NUTRIENTS IN NATURAL WATERS
Herbert E. Allen and James R. Kramer, Editors

pH AND pION CONTROL IN PROCESS AND WASTE STREAMS
F. G. Shinskey

INTRODUCTION TO INSECT PEST MANAGEMENT
Robert L. Metcalf and William H. Luckman, Editors

OUR ACOUSTIC ENVIRONMENT
Frederick A. White

ENVIRONMENTAL DATA HANDLING
George B. Heaslip

THE MEASUREMENT OF AIRBORNE PARTICLES
Richard D. Cadle

ANALYSIS OF AIR POLLUTANTS
Peter O. Warner

ENVIRONMENTAL INDICES
Herbert Inhaber

URBAN COSTS OF CLIMATE MODIFICATION
Terry A. Ferrar, Editor

CHEMICAL CONTROL OF INSECT BEHAVIOR: THEORY AND APPLICATION
H. H. Shorey and John J. McKelvey, Jr.

MERCURY CONTAMINATION: A HUMAN TRAGEDY
Patricia A. D'Itri and Frank M. D'Itri

Air Pollution Control, Part III

AIR POLLUTION CONTROL

Part III

MEASURING AND MONITORING AIR POLLUTANTS

EDITED BY

WERNER STRAUSS

University of Melbourne
Victoria, Australia

A Wiley-Interscience Publication

JOHN WILEY & SONS

New York • Chichester • Brisbane • Toronto

Library of Congress Cataloging in Publication Data:
Main entry under title:

Air pollution control.

Includes bibliographical references.
1. Air—Pollution. I. Strauss, Werner, ed.
TD883.A474 628'.53 79-28773

ISBN 0-471-83323-1

Printed in the United States of America

10 9 8 7 6 5 4 3 2 1

SERIES PREFACE

Environmental Science and Technology

The Environmental Science and Technology Series of Monographs, Textbooks, and Advances is devoted to the study of the quality of the environment and to the technology of its conservation. Environmental science therefore relates to the chemical, physical, and biological changes in the environment through contamination or modification, to the physical nature and biological behavior of air, water, soil, food, and waste as they are affected by man's agricultural, industrial, and social activities, and to the application of science and technology to the control and improvement of environmental quality.

The deterioration of environmental quality, which began when man first collected into villages and utilized fire, has existed as a serious problem under the ever-increasing impacts of exponentially increasing population and of industrializing society, environmental contamination of air, water, soil and food has become a threat to the continued existence of many plant and animal communities of the ecosystem and may ultimately theaten the very survival of the human race.

It seems clear that if we are to preserve for future generations some semblance of the biological order of the world of the past and hope to improve on the deteriorating standards of urban public health environmental science and technology must quickly come to play a dominant role in designing our social and industrial structure for tomorrow. Scientifically rigorous criteria of environmental quality must be developed. Based in part on these criteria, realistic standards must be established and our technological progress must be tailored to meet them. It is obvious that civilization will continue to require increasing amounts of fuel, transportation, industrial chemicals, fertilizers, pesticides, and countless other products and that it will continue to produce waste products of all descriptions. What is urgently needed is total systems approach to modern civilization through which the pooled talents of scientists and engineers, in cooperation with social scientists and the medical profession, can be focused on the development of order and equilibrium to the presently disparate segments of the human environment. Most of the skills and tools that are needed are already in existence. Surely a technology that has created such manifold environmental problems is also capable of solving them. It is our hope that this Series in Environmental Sciences and Technology will not only serve to make this challenge more explicit to the established professional but that it also will help to stimulate the student toward the career opportunities in this vital area.

Robert L. Metcalf
James N. Pitts, Jr.
Werner Stumm

PREFACE TO PART III

In discussing the need for developing precise methods for measuring air pollutants at the source and in the environment, D. H. Lucas* has summarized the present situation as follows:

The aim of pollution control is to minimize the adverse effects of pollutants on the environment. Pollutants in the atmosphere move and produce their effects in such a complex way that basic scientific understanding of high quality is essential to good pollution control.

In the past inadequate understanding led to control philosophies that were successful in some cases, but not in others, but we are now at a stage where there is sufficient scientific knowledge for an increase in the soundness of base for control philosophies.

One of the essential criteria for good control must be based on this understanding. Field measurements, if properly devised, may give an accurate indication of the pollutant effects being produced, but they are of little value in themselves in identifying the primary sources of the effects or of controlling them.

Knowledge and techniques have now improved to such an extent that it is possible to model the contributions of various sources to an overall effect, and conversely to specify the control needed at each source. It is now possible not only to monitor effects in the environment but also to make measurements at the sources which anticipate a given effect, and enable the effect to be avoided.

This part of *Air Pollution Control* is therefore concerned both with ambient and source measurement of air pollutants. Like its predecessors it attempts to impart a basic understanding of the technologies in current practice and in advanced systems. The authors, who are leaders in the fields of their particular specialties, were selected for their practical understanding and their own original contributions, placed in context.

The book deals with source and ambient sampling and monitoring. The sampling, analysis, and monitoring of gaseous pollutants are discussed first,

* D. H. Lucas, Central Electricity Generating Board (London), private communication.

followed by a series of chapters on particulates covering monitoring, sampling, and the physical and chemical analysis of pollutants, including particle size. The chemical analysis material concentrates on the new and critical field of trace elements.

Ambient monitoring is concerned essentially with the physicochemical methods that are available and either are currently in use or are being developed for instrumental systems that are fundamental to regional control. The integration of these monitors into a network is discussed, along with the basis for selecting a particular network. The concluding chapter deals with hydrogen sulfide, which can occur not only in pollutant concentrations from anthropogenic sources (paper mills, oil refineries, coke ovens, sewage plants, etc.), but also naturally in geothermal areas, in surprisingly high concentrations. In the latter case it cannot be controlled by modified technology or source control systems. With the increasing use of geothermal energy as a source of power, the problems of hydrogen sulfide measurement (and protection from its corrosive effects) are of critical importance.

I thank the authors of the chapters, who have been most cooperative in following rather stringent guidelines designed to ensure that the book becomes a useful and comprehensive work, in which there is very little overlap between the chapters. I also express my appreciation to Miss V. Carter, who retyped many of the chapters from very poor notes, and to my colleagues and friends at the universities of North Carolina, Karlsruhe, and Melbourne for their help, advice, and encouragement.

WERNER STRAUSS

Melbourne, Australia
August 1977

CONTENTS

* Professor Natusch, now at Colorado State University, carried out this work while at the University of Illinois.

Air Pollution Control, Part III

1

SOURCE SAMPLING, ANALYSIS, AND MONITORING OF GASEOUS EMISSIONS

Richard W. Gerstle

Pedco Environmental Specialists, Cincinnati, Ohio

1.1. BASIC CONCEPTS, PRINCIPLES, AND EQUIPMENT

The methods used for measuring gaseous emissions from a stack or vent depend on the nature of the compound and the purpose for making the measurement. In addition, the composition and the temperature of the carrier gas stream affect the selection of a sampling technique and analytical method, and the sampling plan.

1.1.1. Emission Measurement Plans

The plan, or test protocol, for measuring gaseous emissions involves the sampling site; that is, the location in the vent system and within the duct itself at which the samples will be taken, the number of samples required, the time period covered by each sample, the sampling technique that should be used, and the process operating conditions. Thus a specific source must be examined in detail and the results desired from the testing program specified in planning the measurements.

In some cases gas streams are sampled in the final exit vent before the gas enters the atmosphere, whereas in others an intermediate gas composition is required. For steady state processes the exact length of sampling is not critical, since the emissions do not vary with time. When a process exit gas stream is well characterized from previous measurements or from examination of the process itself, the sampling method can be preselected with confidence. This is often the case when the carrier gas stream is air that contains a low level of some compound to be measured. In other cases, however, the process emission stream is not well characterized: its composition is not known, and the emission rate may be variable because of changes in gas flow rate, concentration, or both. When a completely unknown process stream is to be sampled, trial sampling runs may be required to make preliminary determinations of concentration ranges, and possible interfering factors, before a final sampling technique can be selected.

1.1.1.1. Sampling Location

A suitable location for sampling gases is one at which the gases are well mixed (i.e., almost any location more than 2 to 3 duct diameters after a flow disturbance—a bend or a T). For large ducts (greater than approximately 10 m^2 in area), two or three points within the duct should be checked to obtain average concentrations. Stratification of gases in large ducts has been observed and must be considered in sampling.[1] When *in situ* instrumental methods are used (see Section 1.3.3), average concentrations across the duct area are obtained.

1.1.1.2 Sampling Time and Rate

The length of the sampling period is determined by the amount of sample needed for accurate analysis and the variations in process operation. When sampling for legal compliance, the time period (or volume) specified in the regulation must be used. The sampling rate depends on emission variations, the quantity of sampling required for analysis, and the sampling equipment used. Typically, rates in the range of 1 to 30 l min^{-1} are used.

Table 1.1 gives various combinations of gas concentrations and stack gas flow that can occur and must be considered in gas sampling. When either the gas concentration or the gas flow is steady (uniform with respect to time), as in categories 1, 2, and 3, a constant sampling rate will yield a true average

Table 1.1. Concentrations and Flow Conditions for Well-Mixed Gaseous Emissions

	Time Variation of Conditions	
Category	Concentration	Velocity
1	Steady	Steady
2	Steady	Unsteady
3	Unsteady	Steady
4	Unsteady	Unsteady

concentration for the sampling period. When both the carrier gas flow and the compound concentration are unsteady, as in category 4, the sampling rate should be proportional to the carrier gas flow variations at the sampling point.[2] When the variation in pollutant concentration is not known, proportional sampling is used by simultaneously measuring the velocity at the sampling point and adjusting the sampling rate accordingly. Thus if the velocity at the sampling point increases by 30%, the sampling rate should also be increased by 30%

1.1.1.3. Sampling Method

Figure 1.1 illustrates different techniques available for gaseous sampling and analysis. Extractive sampling covers the techniques in which a sample of gas is taken from the stack and passed through an instrument, a collecting medium, or into a container, either continuously or intermittently. *In situ*

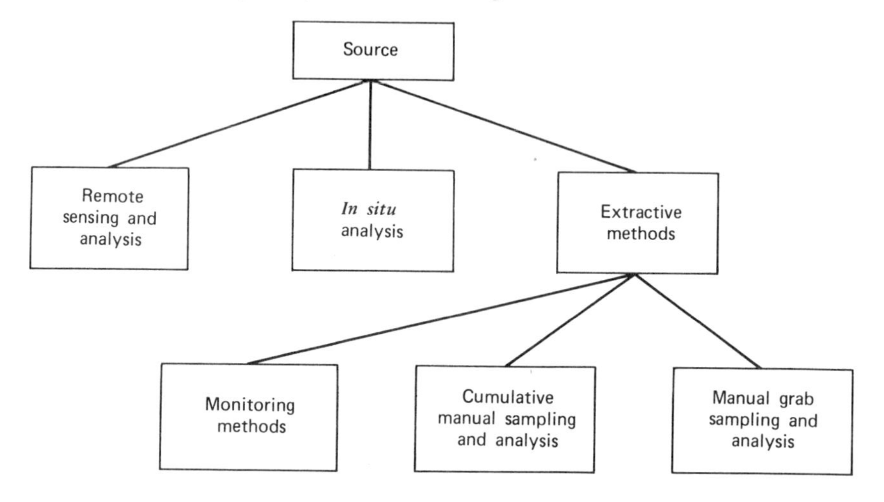

Figure 1.1. Techniques for sampling gaseous components.

monitoring and analysis, and remote sensing of the stack gases, involve instrumental analytical techniques and can be performed either continuously or intermittently to provide real-time data.

Historically, extractive sampling techniques using a liquid or solid sorption medium and subsequent analysis have been most widely used. They are prescribed by many regulatory agencies. However instrumental methods combined with extractive sampling have been developed to provide continuous readings of concentrations for a number of commonly encountered gases.[3,4] Instruments for *in situ* and remote measurement of certain stack gases are available. In the former case, energy is beamed across the stack to determine a gas concentration. In the latter case, the energy emitted by the plume is monitored by a detector at a remote location (100 to 300 m away).

1.1.1.4. Number of Samples

The number of samples required to determine an average gas concentration varies with the nature of the process (i.e., a constant or a variable process), the purpose of the test, the required precision, and the personnel and equipment available to perform the tests. With extractive manual sampling methods, three to nine samples is usually sufficient in practice to obtain a reliable average concentration under a single process operating condition. From a number of samples, the mean and standard deviation of the measurements can be determined by using statistical methods. For instrumental methods of sampling and analysis, data may be obtained continually or for a number of representative time periods.

1.1.2. Extractive Manual Sampling Techniques

Manual techniques that withdraw the sample from the stack for subsequent analysis, including both cumulative and grab sampling methods, are called extractive techniques.

Cumulative manual sampling methods have proved most useful for making short-term measurements at a site. Because of their relative simplicity and flexibility, they can be applied to almost any source and to many compounds. Basically it involves contacting a measured amount of gas with an absorption medium that is then analyzed for the desired compound. The absorption medium (usually a liquid) depends on the gas being sampled. Equipment of the type illustrated in Figure 1.2*a* provides intimate contact between the sampled gas and the liquid that is contained in the impingers (shown) or other adsorption unit, such as bubblers or gas washing bottles. A wide variety of gas–liquid contactors are available, including large (1000 ml) impingers (Greenburg–Smith), gas washing bottles, bubblers with fritted glass, gas distributors, and midget impingers. Plugging by particles may occur, and a probe filter or initial trap should be used. The efficiency of collection of any specific compound in a liquid depends mainly on its solubility and/or reaction rate with the liquid. When the gas reacts chemically with the liquid, absorption efficiency tends to be higher than with purely physical absorption.[5] The collection efficiency decreases as the sampling proceeds because of the increased concentration of the gas in solution. Gas solubility also decreases rapidly with increasing temperature of the solution; therefore the absorbing liquid should be kept cool.

Each contacting device is essentially a continuous stirred tank reactor, which rapidly builds up a concentration of the dissolved gas, and becomes less effective. It is good practice to use two or three absorbers in series to maintain high (99%) efficiency of collection. Figure 1.3 plots collection efficiencies obtained in single stages of common types of absorber when collecting ammonia in equal amounts of water kept at 27°C. The absorption rate varies inversely with the sampling rate and the gas bubble size in the device, and directly with the height of the liquid in the absorber.

After completion of sampling, the absorption medium is carefully removed from the sampling train and analyzed promptly to prevent possible deterioration. The amount of pollutant found in the sample is then related to the amount of gas sampled to determine the concentration.

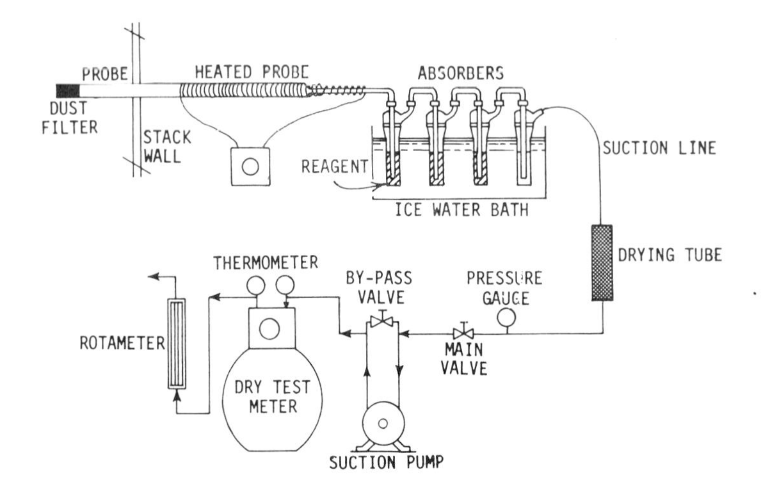

A. GAS ABSORPTION SYSTEM

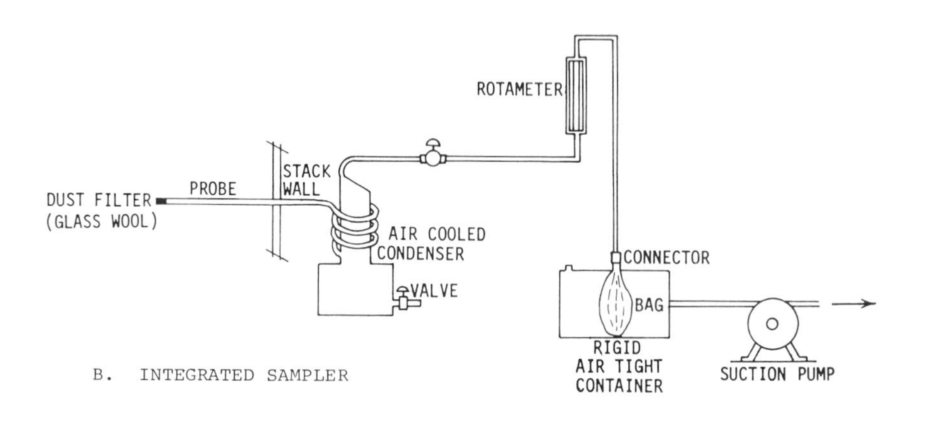

B. INTEGRATED SAMPLER

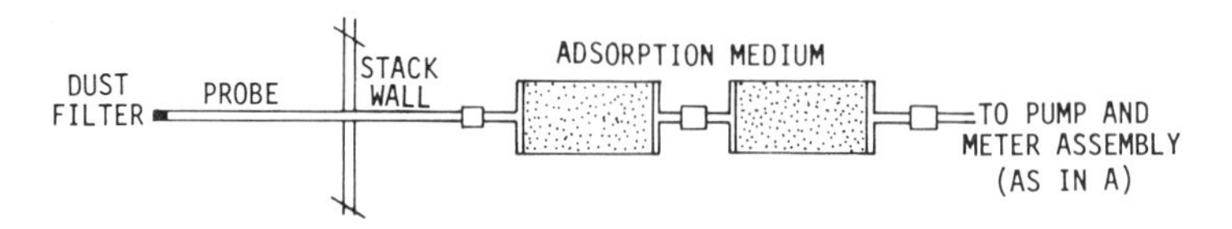

C. GAS ADSORPTION SYSTEM

Figure 1.2. Cumulative manual gas sampling train.

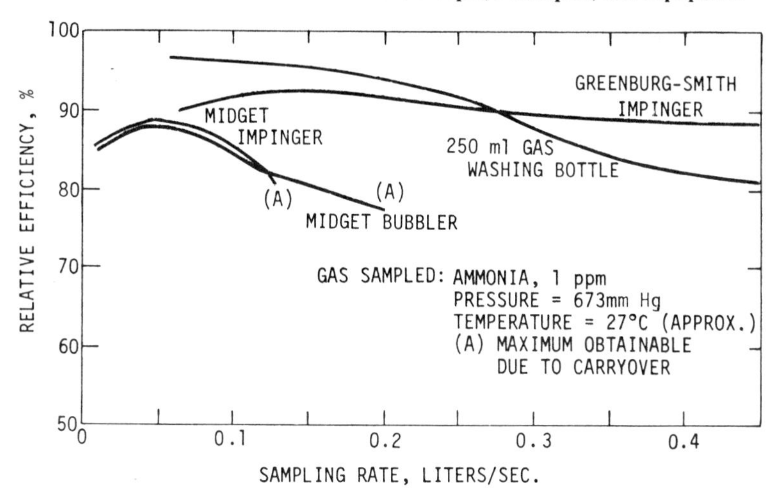

Figure 1.3. Effectiveness of various gas sampling devices.[6]

The sampling train appearing in Figure 1.2*b* collects a total sample in a bag, which can be analyzed in the plant or returned to the laboratory for further analysis. This type of train is especially useful for collecting such nonreactive gases as carbon monoxide or methane, which are neither readily adsorbed on solids nor absorbed in liquids. The bag is filled slowly over a 30 to 40 min period to obtained a cumulative average concentration.

Collection by adsorption on a solid medium is also possible for some gases (especially organic compounds), as in Figure 1.2*c*. Activated carbon and various molecular sieves are used for this purpose. The adsorbent with the collected sample is returned to the laboratory, where it may be carefully weighed or desorbed and chemically analyzed. The degree of adsorption is inversely dependent on temperature and on the amount of gas adsorbed. A number of preliminary experiments usually is required to determine the amount of adsorbent needed to sample a very large proportion of a specific gas of unknown concentration. Two or three adsorption canisters are usually placed in series to ensure high collection efficiency. Special precautions should be taken to keep the canisters above the dew point of water vapor.

Manual grab sampling techniques are also used for extractive sampling. These differ from the simulative methods in that the sampling period is much shorter, extending over only a few seconds or minutes, and only a short-term concentration can be obtained.

Since grab samples yield the gas concentration over a short time period, a number of samples usually is taken consecutively to obtain an average concentration over a longer time period. Short time samples of this type are useful in determining concentration variations in a cyclical process, and variations in concentration across a duct. Equipment used for grab sampling includes glass or metal flasks (Figures 1.4*a* and 1.4*b*) and flexible bags (Figure 1.4*c*). In addition, freezeout traps, large syringes, and adsorption tubes (Figure 1.2*c*) may be used for obtaining a grab sample.

Grab sampling equipment may contain chemical sorbents or liquids to react with the pollutant in question, or it may be used to provide a bulk sample for further analysis.

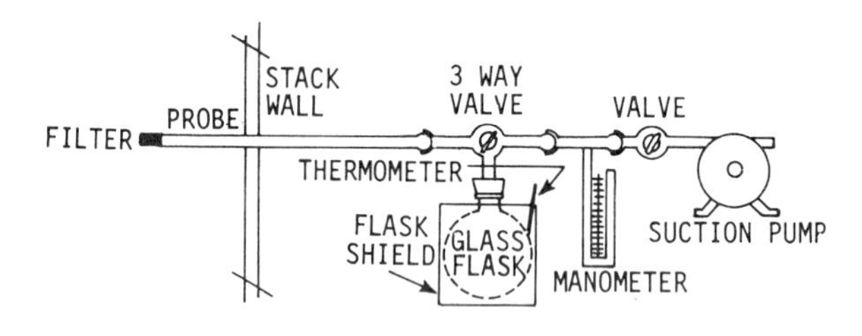

A. EVACUATED FLASK

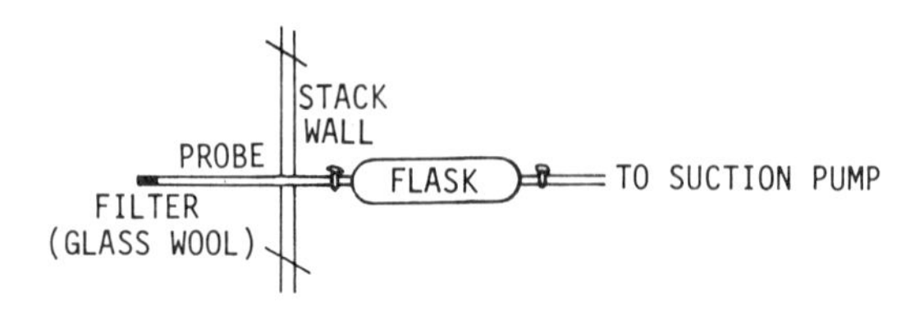

B. DISPLACEMENT OR PURGING SYSTEM

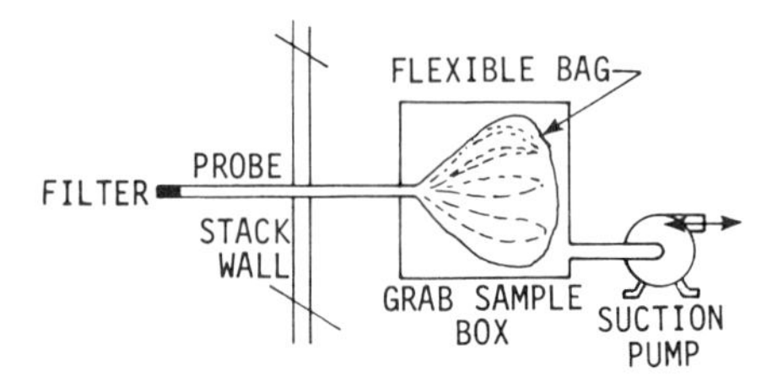

C. FLEXIBLE BAG

Figure 1.4. Examples of grab sampling methods.

When evacuated flasks are used (as in Figure 1.4*a*), the sample volume may be determined according to the ideal gas law and its pressure-volume relationships. Thus the volume of gas sample V_s is

$$V_s = (V_f - V_r)\frac{T_{sc}}{P_{sc}}\left(\frac{P_i}{T_i} - \frac{P_f}{T_f}\right)$$

where V_s = volume of gas sampled at standard conditions
V_f = volume of flask
V_r = volume of reagent, if any
T_{sc} = temperature at desired standard conditions
P_{sc} = pressure at desired standard conditions
P_i = initial vacuum in flask (mm Hg)
P_f = final vacuum in flask (mm Hg)
T_i = initial temperature in flask (K)
T_f = final temperature in flask (K)

Table 1.2. Examples of Gas Sorption Colorimetric Indicator Tubes[7]

Acetaldehyde	Hydrogen cyanide
Acetone	Hydrogen sulfide
Acrolein	Isopropyl alcohol
Ammonia	Mercury
Arsine	Methyl bromide
Beryllium	Methyl chloride
Butyl alcohol	Methyl ethyl ketone (MEK)
Butyl cellosolve	Methyl mercaptan
Carbon disulfide	Nickel carbonyl
Carbon monoxide	Nitrogen dioxide
Carbon monoxide (in blood)	Nitrogen oxides
Carbon tetrachloride	Oxidants
Chlorine	Phenol
Chlorobenzene	Phosgene
Chloroform	Propyl alcohol
Cyclohexane	Styrene (monomer)
Dibromoethane	Sulfur dioxide
Dichloroethylene	Tetraethyl lead
Ethyl alcohol	Tetrahydrofuran
Ethylene	Toluene
Ethyl mercaptan	Trichloroethylene
Formaldehyde	Vinyl chloride
Hydrocarbons	Xylene
Hydrogen chloride	

This equation holds whenever the quantity of gas absorbed in the reagent or condensed in the flask is small (less than a few percent).

When a displacement or purging technique is used, as in Figure 1.4*b*, the volume of gas sampled should be at least 5 times the volume of the flask, to ensure thorough purging. It is theoretically impossible to displace all the gas by purging, but by sampling a volume at least 5 times the volume of the flask, virtually a true concentration will be obtained.

Gas adsorption tubes are available for a wide range of gases (Table 1.2). The tubes provide an indication of the gas concentration by developing a color upon exposure. The length of material colored or the depth of the color is a measure of concentration.[7] The tubes cannot be used for measuring exact concentrations, but they are nevertheless useful for indicating concentration ranges. Some adsorption tubes respond to more than one contaminant; the manufacturer's literature indicates possible interferences. The useful life of many adsorption tubes is limited, and they should not be used after their expiration date.

1.1.3. Analytical Techniques

Analytical techniques covering the entire range of chemical analysis may be used to determine a specific constituent after collection in a container or in a reagent. The exact technique to be used will depend on the equipment available, the compound to be measured, the collecting medium, possible interfering compounds that might be present, and the ingenuity of the analyst.

Analytical techniques can be divided broadly into the following categories:

- **Wet Chemical**
  Gravimetric
  Titrametric
  Colorimetric
  Turbidimetric
- **Instrumental Methods**
  Gas chromatography
  Atomic Absorption
  Electromagnetic
  X-Ray Diffraction
  Mass Spectrometer

In all cases standards must be determined; this permits the analyst to establish a method's response to known amounts of the pollutant of interest.

1.1.4. Monitoring Methods

Instrumental equipment for measuring certain stack gases is commercially available, and such methods are in use with various extractive sampling systems. In addition, there are a number of electrooptical monitoring methods for *in situ* measurements and remote sensing.

The extractive methods all use a probe, a stack gas conditioning section, and a sensor–measurement device (Figure 1.5). The probe usually contains a filter to remove particulates and conveys the stack gases from within the stack. The gas conditioner consists of a cooler, a condenser, and a dilution system to lower the gas temperature and to reduce the moisture content. Great care must be taken to ensure the pollutant gas is not absorbed in the condensate and removed from the gas stream. The conditioned gas is conveyed to the sensor that measures the pollutant gas concentration by a sample line of an inert material, which may be heated. This type of sampling system operates at a constant sampling rate at a single point within the stack. The sampler may be operated continually or intermittently, depending on the information required. A means of injecting calibration gases must also be included in the sampling system. Calibration gas should be injected before the gas conditioner, to expose it to the same environment as that of the sample gas.

When *in situ* monitors are used, the light or energy source is generally mounted on one side of the stack or duct, and the detector on the other side, directly opposite the energy source. In some systems a mirror reflects the energy beam back across the stack diameter. Since the energy beam traverses the mixture of stack gases including particulate, moisture, and other carrier components, it must be capable of discriminating against the unwanted compounds.

The *in situ* instruments have the advantage of measuring the gas at its source with no probes, tubes, or other interfacing requirements that could affect the gas concentration. Since they are mounted directly on the stack,

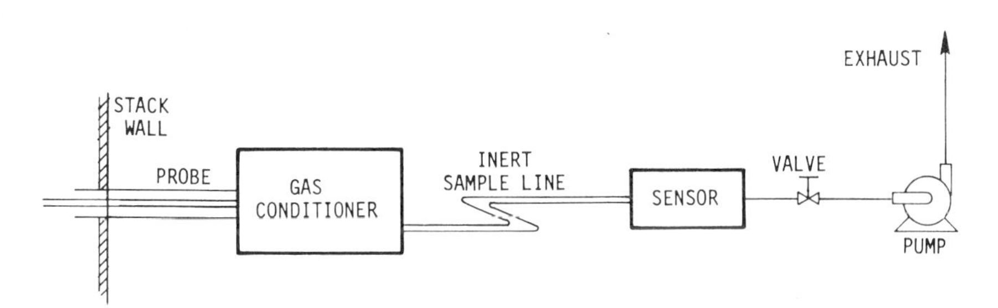

Figure 1.5. Gas monitoring with extractive sampling system.

Table 1.3. Example Monitoring Instruments Available in the United States for Gaseous Emission Measurements[4]

Technique and Company	SO_2	NO/NO_2	CO	HC
Conductometric				
Calibrated Instruments, Inc.	X			
Colorimetric				
F & J Scientific		X		
Amperometric (coulometric)				
Barton ITT	X			
Philips	X	X		
Paper tape				
Houston Atlas, Inc. (via conversion to H_2S)	X			
Catalytic oxidation				
Devco Engineering, Inc.			X	
Matheson Gas Products			X	
Mine Safety Appliances Co.			X	
Chemical sensing electrode				
Geomet, Inc.	X	X		
Orion Research, Inc.	X			
Electrochemical cell				
Dynasciences Corp.	X	X	X	
EnviroMetrics, Inc.	X	X	X	
Theta Sensors, Inc.	X	X		
Chemiluminescence				
Aerochem Research Laboratories, Inc.		X		
Beckman Instruments, Inc.		X		
Bendix Corp., Process Instruments Division		X		
Intertech Corp.		X		
LEDCo Corp.		X		
McMillan Electronics Corp.		X		
REM Scientific, Inc.		X		
Scott Research Laboratories, Inc.		X		
Thermo Electron Corp.		X		
Meloy Laboratories		X		
FID (flame ionization detector for hydrocarbons)				
Beckman Instruments, Inc.				X
Mine Safety Appliances Co.				X
Process Analyzers, Inc.				X
Scott Research Laboratories, Inc.				X
Teledyne Analytical Instruments				X
Thermo Analytical Instruments				X
Wemco Instrumentation Co.				X
Correlation Spectroscopy				
Barringer Research Ltd., U.S. distributor, Environmental Measurements, Inc.	X			
CEA Instruments	X			

FPD (flame photometric devices)				
Meloy Laboratories, Inc.	X			
Bendix Corp.	X			
GC–FID (gas chromatography–flame ionization detection)				
Beckman Instruments, Inc.			X	X
Byron Instruments, Inc.			X	X
Hewlett-Packard Co.			X	X
Bendix Corp.			X	X
NDIR (nondispersive infrared)				
Barnes Engineering Co.			X	X
Beckman Instruments, Inc.	X	X	X	X
Bendix Corp., Process Instruments Division	X	X	X	X
Calibrated Instruments, Inc.	X	X	X	
Ecological Instrument Corp.			X	X
Horiba Instruments, Inc.		X	X	
Infrared Industries	X	X	X	
Intertech Corp.	X	X	X	
Leeds & Northrup	X		X	
Mine Safety Appliance Co.	X	X	X	X
Peerless Instrument Co., Inc.		X	X	X
Scott Research Laboratories, Inc.		X		
Nondispersive UV and visible absorption				
Beckman Instruments, Inc.		X		
Canadian Research Institute, U.S. distributor, Sara Scientific Co.	X			
E. I. du Pont de Nemours & Co.	X	X		
Intertech Corp.	X	X		
Peerless Instrument Co., Inc.	X	X		
Teledyne Analytical Instruments	X	X		
Mercury substitution UV absorption				
Bacharach Instrument Co.			X	
Dispersive IR and UV absorption				
Environmental Data Corp. (IR/UV)	X	X	X	
Wilks Scientific Corp. (IR)	X		X	

however, they tend to be difficult to maintain and calibrate, and careful attention is required to obtain reliable data. In contrast, the extractive instruments are generally easier to maintain and calibrate because they can be sited at a more convenient location, and they are not subject to the direct action of hot, dirty, corrosive stack gases. They do, however, present potential problems in the area of probe reactivity, and reactions of gas with condensed moisture.

Table 1.3 lists stationary source monitors available in the United States. This area of measurement is under active development, and new devices frequently appear on the market.

Sensors currently in use for these systems can be divided into three broad classes, namely:

- **Electrometric.**
- **Spectrometric or electrooptical.**
- **Chemiluminescent.**

Electrometric methods include the sensors that operate by polarographically sensing the gases that diffuse through a selective membrane into an electrochemical cell. These instruments are available for sulfur dioxide (SO_2), hydrogen sulfide (H_2S), nitric oxide (NO), nitrogen dioxide (NO_2), and carbon monoxide (CO).

A number of conductometric and colorimetric analyzers have been adapted for stack gas sampling. These devices, which measure the current in an absorbing solution, or the amount of electrical current needed to regenerate a reagent, have found wider application for ambient air analysis. Colorimetric analyzers, which measure the color of a reagent after exposure to a gas, have also been adapted for semicontinuous stack gas measurement.

Flame ionization detectors (FID) and thermal conductivity monitors are used in the electrometric measurement of hydrocarbons, and in the case of thermal conductivity, other gases. In the flame ionization detector, compounds are passed through a flame that combusts hydrocarbon compounds and ionizes the carbon atoms. The degree of ionization is measured electrically and is related to the hydrocarbon content of the gas stream. With the thermal conductivity detectors, the thermal conductivity of the gas stream is used to cool a branch of an electrical Wheatstone bridge. The resulting electrical signal is related to the thermal conductivity of the gas stream, which in turn is related to its composition.

Flame photometric devices have been used to detect sulfur compounds. The pollutants are burned in a hydrogen-rich flame, in which the sulfur compound emits a characteristic wavelength that is measured and related to the amount of sulfur in the sample.

Spectrometric or electrooptical methods have found very wide applications for stack gas monitoring because different gases absorb electromagnetic waves at specific wavelengths in the infrared, visible, and ultraviolet ranges. This absorbance can be related to the concentrations of specific gases, but interferences occur when other gases present absorb over the same wavelength. Water vapor frequently presents this problem.

Sulfur dioxide exposed to short wavelength ultraviolet light fluoresces. The light created by this fluorescence, after being filtered, is a measure of the amount of SO_2.

Chemiluminescence monitors use the principle of reacting certain gases with a substance, causing the emission of a quantity of light that is proportional to the amount of pollutant present. For example, the reaction of ozone with nitric oxide causes the release of radiation, which measures the amount of NO present. Many other gases also react with ozone to yield measurable radiation, but in spite of development work, only nitric oxide is commonly measured by this method.

1.2. MANUAL SAMPLING TECHNIQUES

Manual sampling techniques for selected atmospheric pollutants are presented to introduce the approaches, the equipment, the analytical procedures, and some of the problems. Detailed descriptions of the steps required to conduct the emission test are not provided, but references are given.

1.2.1. Sulfur Oxides[8,9]

The existence of numerous sources of sulfur oxides (SO_x) has stimulated considerable research in developing measurement methods, not only for sulfur dioxide in the presence of other sulfur oxides, but also sulfur trioxide, and sulfuric (H_2SO_4) and sulfurous (H_2SO_3) acid mist. This has led to the development of a number of modified sampling methods.

The collection of acid mists involves specific procedures including isokinetic sampling and filtration. Being liquid droplets, the acid mists are sampled in the same way as particulate matter, which is discussed further in Chapter 3. Analytical techniques similar to those used with gaseous sulfur oxide, however, apply to acid mists.

Total sulfur oxides are most easily sampled by drawing a measured amount of stack gas through a set of impingers containing an aqueous solution of 3% hydrogen peroxide. This is titrated with barium perchlorate or barium chloride, using a Thorin indicator to determine the amount of sulfate formed. If the gas stream contains particulate matter, it becomes necessary to insert a filter—usually fiberglass—at the probe inlet, but this will also remove some acid mist. Some SO_2 may be oxidized to SO_3 as it passes through the glass fiber wool and/or the collected particulates when oxygen is present and the flue gases are very hot, but this oxidation is usually negligible for conditions found in most source sampling applications.

When only SO_2 is to be measured, the SO_3 and H_2SO_4 must be first separated from the gas stream by initially passing the gas through an impinger containing an 80% aqueous isopropanol solution (Figure 1.6). This solution

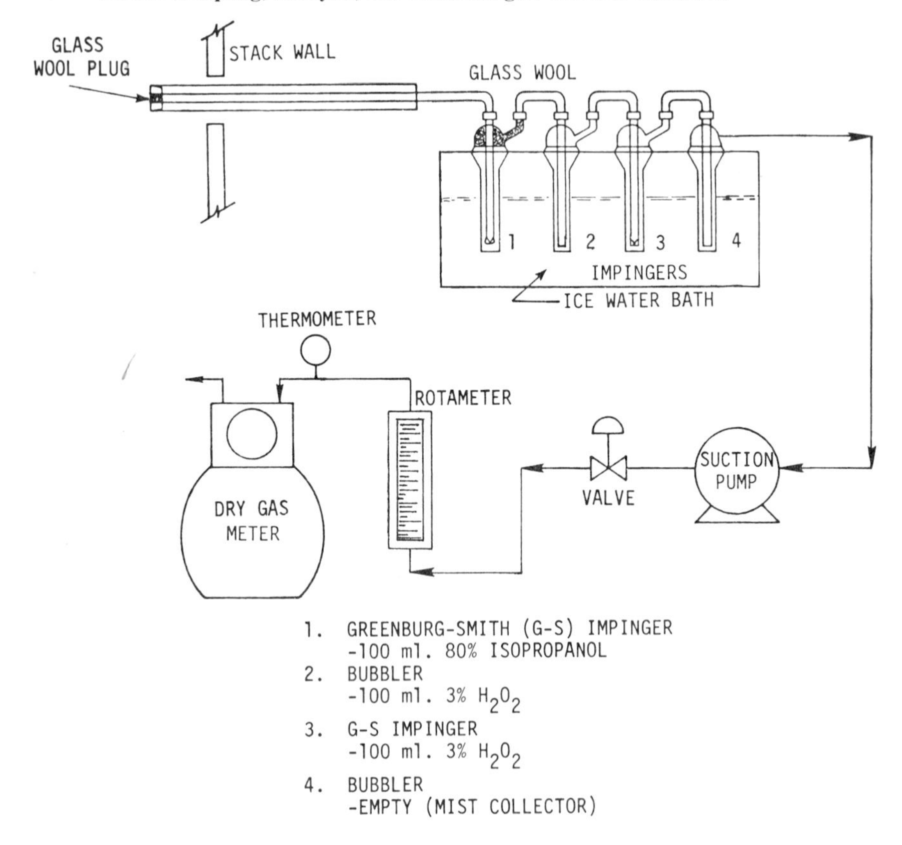

Figure 1.6. Typical sulfur oxides sampling train.

retains the SO_3 and inhibits the oxidation of SO_2. A glass probe heated to at least 260°C should be used to avoid condensation. Upon completion of the sampling period, clean air is drawn through the sampling train for 10 to 15 min to complete the carryover of any entrained SO_2 to the impingers containing the hydrogen peroxide (H_2O_2).

The sampling train should always be checked for leaks before testing by plugging the first impinger and drawing a vacuum: 20 to 30 cm (8 to 12 in.) Hg. Any leakage will be indicated on the gas meter, and of course leakage must be remedied before sampling. A similar sampling train, with midget impingers containing 10 ml of solution, may also be used.

The sensitivity of the method can be increased by increasing the sample volume. For most industrial sources with SO_x concentrations of 50 parts per million (ppm) and higher, a sample volume of approximately 0.6 m^3

(20 ft^3) is suitable. If a midget impinger train is used, 20 to 30 l (1 ft^3) is suitable.

Analysis of the collected sample consists of mixing an aliquot of the peroxide impinger solutions with 80% isopropanol, adding 4 drops of Thorin indicator, and titrating with 0.01 *N* barium perchlorate to a pink end point. Alternatively, barium chloride may be used to avoid the danger involved in working with perchlorates. The perchlorate is preferred, however,

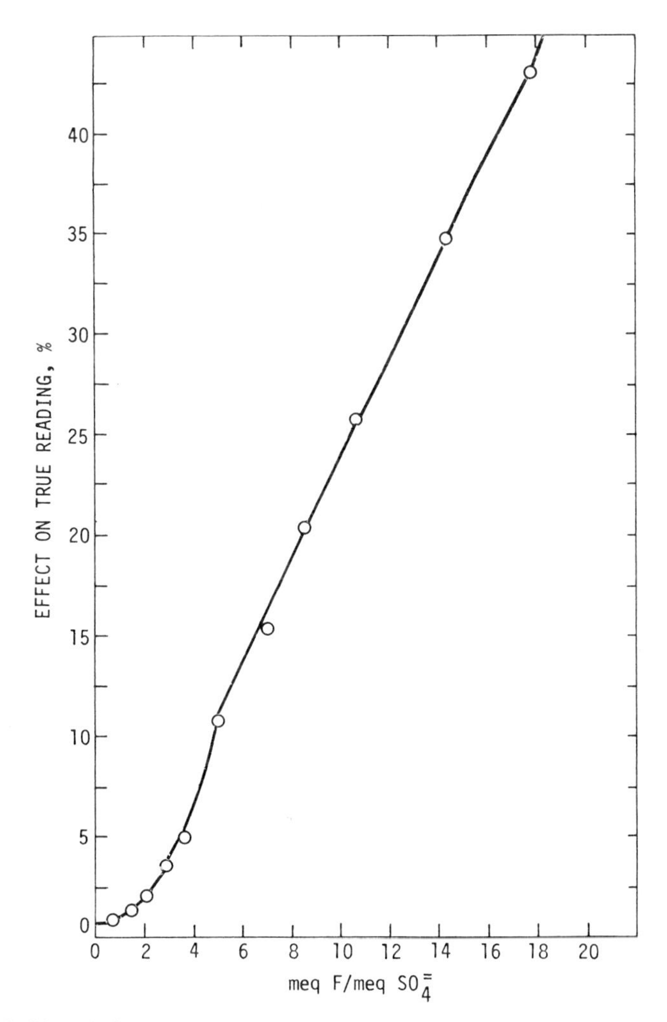

Figure 1.7. Fluoride interference in sulfate determination using Thorin and barium chloride titration.[10]

since it is more soluble in alcohol. The isopropanol impinger solution is titrated directly after the Thorin indicator has been added. A blank should always be run in parallel with the sample. Each milliequivalent of barium compound titrant (normality times milliliters) reacts with one milliequivalent weight of sulfate (32 mg); thus the total weight of sulfate in the sample can be calculated.

The barium titration method is subject to interferences from fluorides (Figure 1.7), which will yield lower results. In addition, ammonia and its compounds, metal sulfates, and phosphates interfere by forming complexes with the Thorin indicator.

Another method for determining total sulfur oxides involves collection of the stack gases in a 5% sodium hydroxide solution utilizing the previously described sampling trains. When only sulfur oxides are present in the gas stream, the resulting solution can be subjected to an acid-base titration. With most gas streams, however, the collecting solution must be treated with bromine, then with barium chloride, to form barium sulfate insoluble crystals. The sulfate crystals are aged and carefully filtered and weighed.[11]

1.2.2. Nitrogen Oxides

The oxides of nitrogen (NO_x) emitted by combustion processes and some chemical processes are usually measured by collecting samples in evacuated flasks as in Figure 1.4*a*. The sample is drawn through a heated glass probe into the evacuated flask, which contains an absorbing reagent that is allowed to remain in contact with the sampled gas after the sampling period is completed. Nitrous oxide (N_2O), being very inert, is not detected by any of the NO_x sampling methods.

Continuous sampling methods for NO_x have not yet been developed, although some progress has been made.[12] Nitric oxide is a relatively inert gas with only limited solubility, and collection techniques using bubblers have yielded poor results. The evacuated flask sampling technique provides only a short-term sampling period (up to about 3 min); thus a number of samples are required to obtain an average value because a single sample may yield a nonrepresentative concentration. Separation with manual methods of the various nitrogen oxides, especially NO and NO_2, has not been very successful, but it can be accomplished with a gas chromatograph.

One of the most widely used NO_x gas analysis methods, for concentrations in the range of 30 to 3000 ppm, is with phenoldisulfonic acid. Lower concentrations down to about 5 ppm can be measured, but accuracy is very limited in the lower range. An absorbing solution of 3% hydrogen peroxide and 0.1 *N* sulfuric acid is used in a 2 l flask evacuated to 50 to 75 mm Hg absolute pressure. The stack gas is drawn into the flask through a glass probe until the flask vacuum is reduced to about 25 mm Hg below atmospheric pressure. A filter at the probe end is required when particles are present, and the probe should be heated to prevent condensation. The flask is then sealed, and the sampled gas is allowed to remain in contact with the reagent for 24 hr.[13] Analysis consists of reacting the absorbing solution and flask washings with phenoldisulfonic acid, which is subsequently treated with ammonium hydroxide to form a yellow compound. The intensity of the color is read spectrophotometrically at 410 nm and compared to the color developed by standard nitrite solutions similarly analyzed.[14]

The phenoldisulfonic acid test can be used for all emission measurements because it is very specific for nitrates, but chlorides interfere (Figure 1.8), and a correction factor should be applied when the chloride content of the gas stream is known. Particulate matter containing nitrate or organic nitrogen compounds will also interfere and must be filtered out before the undesirable substances enter the flask. Additional peroxide absorbing solution should be used if large quantities of oxidizable compounds such as SO_2 are present in the sample stream. Considerable practice is required

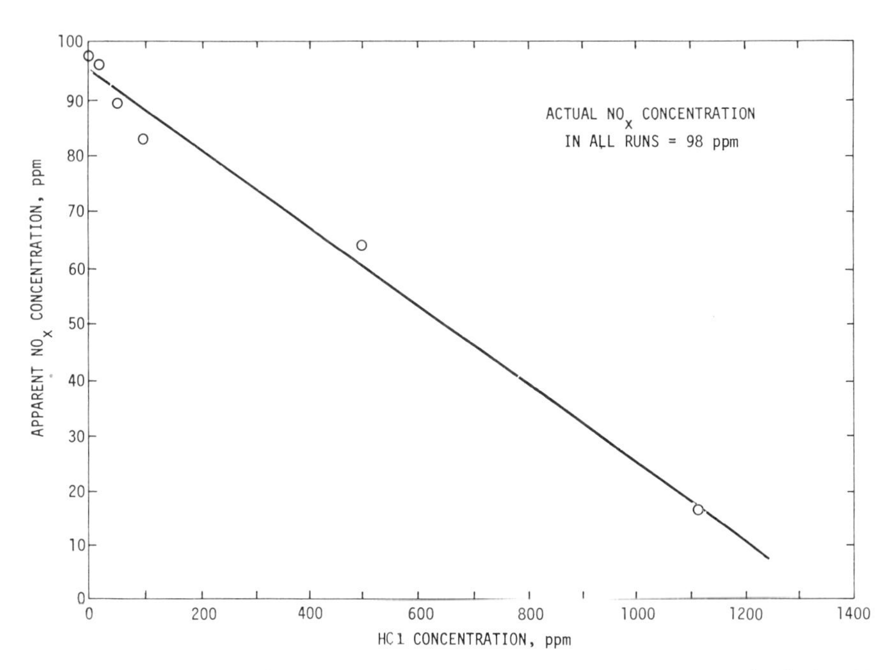

Figure 1.8. Interference of HCl with the determination of NO_x by the phenol disulfonic acid method.[15]

in performing the analysis, since a number of steps are involved and the procedure must be carefully followed to minimize errors.

A simpler analytical procedure can be used when unacidified 3% hydrogen peroxide solution is used to absorb NO_x. An acid-base titration is then performed on the resulting solution. However this method can be used only on gas streams that do not contain any other acidic or basic gases (chiefly CO_2), and this condition is seldom encountered in practice except with some chemical processes.

1.2.3. Fluorides

Fluorides are collected by drawing a gas sample continuously through a set of Greenburg–Smith impingers or bubblers (Figure 1.9). The basic sampling technique is identical to that used for SO_2 sampling with a series

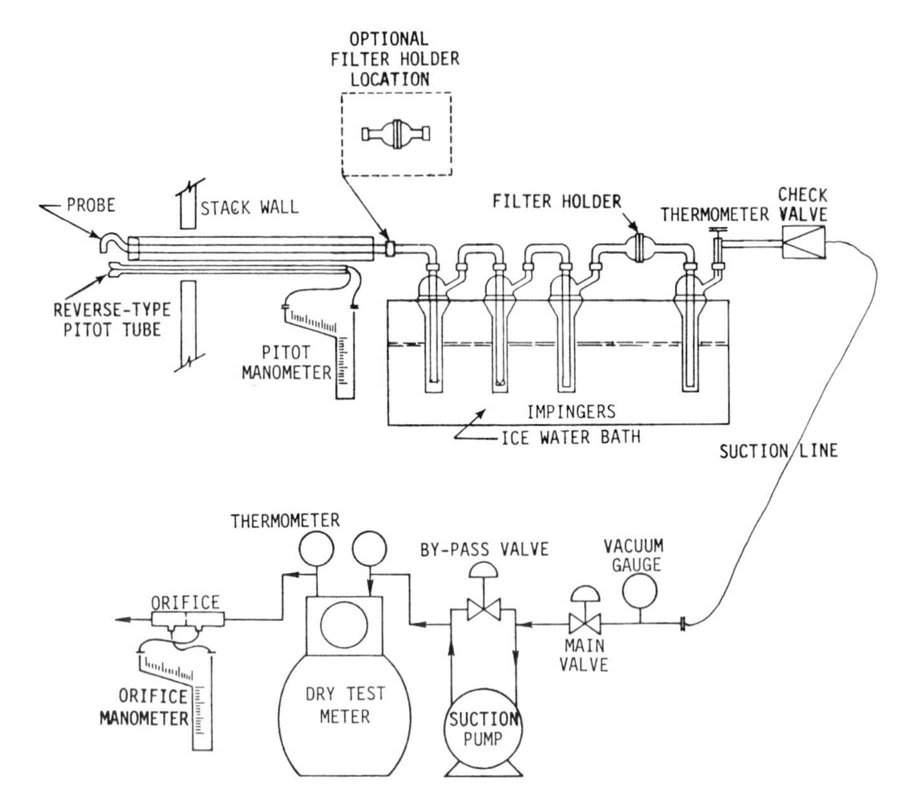

Figure 1.9. Fluoride sampling train.[16]

of three impingers and a sampling rate of approximately 15 l min^{-1}. Fluorides, usually found as hydrogen fluoride (HF), silicon tetrafluoride (SiF_4), and fluorosilicic acid (H_2SiF_6), are very reactive and easily absorbed.

Both gaseous and solid fluorides are usually collected in the impingers, but a reasonable separation between gaseous and solid fluorides can be achieved by filtering the gas with a paper filter before it enters the impingers. Glass fiber filters cannot be used because they react with fluoride compounds. However the reactivity of fluorides with the sampling equipment and any particulate on the filter makes an exact separation difficult. When particulate fluorides are present, sampling must be performed isokinetically and the stack cross section must be traversed. Water or a 5 to 10% sodium hydroxide (NaOH) solution is used for the absorption medium.

Most of the analytical techniques available to determine the fluoride content of the absorbing reagent after sampling involve a neutralization and

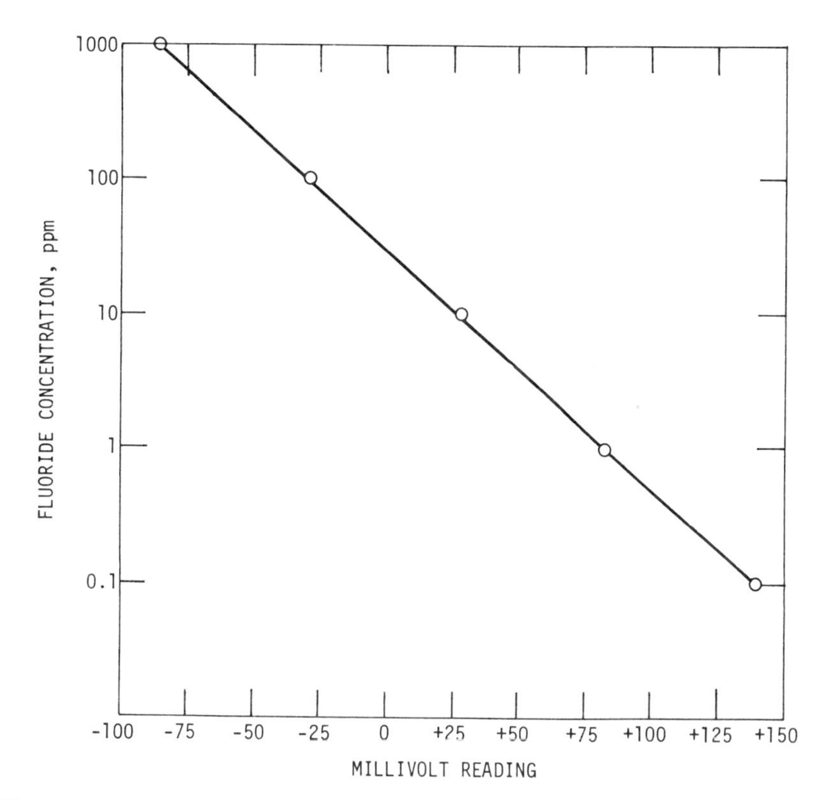

Figure 1.10. Example of calibration curve of fluoride concentration versus specific ion, millivolt readings (buffered to pH 7.5 with 0.1 *M* sodium citrate).[19]

distillation step to separate the fluorides from other compounds. An aliquot of the distillate is then treated with SPADNS reagent [sodium 2-(para-sulfophenylazo)-1,8-dihydroxy-3,6-naphthalene disulfonate], and the resulting color is read at 570 nm with a spectrophotometer.[17] Alternatively, an aliquot is titrated with thorium nitrate with alizarin red sulfur indicator to a pink end point.[18] In both cases, blanks of the absorbing solution should be analyzed along with the sample, and a set of calibration standards must be run. Ions, including aluminum, chloride, iron, magnesium, phosphate, and calcium, interfere with SPADNS reagent, but distillation removes these.

Specific ion electrodes have been developed for directly measuring the fluoride content of a solution after buffering with sodium citrate. This method compares well with analyses by the SPADNS reagent method and is much simpler to perform, since an aliquot of the absorbing reagent is diluted 1:1 with the buffer solution. The specific ion electrode is then immersed in the solution, and the electrical output is measured. A typical calibration curve for the specific ion electrode appears in Figure 1.10.

1.2.4. Chlorides

To measure chlorine and chloride concentrations, gases are collected in a set of impingers containing sodium hydroxide solution and set in an ice bath (Figure 1.6).[20]

For concentrations less than 1000 ppm, a 1 *N* NaOH solution is used; for higher concentrations, a 2.5 *N* solution should be used. For percentage quantities, samples should be collected in a 2 l grab sampling flask (see Figure 1.4*a*). Total chlorides are determined in this solution by the Volhard titration,[21] which involves addition of excess silver nitrate to precipitate silver chloride, and back titration with ammonium thiocyanate with ferric alum indicator to determine the quantity of silver nitrate that reacted. If free chlorine is also present, alkaline arsenite solution should be used for the absorbing reagent. The unconsumed arsenite is then measured by titrating with standard iodine solution. The total chloride content is again determined by the Volhard titration.

The hydrogen chloride content is obtained by subtracting the chlorine value from the total chloride content. To prevent contamination of the absorbing reagent, particles should be filtered out in the probe and they may be analyzed separately.

Sulfur oxides and nitrogen dioxide interfere with the iodine titration for chlorine determinations and yield low results, but they do not interfere with the total chloride determination by the Volhard method. Bromides and iodides, but not fluorides, are included in the Volhard method, but these compounds are not commonly found in vent gases. Bromides, however, are a constituent in automobile exhaust gases. Similar methods, giving like results, are also available for chlorine-chloride analysis; one involves collection in potassium iodide and analysis for free iodine.[22]

1.2.5. Ammonia[23]

Ammonia (NH_3) is collected by using a set of impingers containing distilled water absorbing reagent. The resulting solution can be analyzed by the Kjeldahl distillation procedure or, for low concentrations, by Nessler's reagent. These methods are sensitive to ammonia concentrations of less than 1 ppm when sampling approximately 2 m^3 of gas. A set of two Greenburg–Smith impingers containing 100 ml each of distilled water, followed by a third impinger, initially dry, is used for sampling at a rate of approximately 18 $l\ min^{-1}$. Upon completion of the sampling period, the probe and impingers should be rinsed out with distilled water and added to the impinger contents. An aliquot of the absorbing reagent is distilled in a Kjeldahl apparatus, and the distillate is titrated with 0.1 *N* sodium hydroxide. Blank corrections must be subtracted from the quantity of NaOH used to titrate the sample.

1.2.6. Gaseous Organic Compounds

A variety of methods are being used to measure organic emissions. No one single sampling or analytical method can be specified for organic compounds, since there are so many compounds with different characteristics. Grab samples are taken in bags of nonreactive plastics, 20 to 30 l capacity, suitable for the lighter noncondensable compounds such as methane and ethylene. This sampling system is illustrated in Figure 1.4*c*. The bags must be purged with nitrogen before sampling and then fully evacuated. Table 1.4 shows the reactivities of various plastics with organic compounds.

After sampling, the bags should be promptly analyzed. Generally chromatography is chosen for determining the compounds. Detectors commonly used for air pollutants include flame ionization, hot wire, and electron capture. Each column and detector must be calibrated for the gas to be measured.

When only total hydrocarbons are to be measured, the concentration present can be determined with a flame ionization detector alone. Careful calibration is required. These detectors do not respond quantitatively to oxygenated compounds,[25] and a more precise measure may be obtained by

Table 1.4. Storage Properties of Vapors and Gases in Plastic Bags[24]

Plastic Film	Gas or Vapor Stored	Concentration	Remarks
Esters of dehydric alcohols and terephthalic acid; trade name, Mylar (Dupont)	Olefins	20–400 ppm	5 to 10% loss in 24 hr
	Formaldehyde	2–3 ppm	$<5\%$ in 24 hr in air mixture
	Formaldehyde	Irradiated car exhaust	5 to 10% loss in 2 hr
	Ozone	70 ppm	10% loss in 5 hr in synthetic air
	Nitrogen dioxide	0.2–0.5 ppm	5% in 8 hr in synthetic air
	Sulfur dioxide	0.5 ppm	Stable for 4 hr in synthetic air
	Acrolein	Car exhaust	$<10\%$ loss in 24 hr
	Acrolein	0.1–10 ppm	Stable in air mixture
	Aliphatic hydrocarbon	0.1–130 ppm	$<10\%$ loss for 3 days in air
Polyvinyl chloride	Carbon monoxide	1–100 ppm	Storage variable with source of supply
Trade name, Scotch Pak (3M Co.)	Carbon monoxide	1–100 ppm	Stable several days in expired air
	Ethanol		$<10\%$ loss in 24 hr in expired air
	Aliphatic hydrocarbon	0.1–20 ppm	$<10\%$ loss in 3 days in air
	Acrolein	Car exhaust	$<10\%$ loss in 24 hr
Polyvinylidene chloride; trade name, Saran (Amtech, Inc.)	Chlorinated hydrocarbon	200 ppm	Expired air and synthetic air standards
Fluorinated ethylenepropylene; trade name, Teflon (Dupont)	Hydrocarbons	Irradiated car exhaust	Stable for several hr
Chlorotrifluoroethylene; trade name, Kel-F (3M Co.)	Nitrogen dioxide	1 ppm	Stable for 120 hr

combusting the compounds and measuring with an infrared analyzer the carbon dioxide that is formed.

Larger molecules such as benzene, xylene, and methanol, which are present in quantities so small that their gaseous partial pressure does not exceed their vapor pressure at the sampling temperature, can also be sampled with this technique. The danger in sampling for the compounds with larger molecules is that one does not normally know what concentration is present before sampling, and after the stack gas sample is cooled in the bag, these gases may condense or may react with condensed water. The bag sampling technique thus has limited usefulness, but it is easy to use and provides quick results.

Low levels of various organic compounds can be adsorbed on activated carbon or packed tubes[26] using a sampling train of the type shown in Figure 1.2*c*. A number of trial runs are usually required to determine the correct sample volume and the desorption procedure. A nonorganic solvent such as carbon disulfide or an inert gas may be used to desorb the sample. The resultant solution is then injected into a gas chromatograph or direct into a detector.

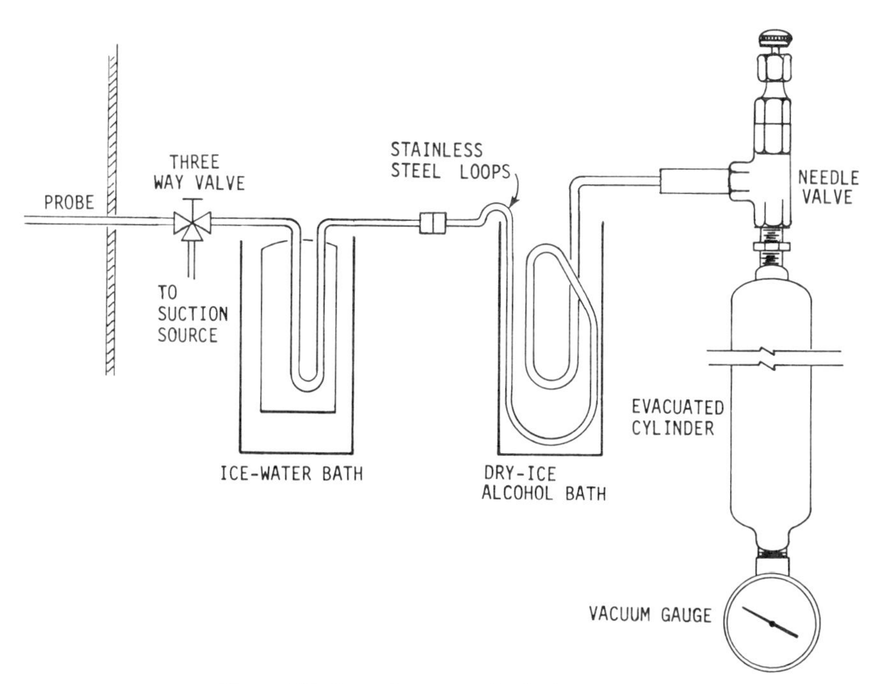

Figure 1.11. Sampling system for hydrocarbons.

When compounds that condense at ambient temperatures are present, the sample must be cooled in the sampling train. Figure 1.11 shows a freeze-out train that combines collection of both condensable and noncondensable compounds. The stainless steel tubing sections are immersed in an ice water and dry ice–alcohol bath before and during sampling. The collection tank is attached, and the coils and tank are evacuated to a measured vacuum. The probe is attached and purged, the three-way valve is opened to the probe, and the needle valve is opened slowly to allow the gas to enter the train until the pressure is near the stack gas pressure. The valves are then closed and the entire train is transferred to the laboratory. The freeze-out portions are heated to selected levels, and the organic compounds vaporized at the different temperatures are measured. The gases in the evacuated cylinder which are the noncondensable compounds are analyzed separately.[27]

1.2.7. Carbon Monoxide

The techniques used for sampling and analysis of carbon monoxide in stack gases depend on the expected concentrations and desired degree of accuracy. In many cases a grab sample can be taken from the stack into a flexible plastic bag (see Fig. 1.2*b*), then analyzed in an Orsat apparatus or by instruments utilizing nondispersive infrared (NDIR) absorption or hot-wire detectors. The CO can also be catalytically converted to methane (CH_4) and measured with a flame ionization detector. Or, the Orsat apparatus, and a number of portable electrical CO monitors can be carried directly to the stack and samples taken from the duct.

The Orsat apparatus is useful when expected concentrations are greater than approximately 0.2% by volume and a high degree of accuracy is not required. This device functions by absorbing the CO in a solution of ammoniacal cuprous chloride or other CO absorbent, then measuring the volume of the absorbed gas by liquid displacement. Glass indicator tubes that change color (Section 1.1.2) are also useful for determining approximate concentrations of low levels of CO.

Infrared absorption (see Section 1.3.1.8) is a reliable and accurate technique for measuring CO, but water vapor and CO_2 interfere with this technique and should be removed from the gas stream or corrections applied to the measured CO concentration. Optical filters in the instrument will also minimize these interferences.

Also available for measuring CO are a number of wet chemical techniques such as absorption in the silver salt of *p*-sulfaminobenzoic acid.[28]

1.2.8. Hydrogen Sulfide

Aqueous cadmium hydroxide [$Cd(OH)_2$] or aqueous zinc acetate [$Zn(CH_3CO_2)_2$] in a series of impingers as shown in Figure 1.6 can be used to measure concentrations of hydrogen sulfide in stack gases. This method has a wide range of sensitivity depending on the volume sampled and the amount and concentration of absorbent used.[29–31] Sulfur dioxide, if present, should be removed from the sampled gas by inserting an impinger containing 3% hydrogen peroxide before the impingers containing the $Cd(OH)_2$ solution. Other reduced inorganic sulfur compounds such as carbon disulfide may interfere with this method.

When cadmium hydroxide is used, it must be thoroughly mixed before sampling to prevent settling of the hydroxide. A measured volume of stack gas is then bubbled through the impingers. Care must be taken not to exceed the capacity of the absorbing solution. The appearance of the yellow color of cadmium sulfide (CdS) in the impingers indicates that the solution is saturated. Sampling higher concentrations of hydrogen sulfide will require more absorbing reagents and larger bubblers. With the larger bubblers or impingers (Greenburg–Smith type), a sampling rate of less than 10 l min^{-1} should be used. Upon completion of sampling, the impinger contents are transferred to a flask and the impingers and connecting glassware are rinsed with an excess of acidified standardized iodine solution, which is added to the flask. Care should be exercised to prevent exposure of the iodine solution to air. The iodine solution is then back titrated with standardized sodium thiosulfate solution to a light yellow color. Starch indicator solution is used, changing the color to blue, and titration is continued until the blue color disappears. The amount of iodine consumed by the sulfide is then calculated from the titration. Phenylarsine oxide, a more stable compound, may be used for the titration instead of sodium thiosulfate.

1.3. GASEOUS EMISSION MONITORING*

Monitoring of gaseous emissions can be performed by (1) attaching an analytical instrument to an extractive sampling system, (2) placing an instrument in the gas stream, and (3) remote measurement of the vent gases after they leave the stack. Of these methods, the first is most commonly used and has reached the highest state of development. Continuous monitoring offers the advantage of providing real-time measurement of the

* More details concerning the principles of these methods are given in Chapter 6.

compound in question, thus allowing the analyst to determine short-term fluctuations.

In the long run, moreover, the use of automatic instruments reduces the amount of manual labor required to obtain emission data. In all cases, an instrument must be carefully installed and calibrated to yield reliable data. Periodic calibration on a routine basis is also required. Monitoring instruments measure only the gas concentration and must be used with a flow measuring device, to calculate total emissions on a mass per unit time basis.

1.3.1. Principles of Detection

A wide range of analyzers is available for measuring the gas stream after it has been removed from the stack and properly conditioned to the instrument manufacturer's requirements. Regardless of which type of

Table 1.5. Example Performance Specifications for Instruments Monitoring Gaseous Emissions[32]

Parameter	Specification
Accuracy (relative)	≤ 20% of mean reference value[a]
Calibration error	≤ 5% of each test gas value[a]
Zero drift (2 hr)	≤ 2% of emission standard[a]
Zero drift (24 hr)	≤ 4% of emission standard[a]
Calibration drift (2 hr)	≤ 2% of emission standard[a]
Calibration drift (24 hr)	≤ 5% of emission standard[a]
Response time	≤ min
Operational time	168 hr continuous

[a] Absolute mean value, 95% confidence interval.

instrument is used, the specifications listed in Table 1.5 are useful as a guide to instrument performance, they apply to both extractive and *in situ* analyses.

Principles of measurement used in stack gas monitoring instruments have been adapted from ambient air measuring instruments and from laboratory analytical instruments.

A general knowledge of the measurement techniques employed by the various monitors is useful in selecting instruments and in understanding their capabilities.

1.3.1.1. Conductivity

The principle of electric conductance by soluble electrolytes has had wide application as an analytical procedure. The conductance of electrolytes is proportional to the number of ions present and their mobility. The measured conductivity can be directly related to the concentration of ions present, thus to the concentration in the sampled gas stream.

Sulfur dioxide in ambient air is commonly measured by this technique. These analyzers use distilled or deionized water reagent modified by the addition of hydrogen peroxide and a small amount of sulfuric acid (to pH 4), the latter reducing the solubility of carbon dioxide and minimizing this interference. The modified reagent forms sulfuric acid upon reaction with sulfur dioxide. This method is not widely used for emission monitoring, however, because despite the precautions, it remains subject to gross interferences from other soluble compounds.

1.3.1.2. Colorimetry

In colorimetry, the quantity of a substance in solution is determined by measuring the relative amount of light passing through that solution. The substance being studied may itself be colored, thus measured directly, or it can be reacted with a reagent, forming a colored compound, whereupon it is determined indirectly. The Beer–Lambert law, which underlies colorimetric analysis, states that the degree of light absorbed by a colored solution is a function of the concentration and the length of the light path through the solution.

Colorimetry is employed, for example, to measure sulfur dioxide and nitrogen dioxide.

1.3.1.3. Coulometry

The quantity of electrons, expressed as coulombs, required to oxidize or reduce a substance, is proportional to the mass of the reacted material according to Faraday's law. Coulometric titration cells for the continuous measurement of sulfur dioxide and nitrogen dioxide have been developed. Another common coulometric technique responds to materials that are oxidized or reduced by halogens and/or halides. Introducing a reactive material shifts the halogen–halide equilibrium. The system's equilibrium is returned to its original value by a third electrode, which regenerates the depleted species. The current required for this generation is measured and is directly proportional to the concentration of the depleted species, and in turn, to the quantity of desired constituent. This is classified as secondary coulometry, usually employing a dynamic iodimetric or bromimetric titration. Most commercial coulometric systems can detect a specific compound by the use of prefiltration devices.

1.3.1.4. Electrochemical Cells[6,33]

In electrochemical cell analyzers, the gas to be measured diffuses through a semipermeable membrane into the cell, the rate of diffusion being proportional to the gas concentration. In the case of electrochemical oxidation, electrons are released at the counterelectrode, and the current generated is proportional to the concentration of SO_2, NO_x, or CO. Cell selectivity is determined by the semipermeable membrane, the electrolyte, the electrode materials, and the retarding potential, the last being adjusted to retard oxidation of the species that are less readily oxidized than the gas of interest.

1.3.1.5. Chemiluminesence

Chemiluminescent analyzers are based on the emission characteristics of a molecular species formed in the reaction between the compound being

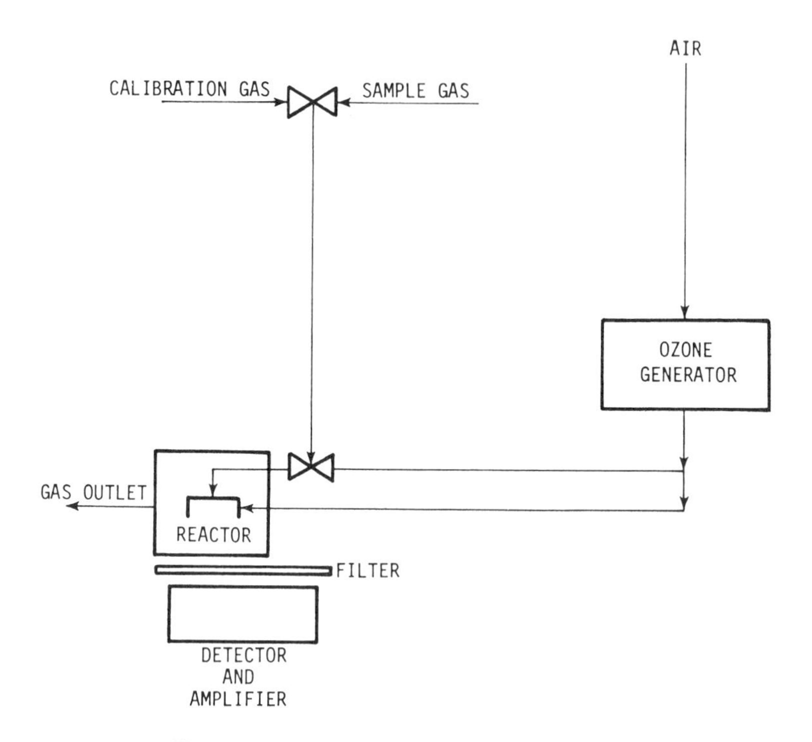

Figure 1.12. Chemiluminescent NO analyzer.

monitored and a gas or solid species. To monitor NO, the sample is reacted with ozone to give excited NO_2, which proceeds to the ground state with the emission of radiant energy, measured with a photomultiplier tube (Figure 1.12). The chemiluminescent reaction occurs according to the equation

$$NO + O_3 = NO_2 + O_2 + h\nu$$

where h is Planck's constant and ν is the frequency for the photon of light emitted by the reaction. Since the ozone is kept in excess, the amount of light generated is proportional to the amount of NO metered into the reaction chamber.

1.3.1.6. Flame Photometry

Flame photometry consists of measuring the intensity of specific spectral lines resulting from the quantum excitation and decay of elements in a flame. Gases are introduced into the flame by mixing them with the flammable gas or with the air supporting the flame. The specific wavelength can be isolated by narrow band optical filters, diffraction gratings, or a prism.

Development of flame photometric detectors having a semispecific response to sulfur compounds has led to their use in continuous monitoring of gaseous sulfur compounds. This flame photometric detector consists of a photomultiplier tube viewing a region above the flame through narrow band optical filters. When sulfur compounds are introduced into the hydrogen-rich flame, they produce strong luminescence at a wavelength of 394 nm.

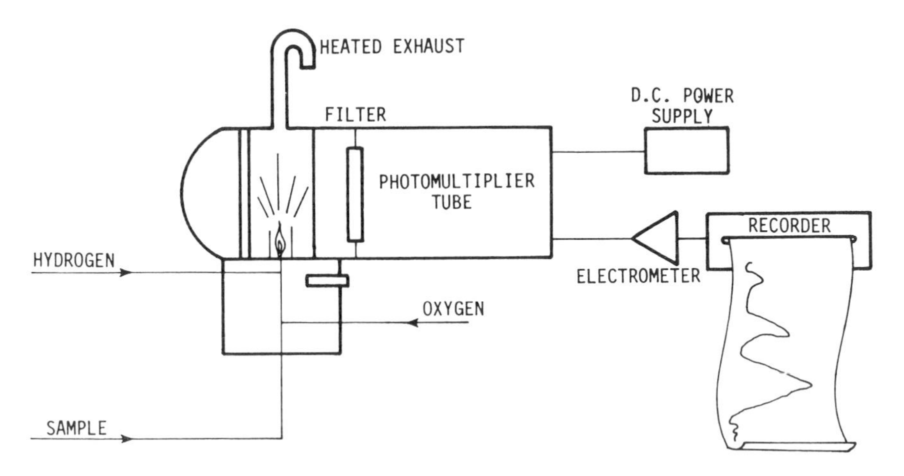

Figure 1.13. Schematic diagram of a typical flame photometric sulfur monitor.[33]

Other sulfur compounds result in positive interferences while monitoring for SO_2, since the detector responds to all sulfur compounds, and discretion must be used in interpreting the data. Figure 1.13 is a schematic diagram of a typical flame photometric sulfur monitor.

Separation of low levels of SO_2, H_2S, CS_2, and CH_3SH has become feasible through the use of chromatographic techniques coupled with flame photometric detection. Typical stack gas concentrations must be diluted for this technique.

1.3.1.7. Flame Ionization

Instruments employing flame ionization detectors, originally designed for the detection by gas chromatography of organic compounds, have had wide application in continuous monitoring of gaseous hydrocarbons. The sample is mixed with hydrogen and passed through a small jet with air supplied to the annular space around the jet to support combustion. Carbon-containing compounds carried into the flame form ions. An electrical potential across the flame jet and an ion collector electrode measure the ion current that is proportional to the number of carbon atoms in the sample (Figure 1.14).

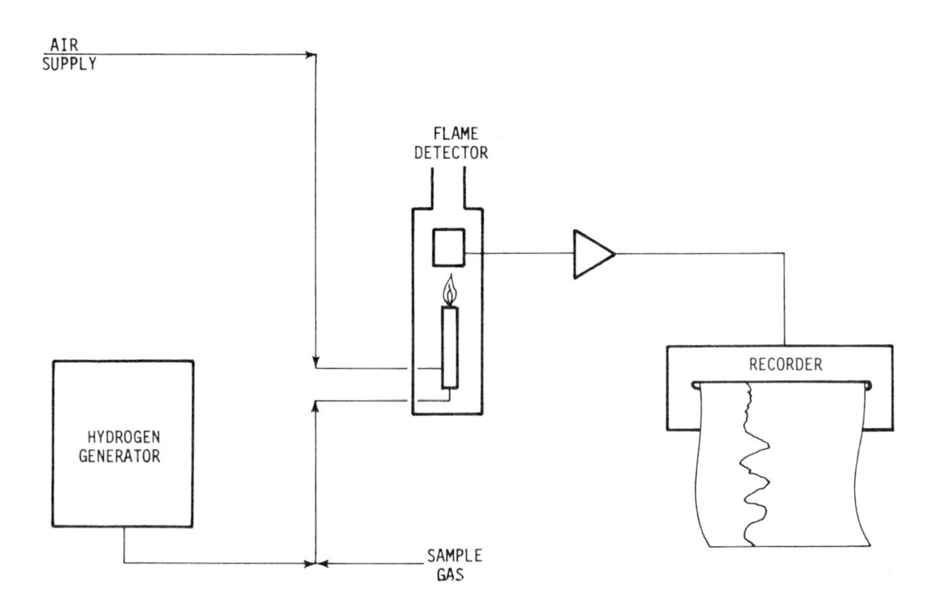

Figure 1.14. Schematic diagram of a typical flame ionization monitor.[33]

*1.3.1.8. Spectrometry**

The infrared and ultraviolet absorption characteristics of several common gases make possible their detection and analysis.

A typical analyzer (Figure 1.15) consists of a sampling system, two infrared sources, sample and reference gas cells, detector, control unit and amplifier, and recorder. The reference cell contains a non-infrared-absorbing gas, such as nitrogen, and the sample cell is continuously flushed with the sample atmosphere. The detector consists of a two-compartment gas cell separated by a diaphragm whose movement causes a change of electrical capacitance in an external circuit, and ultimately an amplified electrical signal that is suitable for input to a controller and/or a recorder.

During operation an optical chopper intermittently exposes the reference and sample cells to the infrared sources. At the frequency imposed by the chopper, a constant amount of infrared energy passes through the reference cell to one compartment of the detector cell, while a varying amount of infrared energy, inversely proportional to the pollutant concentration in the sample cell, reaches the other detector cell compartment. These unequal amounts of residual infrared energy produce unequal expansion of the detector gas, causing the detector cell diaphragm to move and resulting in an electrical signal.

With appropriate optical filters and gas in the detector, the instrument may be made specific in response to SO_2, NO, NO_2, hydrocarbons, CO_2, and CO. Infrared analyzers used in source sampling are susceptible to corrosion of the optics and internal optical coatings by SO_2, NO_2, and chlorides, which are frequently present. Infrared analyzers have reached a high state of development and reliability. Interferences from water vapor and other gases that absorb in the same bands as CO_2 can be reduced by removing them or by trying to correct for the interference.

Spectrophotometric gas analyzers that operate by responding to the second derivative of the absorption band peaks are also available. Figure 1.16 illustrates the difference between the absorption spectrum of NO_2 and the second derivative of the absorption spectrum, the latter showing a more detailed structure than the direct absorption spectrum. By choosing a sharp peak in the second derivative spectrum and scanning a narrow band of it (necessary to generate the second derivative), the sensitivity and the specificity of the analyzer are improved over the same quantities of the direct absorption spectrophotometric analyzers.

Sulfur dioxide analyzers are also available that continuously irradiate a sample of stack gas with short wavelength ultraviolet light, resulting in a

* More details are supplied in Chapter 6.

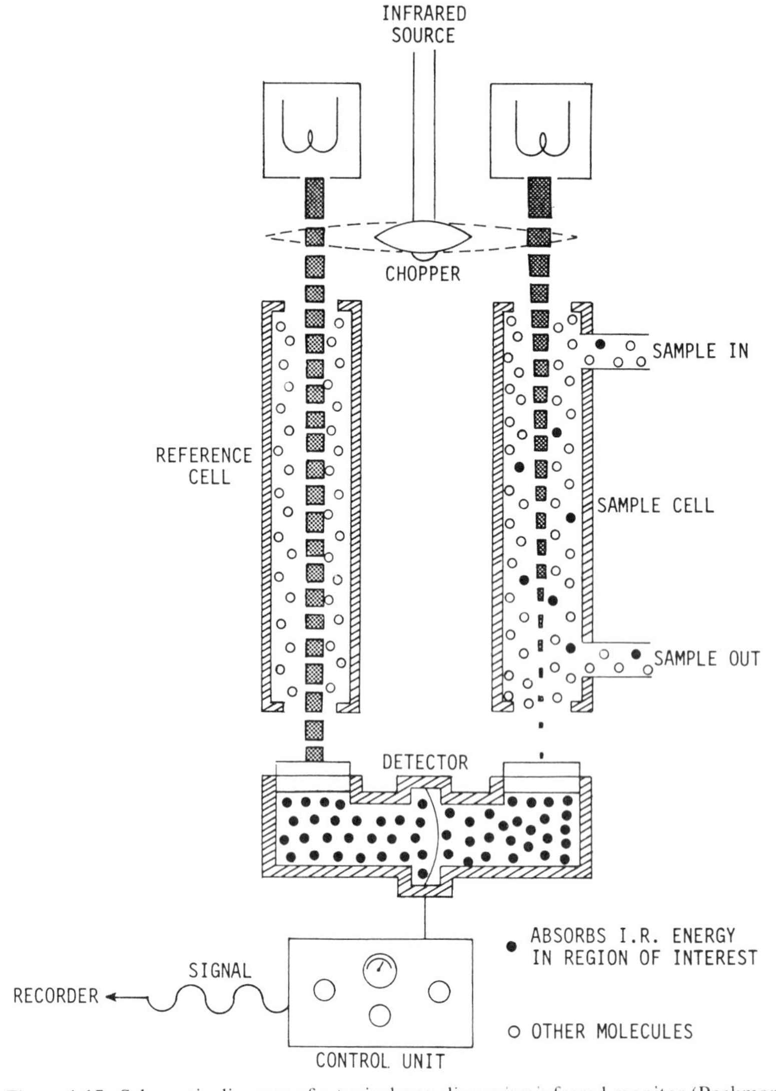

Figure 1.15. Schematic diagram of a typical non dispersive infrared monitor (Beckman).

longer wavelength UV fluorescence, which is passed through a narrow band filter to a photomultiplier tube that measures its intensity. In the lower range of SO_2 concentration (0 to 500 ppm) the response is linear; above 500 ppm the high concentration of SO_2 reabsorbs some of the fluorescent light, and the response becomes logarithmic. Particulate matter and water vapor and probably other common gases interfere with this method.[34]

Ultraviolet analyzers for SO_2 and NO_2 monitoring have been developed

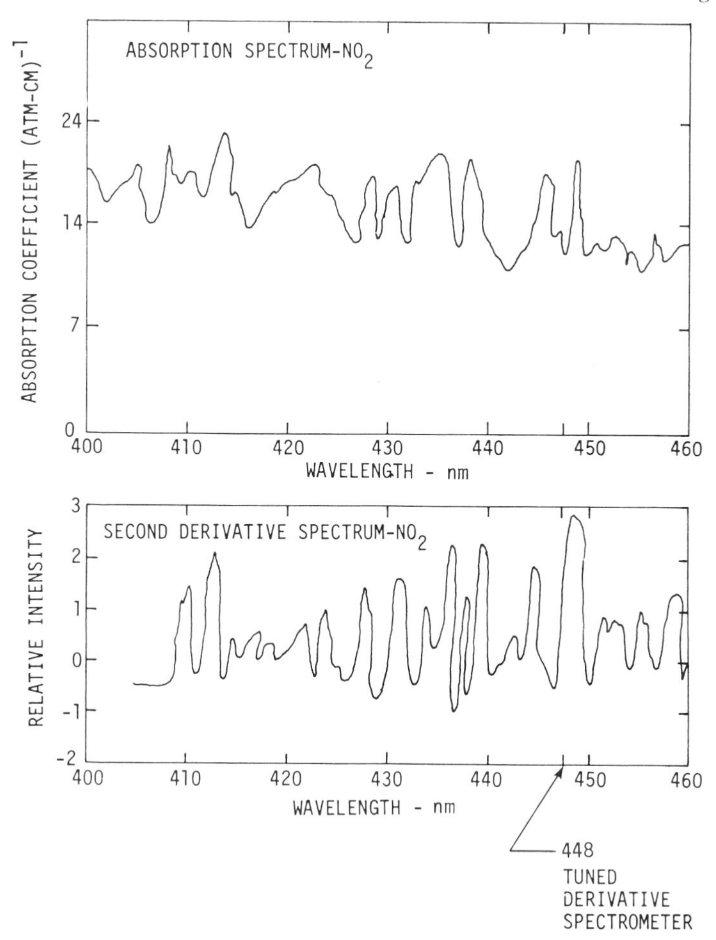

Figure 1.16. Direct and second derivative absorption spectra for NO_2.

in which the sample gas is exposed to selected wavelengths of nondispersive ultraviolet light and focused on the entrance slit of a monochrometer. A plane grating disperses the incident radiation and two adjacent wavelengths (one where the sample absorbs and one where it does not). The ratio of the energy in these two wavelengths in proportional to the concentration.

1.3.2. Extractive Monitoring

Instrumental techniques can be adapted to almost any gas that can be analyzed by manual techniques, but a critical factor is handling the sample between the vent and the analytical instrument. With manual methods, the

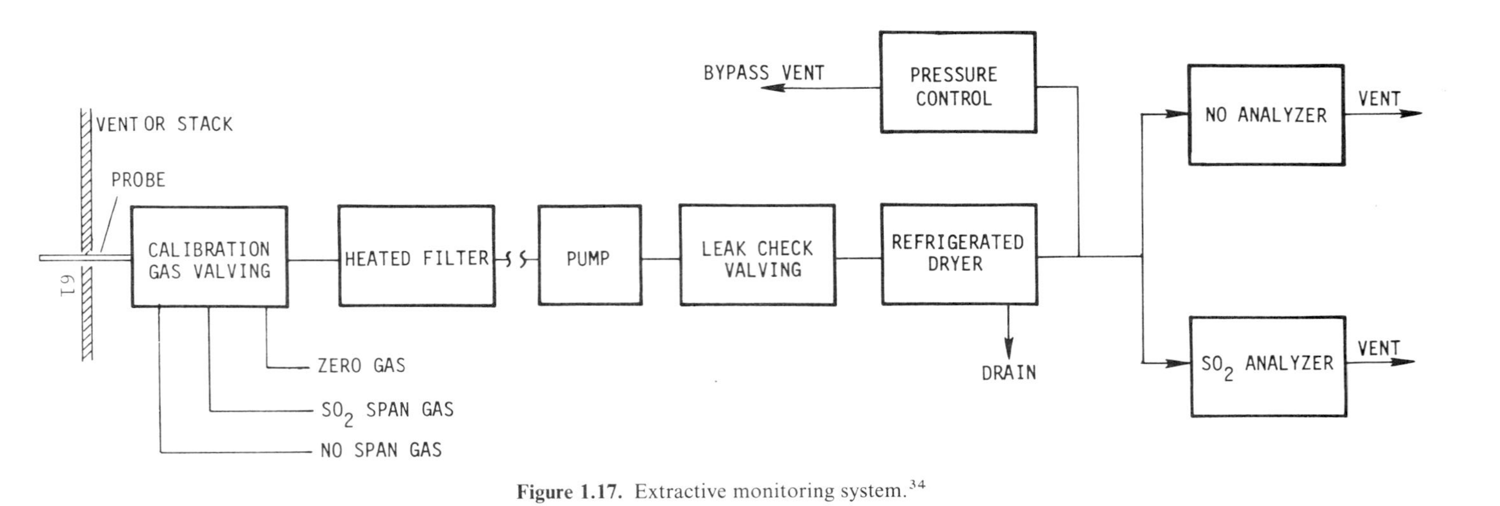

Figure 1.17. Extractive monitoring system.[34]

equipment is placed near the vent and the sample is collected immediately for subsequent transfer and analysis. This close coupling of the vent and the sample collection equipment reduces potential reactions between sample lines and gases, condensation in lines, and gas interactions. Instrumental analyzers are usually located farther from the actual vent to facilitate instrument maintenance and observation. Some instruments require sample conditioning (i.e., cleaning, cooling, and/or drying before analysis), and this requires additional sample handling and treatment. Limited quantitative work has been done in evaluating the effect of the sample handling system on the measured gas composition. Thus there may be some question about the accuracy of the readings obtained by a monitoring instrument, even though the instrument itself is working perfectly.

To reduce these potential problems, the gas conveying system should be kept free of all particulate matter; it should be airtight, and it must be fabricated of nonreactive and noncorrosive materials. A heated filter placed near the vent may be used to keep the sample line clean and to avoid condensation within the filter. The filter itself must be nonreactive and should be easily changed to prevent particulate buildup. Calibration gases should be injected near the stack, to expose these gases to the same environment as the sample gas. Stable gas mixtures in compressed gas cylinders are used for calibration. Purified air may be used as a "zero" gas. Figure 1.17 shows a typical extractive monitoring system as applied to a combustion process.

Water, collected in the refrigerated dryer (condenser), presents one of the major sampling problems. This is especially true in high humidity gas streams, since the interaction of water with the gas must be minimized; condensed water should be removed from the gas stream as quickly as possible.

1.3.3. *In situ* Monitoring

Instruments that analyze the stack gases as they flow past a measurement site in the stack itself are referred to as *in situ* or in-stack monitors. These devices, which employ electromagnetic energy to determine concentrations of SO_2, NO, CO, NH_3, O_2, CO_2 and particulates, have the inherent advantage of continually scanning an entire cross section of stack gases, thus reading a truly average concentration. In addition, since the *in situ* monitors require no sample extraction or conditioning system, possible sample interaction is eliminated. *In situ* devices for particulates are discussed in Chapter 2.

These instruments have the disadvantage of being difficult to calibrate dynamically and to maintain because they are frequently mounted near an inaccessible part of a vent or stack exit, and frequently they are subjected to very corrosive gases, high temperatures, and particulate matter. The instruments are usually too big to be readily portable. Provisions must be made to prevent deposition of particulate on the optical surfaces and to maintain alignment of energy source and detectors.

In situ monitors using dispersive infrared and ultraviolet light, and correlation spectrophotometers have been developed mainly for SO_2, NO, and CO.

1.3.4. Remote Sensing

In remote sensing, as the name implies, the analyzer is not near the stack but is mounted at ground level at a convenient point. The instrument is aimed at the stack exit and senses the temperature and concentration of SO_2, NO, CO_2, or CO in the stack gases as they leave the stack. Some allowance must be made for dilution of the gases after they leave the stack. Correlation spectroscopy has been mainly used for this purpose, and work is in progress to use laser excitation and Raman back scattering, as well as a number of other potential techniques (Table 1.6).

In the correlation spectrometer, an optical system focuses the area under observation on a mirror grating that generates the absorption spectrum of the plume.[37] An oscillating mask, cut to conform to the absorption spectrum

Table 1.6. Development Status of Techniques for Remote Monitoring of Gas Emissions[36]

Analytical Scheme	Pollutants	Development Status
Dispersive correlation, UV and visible	SO_2, NO_2	Commercially available
Dispersive, IR emission	SO_2, NO, CO, CO_2, HF, HCl	Research prototype
Gas filter correlation, IR emission	SO_2	Field prototype
Diode laser, IR absorption	SO_2, CO	Research prototype
Gas filter correlation, IR emission	NO, H_2S	Feasibility
Vibrational Raman, UV	SO_2, NO	Research prototype
Laser heterodyne, IR	SO_2, CO_2	Feasibility
Rotational Raman, UV	NO	Feasibility
Fluorescence, UV	SO_2	Feasibility
Differential absorption lidar (DIAL), UV	SO_2, NO	Feasibility

of the desired gas, alternately passes the light of the absorption spectrum or the background light to a detector. The electronics following the detector are phased with the oscillating mask and generate two signals, one correlating with the absorption spectrum and one with the background light. The two signals are compared to obtain the total amount of pollutant "seen" by the spectrophotometer. The method is quite specific and can be calibrated by use of a sample cell filled with a known concentration of the pollutant being measured.

REFERENCES

1. Zakak, A. et al., Procedures for Measurement in Stratified Gases, Vol. 1, U.S. Environmental Protection Agency, Publication No. EPA-2-74-086-a, Washington, D.C., 1974.
2. Achinger, W. C. and R. T. Shigehara, *J. Air Pollut. Control Assoc.*, 18, 605 (1968).
3. Nader, J. S., *J. Air Pollut. Control Assoc.*, 23, 587 (1973).
4. Hollowell, C. D. and R. D. McLaughlin, *Environ. Sci. Technol.*, 7, 1011 (1973).
5. Calvert, S. and W. Workman, *Ind. Hyg. J.*, 318–324 (August 1961).
6. Roberts, L. R. and H. C. McKee, *J. Air Pollut. Control Assoc.*, 9, 51 (1959).
7. Air Sampling Instruments, American Conference of Governmental Industrial Hygienists, Cincinnati, Ohio, S-11, 1972.
8. Shell Development Company, Emeryville, Calif., Method Series 4516/59a.
9. Corbett, P. F., *J. Inst. Fuel*, 24, 237 (1961).
10. Shigehara, R. T., private communication, U.S. Environmental Protection Agency, Research Triangle Park, N.C.
11. Devorkin, H. et al., *Air Pollution Source Testing Manual*, Los Angeles, 85, 1965, p. 85.
12. Dee, L. A. et al., An Improved Manual Method for NO_x Emission Measurement, U.S. Environmental Protection Agency, Publication No. EPA-R2-72-067, Washington, D.C., 1972.
13. Margolis, G. and J. N. Driscoll, *Environ. Sci. Technol.*, 6, 727 (1972).
14. American Society for Testing and Materials (ASTM), Designation D 1608-60, Philadelphia.
15. Hamil, H. F., Southwest Research Institute, San Antonio, Texas, Publication 01-3462-001, 1973, p.44.
16. *Fed. Reg.*, 39 (October 23, 1974).
17. *Standard Methods for the Examination of Water and Waste Water*, American Public Health Association, 12th ed., 1965, New York, p. 144.
18. Devorkin, H. et al., *Air Pollution Source Testing Manual*, Los Angeles, 1965, p. 82.
19. Elfers, L. A. and C. E. Decker, Paper presented at the 154th Meeting of the American Chemical Society, September 15, 1967.
20. *Atmospheric Emissions from Hydrochloric Acid Manufacturing Processes*, Public Health Service, U.S. Department of Health, Education and Welfare, No. AP-54, 34, 1969, Washington, D.C.

21. *Standard Methods for Chemical Analysis*, 6th ed., Vol. 1, Van Nostrand, New York, 1962, p. 329.
22. Devorkin, H. et al., *Air Pollution Source Testing Manual*, Los Angeles, 1965, p. 100.
23. *Ibid.*, p. 66.
24. *Air Sampling Instruments*, American Conference of Governmental Industrial Hygienists, Cincinnati, Ohio, 1972, p. R-7.
25. Carruth, G. F. and R. Kobayashi, *Anal. Chem.*, 44, 1047 (1972).
26. Russell, J. W., *Environ. Sci. Technol.*, 9, 1175 (1975).
27. Gadomski, R. R. et al., *J. Air Pollut. Control Assoc.*, 24, 484 (1974).
28. Intersociety Committee, *Methods of Air Sampling and Analysis*, American Public Health Association, Washington, D.C., 1972.
29. American Petroleum Institute, Method 772-54, Washington, D.C., 1954.
30. *Fed. Reg.*, 39, No. 47 (March 8, 1974).
31. Flamm, D. L. and R. E. James, *Environ. Sci. Technol.*, 10, 159 (1976).
32. Nader, J. S. et al., U. S. Environmental Protection Agency, Publication EPA-650/2-74-013, 14, Washington, D.C., 1974.
33. *Field Operations Guide for Automatic Air Monitoring Instruments*. PEDCo–Environmental Specialists, Cincinnati, Ohio. U.S. Environmental Protection Agency Contract EPA-70-124, 1971.
34. Homolya, J. B., *J. Air Pollut. Control Assoc.*, 25, 809 (1975).
35. Chapman, R. L., *Environ. Sci. Technol.*, 8, 521 (1974).
36. Chemistry and Physics Laboratory. U.S. Environmental Protection Agency, Research Triangle Park, N.C., November 1974.
37. Moffat, A. J. and M. M. Millan, *Atmospheric Environment*, Vol. 5, Pergamon Press, New York, 1971, p. 677.

2

MONITORING OF PARTICULATES

D. H. Lucas

Environmental Section, Planning Department,
Central Electricity Generating Board, London.

2.1. THE NATURE OF DUST POLLUTION

2.1.1. The Possible Effects

There is no effect produced by dust as a pollutant that is a function of the mass emission alone. All effects are functions of the dust—its mass, size grading, and chemical activity, as well as meteorological variables and other factors such as emission height. Even if the adverse effect being considered was the possibility that a structure might be crushed by the sheet weight of airborne dust lying on it, (an effect that does not need to be seriously considered), the amount of dust that would arrive on the structure in a particular place would be a function of the grading of the dust being emitted.

There are three main possible adverse effects of dust emission. The first is that dust will be inhaled by human beings and cause an adverse medical effect in the lungs. It is widely recognized that only particles smaller than a few micrometers (μm) are readily able to enter the lungs through the body's normal defenses. Larger particles may enter the nose but are ultimately rejected by biological action.

Thus the health effect is very much a function of particle size. It is usually considered to be very small for particles 10 μm or more, but rises to a peak at about 1 μm. By coincidence this function is probably very similar to the size function for the effect of dust on atmospheric obscuration (Section 2.1.12). Here monitoring requirements as related to size are similar to those for preventing obscuration of the atmosphere. It seems likely that if the other adverse effects considered below are dealt with, the effect of dust on health will be more than adequately minimized apart from toxic dusts. Health effects therefore are not considered further.

The adverse effect that has been most positively associated with dust emission is the tendency for dust to deposit where it produces a deterioration in visual appearance. This deposition was considered to be caused by gravitational settlement when each particle fell to the earth at its own characteristic free falling speed, although it was soon recognized that atmospheric turbulence had a strong statistical effect on the process. The result of "deposition" referred to as a "deposit" has been traditionally measured in Britain by the (British) standard "deposit" gauge, and worldwide by an instrument of similar principle having a generally cylindrical shape. The dust was intended to settle through the upper open end of the cylinder. Instruments of this type have been very widely used, but accuracy of the measurements is limited (see Appendix, Section A2.1.1).

This simplistic approach has been modified by the recognition that dust particles in the atmosphere are in general moving in a direction much closer to horizontal than to vertical. Unless the surface of the earth is perfectly smooth, they tend to impact on projections from the earth's surface, such as trees, houses, and blades of grass. The material arriving on a surface by this mechanism is referred to as an "imposit" by analogy with "deposit." A device to measure impaction known as the directional dust gauge has been in use in recent years (see Appendix, Section A2.1.2).

2.1.2. Impaction

The scientific study of the process of impaction is complex. In general if dust particles are being carried by a gas stream toward an object, only a

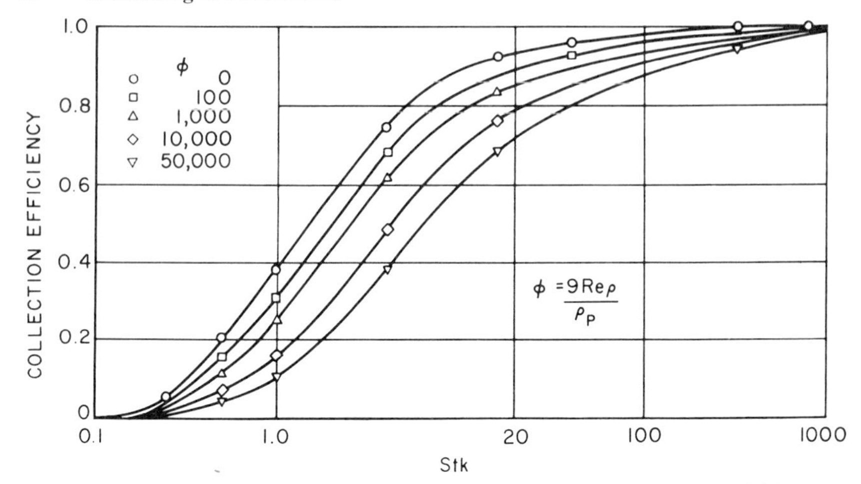

Figure 2.1. Dependence of target efficiency on Stokes number and other variables.

fraction—the target efficiency—of those aimed at the cross section of the object will in fact reach its surface. The target efficiencies for various shapes have been extensively studied.

Figure 2.1 shows how the target efficiency of a cylindrical shape varies under different circumstances. The abscissae for the curves are Stokes numbers, which are defined as follows: if a particle is moving in a gas stream with a certain velocity, it would continue to travel forward for a certain distance if the gas velocity were abruptly brought to zero. This is the stopping distance L for the particle. If the stopping distance is divided by the characteristic length for the obstacle (for a cylinder, the characteristic length will be the radius), this ratio is the Stokes number (Stk $= 2L/D$).

The stopping distance of a particle as a function of its initial Reynolds number and the other relevant variables are given in Figure 2.2.

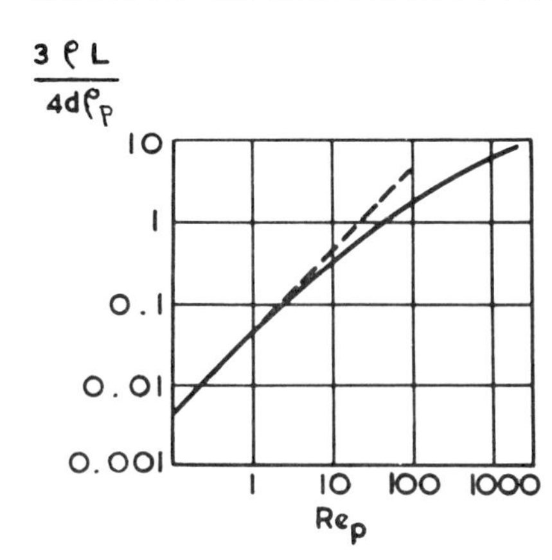

Figure 2.2. Dependence of stopping distance on particle size and particle Reynolds number.[1]

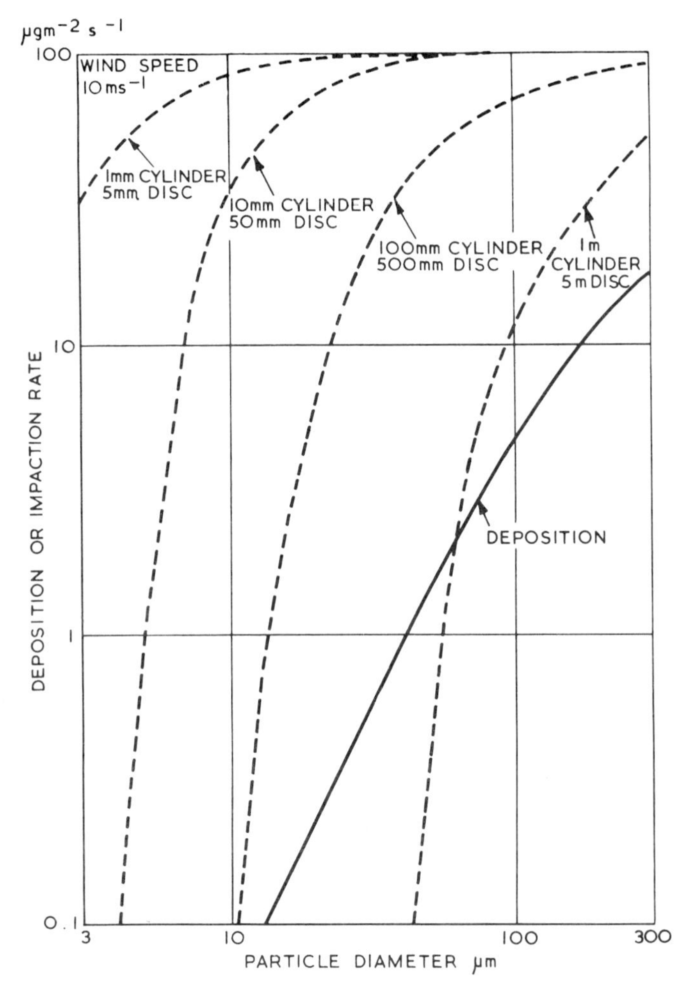

Figure 2.3. Dependence of impaction and deposition on particle size and target size, high wind speed.

It is possible to evaluate target efficiencies for various particle sizes, gas velocities, and targets by the following steps:

1. Using Figure 2.2 to obtain the stopping distance L

$$Re_p = \frac{vd\rho}{\mu} \qquad \frac{\rho}{\mu} = 6.8 \times 10^{-4}\ \text{sec m}^{-1} \qquad \text{for air at } 15^\circ\text{C} \tag{1}$$

2. Evaluating Stokes' number as

$$\frac{2L}{D_o} \tag{2}$$

3. Evaluating

$$\varnothing = \frac{9\text{Re}\rho}{\rho_p} = \frac{9vD_o\rho}{\mu} \times \frac{\rho}{\rho_p} \tag{3}$$

4. Reading off the target efficiency equivalent to the Stokes number from Figure 2.1, using $\varnothing$ to interpolate between the curves.

Some representative results appear in Figures 2.3 and 2.4 for particles of specific gravity 2. Here the simplifying assumption is made that there is an arbitrary horizontal flux of dust of 100 μg m^{-2} sec^{-1}.

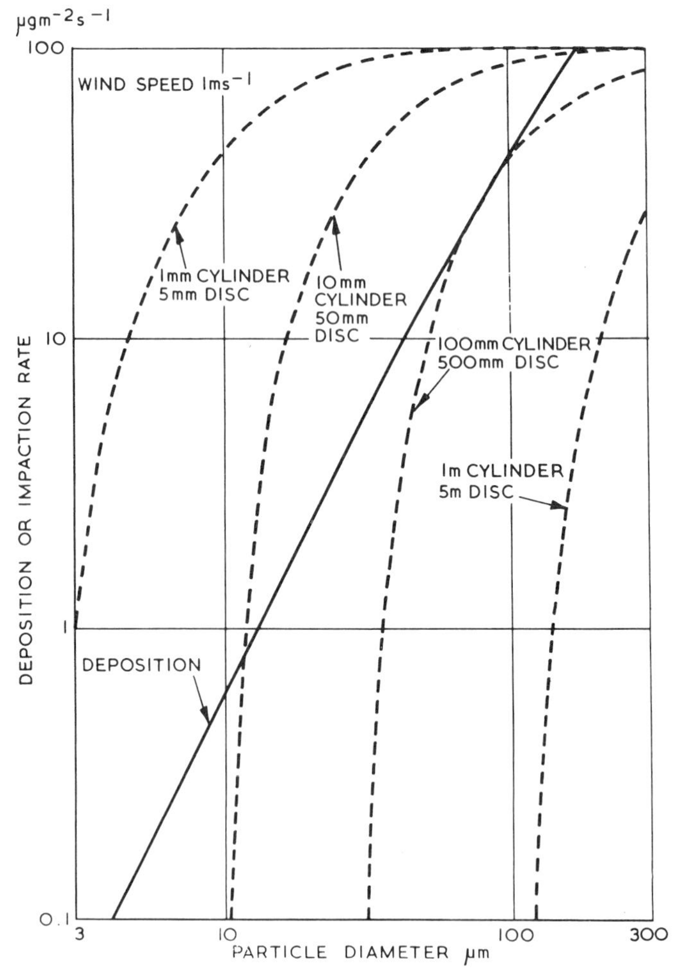

Figure 2.4. Dependence of impaction and deposition on particle size and target size, low wind speed.

Since the cylinder is one of the least efficient shapes for intercepting particles, other shapes are of importance. No simple, general comparison is possible for low collection efficiencies. However at fairly high efficiencies (> 30%) a sphere has a target efficiency equal to that of a cylinder when its diameter is $\frac{4}{3}$ of the cylinder diameter. For a ribbon, the width is 2.5 times the cylinder diameter, and for a disk the diameter is 5 times the cylinder diameter.[2]

2.1.3. Deposition

The movement of dust in the atmosphere is a combination of the movement of a gaseous effluent and the effect of its free falling speed under gravity. For smaller particles their behavior in the atmosphere is similar to the behavior of a gas. If a concentration of dust is postulated *near to ground level*, the rate of deposition of the dust on a smooth surface is given by the concentration times the free falling speed. If it is assumed arbitrarily that the flux is 100 $\mu g\ m^{-2}\ sec^{-1}$ the concentration will be inversely proportional to the wind speed. In practice this will be approximately correct because emission rates, hence fluxes, tend to be independent of wind speed. The net result is that for smaller particles, *maximum* deposition rates near a source are inversely proportional to wind speed and directly proportional to particle free falling speed. For larger particles, where the free falling speed dominates movement by diffusion, the maximum deposition near a source moves nearer to the source as the wind speed falls. Thus the maximum deposition rates increase more rapidly with fall of wind speed and increase of particle free falling speed than for smaller particles. *For the arbitrary flux specified*, however, the curves shown are approximately correct.

On this basis, Figures 2.3 and 2.4 give deposition rates for a range of particle sizes (specific gravity 2) for two wind speeds.

2.1.4. Interdependence of Impaction and Deposition

It is apparent from Figures 2.3 and 2.4 that impaction rates are generally much higher than deposition rates except for low wind velocities, with large obstacles and/or large particles. This is clearly shown in Figure 2.5, where the impaction rate on a vertical disk 50 mm in diameter (analogous to a leaf) is compared with deposition rate on horizontal ground. Even at 1 m sec^{-1} wind speed, deposition rate exceeds impaction rate only when the particle size is greater than 180 μm.

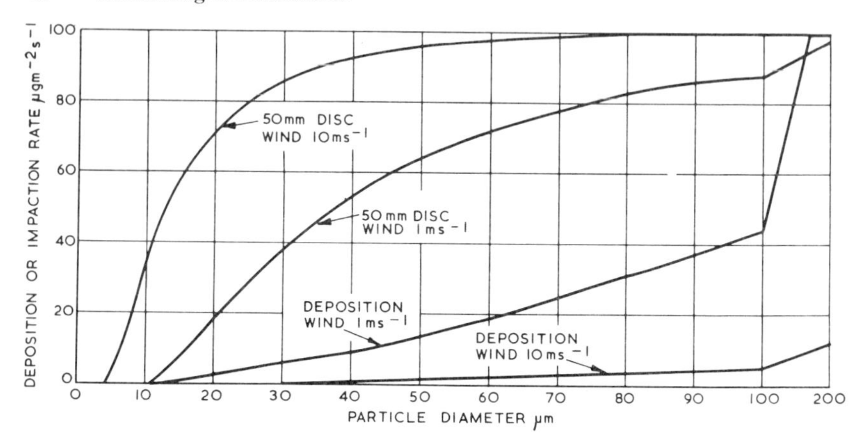

Figure 2.5. Comparison of impaction on a 50 mm disk with deposition on smooth ground.

This implies that under typical conditions, leaves of a tree can be covered with dust to a much greater degree than is possible for horizontal ground. Similarly the vertical surfaces, such as fence posts, will be more heavily coated than the ground.

In a grass-covered area the impaction rate on the tips of the blades of grass (Figure 2.6) will be much higher than the deposition rate on the ground. Indeed all the dust will be stripped from the atmosphere by impaction. However each blade of grass will protect an area behind it, and the *average* rate at which dust is taken up by an area covered with grass will be equal to the deposition rate.

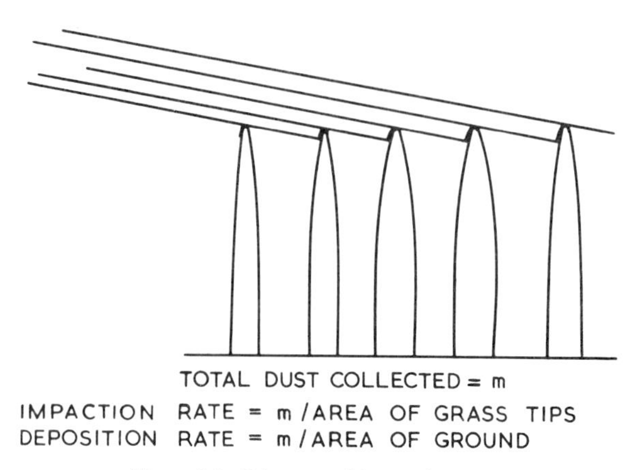

Figure 2.6. Diagram of impaction on grass.

2.1.5. Visual Effects of Deposition and Impaction

For toxic dust the mass rate of deposition is probably directly related to the hazard, for instance, to grazing animals. For nontoxic dusts the most serious effect is visual. The appearance of a surface is much more closely related to the surface area of the settled dust than to its mass. In allowing for this, the curves of Figures 2.3 and 2.4 are modified to give the curves of Figures 2.7 and 2.8, which indicate that apart from the most efficient targets and strong winds, visual effects due to impaction increase with particle size,

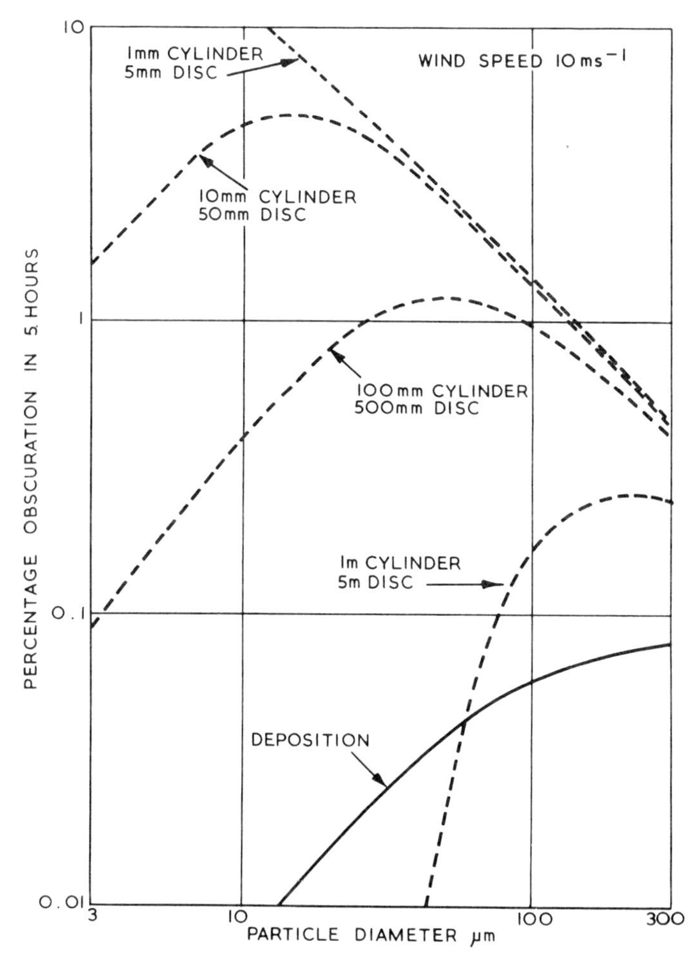

Figure 2.7. Dependence of percentage obscuration due to impaction and deposition on particle size and target size, high wind speed.

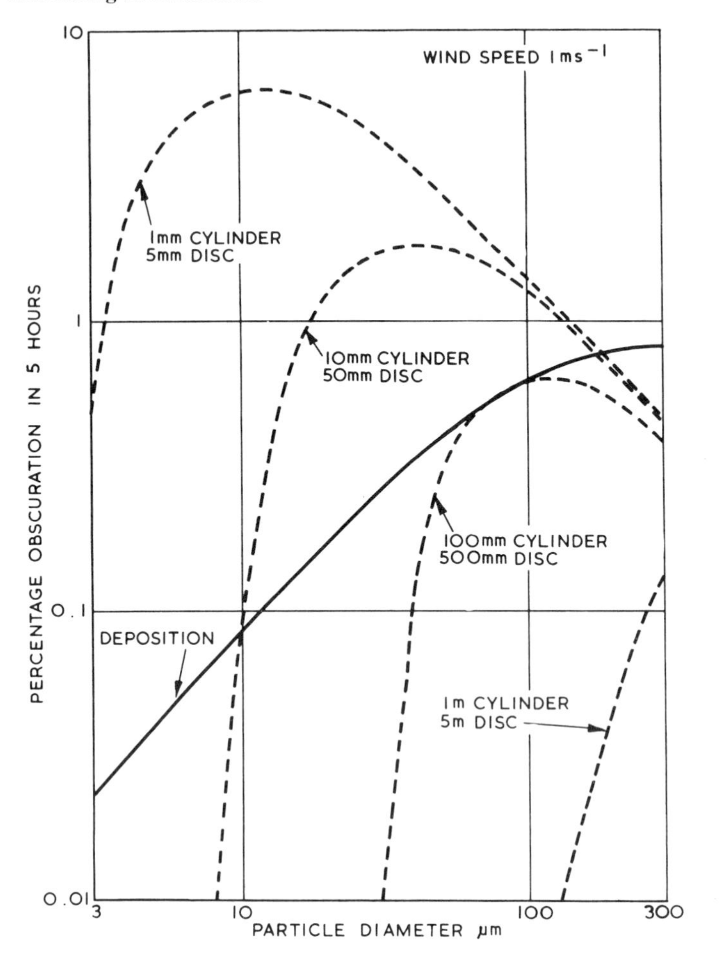

Figure 2.8. Dependence of percentage obscuration due to impaction and deposition on particle size and target size, low wind speed.

at least up to 10 μm. For larger and less efficient targets and lighter winds, visual effects increase up to approximately 100 μm. They then decrease, because similar masses of dust have a smaller area for larger particles sizes.

When particle sizes are larger than 100 μm, however, it is possible to see individual particles, and it is questionable whether the subjective impact decreases with particle size. The objection to these larger particles probably rises with particle size. Where dust settles by deposition, the effect continues to rise with particle size even after allowance for surface-mass effects.

2.1.6. Other Methods of Settlement

Thermal Deposition. When warm air is passing over a cold surface, very small particles are driven down the temperature gradient by differential molecular bombardment.

Inertial Deposition. When turbulent air is passing over a surface, particles larger than a certain size obtain enough lateral movement from the turbulence to pass through the boundary layer.

Brownian Movement Deposition. Particles move in a random way because of the statistical variation of molecular bombardment.

Static Charge Attraction. Electrically charged surfaces of uneven shape attract even uncharged dielectric particles.

All these forces cause particles to settle on surfaces. Typically all the effects have settlement rates *much* lower than impaction and deposition rates.

2.1.7. Chemically Active Dusts

Although most dusts are inert, some, such as cement dust or black smoke, are able to bond themselves to surfaces and can subsequently resist erosion by rain or wind. Soluble dusts can bond, and bonding resists wind, but not rain.

2.1.8. Integration Times

The effects of dust are usually cumulative, and the worst effect produced is the integral of the rate of settlement over a period of time. This period is limited by the occurrence of rain, strong winds, or manual cleaning. The integration time varies enormously depending on place and time. It is convenient to assume that its order of magnitude is 10 days in outdoor situations (apart from chemically active dusts), although there are many special cases.

2.1.9. The Pollution of Buildings

A building is so large a target that on the basis of data in Figures 2.3 and 2.4, we might expect it to be immune from impaction. However every building has features with quite small dimensions. It is these, referred to local velocities, which are soiled by impaction. Furthermore, buildings are subject to the "other methods" of settlement, and although these settling rates are very low, when chemically active dusts are involved, the integration time becomes years or even decades. Hence buildings have in the past been heavily polluted, but the long-term pollution is restricted to chemically active dusts. Because of this long integration time, chemically active dusts need special consideration in determining their allowable emission rates, when compared with inert dusts.

2.1.10. Dust in the Eyes

One of the more serious complaints made about dust pollution is usually associated with strong winds. The "target"—the human head—is roughly a sphere of diameter 150 mm and corresponds to a cylinder of about 100 mm. An increasing chance of impaction as particle size increases can be deduced from Figure 2.3. The chance of eye irritation also increases with particle size.

2.1.11. Dependence of Deposition and Impaction on Particle Size

In all the cases considered thus far the effect increases with particle size. In the case of impaction there *may* subsequently be a decrease in the effect because of the decrease of surface to volume ratio in larger particles. Where this happens in the range above 100 μm, it is doubtful whether the subjective response actually decreases with size, because the particles become individually visible.

2.1.12. Effects on the Atmosphere

When dust or other particulate matter is emitted from a chimney, it may produce a visible plume. This is sometimes objected to because the observer tends to impute toxicity to the emission. When first emitted, the material can cause an obvious reduction in sunlight. When the emission is diffuse and when many plumes have mingled, a loss of sunlight is produced that may be real but not obvious. It may also produce a hazy appearance, and a reduction of the distance of clear visibility that is obvious and objectionable. Whether the optical effect is by light absorption or scattering, it is a function of the area, not of the mass of the particles. On an equal mass basis, therefore, fine dust is more effective in affecting light than is coarse dust, and the net optical effect is inversely proportional to particle diameter.

The optical effects are very complex when the particle diameter is near the wavelength of light, which is of importance with monodisperse dusts;

with typical size distributions of dust, however, these are of secondary importance.

The size dependence of effects in the atmosphere is both simpler than, and different from, the size dependence of effects on the ground. Other things being equal, it is inversely proportional to the diameter of the particles down to about 1 μm. Below this the effect decreases with particle size. In typical situations fine particles in the micrometer range predominate in atmospheric effects.

2.2. THE MONITORING OF DUST EMISSION

In most cases it is practical and economic to reduce dust emission by careful control of industrial plants and by the use of dust collectors. Particularly on large installations, dust collectors tend to be unreliable unless they are carefully maintained and skillfully operated, even when the original design is excellent. It is therefore necessary to check dust emission levels. In the past this has been done by manual sampling from the flue gas followed by weighing (which is fully described in Chapter 3). This indicates the best performance the plant can make under the test conditions, but it may have little relevance to the actual performance of the plant during the 8760 hours of each year. The manual test was, and a large extent still is, the only recognized check on emission. Since it measures mass burden, most of the legal controls of dust emission are in terms of this variable. However mass burden gives no indication of the effects caused by dust emission, either on the ground or in the atmosphere, unless the collected dust is also graded. Even with this precaution results are sometimes misleading because the dust collected in flue ducts is often agglomerated. Thus its effective grading is different from its grading after sampling. Continuous monitors that also respond to size have therefore been developed.

2.2.1. Continuous Monitoring by Opacimeter

In principle it was easy to shine a light through the flue gas and measure the amount of light that penetrated, and this rapidly became a method of continuously monitoring dust emission.

2.2.1.1. Principles of Operation and Sources of Error

The light beam is directed onto a photocell. As the dust concentration in the flue gas increases, the reading on the photocell decreases. When the concentration is zero, the photocell gives its maximum reading. This is *indicated* as zero. This inverse relationship between light intensity and indicated reading is a characteristic feature of the instrument. If any variation of the light intensity occurs for reasons other than a change of dust quantity, it causes a zero error, which tends to be large compared with the zero error of instruments having a direct and not an inverse relationship.

In spite of its apparent simplicity, the opacimeter is a difficult instrument to operate accurately. Zero errors can be caused by the following conditions:

1. A change in optical alignment between light source and photocell.
2. Any dirtying of the windows, lenses, reflectors, and other parts of the optical system—particularly, of course, the windows, which isolate the instrument parts from flue gas.
3. Any "ageing" of lamp or electronic components.
4. Any variation of supply voltage.

Many commercially used instruments have been severely affected by some or all these conditions. The errors have been largely undetected because, on a basic instrument, a clean flue gas duct is required to reveal a zero error. In modern instruments there are varying degrees of sophistication to minimize zero error. The following progression gives increasing accuracy:

1. The stabilization of supply voltage.
2. The preaging of components.
3. The use of geometrical complexes to prevent dust reaching observation windows (Figure 2.9).
4. The use of filtered air blowing over the windows to keep them clean.
5. The splitting of the light beam—one part of the beam being used as a comparison with the other part, which passes through the flue gas. This reduces the effect of any changes of light emission from the light source.
6. The modulation of the light beam to improve amplifier dc stability—even the modulation of the two light beams at different frequencies, to permit the use of a single amplifier and a single photocell can be used.
7. The modulation of the light beam followed by presenting the main beam and the reference beam alternately to a single photocell.
8. The use of a reflector to fold the beam so that light source and photocell are mounted rigidly side by side. The reflector always reflects light back on its original path, even though its orientation changes slightly.

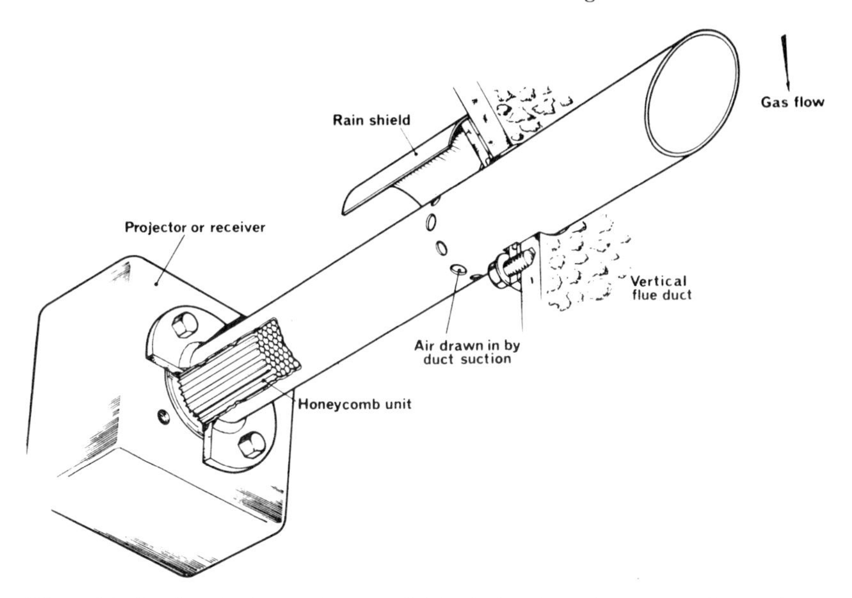

Figure 2.9. Ventilated tube and honeycomb labyrinth to keep observation window clean.

9. The automatic check of zero (for an instrument as in 8) by reflecting the light beam back to the photocell without traversing the flue gas (the windows and the mirror escape this check).
10. The mounting of the light source and the photocell on a single rigid framework that is cantilevered from one side of the duct. This permits the checking of zero by removing the whole instrument from the duct (Figure 2.10).

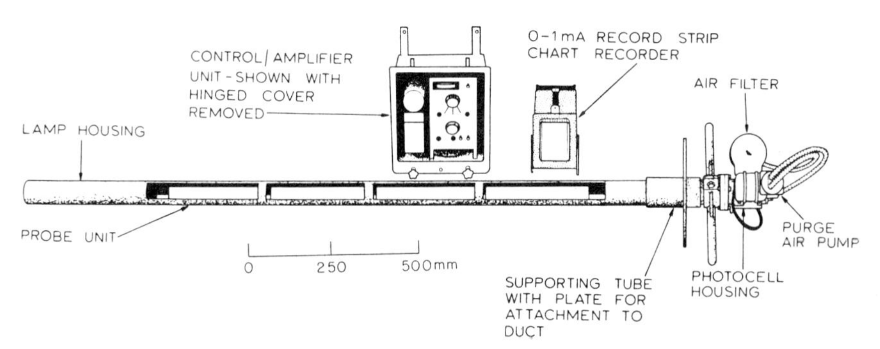

Figure 2.10. Removable opacimeter with rigid connection between light source and photocell.

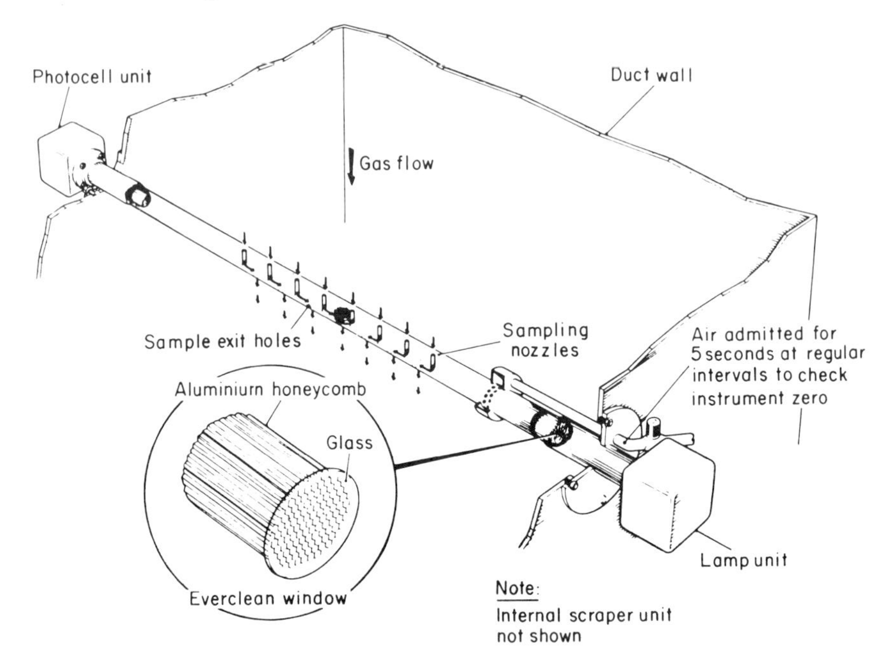

Figure 2.11. Rigid opacimeter with automatic zero checking.

11. The mounting of the light source and the photocell on a simple rigid tube that traverses the flue duct and is pierced by holes, to ensure that it is normally ventilated with flue gas. This permits the automatic checking and recording of zero by occasionally ventilating the tube with clean air sucked or blown into the tube (Figure 2.11).
12. Either automatically or manually introducing a known neutral filter into the light beam to check the span error as distinct from the zero error. The span error—the change of calibration of the instrument—is not usually a serious problem.

2.2.1.2. Monitoring by Reflectometer

Instruments are in commercial use, though they are few compared with opacimeters, in which the light source and the photocell are mounted side by side and the photocell measures the light reflected by the dust. Zero errors are less likely on this type of instrument because it does not have the inverse response of the opacimeter. The basic difficulty of keeping an observation window clean remains, and the instrument has the further disadvantage, in combustion monitoring at least, of being insensitive to black smoke, giving less response as the dust becomes darker. The reflectometer has the same particle size response as the opacimeter.

2.2.1.3. Application of the Opacimeter

It is sometimes claimed that the opacimeter can be calibrated in terms of mass burden, but this is only justified if the particle size is constant. In most cases of dust emission, particle sizes change appreciably from time to time and opacimeters cannot be used to measure dust burden.

However they have the great merit of giving a response that is directly related to dust effects in the atmosphere, and they are ideal for control measures aimed at improving atmospheric visibility. They are particularly suitable for giving warning of the emission of black smoke—a not uncommon failing of combustion plants.

The opacimeter is, however, very insensitive to dust at the coarser end of the spectrum and is not suitable for measuring dust likely to cause complaint because of impaction or deposition effects at ground level. For this reason an instrument aimed particularly at this cause of complaint has been developed and is named an impacimeter.

2.2.2. Monitoring by Impacimeter: Principles of Operation

Whereas the opacimeter is an analog of dust-obscuring light in the atmosphere, the impacimeter is an analog of the impaction of dust on surfaces and the obscuring of light illuminating the surfaces. It measures the *im*paction of dust and the subsequent surface *opacity* caused, hence its name.

The impacimeter[3] has a sampling nozzle that is not aspirated (Figure 2.12); that is, it has a sampling gas velocity of zero. Dust enters the nozzle area by an inertial process almost identical to impaction; instead of settling on this area, however, it settles on an identical area at the bottom of the vertical column, to ensure that it is not reentrained. The collecting area is of glass, and the opacity of the dust as it collects is continuously recorded using a light source and a photocell located outside the duct (Figure 2.13). The dust is allowed to collect for a specified time, usually 15 min, during which the recorded opacity increases on the recorder chart. The dust is then blown out of the collecting chamber by compressed air (assisted if necessary by a scraper), and the recorded opacity falls to a level that is the effective zero and the starting point for the next cycle. The reading appropriate to the cycle is the peak reading in percentage obscuration per

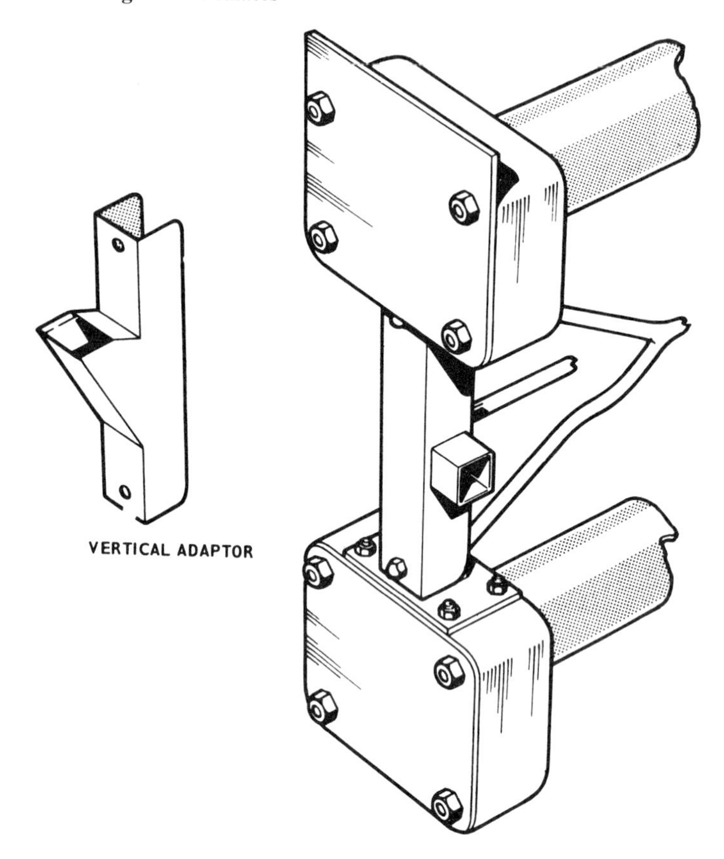

Figure 2.12. Sampling nozzle of impacimeter with alternative nozzle for vertical flue gas flow.

minute (popm) minus the zero reading, if any, at the start of the cycle. Hence the reading is intrinsically free from zero error. The span error can be checked manually at will.

2.2.2.1. Application of the Impacimeter

The response curve for the impacimeter appears in Figure 2.14, contrasted with the curve for an opacimeter. It exhibits an increase in response with particle size up to about 100 μm, then a decline. Often particles larger than 100 μm are agglomerated, particularly in the outlet from electrostatic precipitators. These tend to break up or spread on the glass surface, and the size response for agglomerates may well continue to increase above 100 μm.

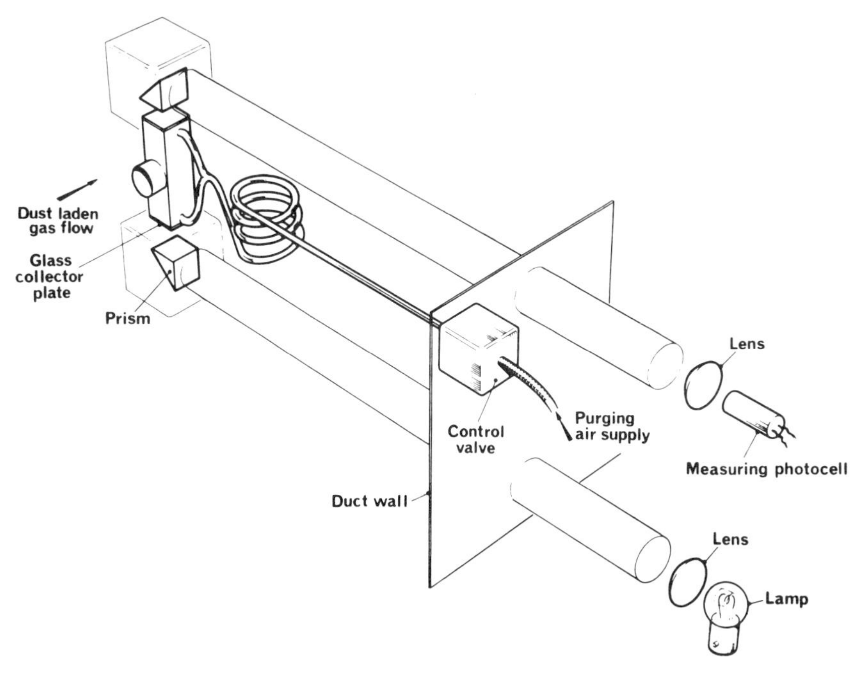

Figure 2.13. Optical system of impacimeter.

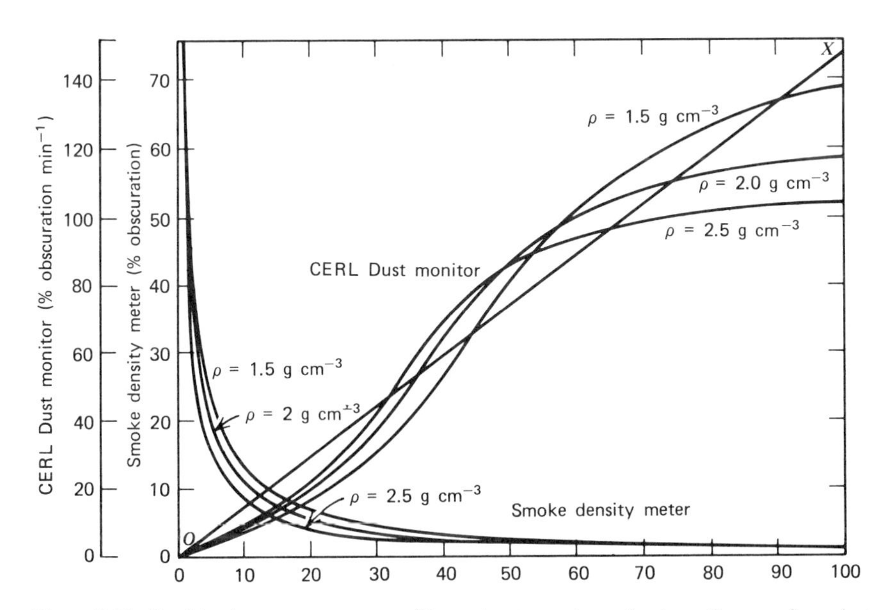

Figure 2.14. Particle size response curves of impacimeter and opacimeter: all curves for a dust concentration of 0.23 g m^{-3} (0.1 g ft^{-3}) actual at a gas velocity of 12.2 m sec^{-1} (40 ft sec^{-1}).

The size response at normal duct velocities corresponds approximately to the size response of a leaf at average wind speeds. The size of the sampling nozzle or the aerodynamic geometry of its situation can be designed to meet a wide range of simulations of any other process of pollution with a different size response.

The size response normally used is typical of the average impaction situation. It also increases steadily over the range of 0 to 100 μm and corresponds well with the size function of the deposition dust in this range. Hence although no single function will represent both the pollution of a blade of grass and the pollution of a house, the function used probably correlates well with the average dust pollution caused in a neighborhood containing grassy surfaces, bushes, trees, fence posts, human beings, houses, and so on. The impacimeter can therefore be used to control pollution of surfaces at ground level.

2.2.3. Mass Measuring Instruments

The two instrument types so far described can be justified because they relate to the two kinds of pollution caused by dust, namely, pollution of the atmosphere and pollution of the earth's surface. They can be applied according to whether one or other problem, or both, are relevant.

Since historically manual sampling was the chief means of expressing the contract terms for the supply of dust collecting plant and also, in many cases, of applying legal limits to emission, there is a continuing interest in mass burden measurement regardless of size grading and of extending it to continuous measurement.

No commerically available method is as cheap and reliable as the best opacimeters and impacimeters. However instruments are in service that use β-ray absorption to measure the mass of dust. A tape of filter paper passes at set intervals through the instrument. It is first subjected to β-rays to measure its absorptive power while clean. This portion of tape is then moved to where a stream of isokinetically sampled flue gas is passed through it for a preset period. The increase in its β-ray absorptive power is a measure of the mass of dust.

There is a limit to the amount of dust that can be retained and measured. When a wide range of dust burdens needs to be measured; the instrument can be adapted to measure the time needed to reach a given mass of dust on the filter. The mass burden is proportional to the reciprocal of the time interval. This type of instrument is more complicated than existing opacimeters and impacimeters and is more expensive and more difficult to maintain.

2.2.4. Practical Problems

Some problems are common to all instrument methods described here, since the instruments need to be able to withstand the heat, dust, acidity, and vibration typical of dust sampling situations. All these problems can be overcome by knowledge or experiment, provided the degree of difficulty is within accepted operational practice. For example, if the degree of acidity is such that the ductwork has a reasonable life, the instrument can be given an equal life.

The major problem of continuous measurement is to find a sampling point that is representative of the total emission. The sampling point should be sufficiently far from an electrostatic precipitator outlet, for example, for the dust from the left-hand side and the right-hand side of the precipitator to be reasonably well mixed. Beyond this point fine dust is usually well distributed, and opacimeters can be used satisfactorily. Since they utilize a beam of light, it is possible to integrate across from left to right. However the devices must be well downstream from precipitators if there is a need to allow mixing between top and bottom.

A further problem is that particles greater than about 20 μm tend to stratify, giving nonuniform distribution at bends in the ductwork and in passing through fans, through centrifugal effects. They redistribute themselves slowly downstream, but typically a further bend occurs before a satisfactory distribution is achieved. The chimney usually gives redistribution a sufficient length of straight run to make a sampling point well up the chimney attractive, although on other considerations it may be difficult or impossible.

The maldistribution at any particular cross section of duct can be measured by manual sampling, and unsatisfactory sampling points can be avoided. However the maldistribution tends to vary with circumstance, and "best" sampling points can be established only after studying a range of circumstances. The departure of the "best" from the ideal may be considerable. Future developments in simple methods of measuring maldistribution will speed up detailed studies of sampling positions.

Sometimes the study reveals that the "best" sampling point is unsatisfactory for all possible sampling positions and automatic duct scanning can be applied. It is available for at least one design of mass burden measuring instrument, but it is expensive and gives only an adequate average over fairly long time periods.

Information on the maldistribution patterns of coarse dust is not well documented or well related to specific causes of maldistribution. However dust in the range of 20 to 100 μm often varies in concentration by at least an order of magnitude at different points in a typical cross section. Still more marked maldistribution is possible. This problem affects impacimeters and mass measuring instruments but not opacimeters, which do not "see" the coarser dust.

Another likely development is the use of turbulence-promoting assemblies designed to bring a maldistribution of dust rapidly into a more uniform distribution, allowing single sampling points to be used with greater accuracy.

2.3. ALLOWABLE INSTRUMENT READINGS

2.3.1. Control by Mass Emission

The legal or official control of dust emission in many countries is at present covered by two kinds of standard. One is based on the appearance of the plume (e.g., Ringelmann number), and the other is based on the mass burden in the flue gases. The existence of the two standards is a tacit recognition that mass burden alone is an insufficient control measurement. But neither separately nor together do they form a consistent or complete basis for emission control.

However the legal control of dust emission by specifying a dust burden without regard to chimney height, sometimes with a limited consideration of particle size, is common in many countries. It is sometimes expressed in terms of mass emission in proportion either to fuel burnt or to electricity generated, but the final effects are very similar. A standard of 150 mg m^{-3} is common, but lower and, in some circumstances, much higher burdens are allowed. The standard is usually referred to a specified temperature and pressure.

2.3.2. Control Related to Effects

Since the most obvious (and immediate) effect of particulate emission is a visible plume, the Ringelmann number—comparing the darkness of the plume against the sky with the darkness of five shades on a test card—was one of the earliest methods of control. Although primitive, it has some merit because it relates control to some *effect* of the emission.

Control by Ringelmann number has been reassessed[4] recently, with the following conclusions.

1. A higher standard of emission (in terms of mass burden) is required for larger plants than smaller.

2. The total *area* of plume permitted is proportional to the amount of flue gas being emitted (or fuel burnt, etc.).
3. The total area of plume permitted is also proportional to the flue gas emission velocity.

Items 1 and 2 reflect realistic recognition that larger plants can achieve higher technical performance and that adverse effects permitted should be roughly in proportion to the industrial benefits achieved by the plant. Item 3 represents an incidental relaxation in favor of plants with high emission velocities that could with advantage be corrected.

2.3.2.1. *Control by Opacimeter*

Ringelmann numbers as assessed by the standard method vary appreciably with the brightness of the sky.[4] If an arbitrary "average" sky brightness is taken, objective and repeatable Ringelmann numbers can be deduced from the readings of opacimeters, provided a correction is made for the ratio of optical path length to chimney diameter and for temperature. This "objective" Ringelmann scale is set out in an addendum to a British Standard.[5] It agrees to some extent with the usually determined Ringelmann numbers and is repeatable.

It has the disadvantage that Ringelmann 0 does not correspond with a clear plume because a partially obscured sky is often as bright as a clean white card. Furthermore, it is linear with percentage obscuration but not with either the area or the mass of the particles being emitted.

It continues to be sensible to make an optical measure across the diameter of a chimney because of items 1 and 2, just enumerated.

Light extinction (i.e., $\ln I_0/I_1$, where I_0 is the light emitted and I_1 is the light reaching the far side of the chimney) is similar in principle to optical density but is based on base e rather than base 10. If the light is reduced to $1/e$, its extinction is 1. This is very nearly equivalent to Ringelmann 2 as rationalized. A convenient scale, which could be referred to as the *extinction scale*, is given in Table 2.1.

Thus the extinction scale number 2 is closely equivalent to Ringelmann 2, and extinction scale 0 corresponds to clear view. The rest of the scale increases linearly with increase of the total cross-sectional area of the dust emitted.

If objective control by opacimeter is desired, the extinction measured by the instrument should be multiplied by chimney diameter and divided by the length of the optical path; it should be divided by the absolute temperature at the instrument and multiplied by the absolute temperature at the

Table 2.1

Extinction Scale Number	Extinction $\ln I/I_0$	Approximate Ringelmann Number
0	0	−2.5
1	0.5	0
2	1.0	2
3	1.5	3
4	2.0	4
5	2.5	—
6	3.0	—

chimney top. To correct for item 3, it should be multiplied by the full load flue gas velocity at the chimney top and divided by 10 m sec^{-1}. This corrected extinction should be kept below 1.0 for extinction scale number 2, or the equivalent number given in Table 2.1 if a higher or lower standard is required.

2.3.2.2. Control by Impacimeter

Another major effect of dust emission is the fouling of surfaces by deposition and impaction. These effects vary with wind velocity and target dimension but particularly with particle size. For mass impaction, and for deposition whether measured as mass or area, the effect increases with particle size, in contrast to atmospheric effects, which decrease with particle size. For impaction measured by particle surface area, there is an initial increase with particle size followed by a decrease.

It is not possible for any single instrument to have a response that represents all these effects. If we take the size of a leaf as a typical target size, however, an impacimeter has a response very similar to the size dependence of impaction on a leaf. Over the range of 0 to 100 μm, the response is generally similar to the size dependence of deposition. Hence control by opacimeter will deal effectively with the impaction and deposition effects of dust emission.

The allowable reading of an impacimeter has been considered elsewhere.[6] If a simple allowance is made for the dispersion of a plume, the allowable reading on an instrument is given by

$$\frac{0.01H^2}{\Sigma A} \tag{4}$$

where H is the effective chimney height and ΣA is the total area of ducting being monitored (Section 2.3.3.6). If this reading is not exceeded, no surface will be perceptibly polluted by the monitored source in a period of 10 days. This period is arbitrarily taken as a representative period between cleanings by rain or high wind.

2.3.2.3. Averaging Times for Monitor Readings

The first function of a monitor is to give warning when a change in emission occurs so that preventive steps can be taken if necessary.

The second function is to provide a record of emission levels that have—or have not—been satisfactory. It is unnecessarily severe to require that monitor readings be kept below a specified limit for every second of the year. If a single source is producing pollution by deposition or impaction, the effect builds up progressively until there is a change of wind direction. The time during which the effect is integrating will depend on the wind regime. In the European area affected by westerlies, it is unlikely that a particular spot will continue to receive dust from the same source for longer than 5 hr. It is therefore reasonable to assume this integration time in setting the emission limit (as in Sections 2.3.2.2 and 2.3.3.6) and also to accept it as an averaging time in assessing monitor readings. Thus the requirement for satisfactory operation is that the 5 hr average reading of an impacimeter should not exceed the agreed emission limit.

If wind directions are changing rapidly, however, each location receiving the plume from a source receives a total of about 5 hr in the 10 days before rain or strong wind is probable. Coincidentally each consideration leads to an averaging time of about 5 hr. When wind directions persist for very long periods, much lower emission figures are allowable for impacimeters, but correspondingly longer averaging times should be applied.

Effects on the atmosphere occur in times of a different magnitude. An effect is visible within seconds while the visible plume length builds up for several minutes when black smoke is emitted. An averaging time of one minute is therefore appropriate to opacimeter readings.

2.3.3. Control by Joint Use of Opacimeter and Impacimeter

If the readings of opacimeters and impacimeters are kept below specified values, both adverse effects of dust emission can be avoided:

Since mass burden is widely used, traditionally and legally, it will be helpful to consider the relationship between the instrument readings, mass burden, and particle size distribution. We must consider real distributions of dust sizes or at least their idealized representation.

2.3.3.1. *The Size Distribution of Particles*

Size distributions of fine particles vary considerably according to the circumstances of their formation and treatment. However it is recognized that in general they tend to have a log-normal distribution; that is, the fraction of the total mass of the particles with sizes within an infinitesimal range of d,

$$\frac{dM}{M} = \frac{1}{\sqrt{2\pi}\log B}\exp\left[\frac{-(\log d - \log d_1)^2}{2(\log B)^2}\right]d(\log d) \tag{5}$$

where d_1 is the geometric mean diameter (by mass weighting not number) and B is a constant that varies with the total range of particle size.

The arithmetic mean diameter (on a mass not a number basis) is

$$d_2 = d_1 \exp[(\log B)^2/2]$$

Thus any size distribution can be specified by two numbers d_1 and B. In practice this will not be quite accurate, but possible errors are not large. Some examples of actual and idealized dust gradings are given in Figure 2.15.

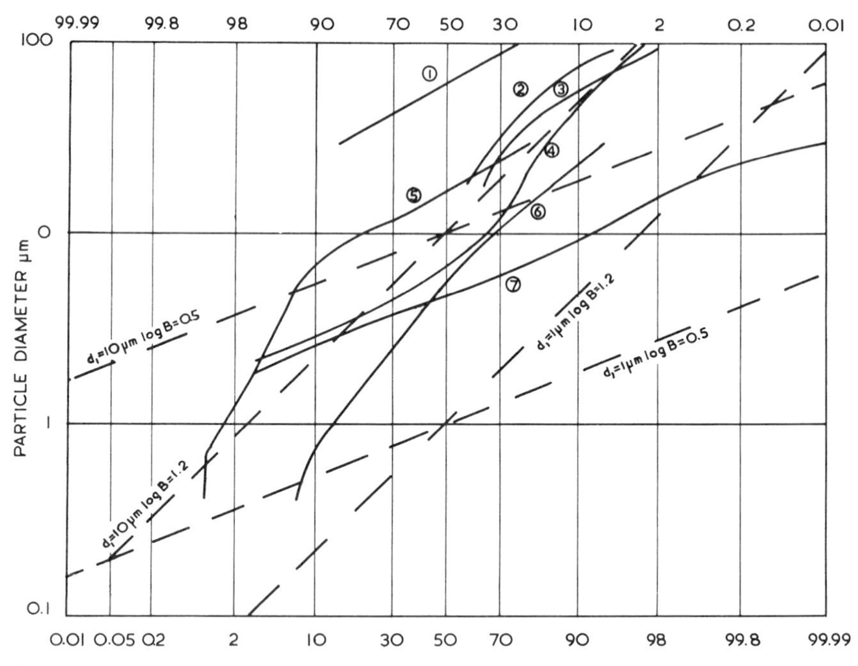

Figure 2.15. Some real and some idealized gradings. (1) Stoker fired dust (coarse) (2) Stoker fired dust (fine) (3) Pulverised fuel ash (coarse) (4) Pulverised fuel ash (medium) (5) Pulverised fuel ash (inlet) (6) Pulverised fuel ash (outlet) (7) Pulverised fuel ash (fine) Dusts (5) and (6) are from the same plant.[9]

2.3.3.2. The Dependence of Opacimeter Response on Size Grading

If some part ΔM of the mass burden of particles in a gas stream has a size d, the number of particles is

$$\frac{6\Delta M}{\pi d^3 \rho_p} \tag{6}$$

The projected area of all the particles is

$$= \frac{6\Delta M}{\pi d^3 \rho_p} \times \frac{\pi d^2}{4} = \frac{1.5\Delta M}{d\rho_p} \tag{7}$$

and the total projected area of all the particles is

$$\int_0^\infty \frac{1.5M}{d\rho_p \log B\sqrt{2\pi}} \exp\left[\frac{-(\log d - \log d_1)^2}{2(\log B)^2}\right] d(\log d) \tag{8}$$

The optical density of path length D is

$$O_D = D \log_{10} e \int_0^\infty \frac{1.5M}{d\rho_p \log B\sqrt{2\pi}} \exp\left[\frac{-(\log d - \log d_1)^2}{2(\log B)^2}\right] d(\log d) \tag{9}$$

$$O_D = \frac{0.65MD}{\rho_p} \exp\left[\frac{(\log B)^2/2}{d_1}\right] \tag{10}$$

2.3.3.3. The Dependence of Impacimeter Response on Grading

The CERL* Mark II impacimeter is assumed to have the response shown by line OX in Figure 2.14, independent of density. This is not strictly true, but because density changes produce positive effects for small particles and negative effects for large ones, the overall effects of small density changes are unimportant. The line OX can be presented by

$$I = 5.2 \times 10^8\ Mdv \text{ for a particular size of particle} \tag{11}$$

For a distribution of particles, we have

$$I = 5.2 \times 10^8\ Mv \int_0^\infty \frac{d}{\sqrt{2\pi} \log B} \exp\left[\frac{-(\log x - \log d_1)^2}{2(\log B)^2}\right] d(\log d) \tag{12}$$

$$I = 5.2 \times 10^8\ Mvd_1 \exp\left[\frac{(\log B)^2}{2}\right] \tag{13}$$

* Central Electricity Research Laboratories of the UK Central Electricity Generating Board.

2.3.3.4. *Deducing Mass Burden and Mean Particle Size from Instrument Readings*

By eliminating d_1 from eqs. 10 and 13, we have

$$M = 5.4 \times 10^{-5} \frac{O_D \rho_p}{D} \times \frac{I}{v} \times \frac{1}{\exp[(\log B)^2/2]} \tag{14}$$

Therefore

$$M = \frac{5.4 \times 10^{-5}}{\exp[(\log B)^2/2]} \left(\frac{O_D \rho_p}{D} \times \frac{I}{v} \right)^{1/2} \tag{15}$$

In principle this is indeterminate because B is unknown, although in practice $\exp[(\log B)^2/2]$ ranges from 1.1 to 2.1.

Taking $(\log B)^2/2$ as 1.5 with an uncertainty factor of 4:3 either way, we write

$$M = 1.14 \times 10^{-3} \left(\frac{O_D}{D} \times \frac{I}{v} \right)^{1/2} \tag{16}$$

when the particle specific gravity is 1. If the particle specific gravity is n, the mass burden must be multiplied by $\sqrt{n}$.

Hence actual dust burden (not at NTP), in kilograms per cubic meter, is

$$1.14 \times 10^{-3} \left(\frac{\text{optical density per meter} \times \text{impacimeter reading (popm)}}{\text{gas velocity (m sec}^{-1}\text{)}} \right)^{1/2} \times \sqrt{n} \tag{17}$$

In milligrams per cubic meter, the expression becomes

$$M = 1.14 \times 10^{3} \sqrt{n} \left(\frac{O_D}{D} \times \frac{I}{v} \right)^{1/2} \tag{18}$$

From eqs. 10 and 13, by eliminating M, it is also possible to obtain

$$d_1 = 3.54 \times 10^{-5} \left(\frac{D}{O_D \rho_p} \times \frac{I}{v} \right)^{1/2} \tag{19}$$

$$d_1 = \frac{1.12 \times 10^{-6}}{\sqrt{n}} \frac{Iv}{O_D/D} \tag{20}$$

Hence the mass burden of flue gas and also the mean particle size can be determined from the simultaneous measurement of impacity and opacity.

The results depend on density and B, but these values may be fairly constant in practice. By using two instruments, not only are the effects adequately controlled but a continuous indication of mass burden and particle size can be deduced.

2.3.3.5. *Emission Control by Opacimeter: Possible Mass Burdens*

If a plant obeys the condition that it must observe Ringelmann 1, as defined in eq. 5, this implies an optical density at the chimney diameter of 0.32. Hence

$$0.32 = \frac{0.65MD}{\rho_p d_1} \exp[(\log B)^2/2], \tag{21}$$

$$M = \frac{0.49\rho_p d_1}{D \exp[(\log B)^2/2]} \tag{22}$$

If ρ_p is converted to specific gravity and M is in milligrams per cubic meter,

$$M = \frac{0.49n \times 10^3 d_1}{D \exp[(\log B)^2/2]} \times 10^6 \tag{23}$$

$$M = \frac{4.9 \times 10^8 nd_1}{D \exp[(\log B)^2/2]} \text{ mg m}^{-3} \tag{24}$$

If M is at 110°C, we have

$$M_{\text{NTP}} = \frac{4.9 \times 10^8 \times 2d_1}{D1.5} \times \frac{383}{273} \quad \text{if } n = 2 \text{ and } \exp(\log B)^2/2 = 1.5 \tag{25}$$

$$= 9.2 \times 10^8 \frac{d_1}{D} \text{ mg m}^{-3} \tag{26}$$

2.3.3.6. *Emission Control by Impacimeter: Possible Mass Burdens*

If a plant obeys the condition[6] that its impacimeter reading must be $I = (0.01H^2)/\Sigma A$, where H is the effective chimney height, ΣA is the total area of ducting being monitored, and a 5 hr integration time is taken, the following may be deduced.

In good practice it is approximately true that $H^2/(v_{\max}\Sigma A) = 66$, where H is taken conservatively as twice the chimney height.

Hence the condition becomes

$$I = 0.01 \times 66v_{\max} \quad \text{for} \quad v_{\max} = 12.2 \text{ m sec}^{-1}, I = 8 \text{ popm} \tag{27}$$

$$I = 0.66v_{\max} \tag{28}$$

If the impacimeter condition is satisfied, eq. 13 yields

$$5.2 \times 10^8\, M v d_1 \exp\left[\frac{(\log B)^2}{2}\right] = 0.66 v_{\max} \tag{29}$$

Hence

$$M = \frac{0.66 v_{\max}}{5.2 \times 10^8\, v d_1 \exp[(\log B)^2/2]} \tag{30}$$

$$= \frac{1.27 \times 10^{-9}\, v_{\max}}{v d_1 \exp(\log B)^2/2} \tag{31}$$

If M is in milligrams per cubic meter, this becomes

$$M = \frac{1.27 \times 10^{-3}\, v_{\max}}{v d_1 \exp(\log B)^2/2};$$

and at full load,

$$M = \frac{1.27 \times 10^{-3}}{d_1 \exp(\log B)^2/2}\ \text{mg m}^{-3} \tag{32}$$

Note that at lower loads a higher dust burden is permitted but not an increase in total mass emission.

$$\text{If} \quad \exp\left[\frac{(\log B)^2}{2}\right] \quad \text{is taken as 1.5,} \tag{33}$$

$$M = \frac{0.85 \times 10^{-3}}{d_1}\ \text{mg m}^{-3} \tag{34}$$

If the flue gas temperature is arbitrarily assumed to be 110°C, the mass burden corrected to NTP is

$$M_{\text{NTP}} = 0.085 \times \frac{383}{273} \times \frac{10^{-3}}{d_1}\ \text{mg m}^{-3} \tag{35}$$

$$= \frac{1.19 \times 10^{-3}}{d_1}\ \text{mg m}^{-3} \tag{36}$$

2.3.3.7. Variation in Opacity Allowed by Mass Burden Control

From eq. 1

$$O_D = \frac{0.65 M D}{\rho_p d_1} \exp[(\log B)^2/2] \tag{37}$$

If by control measures $M = 115 \text{ mg m}^{-3}$ at NTP, at 110°C

$$M = 115 \times \frac{273}{383} \times 10^{-6} \text{ kg m}^{-3} \quad (38)$$

For $n = 2$ and $\exp[(\log B)^2/2] = 1.5$, we have

$$O_D = 0.65 \times \frac{0.115}{2000} \times \frac{273}{383} \times 10^{-3} \frac{D}{d_1} \times 1.5$$

$$= 3.9 \times 10^{-8} \frac{D}{d_1} \quad (39)$$

2.3.3.8. Variation in Impacimeter Reading Allowed by Mass Burden Control

From eq. 13 we have

$$I = 5.2 \times 10^8 \, Mvd \exp[(\log B)^2/2]$$

If by control measures $M = 115 \text{ mg m}^{-3}$ at NTP, then at 110°C we have

$$M = 82 \text{ mg m}^{-3}$$

Hence

$$I = 78 \times 10^4 \times d_1 \quad (40)$$

where $v = 12.2 \text{ m sec}^{-1}$ and $\exp[(\log B)^2/2] = 1.5$.

2.3.3.9. Comparison of Control Methods

Equations 26 and 36 are plotted in Figure 2.16. This shows that for the conditions assumed, if the opacimeter reading for an emission is kept below Ringelmann 1 and the impacimeter reading is kept below 8 popm, the dust burden cannot rise above 300 mg m^{-3} for a 12 m chimney. If the mean dust diameter rises *or* falls from the "optimum," the permitted dust burden falls. For a typical size grading ($d_1 = 10\ \mu$m), the burden permitted is 115 mg m^{-3}.

Under these conditions, both the major effects of dust emission are adequately controlled.

However if for legal or other reasons the dust burden is of interest, a value can be deduced from eq. 18 without further measurement. In addition, the mean diameter of the dust can be deduced from eq. 20. In practice there

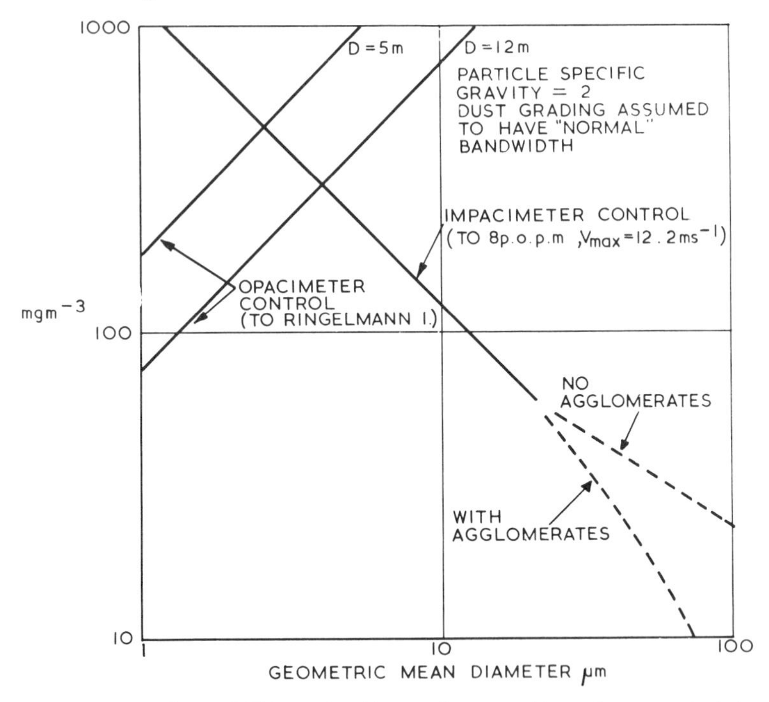

Figure 2.16. Dust burdens at NTP permitted by opacimeter and impacimeter control at certain levels.

will be some uncertainty about the specific gravity of the dust, its departure from a log normal distribution, and its value of *B*. These will to some extent be characteristic of the situation, and the uncertainty can be reduced by calibration.

It is sometimes claimed that opacimeters can be "calibrated" in terms of mass burden. This calibration depends on the factors already mentioned and also the mean size of the dust distribution. In general the total factor of uncertainty will be too large for calibration to be worthwhile.

Equations 30 and 40 are plotted in Figure 2.17, showing that even if mass burden control is rigidly applied, changes in the mean particle diameter can cause unacceptable increases in the adverse effects of dust emission. If 10 μm is taken as a typical mean particle diameter, any increase in dust coarseness causes acceptable levels to be exceeded even if the mass burden is kept at 115 mg m^{-3}. Readings up to an order of magnitude above the acceptable level are possible. On the other hand, optical densities are well within limits and the mean particle diameter must fall to 1.5 μm before Ringelmann 1 is exceeded, even on a large chimney (the mass burden being

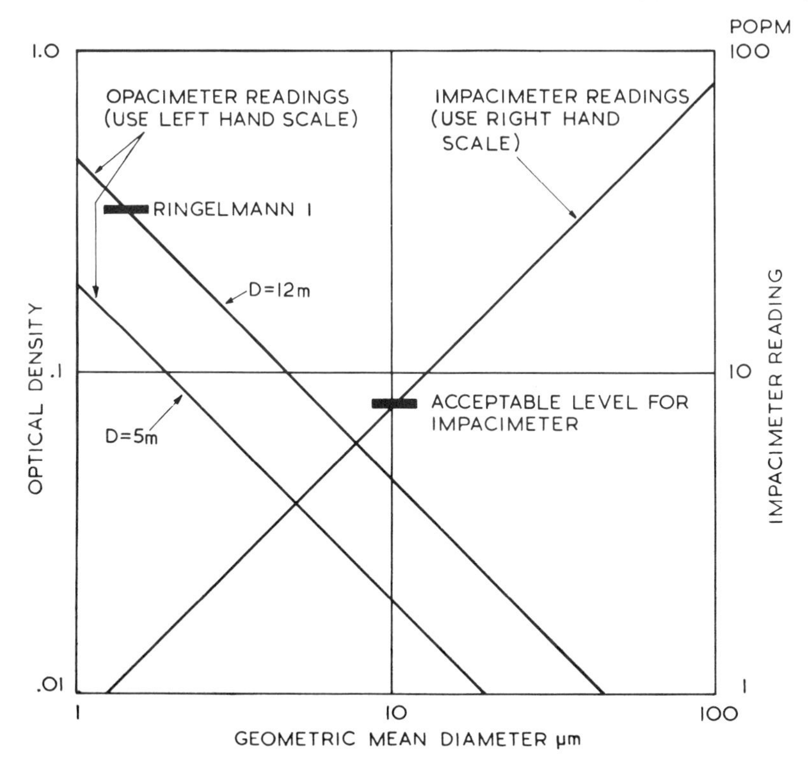

Figure 2.17. Opacimeter and impacimeter readings permitted by mass burden control.

115 mg m^{-3}). Since a decrease in particle size below 1 μm does not increase optical effects, optical densities are not likely to exceed acceptable levels by a large factor, except of course at higher mass burdens.

2.4. CONCLUSIONS

Although emission control by mass burden is well established by tradition and by law, it is neither cheap to apply nor effective in limiting adverse effects. As a method of control, manual sampling is not possible on a day by day basis. The use of continuous mass burden instruments, although by no means technically impossible, is expensive and does not help in dealing with the important effects of the changes in particle size that occur in practice.

Control by opacimeter and impacimeter together is cheap, reliable, and effective in limiting the adverse effects of emissions. The relative values of the two readings, which have marked particle size dependence, give some assistance in diagnosing the reasons for plant malfunction.

Moreover, if for legal or traditional reasons the mass burden is required on a continuous basis, an approximate value can be deduced from the readings of the two instruments. The error of this approximation can be reduced by duct distribution studies, by distribution improvement methods, and by local calibration.

SYMBOLS

d	particle diameter (m)
d_1	mean particle diameter, geometric (m)
d_2	mean particle diameter, arithmetic (m)
D	chimney diameter (m)
D_o	obstacle diameter (m)
H	effective chimney height (m)
I	impacimeter reading (CERL Mark II)
L	stopping distance for particle
M	actual mass burden (kg m^{-3}) (not reduced to NTP)
n	specific gravity
O	optical density
O_d	optical density across chimney diameter
Re	Reynolds number for the fluid flow
Re_p	particle Reynolds number (initial)
Stk	Stokes' number
v	gas velocity (m sec^{-1})
μ	viscosity (N sec m^{-2})
ρ_p	particle density (kg m^{-3})
ρ	gas density (kg m^{-3})

A2.1. FIELD MEASUREMENTS

Measurements in the field by isokinetic sampling are dealt with in Chapter 3. Here we consider measurement directly related to effects.

A2.1.1. The Measurement of Deposition

Deposition has traditionally been measured by a vessel having a circular orifice, 200 or 300 mm in diameter, facing up to receive descending particles. The original British design had a funnel and collecting jar below the orifice.

Most countries in the world today have a basically cylindrical shape below the orifice. The British funnel has the demerit that dust lying in the funnel is likely to be blown out by strong winds. The cylindrical shape is better at retaining dust, especially if it is deep compared with the diameter.

In strong winds the average angle of descent of even 100 μm dust is about 2°. In consequence, deposit gauges catch very little dust in strong winds, and the amount they do catch is appreciably affected by any change in the vertical angle between the wind direction and the plane of the orifice. The catch is still further reduced because the aerodynamic effect of the top of the gauge is to cause the wind to lift as it passes over the orifice.

The deposit gauge gives only a reasonable measure of deposition on the earth's surface when wind speeds are low and particle sizes are high. These are, of course, the conditions when the highest values of deposition occur. Hence the deposit gauge, though not an accurate means of measuring deposition at all times, is probably accurate enough (apart from loss of catch) when deposition is a problem. However it completely ignores impaction and needs to be supplemented.

A2.1.2. The Measurement of Impaction

The British Standard Directional Dust Gauge[8] is illustrated in Figure A2.1. It consists of a vertical cylinder (reproduced four times to cover four cardinal directions) with a slot entrance. Its target efficiency is undoubtedly higher than that of a simple cylinder. It is probably similar to a simple cylinder of diameter 30 mm or a disk of diameter 150 mm.

Its catch is undoubtedly conserved even in strong winds. It has the incidental advantage that the results from the four elements give directional information, and experience has shown that the dominant source in an area (if one exists) can be indicated, provided sufficient results are obtained to give a reasonably representative wind direction frequency.

There is some evidence that the gauge facing away from the wind direction has finite collection efficiency for fine dust because eddies in the lee of the gauge cause a reflux of dust. This efficiency is, however, at least an order of magnitude down on the forward facing efficiencies and does not seriously impair the directional discrimination.

The collected dust can be weighed or its optical absorption can be measured. The latter step can be performed quickly and cheaply, and it gives the appropriate analogy with the obscuration of surfaces by impaction. (The catch from a deposit gauge can also be similarly treated for the same reason.)

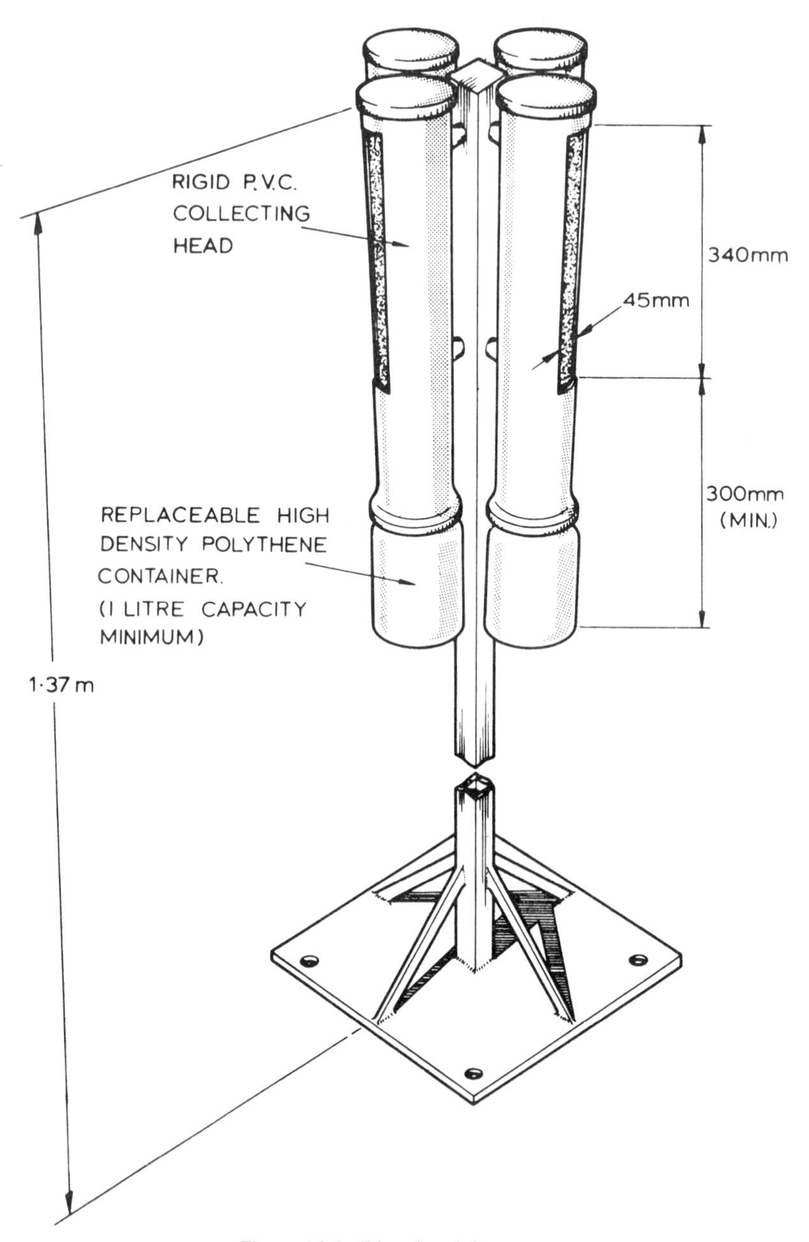

Figure A2.1. Directional dust gauge.

The readings in the plume from a source, of deposit gauge and directional gauge (or imposit gauge) should have consistency with the readings of an impacimeter monitoring the source. They have roughly similar particle size responses. They differ in their response to wind speed, of which the impacimeter reading is, of course, independent.

REFERENCES

1. Fuchs, N. A. *The Mechanics of Aerosols*, rev. ed., C. N. Davies, Ed. Pergamon Press, Oxford, 1964.
2. Hawksley, P. G. W., S. Badzioch, and J. H. Blackett, *Measurement of Solids in Flue Gases*, British Coal Utilization Research Association, Lea Hampstead, 1961.
3. Lucas, D. H., W. L. Snowsill, and P. A. E. Crosse, *Measurement and Control*, 15, 9 (1972).
4. Lucas, D. H., *Atmos. Environ.*, 6, 775 (1972).
5. British Standard No. 2742, Addendum 1, 1972.
6. Lucas, D. H. and W. L. Snowsill, *Atmos. Environ.*, 1, 619 (1967).
7. British Standard No. 1747, Part 1, The Standard Deposit Gauge, 1969.
8. British Standard No. 1747, Part 5, The Directional Dust Gauge, 1969.
9. Lee, R. H., H. L. Crist, A. E. Riley, and K. E. Macleod, *Environ. Sci. Technol.*, 9, 643 (1975).

3

PARTICULATE SAMPLING

M. Bohnet

Institute for Process Technology,
University of Braunschweig, Braunschweig, Germany

3.1. INTRODUCTION

Air pollution is due to changes in the natural composition of the air, by aerosols (smoke, dust, or carbon black) and gases (vapors or odorous substances). Much progress in the field of pollution control has been possible because of the development in recent years, of new and better methods for measuring pollutants in the air and the water, particularly in the area of air pollution control. The measurement of the concentration of gases, solids, or liquids in fluids is necessary both for emission control to reduce air pollution and for control of individual process stages. An example is the measurement of the efficient operation of dust collectors followed by the discharge of dust into the atmosphere.

This chapter deals essentially with the determination of solids in flowing gases, but the methods can also be applied to the measurement of droplet concentrations and their size distribution. One must differentiate between control measurements, which continuously monitor the operating behavior of dust collectors (or of process plant) and of stack emissions, and measurements that give the absolute content of solids in the gas stream. For control, devices using light intensity reduction (as a measure of density), intensity reduction of radioactive radiation, or changes in capacity are widely used. Particle counters based on light scattering are used for control of "clean" rooms because of their excellent sensitivity.

When measuring the absolute solids content in flowing gases, part of the gas stream is drawn off and analyzed. However satisfactory results can be achieved only at measuring points, where there is a representative distribution of the solid or liquid. For equipment control using measurement of dust content, the sites for the sampling points must be fixed during the planning and design stages.

3.2. SAMPLING IN FLOWING GASES

Correct sampling in flowing gases determines the success of the measurement. There are also special conditions when gases, solids, or liquids in the main gas stream are to be analyzed.

3.2.1. Sampling of Gases

Since gases are completely miscible, there are normally no special requirements for the sampling of a partial gas stream. However the measurement must be made at a point where the concentrations are representative. This is difficult when gases are being mixed or if a chemical reaction takes place. Normally the gas samples from ducts, pipes, or vessels are taken with radial nozzles (Figure 3.1). Sampling probes may also be used if a concentration distribution has to be determined.

3.2.2. Sampling of Solids

Solids from flowing gases are generally sampled by suction of part of the gas stream. Figure 3.1 demonstrates the correct and incorrect methods. Sampling with a radial nozzle (Figure 3.1*a*) or an inserted pipe (Figure 3.1*b*) leads to incorrect results. Also sampling immediately after a bend or other obstruction (Figure 3.1*c*) is incorrect because of the uneven dispersion of the solids. Depending on the position of the probe, the gas stream entering the sampling device may be loaded with too many or too few particles. Only if the sampling probe is located in a position parallel to the main stream in the undisturbed flow, and sampling is isokinetic (Figure 3.1*f*), will measurements be correct. Differences between the free stream velocity in the duct and the sampling velocity at the probe inlet lead to incorrect results (Figure 3.1 *d*, *e*).

3.2.3. Sampling of Liquids

The sampling of liquids in flowing gases is very complicated because usually the state of the liquid is unknown. It may occur as film flow at the pipe wall, as dispersed droplets, or as a fine mist. Combinations of continuous and dispersed flow regimes are possible. Sampling of the gas stream with an opening in the pipe wall will collect mainly the liquid film, and sampling with a probe inserted into the duct will only collect droplets; thus the most

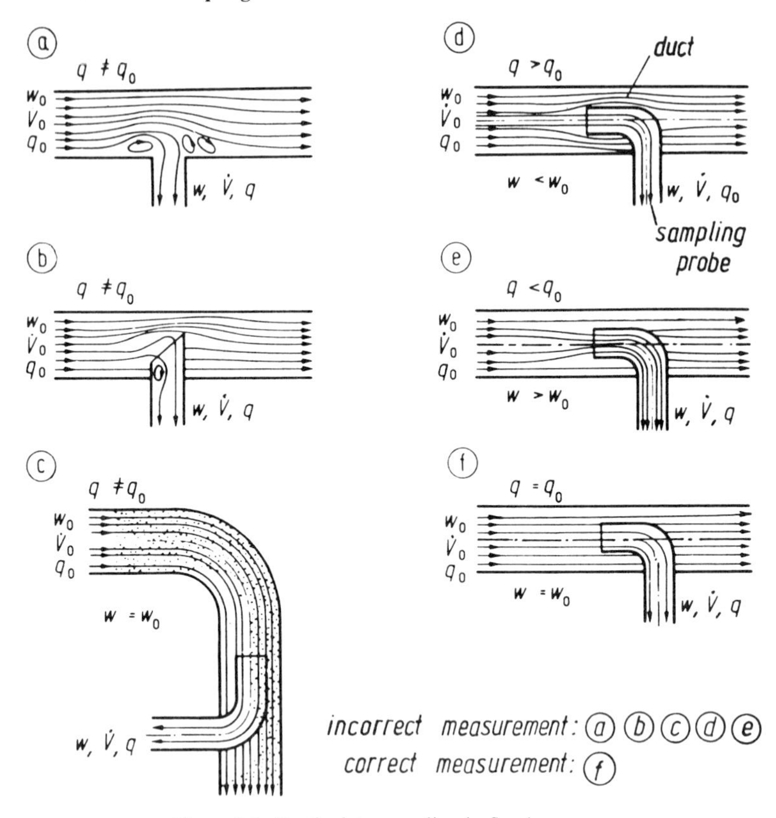

Figure 3.1. Particulate sampling in flowing gases.

effective type of sampling must be determined. Generally sampling a liquid-laden gas stream gives information only about the liquid droplet concentration within the gas stream. However by using special measuring devices (see below), the droplet size distribution can be ascertained. The problem is much easier when solids are being sampled from flowing gases because it is possible to determine the particle size distribution after the measurement of total solids. This requires an adequately sized sample that can be completely removed from the dust collection filter.

3.3. SAMPLING DEVICES

All the foregoing methods for determining the solids content in flowing gases require calibration of the different devices. For total solids, sampling of part of the gas stream with probes is necessary.

3.3.1. Streamlines in the Neighborhood of the Probe Inlet

For accurate measurements, the flow conditions in the neighborhood of the probe inlet are of great importance, and there are several points to be considered. First, the sampling velocity relative to the flow velocity has a decisive influence, and normally only isokinetic sampling gives correct results. Second, the shape of the probe inlet is important: the thicker the probe wall in relation to the probe diameter, the larger the errors that occur, even with isokinetic sampling; thus sharp-edged probes are usually used. For calculating flow conditions in the neighborhood of the probe inlet, it is generally assumed that the probe wall is infinitely thin. Finally, the angle the probe makes relative to the main stream direction influences the result.

Figure 3.2 shows the pattern of the streamlines in the neighborhood of the probe inlet for three cases: (1) isokinetic sampling, (2) sampling velocity lower than free stream velocity, and (3) sampling velocity higher than free stream velocity. In (1) the correct gas and solids mass flow, flowing at the same velocity as in the cross section of the duct, enters the probe.

In (2) the gas stream expands in front of the probe inlet. The streamline pattern shows that a "limiting streamline" exists, ending exactly at the probe wall. The gas stream within both limiting streamlines (denoting a cylindrical cross section due to the three-dimensional flow) reaches the probe; the gas outside the limiting streamlines bypasses the probe.

When the sampling velocity is higher than the free stream velocity (3), the streamlines contract in the vicinity of the probe. Here also limiting streamlines are observed, which show the gas respectively reaching the probe and passing it.

3.3.2. Particle Paths in the Neighborhood of the Probe Inlet

Figure 3.2 indicates that depending on the sampling velocity, the pattern of the streamlines in front of the probe inlet changes. At flow conditions corresponding to (1), all solid particles will move on streamlines, as long as the influence of the gravity is neglected in horizontal pipes. This does not apply, however, to (2) and (3). Because of their inertia, the solid particles are unable to follow the gas flow, but move on trajectories, which are not identical with streamlines. Only the solid particles, which move within the "limiting particle paths" reach the probe. This describes the flight path of a particle, which just contacts the probe wall. With unchanged flow conditions, different limiting particle paths result from the varying inertias for differently sized particles and different solid densities. The limiting particle paths also appear in Figure 3.2. In (1) the solid concentration in the duct

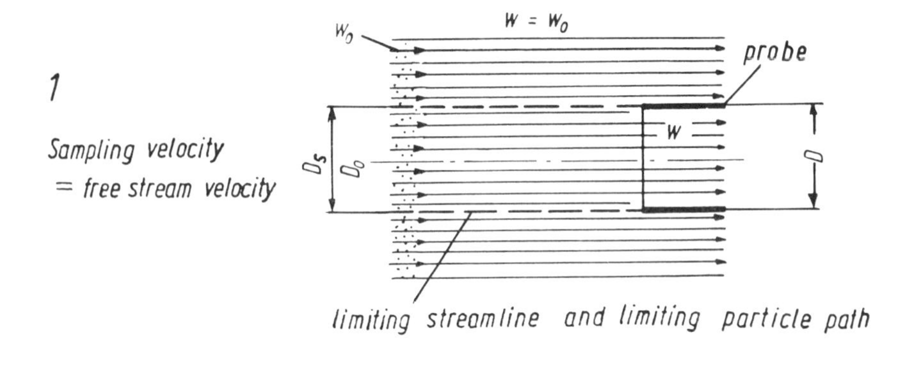

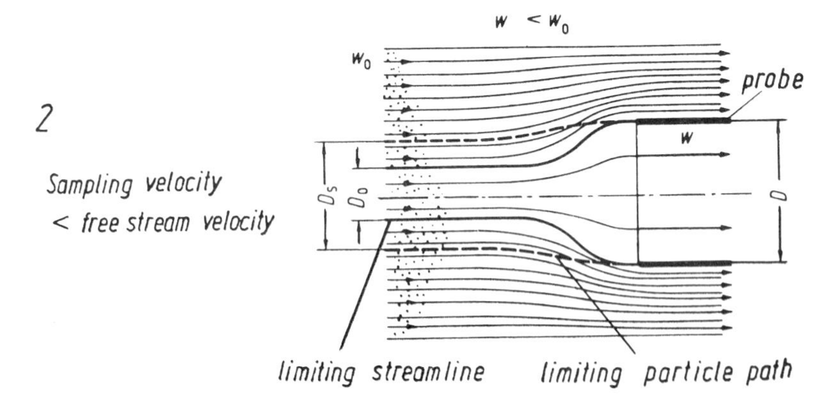

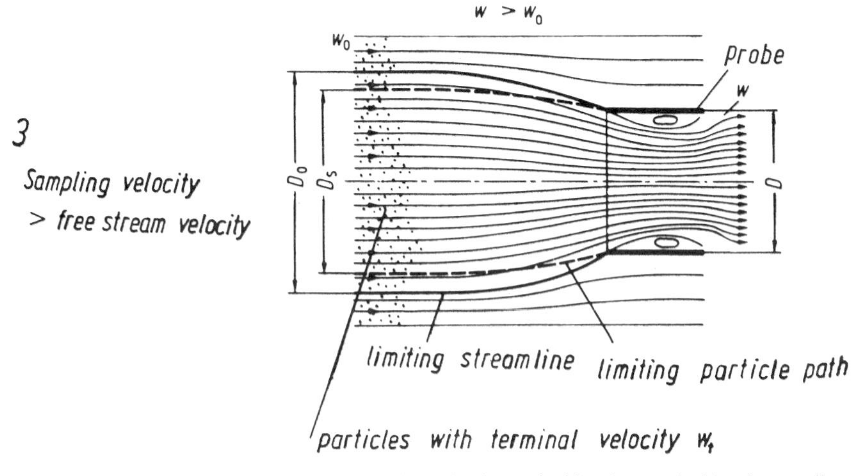

Figure 3.2. Limiting streamlines and particle paths due to isokinetic or anisokinetic sampling.

flow corresponds to that in the probe. Measurements with sampling velocities lower than the free stream velocity lead to an excessively high solids concentration in the sampled part of the gas stream, because the limiting particle paths run outside the limiting streamlines. If the sampling velocity is higher than the free stream velocity, the limiting particle paths run inside the limiting streamlines. The measurement yields too low a solids concentration in the sample.

3.3.3. Error due to Anisokinetic Sampling

For the sampled gas volume stream, the continuity equation is

$$\dot{V}_0 = \dot{V} = \frac{\pi}{4} D_0{}^2 w_0 = \frac{\pi}{4} D^2 w \tag{1}$$

where $\dot{V}_0$ = flow rate in duct ($m^3\ sec^{-1}$)
$\dot{V}$ = flow rate of sampled gas ($m^3\ sec^{-1}$)
D_0 = diameter of cylinder formed by limiting streamlines (m)
D = probe diameter (m)
w_0 = free stream velocity ($m\ sec^{-1}$)
w = sampling velocity ($m\ sec^{-1}$)

For the specific solid contents in the gas stream and in the probe, respectively, we write

$$q_0 = \frac{\dot{S}_0}{\dot{V}_0} \qquad q = \frac{\dot{S}}{\dot{V}} \tag{2}$$

where q = solids concentration in sampled gas ($kg\ m^{-3}$)
q_0 = true solids concentration ($kg\ m^{-3}$)
$\dot{S}$ = flow rate of solids in the probe gas stream ($kg\ sec^{-1}$)
$\dot{S}_0$ = flow rate of solids in the dust gas stream ($kg\ sec^{-1}$)

The solid mass stream sampled with the probe is

$$\dot{S} = q_0 w_0 \frac{\pi}{4} D_s{}^2 \tag{3}$$

where D_s = diameter of the cylinder formed by the limiting particle paths (m)

The relative solid content within the probe—that is, the ratio of the solid content in the probe to the solid content in the duct—is then

$$\varepsilon = \frac{q}{q_0} = \frac{\dot{S}}{\dot{S}_0} = \left(\frac{D_s}{D_0}\right)^2 \tag{4}$$

To specify the relative solids content in the probe, it is necessary to know the patterns of the limiting streamlines and of the limiting particle paths. Two extreme cases can be readily calculated: if the dust particle size is very small, the particles follow the movements of the gas because of their low inertia. Independently of the ratio of the sampling velocity to the free stream velocity, the solids concentration in the sample gas stream corresponds to that in the duct. For this case $\varepsilon = 1$, and measurement error does not occur.

On the other hand, when the inertia of the solid particles is so high that the particles fly straight into the probe without being disturbed by the gas flow, their concentration in the main stream can be calculated.

The solid mass rate reaching the probe is

$$\dot{S} = q_0 w_0 \frac{\pi}{4} D^2 \tag{5}$$

From eqs. 1, 2, and 4, it follows that relative solid content in the probe is

$$\varepsilon = \frac{q}{q_0} = \frac{w_0}{w} \tag{6}$$

That is, the relative solid content in the probe is inversely proportional to the ratio of the sampling velocity to the free stream velocity.

All the other possible cases range between these two limits. For determining dust concentration in flowing gases, the particles whose inertia cannot be neglected are of predominant interest. Clearly the error, which occurs with anisokinetic sampling of particles with different sizes and different terminal velocities, requires a knowledge of the pattern of streamlines and particle paths in the neighborhood of the probe inlet.

3.4. DETERMINATION OF THE FLOW REGION

The flow in the neighborhood of the probe inlet can be determined by experiment or by calculation. Both the calculation and the experimental determination of the streamlines show certain inaccuracies. The calculation uses approximation methods and initially assumes sharp-edged probes. In measuring the streamlines, the usual errors must be taken into consideration.

3.4.1. Theory

Vitols[1] calculated the flow near the probe inlet assuming potential flow. For a ratio of sampling to free stream velocity (w/w_0) between 0.25 and 3.0, the results of Vitols' calculations[2] are given in his thesis. Because a closed

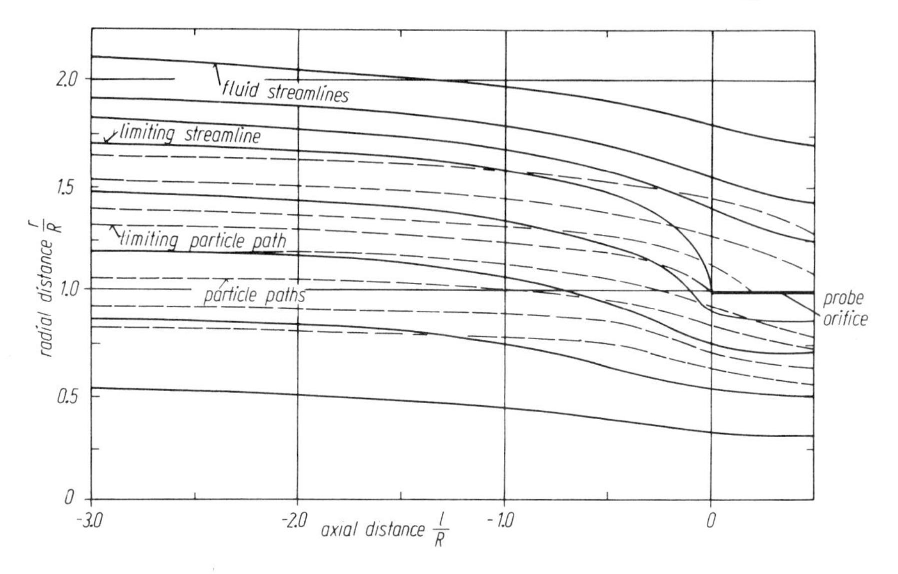

Figure 3.3. Particle paths and fluid streamlines in front of the probe inlet ($w/w_0 = 3.0$, $B = 0.5$, Re = 23.2).[1]

calculation of the flow field is not possible, he uses a step-by-step computation. As an example of his results, Figure 3.3 shows the streamline pattern and the particle paths for the velocity ratio $w/w_0 = 3.0$.

Bartak's[3] calculations are also based on the assumption of potential flow of an incompressible gas. He applies the method of conformal mapping to the flow pattern.

Fernandes and Suter[4] calculate the flow pattern with a rheoelectrical analogy on the basis of which they start the iterative calculation of the particle paths. This no longer requires the assumption of an infinitely thin probe wall, and various thick-walled probes of different shape are considered.

Stenhouse and Lloyd[5] assume potential flow and calculate the flow pattern using a substitution model, namely, the superposition of a point source and a two-dimensional flow.

Bohnet[6] replaced the actual flow pattern near the probe inlet by a more easily calculable flow pattern. When the sampling velocity is less than the free stream velocity, the model for flow around a sphere was used; when the sampling velocity is greater than the free stream velocity, the superposition of parallel and sink-ring flow was used as a model. For these cases the calculation of the flow pattern has the disadvantage that the substitute simplified flow patterns cannot reproduce the true flow patterns faithfully.

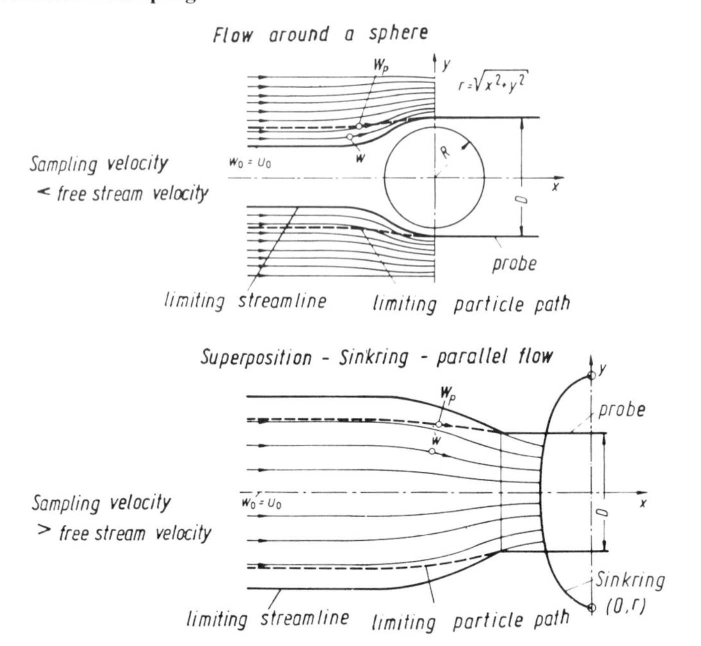

Figure 3.4. Limiting streamlines and limiting particle paths in the substitute flow field.

The patterns of the limiting streamlines and of the limiting particle paths within the substitute flow pattern appear in Figure 3.4. For the flow around a sphere, the calculation of the pattern of the streamlines is simple. By superposing the potential of a parallel flow

$$\phi_p = u_0 x \tag{7}$$

where ϕ_p = potential of parallel flow ($m^2\ sec^{-1}$)
u_0 = velocity in the x-direction (in duct) ($m\ sec^{-1}$)
x = distance in the x-direction (m)

with the potential of a dipole

$$\phi_D = \frac{u_0 x R^3}{2r^3} \tag{8}$$

where ϕ_D = potential of a dipole ($m^2\ sec^{-1}$)
R = probe radius (m)
r = radial length (m)

the potential of the flow around a sphere is

$$\phi = \phi_p + \phi_D = u_0 x\left(1 + \frac{R^3}{2r^3}\right) \tag{9}$$

For the stream function and the components of the gas velocity in the x- and y-directions we get

$$\frac{\partial \phi}{\partial x} = \frac{\partial \psi}{\partial y} = u, \qquad \frac{\partial \phi}{\partial y} = \frac{\partial \psi}{\partial x} = v \tag{10}$$

where ψ = stream function (m sec^{-1})
y = distance in the y-direction (m)

By integrating eq. 10, the equation for the streamlines is deduced. The calculation of the streamline pattern with superposition of a parallel and a sink-ring flow requires the definition of the potential. A three-dimensional sink of point $(0, r', \varphi^0)$ has at an arbitrary point (x, y, φ) the potential

$$\phi^0 = \frac{E^0}{4\pi}[x^2 + y^2 + r'^2 - 2yr' \cos(\varphi - \varphi^0)]^{-1/2} \tag{11}$$

The potential of the sink-ring with the diameter r' results from introducing a sink strength $E^0 = Er'\,d\varphi^0$. Integration yields

$$\phi_s = \frac{Er'}{4\pi}\int_0^{2\pi}[x^2 + y^2 + r'^2 - 2yr' \cos(\varphi - \varphi^0)]^{-1/2}\,d\varphi^0 \tag{12}$$

where E = sink strength constant (m^3 s^{-1}).

From this the potential of the total flow field is gained by superposition of the parallel flow

$$\phi = \phi_p + \phi_s \tag{13}$$

By partial differentiation, according to the position coordinates, the components of the gas velocities in x- and y-directions are

$$u = \frac{\partial \phi}{\partial x} \qquad v = \frac{\partial \phi}{\partial y} \tag{14}$$

These can be calculated at any point of the flow field; the integration of the velocity field leads to a relation for the streamlines.

3.4.2. Experiment

Rüping[7] has measured the flow field around a sharp-edged probe with a wedge-type probe. He then approximated the streamlines upstream of the probe inlet by parabolic curves. Figure 3.5 presents streamlines measured by Rüping for the velocity ratios $w/w_0 = 0.64$ and $= 2.18$. In Figure 3.6

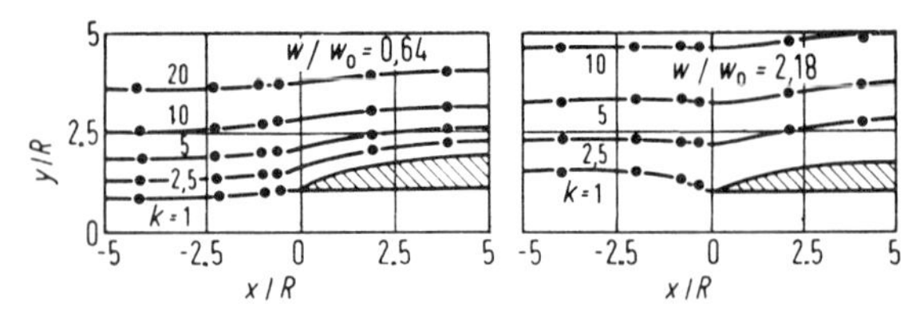

Figure 3.5. Measured streamlines in the neighborhood of the probe inlet.[7]

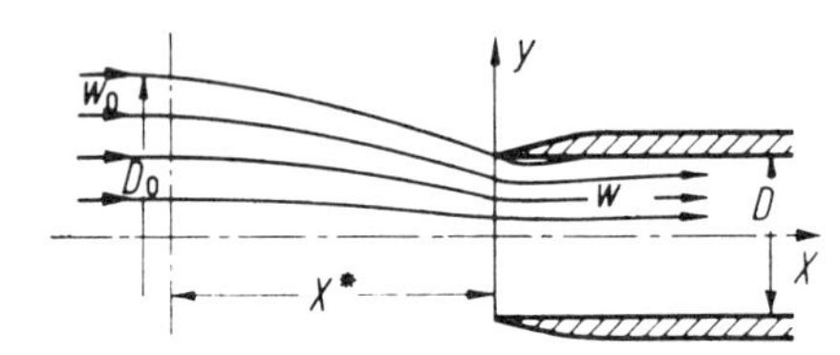

Figure 3.6. Approximation of streamlines by parabolic curves.[7]

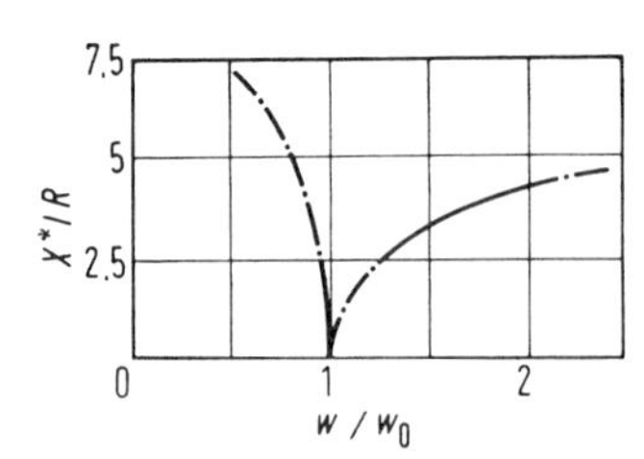

Figure 3.7. Distance of disturbance zone in front of the probe nozzle.[7]

the streamlines are approximated by parabolic curves. The flow index k indicates how much gas flows through the stream pipe compared with the quantity of gas passing into the probe. The measurements also show where upstream, in front of the probe nozzle, the flow pattern is influenced by the probe. The length of this zone of disturbance as a function of the velocity ratio w/w_0 is given in Figure 3.7.

3.5. DETERMINATION OF PARTICLE PATHS

For calculating the error in anisokinetic sampling, the paths of the solid particles are required. Figure 3.2 shows that independent of the terminal velocity of the particles w_t, which essentially is a function of the size of the particles d_p and the density ρ_p, a limiting particle path exists, which ends exactly at the probe wall. All particles outside the limiting particle paths bypass the probe; all particles inside the limiting particle paths enter the probe.

To calculate the path of a solid particle in a gas stream, it is necessary to know the forces acting on it. Neglecting the effect of gravity, the equilibrium of resistance and inertia forces acting on the solid particle is $F = I$, where

$$F = c_w \frac{\pi}{4} d_p^2 (w - w_p)^2 \frac{\rho}{2} \tag{15}$$

where F = flow resistance (N)
c_w = drag coefficient
d_p = particle diameter (m)
ρ = fluid density (kg m^{-3})
w_p = particle velocity (m sec^{-1})

and

$$I = \frac{\pi}{6} d_p^3 (\rho_p - \rho) \frac{dw_p}{dt} \tag{16}$$

where I = inertial force (N)
t = time (sec)

Assuming that the solid particles are spherical, the resistance (drag) coefficient c_w (eq. 8) can be calculated from

$$c_w = \frac{21}{\text{Re}} + \frac{6}{\sqrt{\text{Re}}} + 0.28 \tag{17a}$$

in which

$$\text{Re} = \frac{(w - w_p) d_p}{\nu} \tag{17b}$$

where ν = kinematic viscosity (m^2 sec^{-1}).

Within the Stokes law range, the equation of motion resulting from eqs. 15 and 16 can be solved completely, introducing $c_w = 24/\text{Re}$. For this case a simple relation for the particle velocity is given by

$$\frac{dw_p}{w - w_p} = 18 \frac{\eta}{d_p^2} \cdot \frac{1}{\rho_p - \rho} dt \tag{18}$$

where η = dynamic viscosity of the fluid (Pa sec).

Introducing the terminal velocity of solid particles w_t, we write

$$\frac{dw_p}{w - w_p} = \frac{g}{w_t} dt \tag{19}$$

where g = acceleration of gravity (m sec^{-2}).

After integration, the components of the particle velocity in the x- and y-directions are

$$
\begin{aligned}
w_{p_x} &= u + (w_{p_x} - u)e^{-gt/w_t} \\
w_{p_y} &= v + (w_{p_y} - v)e^{-gt/w_t}
\end{aligned}
\tag{20}
$$

and the components of the particle paths are

$$
\begin{aligned}
x &= ut + \frac{w_t}{g}(w_{p_x} - u)(1 - e^{-gt/w_t}) \\
y &= vt + \frac{w_t}{g}(w_{p_y} - v)(1 - e^{-gt/w_t})
\end{aligned}
\tag{21}
$$

In the relations above the components of gas velocity are introduced according to the methods of Bohnet or Rüping.

For higher Reynolds numbers the empirical drag coefficient shown in eq. 17 must be introduced in the calculation. A closed solution is no longer possible, and the particle paths are calculated step by step using computers.

Results of experiments that define particle paths near the probe inlet have not been published.

3.6. ISOKINETIC SAMPLING

Isokinetic sampling guarantees an exact measuring result only if the probe is placed at points where there is a representative distribution of solid particles. The type of the probe inlet (thin or thick walled) influences the flow pattern, although the effect of the sampling probe on the measuring error is less, the thinner (i.e., the more sharp edged) the probe. For thick-walled probes (Figure 3.8) Fernandes and Suter[4] have calculated the dependence of the streamlines with isokinetic sampling due to the relative thickness of the wall and the face angle. Figure 3.9 gives the streamline pattern in front of thick-walled probes with isokinetic sampling: with decreasing face angle, the deflection of the streamlines directly in front of the probe inlet clearly diminishes. With very little face (large angle α), the streamlines of gas just in front of the probe curve strongly and the particles no longer follow the streamline, resulting in measuring error even with isokinetic sampling.

Sampling probes with very thin walls are difficult to handle in practice; either the probe itself must be very small, or the wall thickness must be increased for stability, to compensate for vibrations in high speed gas streams.

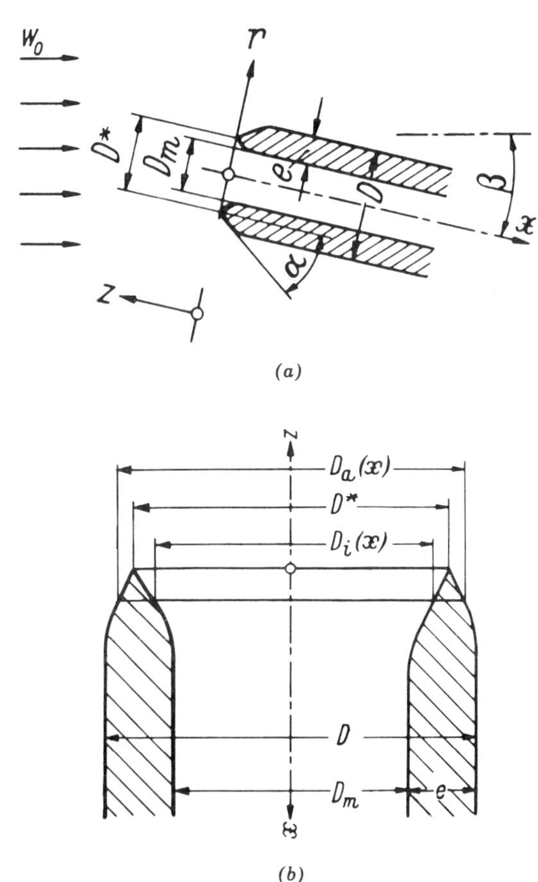

Figure 3.8. Dimensions of the probe.[4] (*a*) Normal probe. (*b*) Probe suggested in Reference 4.

The wall must also be sufficiently thick to allow for abrasion by solid particles. It is therefore desirable to construct thick-walled probes, and minimize the influence of the wall thickness on the streamline pattern. Based on their investigations, Fernandes and Suter propose a probe shape that best meets these requirements. It is characteristic of the probe that the diameter of the sharp edge of the probe D^* divides the outer and the inner annuli of the forward face into equal areas (see Figure 3.8).

$$D^{*2} - \frac{D^2 - D_m^{\ 2}}{2} = D^2\left[1 - 2\left(\frac{e}{D}\right) + 2\left(\frac{e}{D}\right)^2\right] \tag{22}$$

where D is the outer diameter, D_m the internal diameter and e the wall thickness.

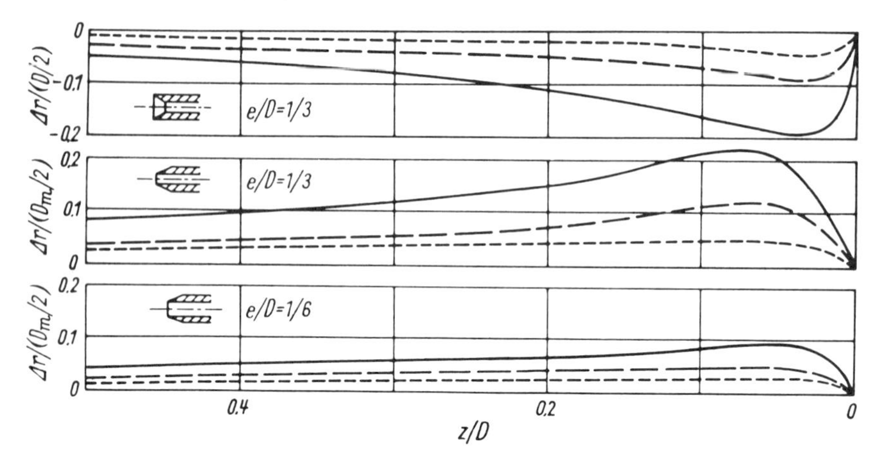

Figure 3.9. Streamlines in front of thick-walled probes and isokinetic sampling. Solid curves, 60°; — — —, 30°, - - -, 15°

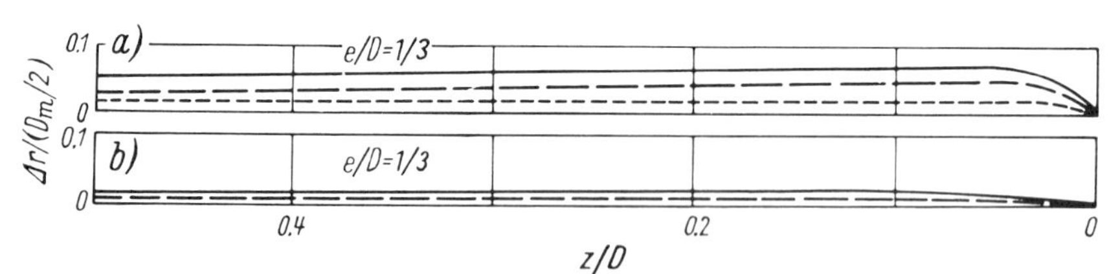

Figure 3.10. Streamlines in front of thick-walled probes and isokinetic sampling. Solid curves, 60°; — — —, 30°; - - -, 15°. (*a*) Point at the middle of the probe wall. (*b*) Probe shape referred to in Reference 22.

At other points of the probe the contours should follow a shape given by

$$D_a^2(x) - D^{*2} = D^{*2} - D_i^2(x) \tag{23}$$

Here $D_i(x)$ can be assumed. It is advisable to choose shapes that can be easily built or cause little turbulent flow within the probe, (e.g., see cross section in Figure 3.8*b*). Figure 3.10 represents the deviation of the streamline at the point of the probe with isokinetic sampling with thick-walled probes, in the case that the point of the probe is situated in the middle of the wall or according to the new proposal.

3.6.1. Measuring Errors due to Anisokinetic Sampling

Different limiting streamlines and limiting particle paths result from various terminal velocities and free stream or sampling velocities. The relative dust content in the probe can be described by introducing the Barth number Ba describing the particle motion

$$\mathrm{Ba} = \frac{w_0 w_t}{gD} \tag{24}$$

where w_t = terminal velocity (m sec^{-1})

This characteristic number permits a dimensionless presentation of calculated or measured results which is generally valid. For different values of Ba, D_s is determined from the particle paths. A simple continuity relation leads to D_0 and with eq. 4 the relative dust content ε of the gas in the probe can be calculated for different values of w/w_0.

A graph of the relative dust content as a function of the ratio of sampling velocity w to free stream velocity w_0; (w/w_0) for various values of Ba is given in Figure 3.11. The foregoing case of very small particles is covered by the limiting curve Ba = 0 and the case of very large particles by the limiting curve Ba = ∞. From Figure 3.11, based on Bohnet's calculations, the error in anisokinetic sampling can easily be found.

Figure 3.12 compares the results of the investigations of Bartak,[3] Stenhouse and Lloyd,[5] Bohnet,[6] Rüping,[7] and Zenker.[9,10] Contrary to the results of

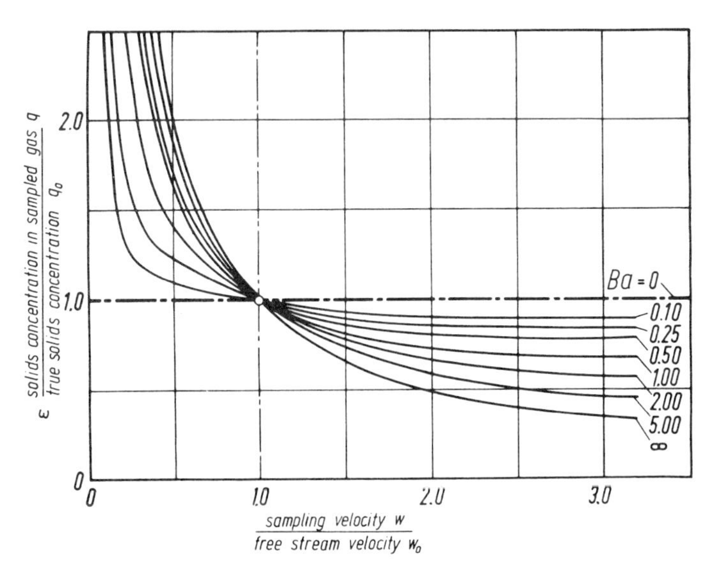

Figure 3.11. Effect of anisokinetic sampling on the solids content sampled dependent on sampling velocity, free stream velocity, particle size, and probe diameter.[6]

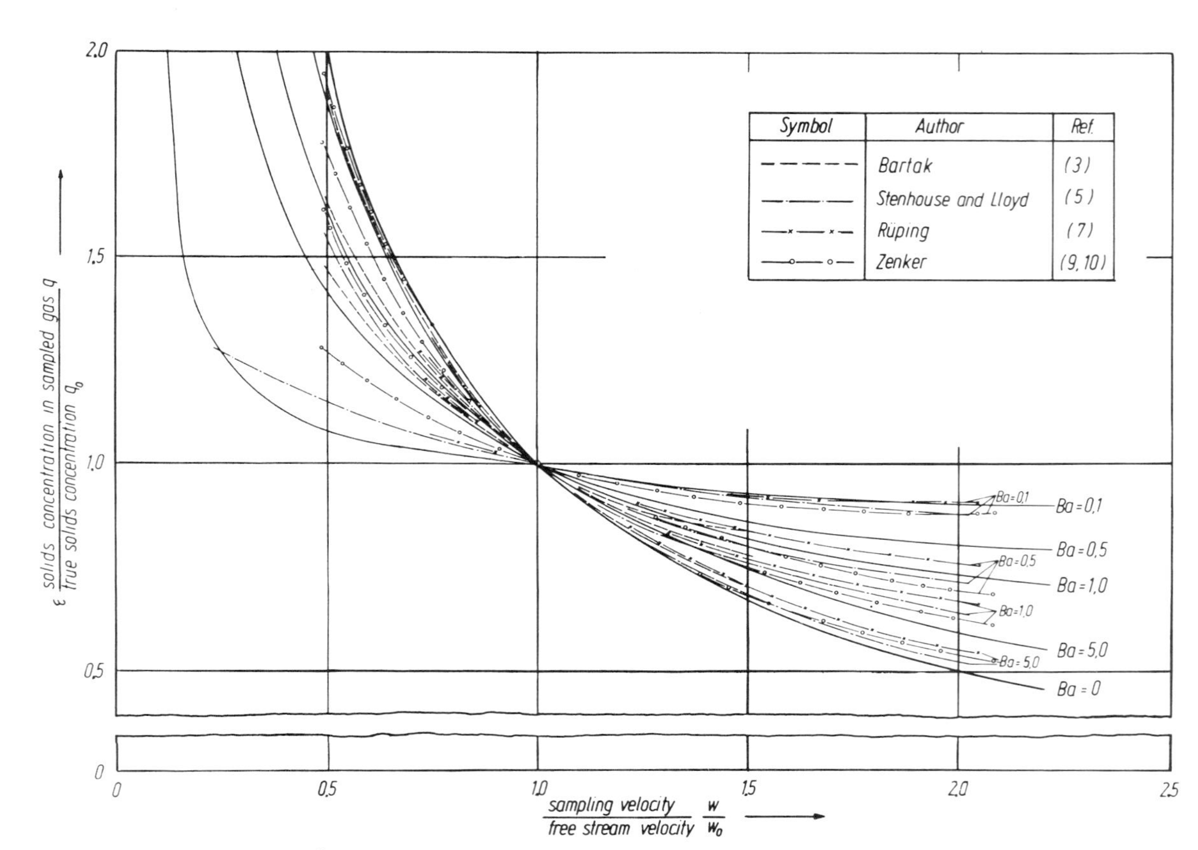

Figure 3.12. Comparison of different methods for the calculation of the relative solids concentration as a function of the velocity ratio and the Barth number.

Bartak, Bohnet, and Stenhouse and Lloyd, which were based on potential theoretical considerations, the particle paths given by Rüping were calculated from measurements of the streamlines. In an experiment Zenker[9,10] measured the relative dust content in the sampled gas stream as a function of the velocity ratio w/w_0. In a circular duct (diameter 411 mm) Zenker determined the measuring error with probes of diameters varying between 15 and 40 mm, at velocity ratios from 0.4 to 2.5, using 10 different particle size fractions. From the measurements Zenker deduced a coefficient K that allowed the relative dust content to be calculated as follows:

$$\varepsilon = \frac{w_0}{w} + K\left(1 - \frac{w_0}{w}\right) \tag{25}$$

The coefficient K depends on the relation l_0/D, where l_0 corresponds to the flight path of a particle induced into a stagnant gas at a given starting velocity. The ratio l_0/D is identical with the factor Ba.

The function

$$K = \{1 + \exp(a + b \log \mathrm{Ba})\}^{-1} \tag{26}$$

was determined by Zenker from his measurements. Replacing the coefficients a and b by 1.04 and 2.06, respectively (these numbers result from a regression analysis), graphs can also be calculated for assumed values of the Ba number. The graphs are based on measured values only. The curves are also shown in Figure 3.12.

Comparison of calculated errors with measured ones does not show much consistency. The calculations proposed by Bartak or Stenhouse and Lloyd both give larger errors for very low and very high sampling velocities, when compared to the work of Bohnet, who based his calculations on substitute flow patterns. Rüping's results, obtained by a combination of experiment and calculation, are close to the curves by Bartak and Stenhouse and Lloyd. There are, however, considerable differences between the results obtained by all authors. Zenker's results, which are based only on experiments indicate the largest measuring errors, except at high Ba values. His curves were included in Figure 3.12 on the basis of a balancing calculation of the experimental results.

The results appearing in Figure 3.12 lead to the conclusion that at present there is no method of calculation that allows adequate precision for results obtained with probes of practical wall thickness. On this basis Figure 3.12 suggests that a diagram based on the mean values of all the published investigations should be drawn. This would provide considerable simplification. Because the curves in Figure 3.12 are later compared with the measured results of other authors, this procedure has been omitted here.

3.6.2. Influence of Probe Inclination

The result of the sampling is influenced not only by the sampling velocity but also by the position of the probe. It is essential that the probe be situated parallel to the free stream. For thin-walled probes Watson[11] considers the deviation caused by inclining the probe to the gas stream at an angle β (see Figure 3.8*a*):

$$\varepsilon(\beta) = \varepsilon(\beta = 0) \cdot \cos \beta \tag{27}$$

This small influence of the angle of incline can be expected only for thin-walled probes (e.g., for $\beta = 10^\circ$, $\cos \beta = 0.985$).

For various types of probe Fernandes and Suter[4] explored experimentally the influence of inclining the probe up to an angle of 10°. They used water droplets with diameters between 5 and 10 μm (corresponding to Ba $= 0.15$ to 0.85). The result indicates that their specially designed probe is less sensitive to sampling error when inclined to the gas stream than the usual design for thick-walled probes.

3.6.3. Influence of Sampling Velocity on the Particle Size Distribution

In all practical cases the solid particles transported in a gas stream do not have uniform particle size, but a particle size distribution. For an exact interpretation of the measurements, it is therefore necessary to calculate separately the error for each particle size. The recommended procedure is to split up the solid particles into different fractions and calculate for each range of sizes the average terminal velocity w_t of the particles, then the characteristic value Ba. The relative dust content ε corresponding to Ba can be taken from Figure 3.11. With these values for ε the dust content in the probe can be calculated using the percentage of the particle size fraction in the total amount of solids.

3.6.4. Determination of Droplet Size Distribution

The method of sampling a gas stream is also applicable to determining the liquid content. As explained before, only liquids transported as drops can be measured. In contrast to solid particles, it is not possible to determine the drop size distribution from the separated and collected liquid. Thus control of the error occurring because of anisokinetic sampling is not possible because the drop size is required for the calculation of Ba, using Figure 3.11.

Here the drop size distribution in the gas stream must be measured with cascade impactors. In recent years these impactors have been successfully used in the United States and in Germany. In comprehensive studies Bürkholz[12–14] described using the cascade impactor and explained the possibility of its application on a number of mist eliminators.[15]

For defining the average drop size of the collected liquid, the proposed sampling procedure can be applied in connection with Figure 3.11. If the actual concentration of the liquid in the duct is known, the relative error can be calculated from the quantity of liquid collected by sampling at a velocity that is higher or lower than the free stream velocity. The preselected sampling velocity and the measured relative error are connected in Figure 3.11 by a specific value of Ba through which the terminal velocity w_t can be calculated using the known free stream velocity. The average drop size can then be found.

3.7 COMPARISON OF THEORY AND EXPERIMENT

Along with the measurements of Zenker, Bartak, and Stenhouse and Lloyd, the results obtained by Badzioch,[16,17] Whiteley and Reed,[18] Zipse,[19] Hemeon and Haines,[20] and Benarie and Panof[21] must be considered. These investigations, however, do not present a uniform picture. Figure 3.13 reveals that the measurements result in larger and smaller sampling errors than those calculated by Bohnet. The measurements of Benarie and Panof were normalized for comparison, because the published results showed a measuring error in the dust content when the sampling velocity was identical with the free stream velocity. Comparing the measured points and the calculated curves and allowing for the fact that the measured points of Benarie and Panof are average values of their own series of measurements, it can be concluded that in almost all cases the measuring error is bigger than the error in the calculation. Stenhouse and Lloyd also obtain larger measuring errors than calculated when Ba become larger. Bartak's results, however, were all below the calculated errors.

The measurements of Badzioch, Hemeon and Haines, and Zipse, using different particle sizes and probe diameters, allow a comparison at constant values of Ba. Figure 3.14 shows very good correspondence between these measurements and Bohnet's calculations.

It is very difficult to decide the criteria that determine the measuring errors. Using measurements alone does not provide the absolute values. Combining measurement of the flow pattern and calculation of particle

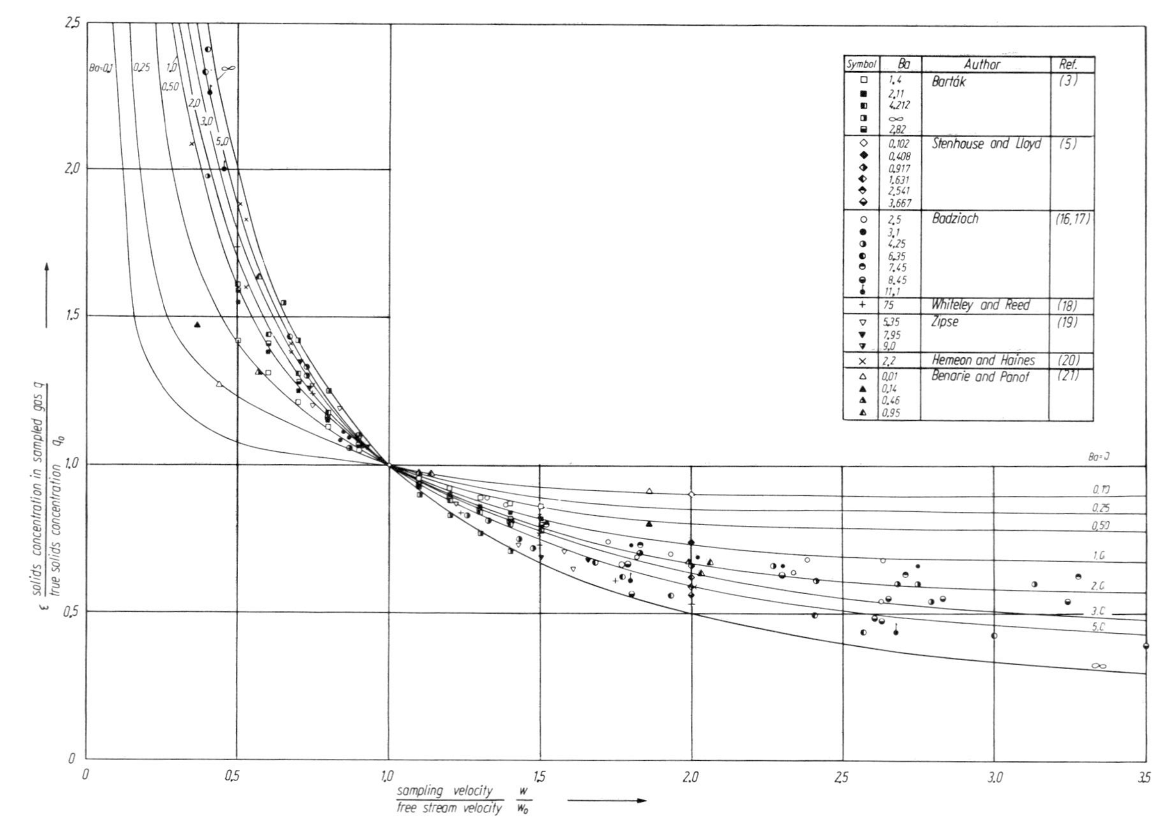

Figure 3.13. Comparison of calculated cruves by the method of Bohnet with experimental data.

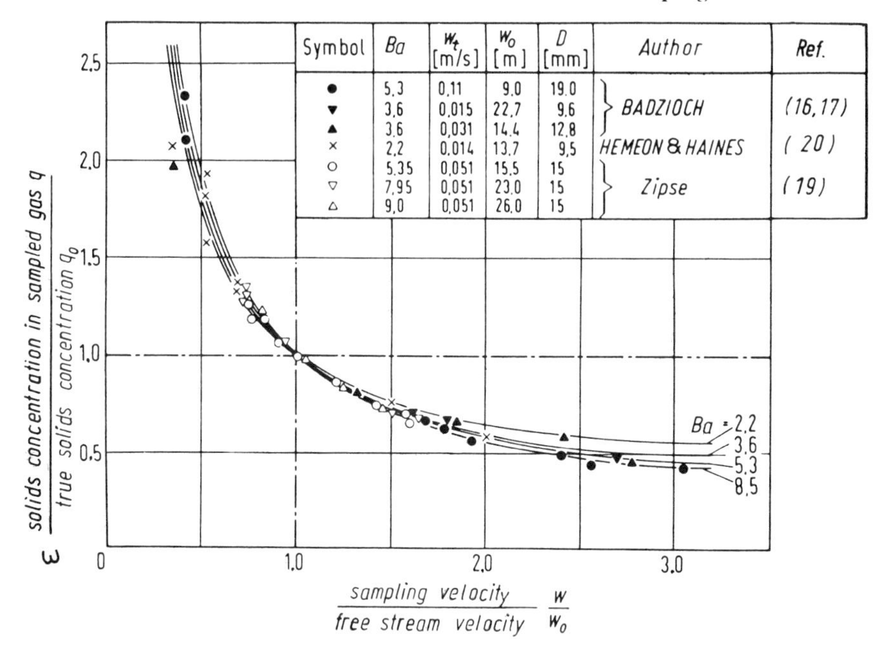

Figure 3.14. Comparison of calculated curves by the method of Bohnet with experimental data.

paths based on them has the advantage that the flow pattern can be calculated with sufficient accuracy. Purely mathematical determination of the flow pattern and the particle paths, thus of the measuring error, remains uninfluenced by measuring inaccuracies. It has the following disadvantages, however: (*a*) the flow pattern near the probe inlet has not yet been calculated exactly, and (*b*) the assumption of substitute flow patterns limits the accuracy. Comparing calculations and measured results obtained by various authors leads to the conclusion that none of the proposed procedures is completely satisfactory. Since the method of calculation using Figure 3.11 suggested by Bohnet agrees closely with measured values, however, its recommendation is justified.

3.8. SELECTION OF SAMPLING POINTS

It has been emphasized that the selection of the sampling points is of great importance. Measurements taken behind bends or other centers of disturbance lead to incorrect results. Generally the number of sample points at a

Table 3.1. Selection of Number of Traverse Points

Number of Stack Diameters Upstream and Downstream of Flow Disturbance		Number of Traverse Points on Each Diameter
Upstream	Downstream	
8+	2+	6
7.3	1.8	8
6.7	1.7	10
6.0	1.5	12
5.3	1.3	14
4.7	1.2	16
4.0	1.0	18
3.3	0.8	20
2.6	0.6	22
2.0	0.5	24

section must be the greater, the closer these are situated to a point of flow disturbance. Such disturbances can be caused by valves, orifices or bends, or other flow interferences. Morrow et al.[22] proposed the minimum number of points to be measured, depending on whether sampling is taking place upstream or downstream from the disturbance (Table 3.1). For rectangular channels and hydraulic diameter d_h is taken as equivalent diameter:

$$d_h = \frac{4A}{U} \tag{28}$$

where A is the cross section of the duct (m^2) and U its circumference (m). Flow disturbances upstream from the sampling point have much greater influence on their number than those downstream. The choice of the sampling sites in the cross-sectional plane is governed by regulations covering both rectangular and circular sections. Generally the measuring section must be divided into equal areas. The gas velocity must be measured in the center of gravity of the areas, usually with a Pitot tube, and the partial gas stream determining the content of solids is drawn off at these points.

Figure 3.15 gives the position of sampling points in circular and rectangular ducts. For rectangular ducts the shape of the partial areas should be similar to the shape of the cross section.

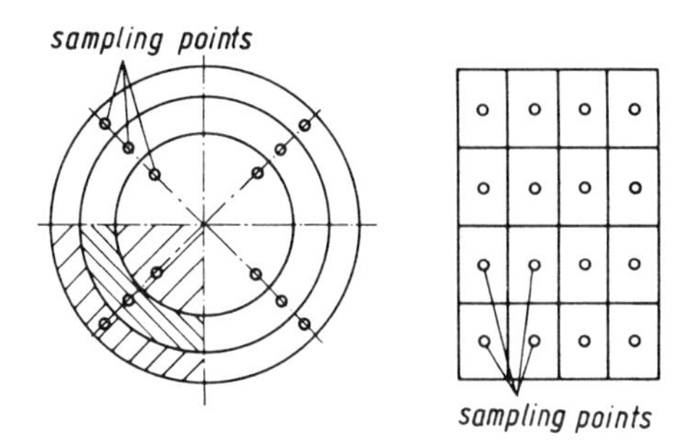

Figure 3.15. Position of sampling points in circular and rectangular ducts.

The position of sampling points in a circular duct is found mathematically by[23]:

$$a_n = \frac{D_d}{2}\left\{1 \pm \left(\frac{2i - 2n + 1}{2i}\right)^{1/2}\right\} = D_d \cdot K_n \tag{29}$$

where i is the number of partial areas and n the ordinal number.

Figure 3.16 presents the values for a division into four parts. The corresponding dimensions for dividing a circular duct into 10 parts are given in Table 3.2, as the distances of the sampling points from the outer wall. The values refer to a circular duct with unit diameter and were always measured from the internal diameter. As a rule for rectangular ducts, a minimum of nine sampling points should be chosen.

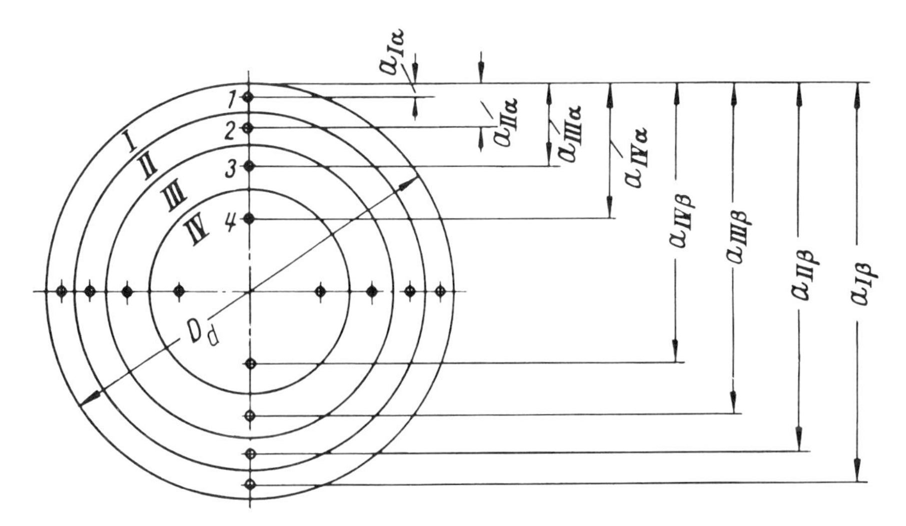

Figure 3.16. Position of sampling points in circular ducts as calculated in Table 3.2.

Table 3.2. **Values of K_n**

						i					
n		1	2	3	4	5	6	7	8	9	10
I	α	0.14645	0.06699	0.04356	0.03229	0.02566	0.02129	0.01819	0.01588	0.01409	0.01266
	β	0.85355	0.93301	0.95644	0.96771	0.97434	0.97871	0.98181	0.98412	0.98591	0.98734
II	α		0.25000	0.14645	0.10472	0.08167	0.06699	0.05679	0.04931	0.04357	0.03902
	β		0.75000	0.85355	0.89528	0.91833	0.93301	0.94320	0.95069	0.95643	0.96098
III	α			0.29588	0.19382	0.14645	0.11812	0.09911	0.08542	0.07508	0.06699
	β			0.70412	0.80618	0.85355	0.88188	0.90089	0.91458	0.92492	0.93301
IV	α				0.32322	0.22614	0.17725	0.14645	0.12500	0.10913	0.09689
	β				0.67678	0.77386	0.82275	0.85355	0.87500	0.89087	0.90311
V	α					0.34189	0.25000	0.20119	0.16928	0.14645	0.12919
	β					0.65811	0.75000	0.79881	0.83072	0.85355	0.87081
VI	α						0.35566	0.26855	0.22049	0.18819	0.16459
	β						0.64434	0.73145	0.77951	0.81181	0.83541
VII	α							0.36637	0.28349	0.23648	0.20419
	β							0.63363	0.71651	0.76352	0.79581
VIII	α								0.37500	0.29588	0.25000
	β								0.62500	0.70412	0.75000
IX	α									0.38215	0.30635
	β									0.61785	0.69365
X	α										0.38819
	β										0.61181

3.9. SAMPLING TRAINS

For performing dust measurements, a train of measuring equipment is required in addition to the actual sampling probes. Figure 3.17 illustrates the train for isokinetic sampling. These must have the following characteristics:

1. Complete collection and loss free recovery of the drawn-off dust. This implies short tube runs to dust collector and use of proper filter material.
2. Completely tight joints in the whole sampling train.
3. The possible need for cooling or heating the connecting tube and the dust collector. This is especially important when measuring hot flue gas streams to ensure that the temperatures in the duct and in the collector do not fall below the dew point.
4. Resistance of the device against chemical attack.
5. Avoidance of electrostatic charges.
6. Low weight, good transportability, and ease in handling, and cleaning of the device.[23]

The sampling devices are all built on the same principles. Figure 3.17 shows that a sampling train for isokinetic sampling needs the following components:

1. Sampling probe (*a*).
2. Sampling tube that can also serve as a support (*b*).

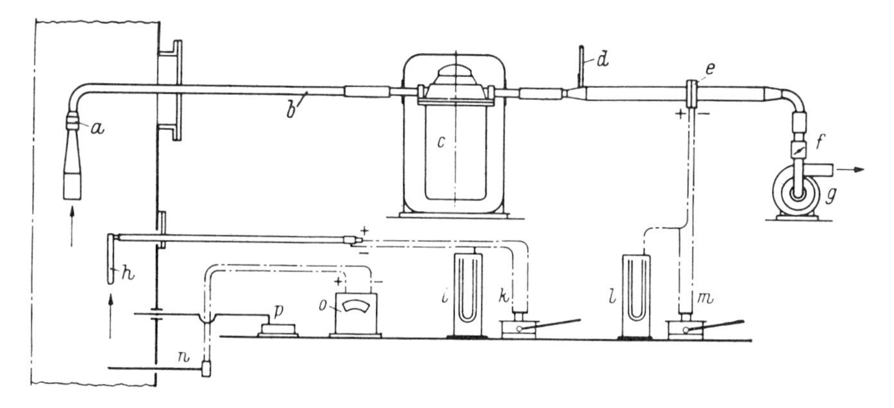

Figure 3.17. Sampling train for isokinetic sampling: *a*, sampling probe; *b*, tube; *c*, dust collector; *d*, thermometer; *e*, orifice meter; *f*, valve; *g*, fan; *h*, Pitot tube; *i*, *l*, measuring device for static pressure; *k*, measuring device for dynamic pressure; *m*, measuring device for orifice pressure loss; *n*, thermocouple element; *o*, voltmeter; *p*, measuring device for gas concentration and humidity.

3. Dust collector (*c*).
4. Device for measuring the volume flow of the sample gas stream (*e*).
5. Suction device (*g*).
6. Check valve for the sample gas stream (*f*).
7. Device for the measurement of the free stream velocity (*h*).
8. Measuring devices for pressure, temperature, and humidity of the gas at the sampling point (*d*), (*k*), (*m*), (*n*), (*p*).

3.10. SAMPLING PROBES

Sampling probes of various types have been built. Like Pitot tubes, they must be positioned so that the probe axis is in the exact direction of the main flow. When the angle to the flow is less than 5°, the error caused by the inclination may be neglected. Figure 3.18 shows calibration curves for Prandtl tubes, obtained from wind tunnel experiments,[24] and Figure 3.19 plots calibration curves of a set of velocity probes of different sizes.[25] The structural details of these probes also appear in Figure 3.19*a*, and additional data are supplied in Table 3.3.

For sampling probes, which do not include special means for measuring gas velocity, it is possible to place a filter at the probe head for dust collection.

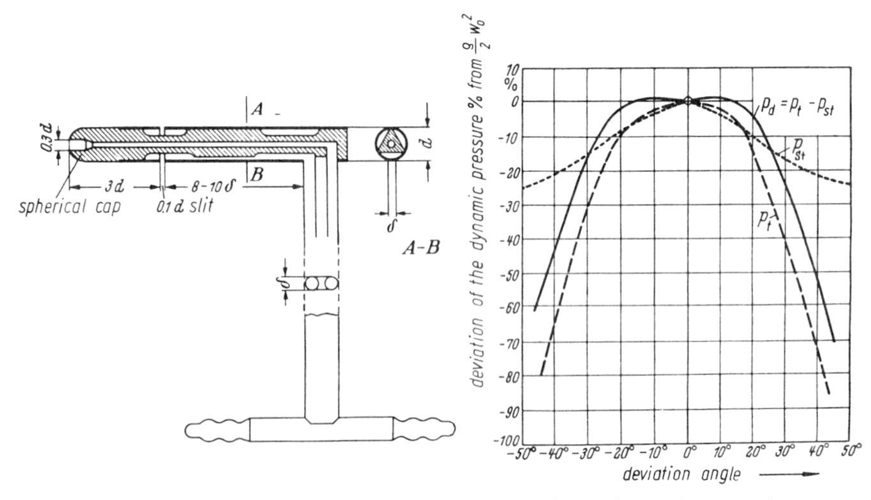

Figure 3.18. Prandtl tube and pressure curves as a function of deviation angle: p_t, total pressure; p_{st}, static pressure; p_d, dynamic pressure.

Table 3.3. Diameters of Probes (Figure 3.19*a*)

Probe[a] Number	Diameter (mm) Inside, D_i	Diameter (mm) Outside, D_a	Effective Diameter, $D \approx (D_i + D_a)/2$ (mm)
1	23.0	53.0	38.0
2	31.0	61.0	46.0
3	38.0	68.0	53.0
4	46.0	76.0	61.0
5	55.0	85.0	70.0
6	65.0	95.0	80.0

[a] Wall thickness of probes, 15 mm.

This arrangement is advantageous for measurements in hot flue gases because the sampling filter has effectively the temperature of the flue gas and it is simple to maintain the temperature above the dew point. Different forms of such probes are illustrated in Figure 3.20. Depending on the temperature, the filter in the probe head is of paper, glass fibers in a mesh basket, or glass wool in stainless steel cartridges. The favorite materials for the filters in the probe heads or the dust collectors are as follows.

1. Paper filters are selected for low dust concentrations and very small particles. They have a small pore width (2 to 5 μm) and a large rough surface. The weight of these filters is small.
2. Special low ash papers are used when it is necessary to incinerate the sample for analysis.
3. Cloth filters are used for large sample gas streams and high dust loads, because they have greater stability and normally lower pressure losses than paper filters.
4. Filters made of glass fibers coated with silicone and filters made from polymeric materials are appropriate for the sampling of corrosive solids or gases.
5. Felted filters made from cotton, asbestos, or glass wool have large surfaces and filtration depth. Essential for good filtration is a uniform but porous felting without breakage of the fibers. These materials are suitable for use at high temperatures.
6. Sintered glass, metal, or ceramic filters (filter stones) can be used at very high temperatures.

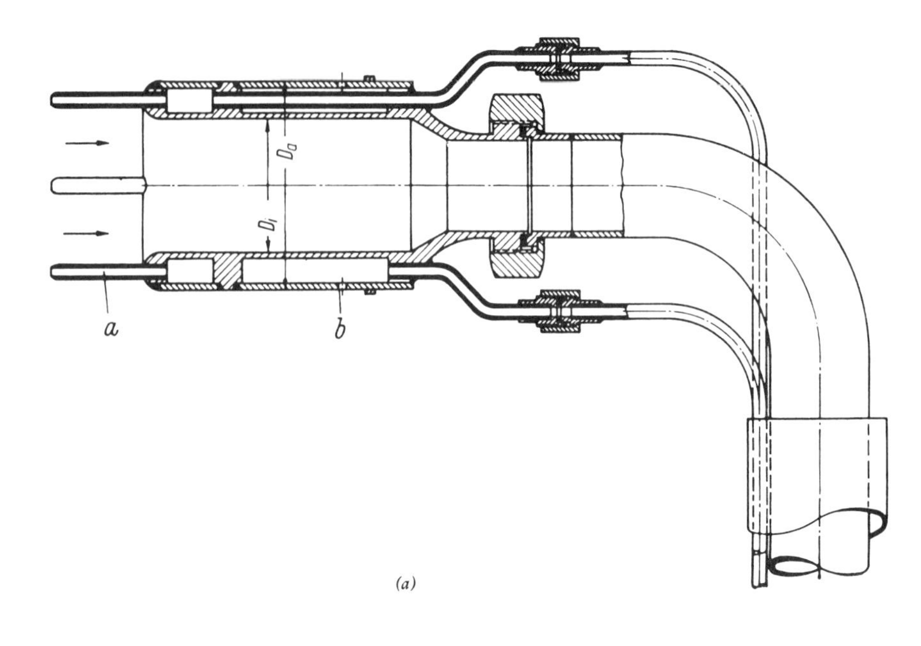

(a)

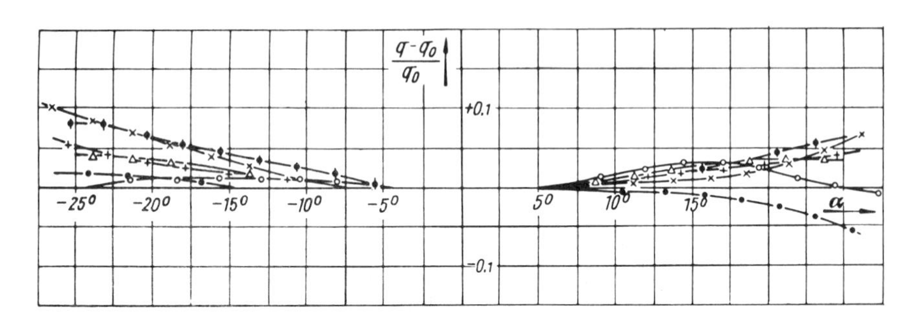

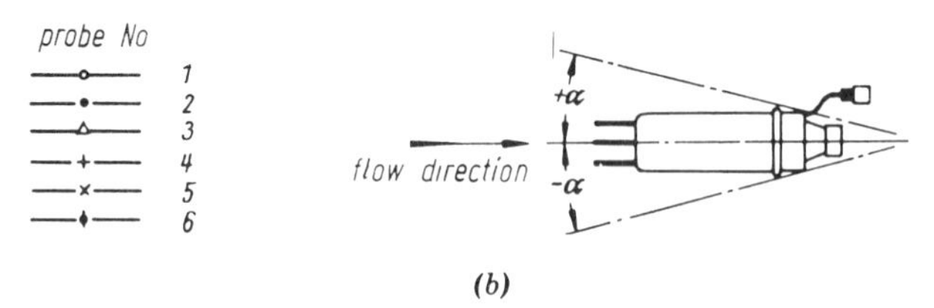

(b)

Figure 3.19. (*a*) Velocity probe: *a*, measuring probes for total pressure; *b*, probe for measurement of static pressure in the main stream. (*b*) α, deviation angle; q_0, dynamic pressure at $\alpha = 0°$; q, dynamic pressure.

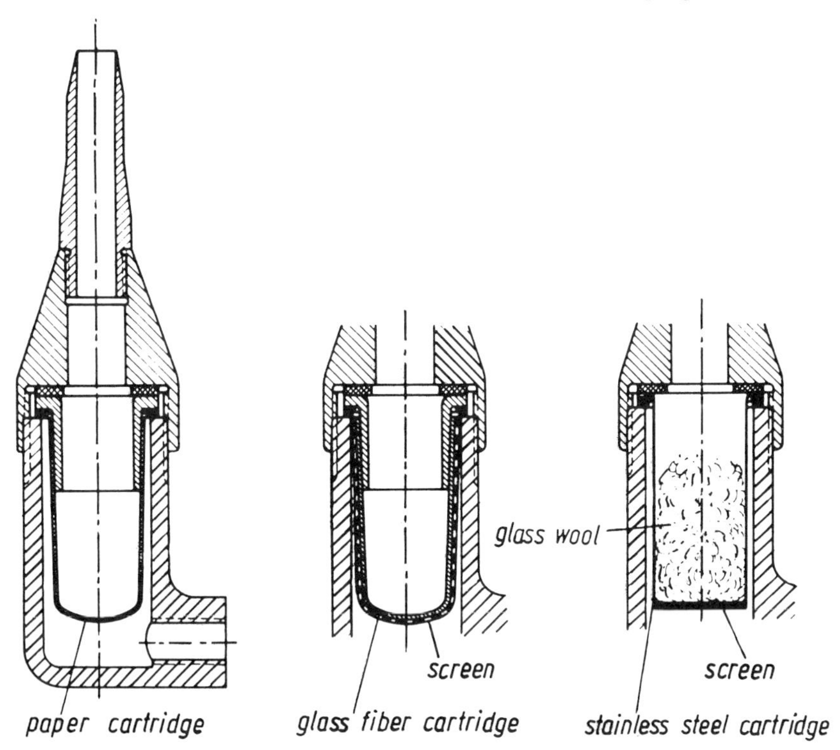

Figure 3.20. Probes with filter.[23]

3.10.1. Standard Probes

Standard probes are sampling probes that do not include special equipment for measuring velocity. Normally these probes are sharp-edged tubes (Figure 3.20). Sampling with these probes presupposes that the velocity of the main gas stream is measured by Prandtl tubes. The advantage of these probes is that it is possible to place a dust-collecting filter into the probe head. Probe heads are exchangeable and can be easily adapted without modifying the measuring device if the flow conditions in the channel change.

3.10.2. Velocity Probes

Thick-walled velocity probes as shown in Figure 3.19 have four tubes at the probe inlet for measuring the total pressure and four holes located in the cylindrical part of the probe for measuring the static pressure in the main stream. Both tubes are connected by way of a liquid manometer; the dynamic pressure of the main stream can be measured directly, for example, with Pitot tubes.

The velocity probe makes it possible to measure the velocity of the main stream at each point during the sampling. This makes it easy to select a sampling velocity identical with the free stream velocity. Changes of the velocity during the sampling are discovered immediately. The probe can also be used for velocity measurement when no sampling takes place.

3.10.3. Null-Type Sampling Probes

With the "null"-type sampling probes, the static pressure, measured at the outer and inner wall of the probe, is balanced. This assumes that with the same static pressure at the outer and the inner wall of the probe, the velocity within the probe is equal to the free stream velocity. This is only true in practice when the probe wall is infinitely thin. In constructing the probe inlet it is important to avoid errors due to the disturbance of the flow just away from the probe inlet. Another type of inaccuracy may result because the static pressure inside and outside the probe should be measured directly at the probe orifice; this is impossible in practice, however, because of structural difficulties. Figure 3.21 presents different types of null-type sampling probe. Two of them (Figure 3.21*a*, *b*) were described by Dennis et al.[26] and have been tested at several places in the United States. The first is used for gas velocities from 5 to 30 m sec^{-1} and sampling volume flows between 2.6 and 14 m^3 hr^{-1}. The sampling volume flows for the probe in Figure 3.21*b* are 17 to 110 m^3 hr^{-1} at gas velocities between 5 and 33 m sec^{-1}. The static pressure inside the probe of Figure 3.21*a* is measured from the probe inlet at a distance of about twice the probe diameter. The measurement of the static pressure of the main stream takes place at a distance from the probe inlet which is about 1.85 times the outer diameter of the probe. The static pressure for the second probe is measured inside at a distance from the probe inlet of about 0.64 times the inner diameter and outside at a distance from the inlet of about 2.73 times the outer diameter of the probe. Proposed by Narjes,[27] the design of Figure 3.21*c* allows the measurement of the static pressure of the inside flow directly at the probe inlet. It is especially suitable for the investigation of gas streams loaded with a high concentration of solids and is widely used for the control of power plants fired by pulverized coal.

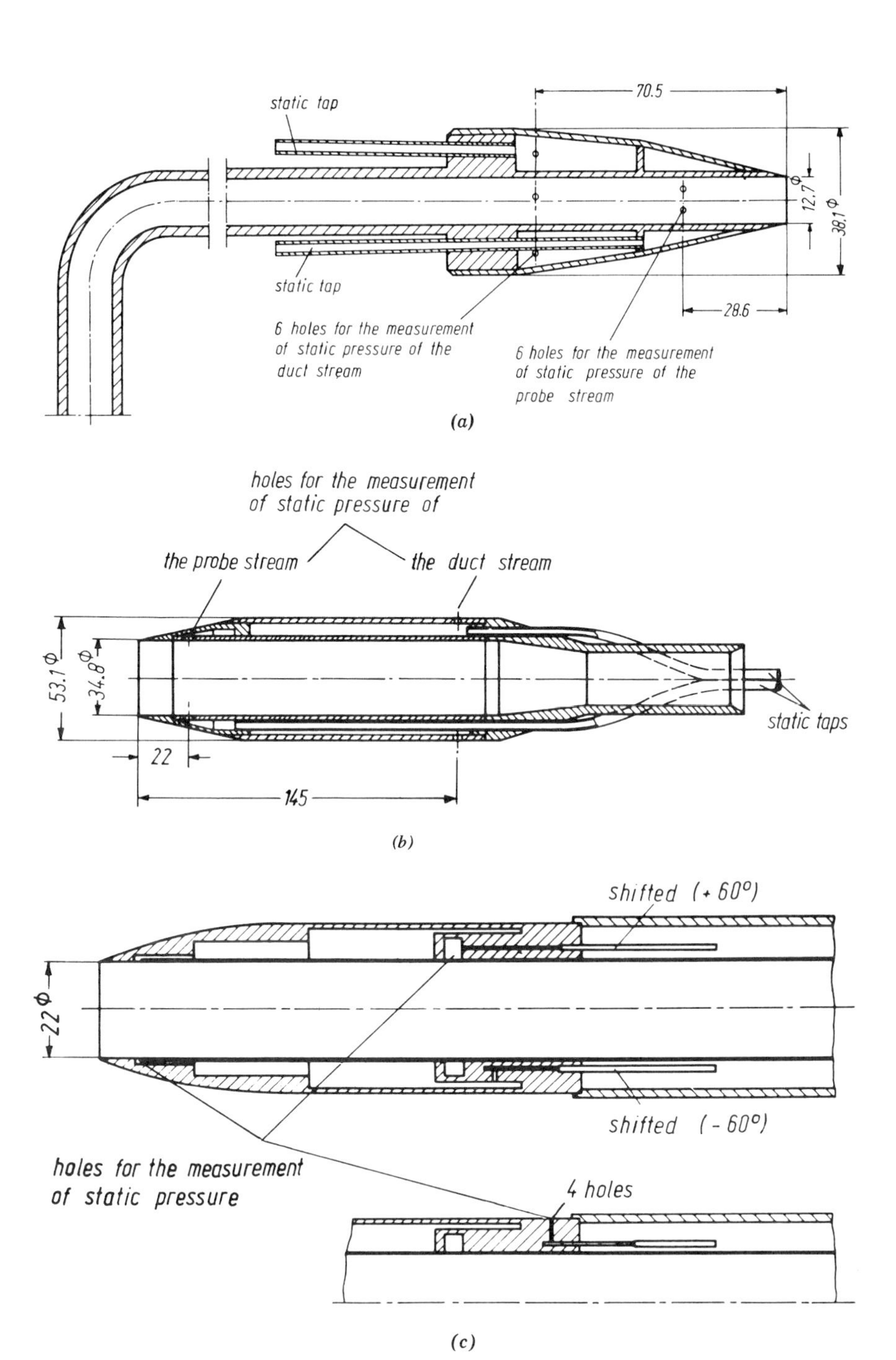

Figure 3.21. Different null-type stack sampler probes. (*a*) Design by Western Precipitation Corp. (*b*) Design by Buell Engineering Co. (*c*) Design by L. & C. Steinmüller GmbH.

3.10.4. Container Probes

A new method for the determination of the distribution of a dust stream in a flow field was proposed by Jung.[28] No longer is part of a gas stream sampled, but the dust is collected by a "container" probe. The container probe (Figure 3.22) consists of a cylindrical tube (diameter d_s) set in the flow transverse to the main stream. The tube contains radial holes (diameter d_H) through which the solid material reaches the tube *a*. With a diversion vane,

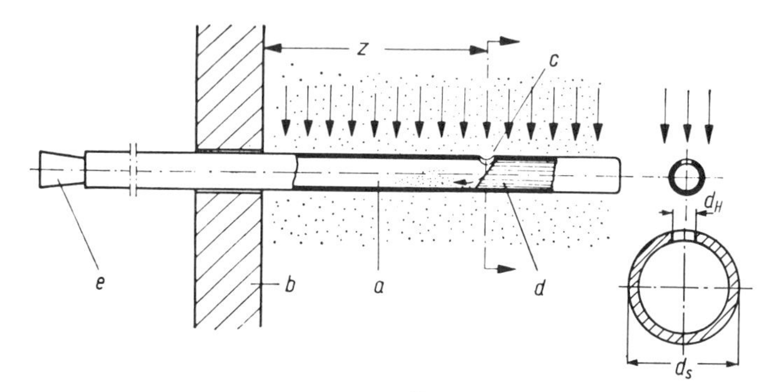

Figure 3.22. Container probe proposed by Jung:[28] *a*, probe shaft; *b*, channel wall; *c*, impaction hole; *d*, impingement sheet; *e*, stopper.

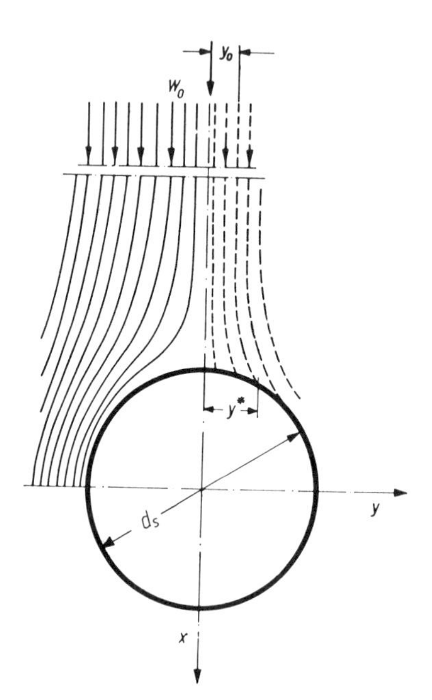

Figure 3.23. Potential streamlines (solid lines) and particle paths (broken lines) in front of a cylinder.

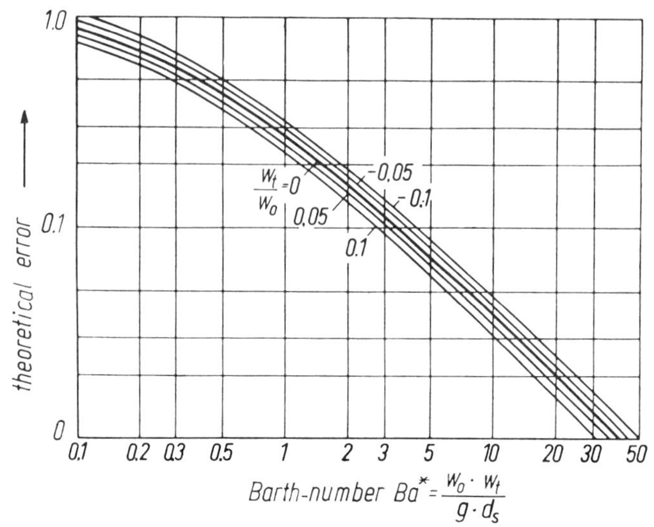

Figure 3.24. Theoretical error as a function of the modified Barth number Ba* and the velocity ratio w_t/w_0 for the container probe.

the solid is transported into the interior of the tube. Simply moving the tube through the duct wall *b* allows a fast scanning of the flow cross section. The limited capacity of the sampling tube for the solid material makes for a very short measuring time. Very short measuring times, of the order of 1 to 10 sec, are obtained because in turning the probe around on its shaft, the measuring holes can be opened or closed. The solid particles entering

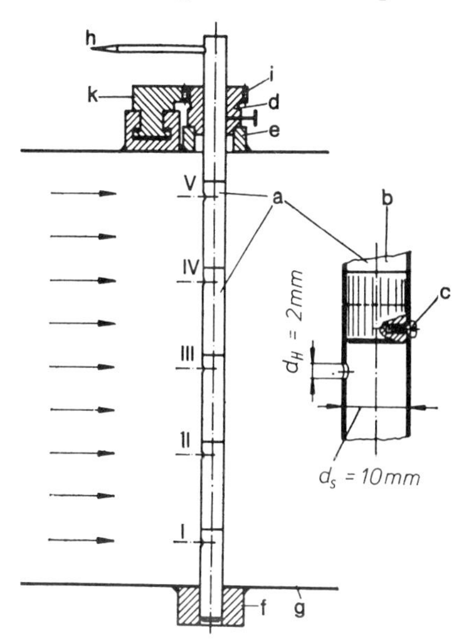

Figure 3.25. Multiple-container probe in position, showing connection point of different probe elements: *a*, container pipe; *b*, bolt; *c*, screw; *d*, adjusting ring; *e*, *f*, holder; *g*, channel; *h*, pointer; *i*, gear wheel; *k*, gear rack; I–V, impaction holes.

a hole of finite diameter can be calculated from a knowledge of streamlines and particle paths around a cylinder that is impacted by the transverse flow. Streamlines and particle paths for the container probe are given in Figure 3.23. The results of Jung's theoretical calculations appear in Figure 3.24, where the theoretical error is plotted as a function of the Barth number for different velocity ratios. To allow for the disturbance of the streamlines, caused by the edge of the collecting holes, the diameter d_H of these should be $d_H \leq 0.2\, d_s$, where d_s is the diameter of the probe tube.

For rapid measurements of the solid mass distribution in a duct, the container probe can be built with multiple openings (Figure 3.25), allowing for simultaneous sampling at several points. The use of this probe has some advantages for process control—for example, in power plants.

3.11. STANDARDS FOR PARTICULATE SAMPLING

Legislative requirements or recommendations of the scientific societies or engineering institutions must be observed in the determination of the solids content in flowing gases. The standards for the measurement of solids in flowing gases of Germany and the United States may be taken as representing the present state of the art. Unless there are conflicting performance standards applying in the country in which the measurements are being carried out, measurements should be performed in accordance with these recommendations.

3.11.1. Germany

In 1974 in the Federal Republic of Germany, specifications were published regarding the limitation of emissions of solid particles. These values, given in the *Technische Anleitung zur Reinhaltung der Luft* ("*TA-Luft*") may not be exceeded. For a number of materials emitted by different industries data, are given in the *VDI-Handbuch Reinhaltung der Luft*.[32]

The German Engineering Society (VDI) published a standard for efficiency measurement on dust collectors (VDI 2066).[33] This publication describes in detail the rules for the preparation and the performance of the dust content measurement and discusses the choice of sampling devices and the measurement of the condition of the gas and of the distribution of the solids. The measuring devices for the determination of the solid content are described, and errors and reasons for errors are analyzed.

3.11.2. United States

In the United States, the specifications for the measurement of the solid content in flowing gases are laid down in the American Society for Testing and Materials Standards, especially ASTM-Standard D 2928.[32] The standard includes details concerning the measurement of the gas stream, the equipment necessary for the measurement of the solid content, and the calculation of the solids concentration based on experimental results, as well as descriptions of sampling trains and different sampling probes.

3.11.3. Great Britain

Specifications are published by the British Standards Institution in British Standards 893[33] and 3405.[34]

3.12. CARRYING OUT A MEASUREMENT

The first step in taking a measurement is to choose the cross section in the duct and the number of sampling points according to the criteria discussed previously.

The diameter of the probe is selected so that in a limited measuring time a sufficient amount of solid material can be collected for an evaluation. If the particle size distribution is to be determined from the sampled solids, the amount of solid material must be large. According to the type of probe, the velocity profile has to be measured by means of a Prandtl tube. When using null-type sampling probes, the measurements of velocity and solid content can be combined. As an example, typical values are given in Table 3.4, and the particle size distribution of the dust transported in the gas stream is plotted in Figure 3.26. Figure 3.27 shows the relative dust content as a

Table 3.4. Typical Values as an Example

Free stream velocity	$w_0 = 20$ m sec^{-1}
Density of gas	$\rho = 1.2$ kg m^{-3}
True solids concentration	$q_0 = 200$ mg m^{-3}
Density of solids	$\rho_p = 2000$ kg m^{-3}
Probe diameter	$D = 20$ mm

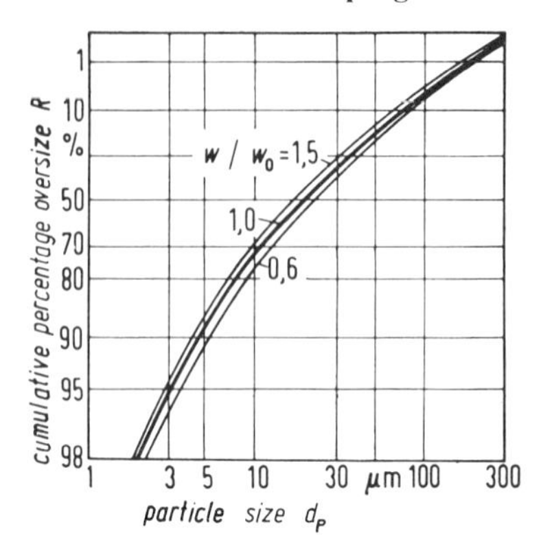

Figure 3.26. Particle size distribution.

function of the sampling velocity. Table 3.5 gives the percentage of measuring error for some sampling velocities. For constant ratio of sampling to free stream velocity, each fraction of particle size has another measuring error. This leads to the result that the sampled solid material has a particle size distribution different from that of the dust transported within the gas stream. The particle size distribution shifts to finer sizes when $w/w_0 > 1$, and to coarser sizes when $w/w_0 < 1$. For two values, $w/w_0 = 2.5$ and $w/w_0 = 0.6$, the shift appears in Figure 3.26. Contrary to this, Benarie and Panof were unable to find differences in the particle size distribution in the probe and the duct for different sampling velocities. This failure can be accounted for only by the limited accuracy of their experiments.

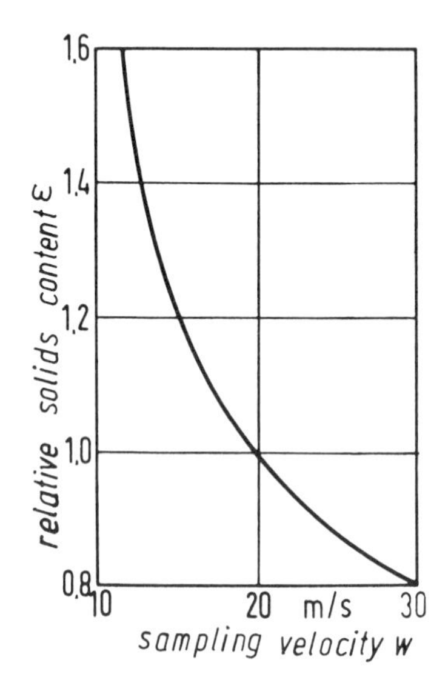

Figure 3.27. Relative solids content in the probe as a function of the sampling velocity.

Table 3.5. Solids Concentration in Sampled Gas and Measuring Error for Different Sampling Velocities

Measurement	I	II	III	IV	V
Sampling velocity, w (m sec^{-1})	12	16	20	24	28
Relative solids concentration in sampled gas, ε	1.54	1.16	1.0	0.895	0.83
Solids concentration in sampled gas, q (mg m^{-3})	308	232	200	179	166
Measuring error, f (%)	+54	+16	+0	−10.5	−17

3.13. CONCLUSIONS

Anisokinetic sampling of a partial gas stream from a dust-laden gas stream leads to incorrect results. The larger particles in the gas stream produce a larger measurement error at equal ratio of sampling to free stream velocity. The error increases at a constant ratio of sampling to free stream velocity with decreasing probe diameter. If the fractional deviation of the sampling velocity is equal to the free stream velocity, velocities higher than the free stream velocity are favored because this results in less error. This chapter shows that in the errors resulting from (1) anisokinetic sampling, (2) clouds of solid material in the duct, and (3) an insufficient number of sampling points, the first type are often overestimated, whereas the errors resulting from cases 2 and 3 are in practice often higher than anticipated.

SYMBOLS

A	duct cross section (m^2)
Ba	Barth number $(w_0 w_t)/(gD)$
c_w	drag coefficient
D	probe diameter (m)
D_0	diameter of the cylinder formed by the limiting streamlines (m)
D_s	diameter of the cylinder formed by the limiting particle paths (m)
D_d	duct diameter (m)
d_h	hydraulic diameter (m)
d_p	particle diameter (m)
e	thickness of the probe wall (m)
E	sink strength constant (m^3 sec^{-1})
F	flow resistance (N)
F_0	duct cross section (m^2)
g	acceleration of gravity (m sec^{-2})
I	inertial force (N)

i	number of incremental area
l_0	flight path of a particle in undisturbed gas (m)
l	length (m)
n	ordinal index
q	solids concentration in sampled gas (kg m^{-3})
q_0	true solids concentration (kg m^{-3})
R	probe radius (m)
Re_0	Reynolds number $(w_0 d_p \rho)/\eta$
Re	Reynolds number $[(w - w_p)d_p \rho]/\eta$
r	radial length (m)
$\dot{S}$	flow rate of solids in the probe gas stream (kg sec^{-1})
$\dot{S}_0$	flow rate of solids in the dust gas stream (kg sec^{-1})
t	time (sec)
U	duct circumference (m)
u	velocity in x-direction (m sec^{-1})
$\dot{V}$	flow rate of sampled gas (m^3 sec^{-1})
$\dot{V}_0$	flow rate in the duct (m^3 sec^{-1})
v	velocity in y-direction (m sec^{-1})
w	sampling velocity (m sec^{-1})
w_0	free stream velocity (m sec^{-1})
w_p	particle velocity (m sec^{-1})
w_t	terminal velocity (m sec^{-1})
z	distance from probe inlet (m)
β	inclination angle (degrees)
$\varepsilon = q/q_0$	relative solids concentration in sampled gas
ν	kinematic viscosity of the fluid (m^2 sec^{-1})
ϕ_p	potential of a parallel flow (m^2 sec^{-1})
ϕ_D	potential of a dipole (m^2 sec^{-1})
ϕ	potential of the flow around a sphere (m^2 sec^{-1})
ϕ_s	potential of a sink ring (m^2 sec^{-1})
ψ	stream function (m sec^{-1})
η	dynamic viscosity of the fluid (Pa s)
ρ	fluid density (kg m^{-3})
ρ_p	particle density (kg m^{-3})

REFERENCES

1. Vitols, V., *J. Air Pollut. Control Assoc.*, 16, 79 (1966).
2. Vitols, V., Determination of Theoretical Collection Efficiencies of Aspirated Particulate Matter Sampling Probes Under Anisokinetic Flow. Ph.D. thesis, University of Michigan, Ann Arbor, 1964.
3. Bartak, J., *Staub-Reinhalt. Luft*, 34, 295 (1974).

4. Fernandes, E. and P. Suter, *Brennstoff-Wärme-Kraft*, 26, 502 (1974).
5. Stenhouse, J. I. T. and P. J. Lloyd, Sampling Errors due to Inertial Classification, American Institute of Chemical Engineers Symposium Series, No. 137, Vol. 70, 1974, p. 307.
6. Bohnet, M., *Chem. Ing. Tech.*, 39, 972 (1967).
7. Rüping, G., *Staub-Reinhalt. Luft*, 28, 137 (1968).
8. Kürten, H., J. Raasch, and H. Rumpf, *Chem. Ing. Tech.*, 38, 941 (1966).
9. Zenker, P., Untersuchung über die Staubverteilung von strömenden Staub-Luft-Gemischen in Rohrleitungen. Dissertation, Technische Universität München, 1975.
10. Zenker, P., *Staub-Reinhalt. Luft*, 31, 252 (1971).
11. Watson, H. H., *Am. Ind. Hyg. Assoc., Quart.*, 15 (March 21–25, 1954).
12. Bürkholz, A., *Staub-Reinhalt. Luft*, 33, 397 (1973).
13. Bürkholz, A., *Chem. Ing. Tech.*, 42, 299 (1970).
14. Bürkholz, A., *Chem. Ing. Tech.*, 45, 1 (1973).
15. Bürkholz, A. and E. Muschelknautz, *Chem. Ing. Tech.*, 44, 503 (1972).
16. Badzioch, S., *Brit. J. Appl. Phys.*, 10, 26 (1959).
17. Badzioch, S., *J. Inst. Fuel*, 33, 106 (1960).
18. Whiteley, B. and L. E. Reed, *J. Inst. Pet.*, 32, 316 (1959).
19. Zipse, G., Die Massenstromdichteverteilung bei der pneumatischen Staubförderung und ihre Beeinflussung durch Einbauten in die Förderleitung. *Fort. Ver. Deut. Ing. Düsseldorf*, Series 13, No. 3 (1966).
20. Hemeon, W. C. L. and F. F. Haines, *Air Repair*, 4, 159 (1954).
21. Benarie, M. and S. Panof, *Aerosol Sci.*, 1, 21 (1970).
22. Morrow, N. L., R. S. Brief, and R. R. Bertrand, *Chem. Eng.*, 24, 84 (1972).
23. Verein Deutscher Ingenieure, *Leistungsmessung an Entstaubern*. VDI-Richtlinien, Nr. 2066, 1966, p. 21.
24. Hengstenberg, J., B. Sturm, and O. Winkler, *Messen und Regeln in der chemischen Technik*, Springer-Verlag, Berlin, 1964.
25. Ibing, R. and G. Meier, *Eichung und Entwicklung von Staubentnahmesonden*. Forschungsberichte des Wirtschafts-und Verkehrsministeriums Nordrhein-Westfalen, No. 474, Köln und Opladen, Westdeutscher Verlag, 1958.
26. Dennis, R., W. R. Samples, D. M. Anderson, and L. Silverman, *Ind. Eng. Chem.*, 49, 294 (1957).
27. Narjes, L., *Staub*, 25, 148 (1965).
28. Jung, R., *Chem. Ing. Tech.*, 41, 620 (1969).
29. Jung, R. and G. Thomas, *Verfahrenstechnik*, 8, 251 (1974).
30. *Technische Anleitung zur Reinhaltung der Luft* (*TA Luft*), August 28, 1974, 25, 427 (1974).
31. *VDI-Handbuch Reinhaltung der Luft*, VDI-Kommission Reinhaltung der Luft, VDI-Verlag Düsseldorf.
32. Standard Method for Sampling Stacks for Particulate Matter. American National Standard Z 257.3-1973, approved March 29, 1973, American National Standards Institute.
33. BSI-894, Method of Testing Dust Extraction Plants and the Emission of Solids from Chimneys of Electric Power Stations, British Standards Institution, 1940.
34. BSI-3405, Simplified Methods for Measurement of Grit and Dust Emission, British Standards Institution, 1971.

4

PARTICLE SIZE MEASUREMENT IN INDUSTRIAL FLUE GASES

W. B. Smith and J. D. McCain

Southern Research Institute
Birmingham, Alabama

4.1. INTRODUCTION

4.1.1. General Discussion

Measurements of particle size distributions in industrial flue gas streams are made for several reasons: the emissions must be characterized as completely as possible, to permit assessment of potential health and environmental effects; emission measurements can be useful as a process monitor; and particle size measurements are required to obtain a complete quantification and better understanding of behavior of gas cleaning devices, which in turn could make particulate removal more efficient.

The emphasis on pollution control has recently shifted toward the fine particle size range, that is, particles having diameters of less than 3 μm. These are most important because they remain in the atmosphere for long periods of time, where they contribute to the scattering of light in the visible range. They can follow the airstream deep into the human respiratory system, where a large fraction may be deposited. Fine particle emissions are more difficult to control than large particles, and large particles are collected with high efficiencies by most pollution control devices; even when large particles pass through the control device, they tend to settle out of the atmosphere more quickly.

Particulate measurement techniques are in a transitory developmental state. Prototype systems often include laboratory components that are unsuitable for the hostile environments frequently encountered by on-line instruments. No single instrument can be used to make size measurements over the entire range of particulate sizes. The measurements are expensive, complicated, and far less accurate than is desirable.

Extensive treatments of aerosol physics and particle sizing have been written by Green and Lane,[1] Davies,[2] Allen,[3] Cadle,[4] and Irani and Callis,[5] but these do not discuss the modern equipment and techniques in enough detail to serve as an adequate preparation for persons engaged in stack sampling.

This chapter discusses techniques that can be used to make measurements of particle size distributions in the range of 0.01 to 10 μm in diameter in process streams where particulate concentration and temperature are usually much higher than ambient. Emphasis is placed on sizing particles before collection, as opposed to the less desirable methods, which require that the aerosol be redispersed from a bulk sample.

4.1.2. Definitions and Fundamentals

4.1.2.1. Particle Size and Shape

The particulate matter suspended in industrial gas streams may be in the form of nearly perfect spheres, regular crystalline forms other than spheres, irregular or random shapes, or as agglomerates made up from combinations of these. It is possible to discuss particle size in terms of the volume, surface area, projected area, projected perimeter, linear dimensions, light scattering properties, or drag forces in a liquid or gas (mobility). Particle sizing work is frequently done on a statistical basis where large numbers of particles are sampled, rather than individuals. For this reason the particles are normally assumed to be spherical. This convention also makes transformation from one basis to another more convenient.

a. Aerodynamic Diameter

The parameter most commonly used to describe particle size is the equivalent aerodynamic diameter. This is the diameter of a sphere of unit density that has the same settling velocity in air as the particle of interest. This convention has found popularity because much of the interest in aerosol research was initiated by meteorologists, health scientists, and environmental scientists, who study among other things the dispersion of plumes, the deposition of particulate matter in the human respiratory system, and the fate of pollutants in the atmosphere. All these depend to a large extent on the aerodynamic behavior of the particles. Many sampling instruments classify particles according to the aerodynamic or Stokes diameter.

b. Stokes Diameter

If the density of the particles is known, particle size may be described by using the Stokes diameter, that is, the diameter of a sphere having the same density and settling velocity as the particle. This holds precisely for particles of moderate diameter (about 20 μm in air, which is the principal range of interest).

An average density for the particles can be obtained from volume-weight data using a helium pycnometer if large enough samples are available. The validity of size information based on an average density depends on the uniformity of the density from particle to particle, particularly with respect to size. Visual inspection of some size-classified samples from flue gases sometimes reveals a variation in color with size that would seem to indicate compositional inhomogeneities. Repeated measurements of the density of bulk samples of fly ash taken from the effluents of coal fired power plants, however, have consistently yielded values near 2.5 g/cm^{-3}.

c. Equivalent Polystyrene Latex (PSL) Diameter

The intensity of light scattered by a particle at any given angle is dependent on the particle size, shape, and index of refraction. It is impractical to measure each of these parameters, and the theory for irregularly shaped particles is not well developed. Sizes based on light scattering by single particles are therefore usually estimated by comparison of the intensity of scattered light from the particle with the intensities due to a series of calibration spheres of very precisely known size. Most commonly these are PSL spheres.* Spinning disk and vibrating orifice aerosol generators can be used to generate monodisperse calibration aerosols of different physical properties.[6,7] Because most manufacturers of optical particle sizing instruments use PSL spheres to calibrate their instruments, it is convenient to define an equivalent PSL diameter as the diameter of a PSL sphere that gives the same response with a particular optical instrument as the particle of interest.

d. Equivalent Volume Diameter

Certain instruments, notably the Coulter counter (see Section 4.3.2), have as the measured size parameter the volumes of the individual particles. Size distributions from such techniques are given in terms of the sizes of spheres having the same volume as the particles measured.

4.1.2.2. Particle Size Distributions

a. General Discussion

Experimental measurements of particle size normally cannot be made with a single instrument if the size range of interest extends over much more than a decimal order of magnitude. Presentations of size distributions covering

* Available from the Dow Chemical Company, P.O. Box 68511, Indianapolis, Indiana 46268.

broad ranges of sizes then must include data points that may have been obtained using different physical mechanisms. Normally the data points are converted by calculation to the same basis and put into tabular form or fitted with a histogram or smooth curve to represent the particle size distribution. Frequently used bases for particle size distributions are the relative number, volume, surface area, or mass of particles within a size range. The size range might be specified in terms of aerodynamic, Stokes, or equivalent PSL diameter. There is no standard equation for statistical distributions that can be universally applied to describe the results given by experimental particle size measurements. However the log-normal distribution function has been found to be a fair approximation for some sources of particulate, and several features make it convenient to use. For industrial sources the best procedure is to plot the experimental points in a convenient format and to examine the distribution in different size ranges separately, rather than trying to characterize the entire distribution by two or three parameters. The ready availability of inexpensive programmable calculators, which can be used to convert from one basis to another, compensates greatly for the lack of an analytical expression for the size distribution.

b. Terminology and Definitions

Figure 4.1 plots generalized unimodal particle size distributions that are used to illustrate graphically the terms commonly employed to characterize an aerosol. Occasionally size distribution plots exhibit more than one peak. A size distribution with two peaks (bimodal) frequently can be shown to be equivalent to the sum of two or more distributions of the types shown in Figure 4.1. If a distribution is symmetric or bell shaped when plotted along a linear abscissa, it is called a "normal" distribution (Figure 4.1*c*). A distribution that is symmetric or bell shaped when plotted on a logarithmic abscissa is called "log-normal" (Figure 4.1*d*).

Interpretation of the frequency or relative frequency (f in Figure 4.1) is very subtle. One is tempted to interpret this as the amount of particulate of a given size. This interpretation is erroneous, however, and would require that an infinite number of particles be present. The most useful convention is to define f in such a way that the area bounded by the curve (f) and vertical lines intersecting the abscissa at any two diameters is equal to the amount of particulate in the size range indicated by the diameters selected. A point on curve f, then, is equal to the *relative* amount of particulate in a narrow size *range* about a given diameter.

The median divides the area under the frequency curve in half. For example, the mass median diameter of a particle size distribution is the size at which 50% of the mass consists of particles of larger diameter, and 50% of the mass consists of particles having smaller diameters. Similar definitions

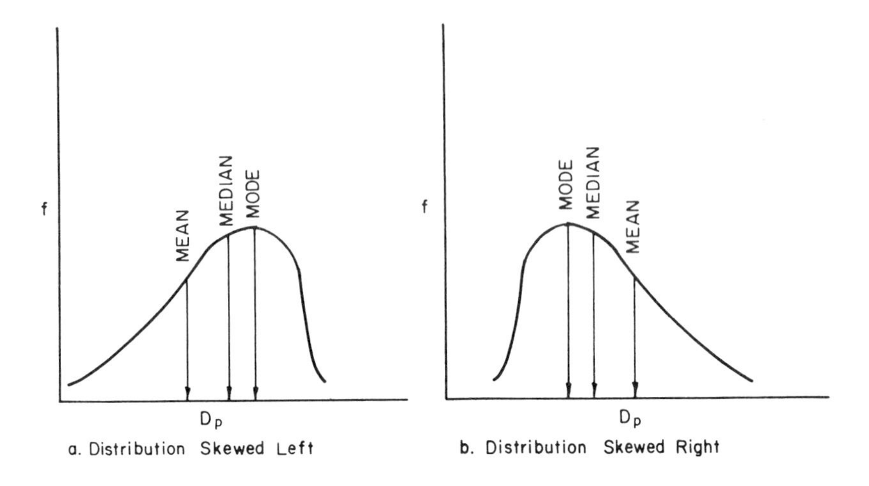

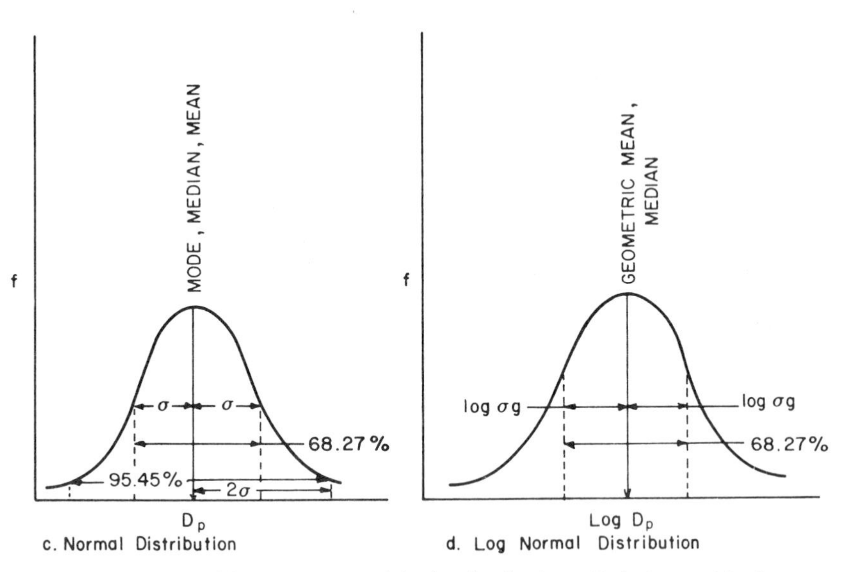

Figure 4.1. Examples of frequency or particle size distributions; D_p is the particle diameter.

apply for the number median diameter (NMD) and the surface median diameter (SMD).

The term "mean" denotes the arithmetic mean of the distribution. In a particle size distribution the mass mean diameter is the diameter of a particle that has the average mass for the entire particle distribution. Again, similar definitions hold for the surface and number mean diameters.

The mode represents the diameter that occurs most commonly in a particle size distribution. The mode is seldom used as a descriptive term in aerosol physics.

The geometric mean diameter is the diameter of a particle that has the logarithmic mean for the size distribution. This can be expressed mathematically as follows:

$$\log D_g = \frac{\log D_1 + \log D_2 + \cdots + \log D_N}{N} \tag{1a}$$

or

$$D_g = (D_1 D_2 D_3, \ldots, D_N)^{1/N} \tag{1b}$$

The standard deviation σ and relative standard deviation α are measures of the dispersion (spread, or polydispersity) of a set of numbers. The relative standard deviation is the standard deviation of a distribution divided by the mean, where σ and the mean are calculated on the same basis (i.e., number, mass, or surface area). A monodisperse aerosol has a standard deviation and relative standard deviation of zero. For many purposes the standard deviation is preferred because it has the same dimensions (units) as the set of interest. For normal distributions, 68.27% of the events fall within one standard deviation of the mean, 95.45% within two standard deviations, and 99.73% within three standard deviations.

Table 4.1 summarizes nomenclature and formulas that are frequently used in practical aerosol research. In practice most of the statistical analysis is done graphically.

c. *Cumulative and Differential Graphs*

Field measurements of particle size usually yield a set of discrete data points that must be transformed to some extent before interpretation. The resultant particle size distribution may be shown as tables, histograms, or graphs. Graphical presentations, the conventional and most convenient formats, can be of several forms.

CUMULATIVE SIZE DISTRIBUTIONS. Cumulative mass size distributions are formed by summing all the mass containing particles less than a certain

Table 4.1. Summary of Nomenclature used to Describe Particle Size Distributions

Name	Symbol	Formula[a]	
Volume or mass mean diameter	D_m	$D_m = \left(\frac{\sum_{j=1}^{N} n_j D_j^3}{N} \right)^{1/3}$	(2)
Surface mean diameter	D_s	$D_s = \left(\frac{\sum_{j=1}^{N} n_j D_j^2}{N} \right)^{1/2}$	(3)
Number mean diameter	D_n	$D_n = \frac{\sum_{j=1}^{N} n_j D_j}{N}$	(4)
Geometric mean diameter	D_g	$D_g = \left(\prod_{j=1}^{N} D_j \right)^{1/N}$	(5)
Surface-volume mean diameter, or Sauter diameter	D_{vs}	$D_{vs} = \frac{\sum_{j=1}^{N} D_j^3}{\sum_{j=1}^{N} D_j^2}$	(6)

Standard deviation	σ	$\sigma = \left(\frac{\sum_{j=1}^{N} f_j (D_j - D_{m,n,s})^2}{N - 1} \right)^{1/2}$	(7)
Relative standard deviation	α	$\alpha = \frac{\sigma}{D_{m,n,s}}$	(8)
Mass median diameter	MMD	Medians are most conveniently determined	
Surface median diameter	SMD	graphically; for many slightly skewed	
Number median diameter	NMD	distributions median = mean + $\frac{1}{3}$(mode–mean)	
Stokes diameter	D	$D = \left(\frac{18\, \mu V_s}{g \rho_p C} \right)^{1/2}$	(9)
Aerodynamic diameter	D_{aer}	$D_{\text{aer}} = \left(\frac{18\, \mu V_s}{g C} \right)^{1/2}$	(10)

[a] In these equations j denotes a particular size interval; N is the total number of intervals; f_j is the relative mass, surface area, or number of particles in the interval; D_j is the diameter characteristic of the jth interval; μ is the viscosity of the gas; V_s is the particle settling velocity; g is the acceleration due to gravity; ρ_p is the particle density; and C is the slip correction factor.

diameter and plotting this mass versus the diameter. The ordinate is specifically equal to $M(j) = \sum_{t=1}^{j} M_t$, where M_t is the amount of mass contained in the size interval between D_t and D_{t-1}. The abscissa would be equal to D_j. Cumulative plots can be made for surface area and number of particles per unit volume in the same manner. Examples of cumulative mass and number graphs are given in Figures 4.2*b* and 4.2*a*, respectively, for the effluent from a coal-fired power boiler. Although cumulative plots obscure some information, the median diameter and the total mass per unit volume can be obtained readily from the curve. Because both the ordinate and abscissa extend over several orders of magnitude, logarithmic axes are normally used for both.

A second form of cumulative plot that is frequently used in the cumulative percentage of mass, number, or surface area contained in particles having diameter smaller than a given size. In this case the ordinate, on a mass basis is

$$\text{cumulative percentage of mass less than size} = \frac{\sum_{t=1}^{j} M_j}{\sum_{t=1}^{N} M_t} \times 100\% \quad (11)$$

The abscissa would be log D_j. Special log-probability paper is used for these graphs, and for log-normal distributions the data set would lie along a straight line. For such distributions the median diameter and geometric standard deviation can be easily obtained graphically. Figures 4.3*a* and 4.3*b* present cumulative percentage graphs for the size distribution in Figure 4.2*a* and a log-normal size distribution.

DIFFERENTIAL SIZE DISTRIBUTIONS. Differential particle size distribution curves are obtained from cumulative plots by taking the average slope over a small size range as the ordinate and the geometric mean diameter of the range as the abscissa. If the cumulative plot were made on logarithmic paper, the frequency (slope), taking finite differences, would be

$$\frac{\Delta M}{\Delta(\log D)} = \frac{M_j - M_{j-1}}{\log D_j - \log D_{j-1}}, \quad (12)$$

and the abscissa would be $D_G = \sqrt{D_j D_{j-1}}$, where the size range of interest is bounded by D_j and D_{j-1}; M_j and M_{j-1} correspond to the cumulative masses below these sizes. Differential number and surface area distributions can be obtained from cumulative graphs in precisely the same way. Differential graphs show visually the size range where the particles are concentrated with respect to the parameter of interest. The area under the curve

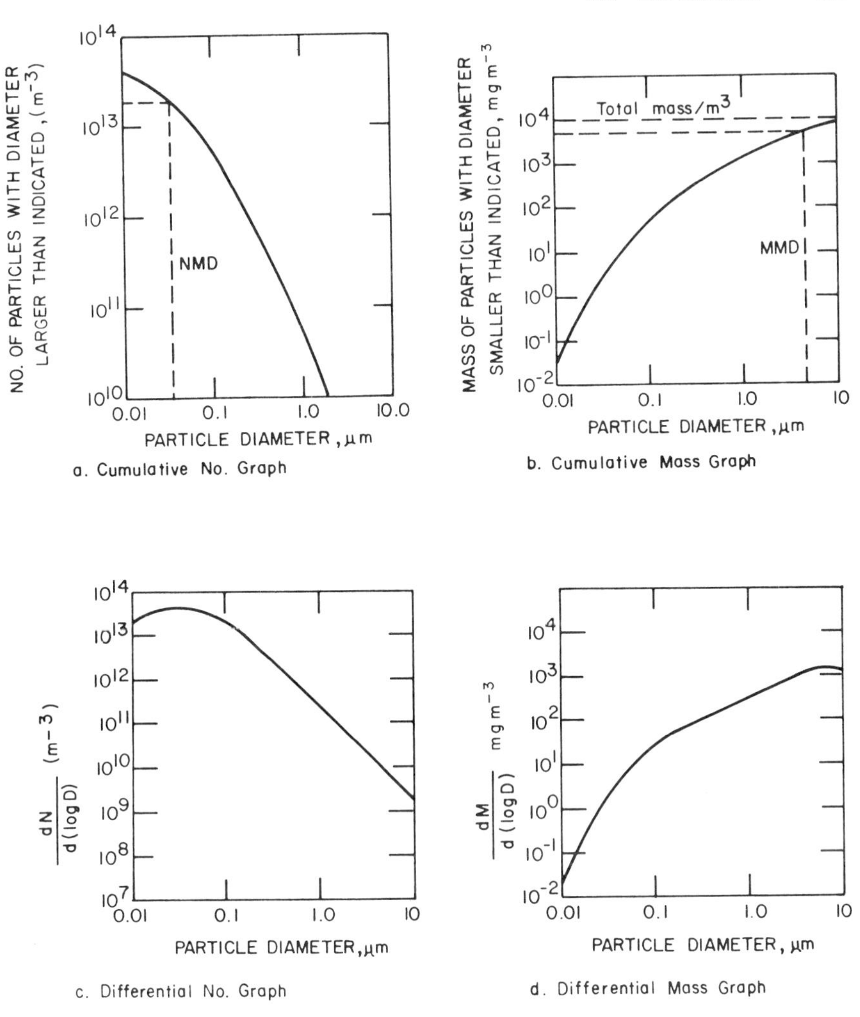

Figure 4.2. A single particle size distribution presented in four ways. The measurements were made in the effluent from a coal-fired power boiler.

in any size range is equal to the amount of mass (number, or surface area) consisting of particles in that range, and the total area under the curve corresponds to the entire mass (number, or surface area) of particulate in a unit volume. Again, because of the extent in particle size and the emphasis on the fine particle fraction, these plots are normally made on logarithmic scales. Figures 4.2*c* and 4.2*d* are examples of differential graphs of particle size distributions.

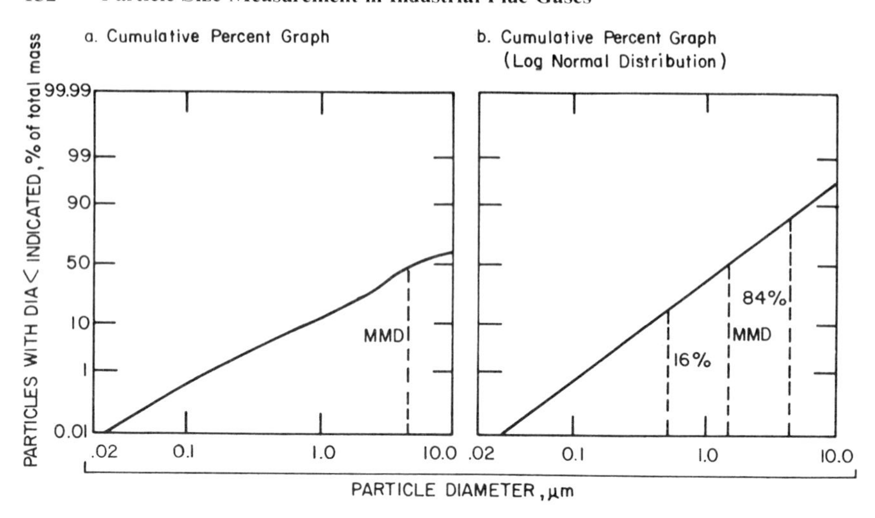

Figure 4.3. Size distributions plotted on log-probability paper.

LOG-NORMAL SIZE DISTRIBUTIONS. Several authors[8–16] have demonstrated that the formation of aerosols by different means frequently results in particle size distributions that obey the log-normal law. For log-normal particle size distributions, the geometric mean and median diameters coincide.

On a mass basis, the normal distribution law is

$$f = \frac{dM}{dD} = \frac{1}{\sigma\sqrt{2\pi}} \exp\left[-\frac{(D - D_m)^2}{2\sigma^2}\right] \tag{13}$$

The log-normal distribution law is derived from this equation by the transformation $D \rightarrow \log D$

$$f = \frac{dM}{d(\log D)} = \frac{1}{\log \sigma_g \sqrt{2\pi}} \exp\left[-\frac{1}{2}\left(\frac{\log D - \log D_{gm}}{\log \sigma_g}\right)^2\right] \tag{14}$$

where $\log \sigma_g$, the geometric standard deviation, is obtained by using the substitutions $D \rightarrow \log D$ and $\sigma \rightarrow \log \sigma_g$ in eq. 7. This distribution is symmetric when plotted along a logarithmic abscissa and has the feature that 68.3% of the distribution lies within one geometric standard deviation of the geometric mean on such a plot. Mathematically, this implies that $\log \sigma_g = \log D_{84.14} - \log D_g$ or $\log D_g - \log D_{15.86}$, where $D_{84.14}$ is the diameter below which 84.14% of the distribution is found, and so on. This can be simplified to yield

$$\sigma_g \simeq \frac{D_{84}}{D_g} \tag{15}$$

$$\sigma_g \simeq \frac{D_g}{D_{16}} \tag{16}$$

$$\sigma_g \simeq \left(\frac{D_{84}}{D_{16}}\right)^{1/2}. \tag{17}$$

When plotted on log-probability paper, the log-normal distribution is a straight line on any basis and is determined completely by the knowledge of D_g and σ_g, as illustrated in Figure 4.3*b*.

Another important feature is the relatively simple relationships among log-normal distributions of different bases. If D_{gm}, D_{gs}, D_{gvs}, and D_{gN} are the geometric mean diameters of the mass, surface area, volume-surface, and number distribution, respectively, we can write

$$\log D_{gs} = \log D_{gm} - 4.6 \log^2 \sigma_g \tag{18}$$

$$\log D_{gvs} = \log D_{gm} - 1.151 \log^2 \sigma_g \tag{19}$$

$$\log D_{gN} = \log D_{gm} - 6.9 \log^2 \sigma_g \tag{20}$$

The geometric standard deviation remains the same for all bases.

4.1.3. Summary of Techniques Available for Particle Sizing

Table 4.2 summarizes techniques available for particle sizing that may be useful in air pollution research. Ideally the measurements of particle size should be made within the flue gas without any perturbation of the sample. Unfortunately the instruments that are theoretically useful for this type of measurement are still in the early stages of development.

In-stream analyses are obtained by the engineering application of certain size-dependent physical mechanisms. The measurements can be made with field-usable prototype systems or, over limited size ranges, with commercially available devices.

Another category of particle sizing techniques, laboratory classification, is useful only in special cases or to supplement field test data. Such techniques

Table 4.2. Status of Particle Size Measurement Techniques[a]

Laboratory Analysis	In-Stream Analysis	*In situ* Analysis (sample unperturbed)
Optical and electron microscopy—C (0.001–100 μm)	Inertial separation—C (0.3–20 μm)	Optical scattering—R
Electrical conductivity—C (0.4–100 μm)	Optical scattering—PC (0.3–3, 3–30, 10–100 μm)	Holography—R
Inertial separation—C (1–30 μm)	Electrical mobility—PC (0.01–1 μm)	Laser interferometry—R
Sedimentation—C (1–100 μm)	Diffusion—PC (0.01–0.2 μm)	

[a] Techniques abbreviated as follows: C, commercially available; PC, prototype system with commercial components; R, research.

are always of questionable value because of the difficulty of redispersing the agglomerated particulate to recreate the original aerosol.

The remainder of this chapter is devoted to specific instruments and techniques of particle sizing. Section 4.2 includes a detailed description of procedures and instruments for in-stream particle sizing as well as results from research and calibration studies that improve the quality of the data obtainable by using these techniques. Section 4.3 is a brief review of established laboratory techniques. Section 4.4 summarizes research and developmental work aimed at achieving the capability of *in situ* analysis.

4.2. IN-STREAM PARTICLE SIZING TECHNIQUES

4.2.1. General Discussion

Measurement procedures can generally be classed in three broad groups: (1) noncontact-nonextractive (e.g., transmissometers), (2) extractive sampling with the measurement device or sample collector located in-stream using minimal sample transport (e.g., in-stack cyclones and cascade impactors), and (3) extractive sampling using probes and transport lines to convey the sample to a collector or sensor located outside the environment of the process stream (e.g., optical particle counters and some impactor systems). Problem areas common to all particulate sampling methods designed to secure a representative size distribution can generally be described as falling into one of three areas: source variations, interferences, and procedural.

Where extractive sampling is used, the sample aerosol flow rate(s) must be adjusted to maintain isokinetic sampling conditions to avoid concentration errors resulting from under- or oversampling of the particles that have too high an inertia to follow the gas flow streams in the vicinity of the sampling nozzle. This is the case for particles having aerodynamic diameters larger than about 5 μm. Because many particulate sizing devices have size fractionation points that are flow rate dependent, the necessity for isokinetic sampling in the case of large particles can result in undesirable compromises in obtaining data—either in the number of points sampled or in the validity or precision of the data for large particles.

In general, particulates within a duct or flue are stratified to some degree, with strong gradients in concentration often being found for larger particles and in some cases for small particles. Such concentration gradients, which can be due to inertial effects, gravitational settling, lane to lane efficiency variations in the case of electrostatic precipitators, and so on, imply that multipoint (traverse) sampling must be used.

Even the careful use of multipoint traverse techniques does not guarantee representative data. The location of the sampling points during process changes or variations in control device operations can lead to significant data scatter. As an example, rapping losses in dry electrostatic precipitators tend to be confined to the lower portions of the gas streams, and radically different results may be obtained, depending on the magnitude of the rapping losses and whether single point or traverse sampling is used. In addition, large variations among results from successive multipoint traverse tests can occur as a result of differences in the location of the sampling points when the precipitator plates are rapped. Similar effects occur in other instances as a result of process variations and stratification due to settling, cyclonic flows, and other conditions. Figure 4.4 illustrates the temporal concentration variations for two particle sizes at a point in space located in a duct immediately downstream of a dry electrostatic precipitator.

Choices of particulate measurement devices or methods for individual applications are dependent on the availability of suitable techniques that permit the required temporal and/or spatial resolution or integrations. In many instances the properties of the aerosol are subject to large changes not only in size distribution and concentration but also in chemical composition (e.g., emissions from the open hearth steel making process). Different methods or sampling devices are generally required to obtain data for long-term process averages as opposed to the isolation of certain portions of the process to determine the cause of a particular type of emission from the process. To extract reasonable samples within fairly short time spans, high

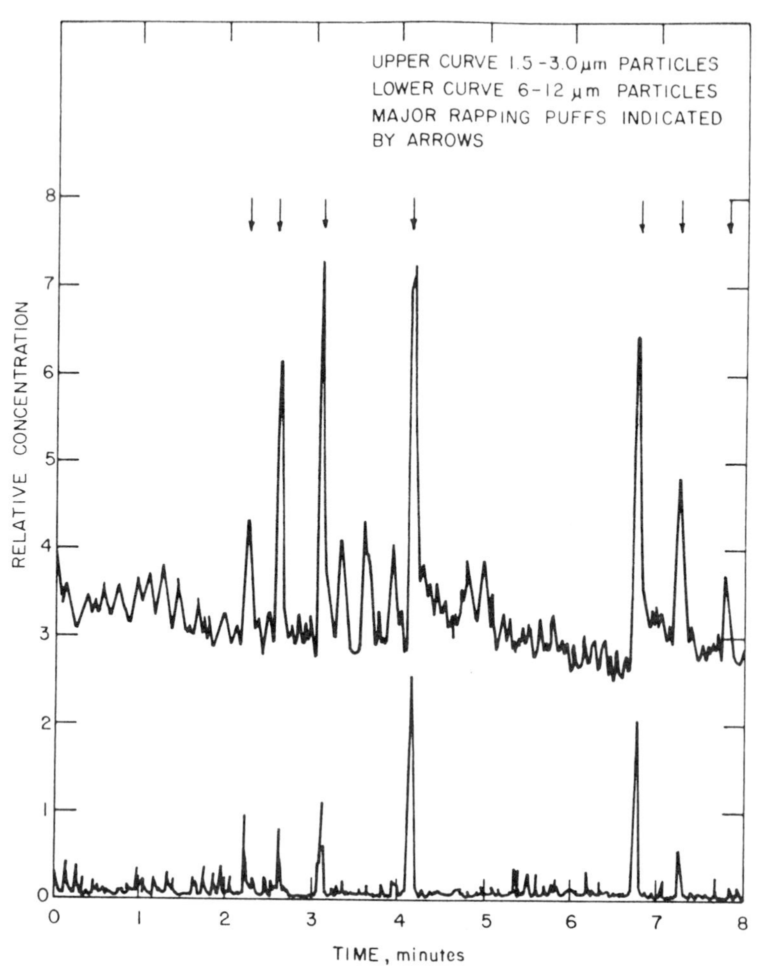

Figure 4.4. Strip chart records showing the temporal variations in particle size and count downstream from an electrostatic precipitator installed at a coal-fired power plant.

volume flow sampling devices must be developed, especially in the area of sizing. Nonsizing devices with high sampling rates are now commercially available.

Interferences exist that can affect most sampling methods. Two commonly occurring problems are the condensation of vapor phase components from the gas stream, and reactions of gas, liquid, or solid phase materials with various portions of the sampling systems. An example of the latter is the formation of sulfates in appreciable (several milligram) quantities on several of the commonly used glass fiber filter media by reactions involving sulfur

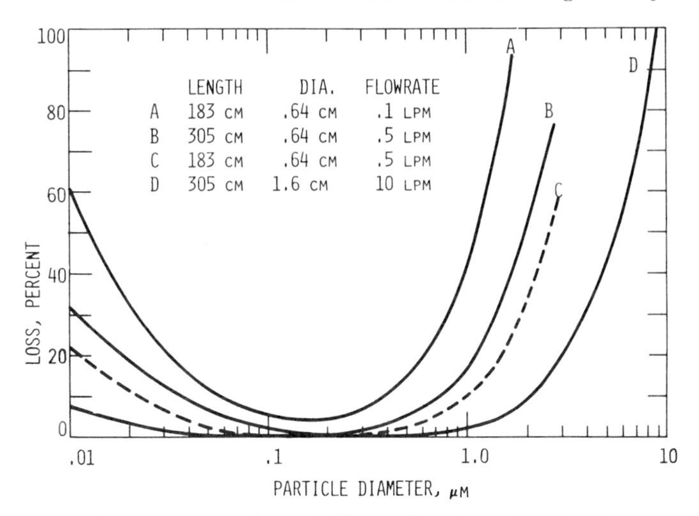

Figure 4.5. Probe lossess due to settling and diffusion for spherical particles having a density of 2.5 g cm^{-3} under conditions of laminar flow.

dioxide and trace constituents of the filter media. Sulfuric acid condensation in cascade impactors and in the probes used for extractive sampling is an example of the former.

If extractive sampling is used and the sample is conveyed through lengthy probes and transport lines, as is the case with several types of continuous mass emission monitors and some particulate sizing methods, special attention must be given to the recognition, minimization, and compensation for losses by various mechanisms in the transport lines. Such losses can be quite large for certain particle sizes. Figure 4.5 shows calculated losses by diffusion and gravitational settling in horizontal probes for several probe geometries and flow rates.

In addition to extremely low particulate mass loading, problems that have recently become significant are sampling under conditions of high pressure and high temperature, and with entrained water.

The following section discusses individual particulate sampling systems and procedures in more detail. These topics are categorized according to the physical mechanism that is used to obtain the data: inertial, optical, electrical, or diffusional.

4.2.2. Inertial Particle Sizing Devices

Devices that fall into the inertial sizing category are impactors, cyclones, centrifuges, and virtual impactors, which are sometimes called cascade centripeters or particle concentrators. In each of these techniques, an aerosol stream is constrained to follow a path of such curvature that the particles tend to move radially outward toward an impaction surface because of their inertia. Subsequent analysis of the particle size distribution may be made by gravimetric means, quantitative chemical analysis, microscopic inspection, attenuation of radiation, or light scattering.

Particle size distribution measurements related to control device research have largely been made using cascade impactors, which are effective in the size range of 0.3 to 20 μm diameter; in some applications, however, hybrid cyclone–impactor units, or cyclones have also been used. The particle size distributions are normally calculated from experimental data by relating the mass collected on various stages to the theoretical or empirical size cut points associated with those stage geometries. Efforts are underway to develop inertial sizing devices with extended size ranges and automatic, real-time readouts of the particle size information; but to date no such device has been used successfully in an industrial flue gas atmosphere.

4.2.2.1. Cascade Impactors

Because of its compact arrangement and mechanical simplicity, the cascade impactor has gained wide acceptance as a practical means of making particle size measurements in flue gases. Frequently the impactors can be inserted directly into the duct or flue, eliminating many condensation and sample loss problems encountered when probes are used for extractive sampling.

Figure 4.6 is a schematic diagram illustrating the principle of particle collection common to all cascade impactors. The sample aerosol is constrained to pass through a slit or circular hole to form a jet that is directed toward an impaction surface. Particles having lower momentum will follow the airstream to lower stages, where the jet velocities are progressively higher. For each stage there is a characteristic particle size that theoretically has a 50% probability of striking the collection surface. This particle size, D_{50}, is called the effective cut size for that stage. The number of holes or jets on any one stage ranges from one to several hundred, depending on the desired jet velocity and total volumetric flow rate. The number of jet stages in an impactor ranges from one to about 20 for various impactor geometries reported in the literature. Most commercially available impactors have between 5 and 10 stages.

Parameters that determine the collection efficiency for a particular geometry are gas viscosity, particle density, jet width, jet-to-plate spacing, and velocity of the air jet.

Most modern impactor designs are based on the semiempirical theory of Ranz and Wong.[17] Modifications of the theory of Ranz and Wong to include effects due to variations in the jet-to-plate spacing were done by

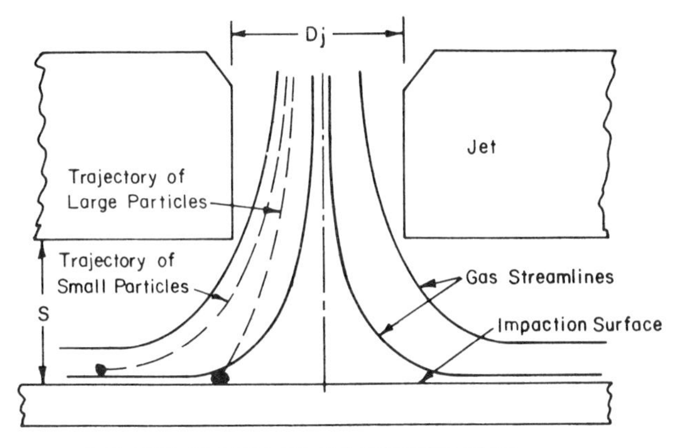

a. Typical Impactor Jet and Collection Plate

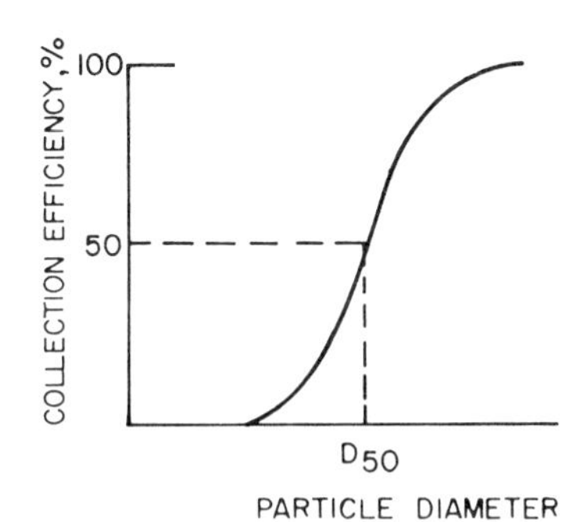

b. Generalized Stage Collection Efficiency Curve

Figure 4.6. Operation principle and typical performance for a cascade impactor.

Mercer et al.[18,19] who also showed that the performance of an impactor is insensitive to the jet-to-plate spacing if the ratio of this distance to the jet width (S/D_j) is between 1 and 3. Similar studies were done in Japan by Yuu and Iinoya.[20] More comprehensive theories have been developed by Davies and Aylward[21] and by Marple.[22] Deviations from ideal behavior in actual impactors dictate that they be calibrated experimentally. Because of the insensitivity to S/D_j, the theory of Ranz and Wong is generally satisfactory for impactor design.

A large number of experimental studies have been published on cascade impactor design and performance in the laboratory environment. Most of these have been reviewed in the dissertations of Marple[22] and Rao.[23] In addition, Brink[24] and Pilat et al.[25] have reviewed impactor operational procedures and described their own impactors.

a. General Considerations for Impactor Design

According to the theory of Ranz and Wong, impactor stage collection efficiencies can be expressed in terms of the dimensionless parameter $\sqrt{\psi}$, where, for round jets,

$$\sqrt{\psi} = D_p(3.075 \times 10^{-7})\left(\frac{\rho_p Q P_0 C}{\mu D_j^3 N_j P_j}\right)^{1/2} \quad (21)$$

and D_p = particle diameter (cm)
ρ_p = particle density (g cm^{-3})
Q = aerosol flow rate (l min^{-1})
P_0 = air pressure at the impactor inlet (atm)
C = the slip correction factor for small particles
μ = gas viscosity (poise)
D_j = jet diameter (cm)
N_j = number of holes per stage
P_j = air pressure at the jet stage (atm)

According to the experiments of Ranz and Wong, the value of $\sqrt{\psi}$ at 50% collection efficiency is 0.57 for infinitely long slits and 0.38 for circular jets. Using these values to invert eq. 21, one finds for circular jets

$$D_{50} = (1.237 \times 10^6)\left[\frac{\mu D_j^3 P_j N_j}{P_0 \rho_p C Q}\right]^{1/2} \quad (22)$$

Since the slip correction factor includes the particle diameter, the D_{50}'s for an actual impactor test must be determined by iteration using eq. 22 and the following equation for C:

$$C = 1 + \frac{2\lambda}{D_{50}}\left[1.2 + 0.4\exp\left(\frac{-0.44 D_{50}}{\lambda}\right)\right] \quad (23)$$

where λ is the mean free path of the gas molecules.

Equation 20 is the basic theoretical tool for impactor design. To vary the cut point one may (1) change the hole size, (2) change the number of holes, (3) change the flow rate, and (4) under some conditions vary the magnitude of the slip correction factor. Practical considerations in impactor design are size, handling ease, flow rate, type of collection substrates, and effective cut size selection. Figures 4.7, 4.8, and 4.9 show three of the commercial impactor designs commonly used in source testing. Table 4.3 summarizes the operating characteristics of each of these three impactors. It is usually impractical to use the same impactor at the inlet and outlet of a pollution control device

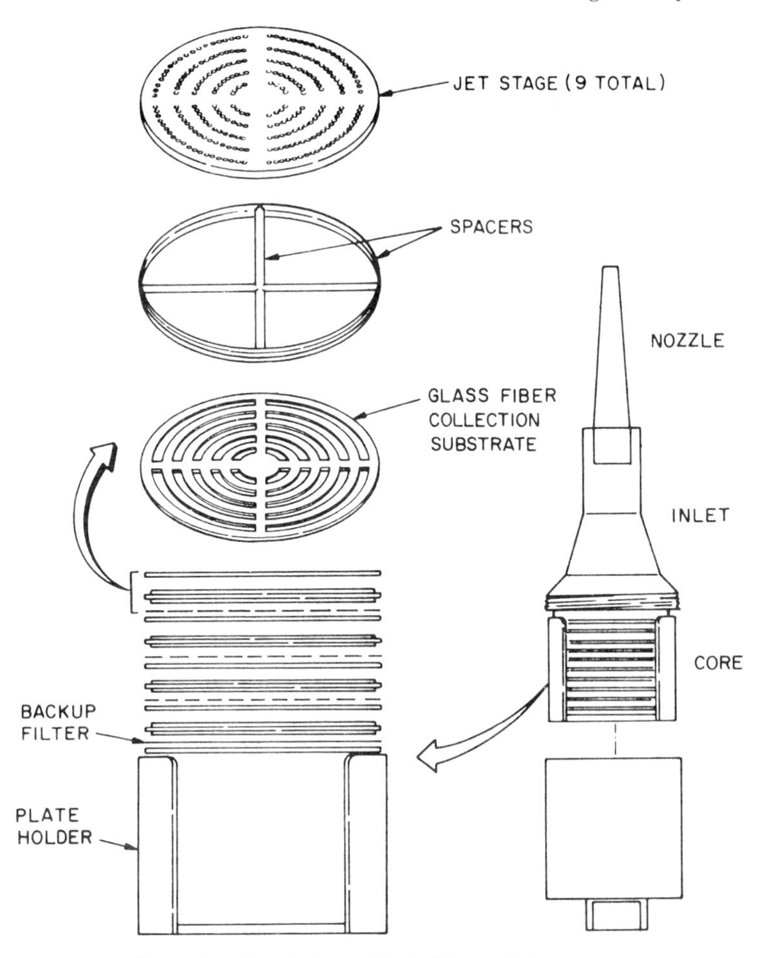

Figure 4.7. The Andersen Mark III cascade impactor.

when making fractional efficiency measurements because of the large difference in particulate loading. For example, if a sampling time of 30 min is adequate at the inlet for the same impactor operating conditions and the same amount of sample collected, approximately 3000 min sampling time would be required at the outlet (control device efficiency of 99% is assumed). Although impactor flow rates can be varied, they cannot be adjusted enough to compensate for this difference in particulate loading without creating other problems. Extremely high sampling rates result in particle bounce and in scouring of impacted particles from the lower stages of the impactor, where the jet velocities become excessively high. When sampling times are short,

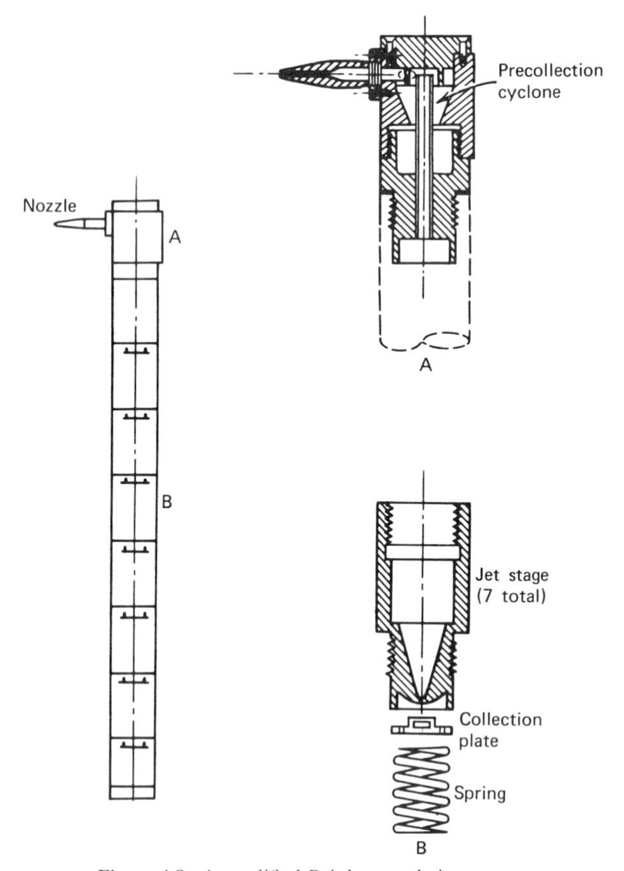

Figure 4.8. A modified Brink cascade impactor.

atypical samples may be obtained as a result of momentary fluctuations in the particle concentration or size distribution within the duct. Normally a low flow rate impactor is used at the inlet and a high flow rate impactor at the outlet. The impactors are then operated at their respective optimum flow rates, and the sampling times are dictated by the time required to collect weighable samples on each stage without overloading any single stage. Table 4.4 lists several cascade impactors that are manufactured in the United States and are commercially available for stack sampling.

b. Impactor Calibration Studies

An extensive literature reports experimental verification of impaction theories for special laboratory impactor designs. Unfortunately this is not true of impactors that are useful for in-stream sampling.

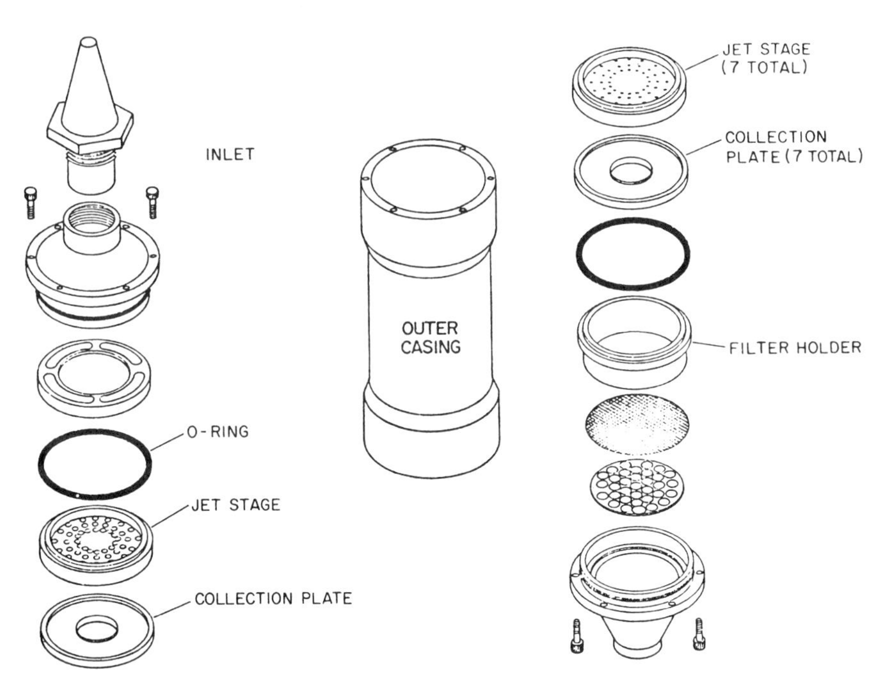

Figure 4.9. University of Washington Mark III cascade impactor.

Table 4.3. Theoretical Size Fractionating Points (μm) of Some Commerical Cascade Impactors for Unit Density Spheres

Stage	Modified Brink, 0.85 l min^{-1}	Andersen Mark III, 14 l min^{-1}	University of Washington (Pilat), 14 l min^{-1}
Cyclone	18.0		
0	11.0		
1	6.29	14.0	39.0
2	3.74	8.71	15.0
3	2.59	5.92	6.5
4	1.41	4.00	3.1
5	0.93	2.58	1.65
6	0.56	1.29	0.80
7		0.80	0.49
8		0.51	

Table 4.4. Commercial Cascade Impactor Sampling Systems

Name	Nominal Flow Rate (cm^3 sec^{-1})	Substrates	Manufacturer
Andersen Stack Sampler	236	Glass fiber (available from manufacturer)	Andersen 2000, Inc. P.O. Box 20769 Atlanta, Ga. 30320
University of Washington Mark III Source Test Cascade Impactor	236	Stainless steel inserts, glass fiber, grease	Pollution Control System Corp. 321 Evergreen Bldg. Renton, Wash. 98055
Brink Cascade Impactor	14.2	Glass fiber, aluminum, grease	Monsanto Envirochem, Inc. 800 North Lindbergh Blvd. St. Louis, Mo. 63141
Sierra Source Cascade Impactor, model 226	236	Glass fiber (available from manufacturer)	Sierra Instruments, Inc.
Sierra Tag Sampler, model 2600	236	Stainless steel plates, grease	P.O. Box 909
Sierra Cyclone Preseparator, model 220 CP (used with both the Sierra samplers listed above)	236	—	Village Square Carmel Valley, Calif. 93924
MRI Inertial Cascade Impactor	236	Stainless steel, aluminum, Mylar, Teflon; Optional: gold, silver, nickel	Meteorology Research, Inc. Box 637 Altadena, Calif. 91001

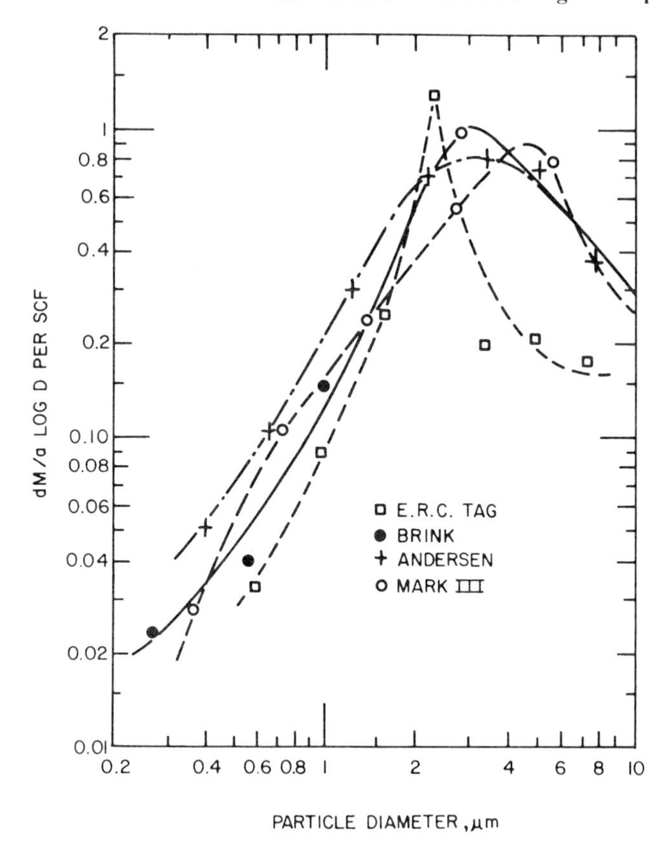

Figure 4.10. Comparative size distributions measured in the effluent from a coal-fired power boiler using four cascade impactors.

Bird and McCain[26] have published results from field tests where several impactors were used to sample a single source (Figure 4.10). Smith, Cushing et al.[27–29] have done laboratory evaluations of the available commercial impactors using monodisperse aerosols. Some of the results of these tests appear in Figures 4.11 to 4.16. These data are all presented as graphs of stage collection versus $\sqrt{\psi}$, with the theory and experimental results of Ranz and Wong (R & W) given as continuous curves. In this way the calibration data for all the stages in an impactor can be displayed on a single plot. All the data were taken at a single flow rate for each impactor (which was chosen to be well within the range recommended by the manufacturers), at which the jet velocities were not believed to be excessively high.

It is possible to draw several conclusions from these calibration studies.

1. Different stages within any multiple-stage impactor do not in general obey the same theoretical equation.
2. Upper stages with large cut points tend to agree with the experimental results of Ranz and Wong and of Cohen and Montan,[30] whereas the lower stage cut points tend to be shifted to smaller sizes.
3. The performance and cut point of a stage is sensitive to the type of substrate used.
4. The maximum efficiency for many stages is less than 100% for all particle sizes.

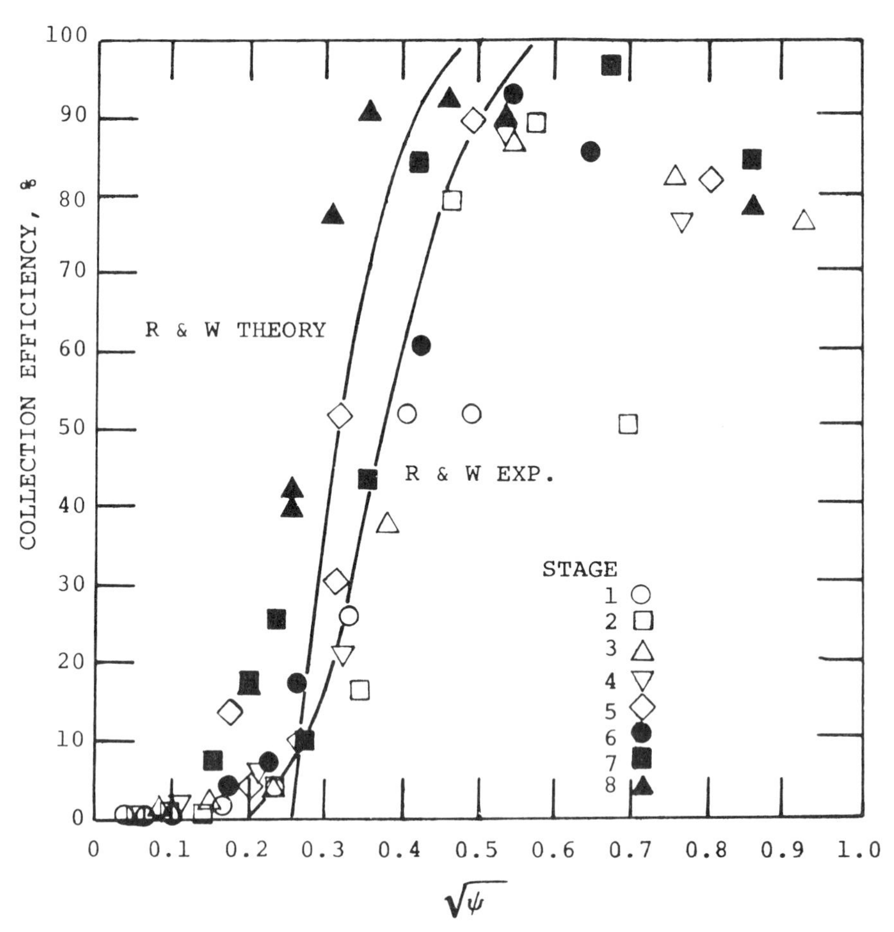

Figure 4.11. Some laboratory calibration data for the Andersen Mark III cascade impactor ($14\ l\ min^{-1}$, 22°C, 29.5 in. Hg, $\rho_p = 1.35\ g\ cm^{-3}$).

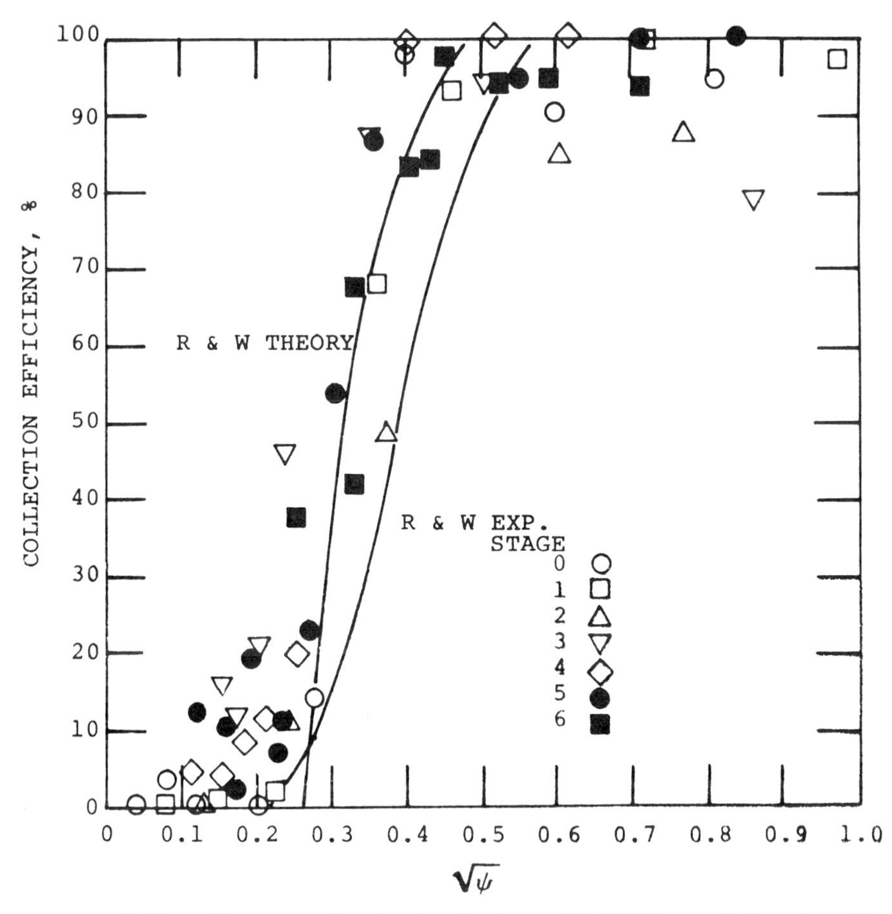

Figure 4.12. Some laboratory calibration data for the modified Brink cascade impactor with greased foil substrates (0.85 l min^{-1}, 22 C, 29.5 in. Hg, $\rho_p = 1.35\ \text{g cm}^{-3}$).

5. The stage efficiency for a given stage *decreases* for particle sizes much greater than the cut point because of high impact momentum and resultant particle bounce.

It is clear that an impactor should be calibrated before use, at test conditions as close as possible to the anticipated sampling conditions.

c. Impactor Data Reduction Techniques

Two methods of calculation have been used to calculate particle size distribution from field test data. In the "D_{50}" method the particles caught on a particular impactor stage are essentially treated as monodisperse and are assigned diameters equal to that stage D_{50}. Picknett[31] has introduced a more sophisticated method that accounts for the nonideal nature of the stage collection efficiencies. Specifically, particles of a given diameter may be caught by two or more stages, and this property is incorporated into the theory. In the Picknett method discrete particle sizes are arbitrarily chosen, using the D_{50}'s as an initial set and calculating their concentrations such that the measured stage loadings would have been obtained if the real particle size distribution had consisted of these chosen sizes. A large number of particle sizes may be chosen to give a very good representation of the real distribution. This technique, however, frequently fails when field test data

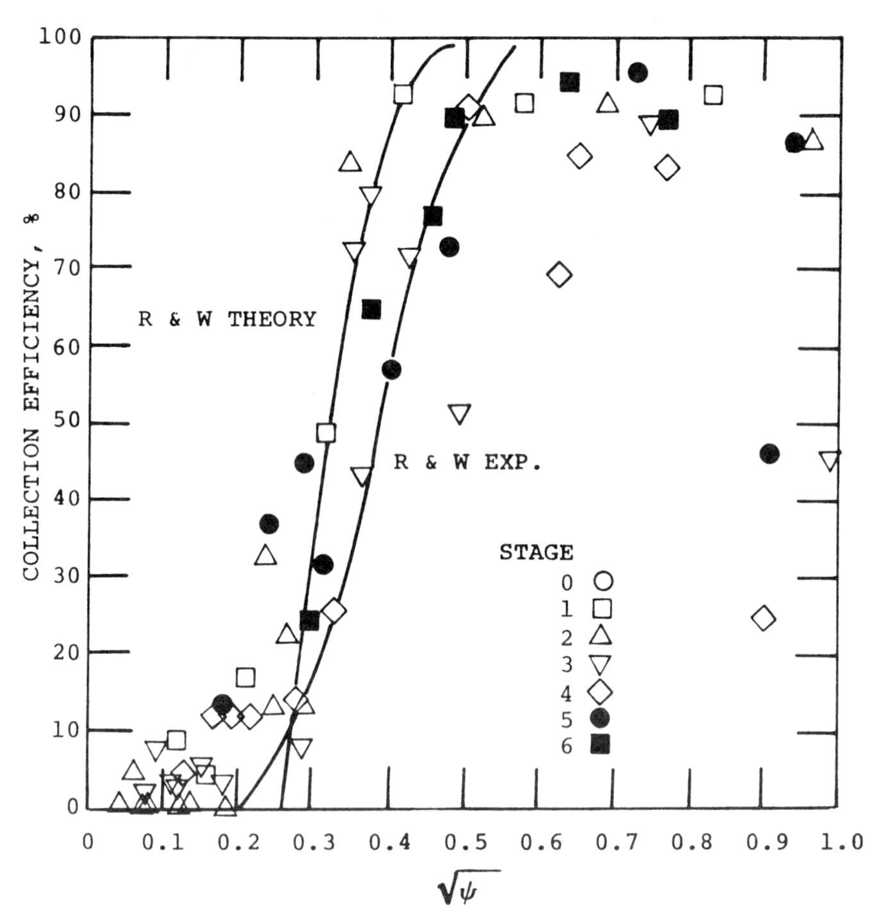

Figure 4.13. Some laboratory calibration data for the modified Brink cascade impactor with glass fiber substrates (0.85 l min^{-1}, 22°C, 29.5 in. Hg, $\rho_p = 1.35$ g cm^{-3}).

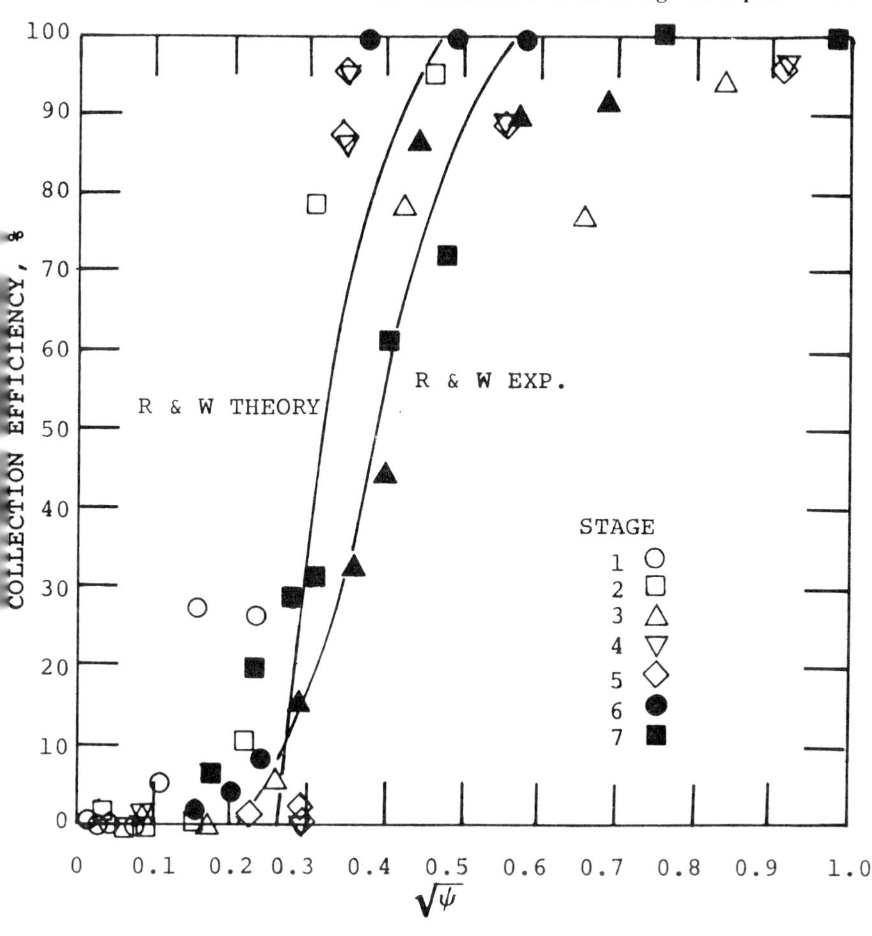

Figure 4.14. Some laboratory calibration data for the MRI model 1502 cascade impactor (14 l min^{-1}, 22°C, 29.5 in. Hg, $\rho_p = 1.35$ g cm^{-3}).

are used in which the observed stage collection weights are not consistent with the theoretical impactor behavior.

In investigations of impactor data reduction techniques it is difficult to separate inaccuracies in the theory and errors introduced by reentrainment, bounce, scouring, and poor calibration of the impactors. This problem can be eliminated by simulating the capture of a fictitious aerosol using the theoretical stage efficiency curves. Once the stage loadings are calculated, the data can be used to recalculate a particle size distribution, using either the D_{50} or the Picknett method, which should, ideally, be identical to the "fictitious" input distribution.

Figure 4.17 is a test of both the D_{50} and the Picknett techniques using for input, or "true" distribution, the sum of two log-normal distributions. Theoretical efficiencies for a Brink impactor with a cyclone were used in these calculations.

d. Wall Losses

Impactors are usually operated in stack to avoid probe losses and to minimize condensation problems. Nevertheless, particles are lost in the inlet nozzle and onto the internal surfaces by diffusion, electrical forces, settling, and impaction. Lundgren[32] and Gussman et al.[33] have shown that the losses per stage can amount to as much material as is collected by that stage. Figure

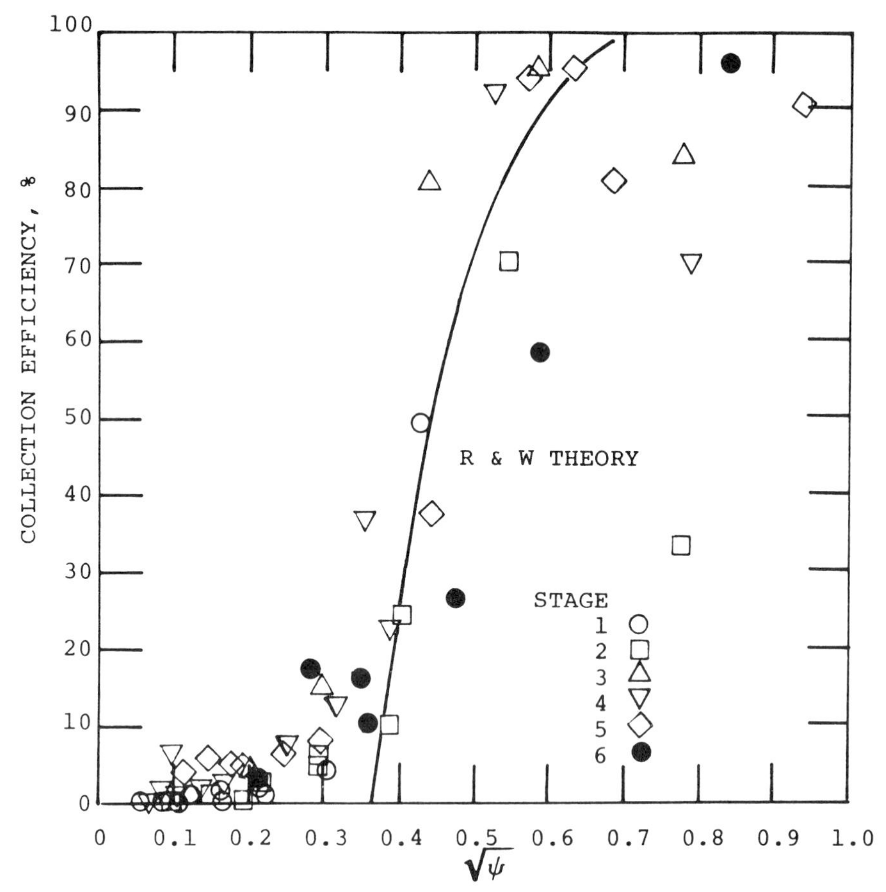

Figure 4.15. Some laboratory calibration data for the Sierra Instrument Co. model 2600 cascade impactor (7 l min^{-1}, 22°C, 29.5 in. Hg, ρ_p = 1.35 g cm^{-3}).

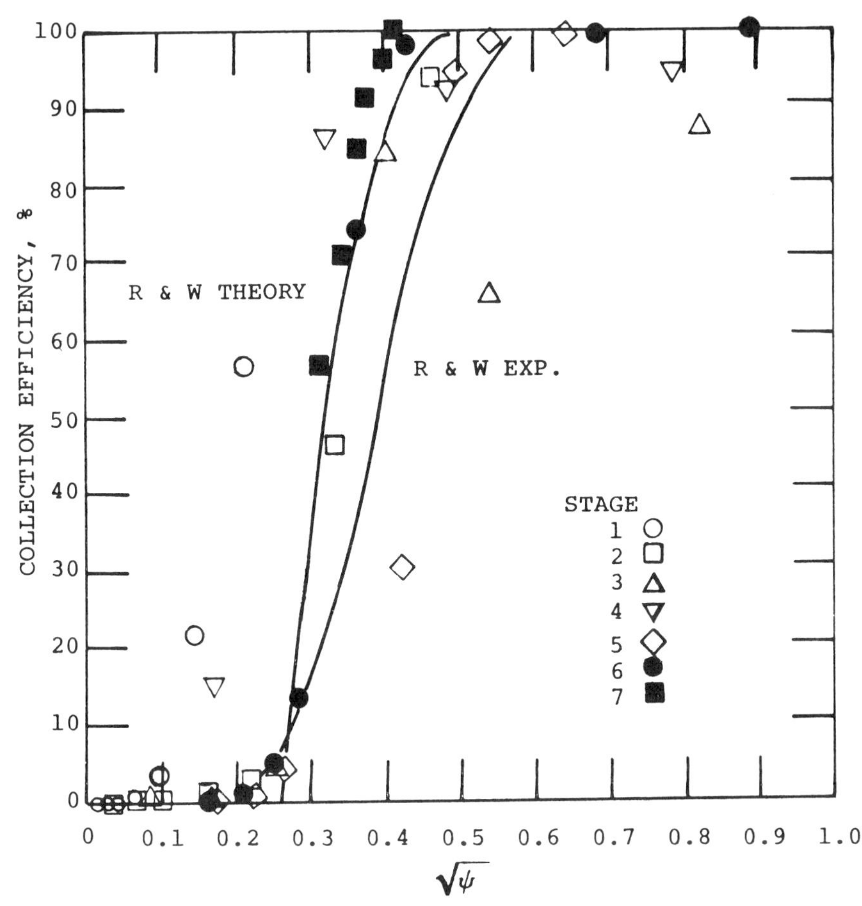

Figure 4.16. Some laboratory calibration data for the University of Washington Mark III cascade impactor (14 l min^{-1}, 22°C, 29.5 in. Hg, ρ_p = 1.35 g cm^{-3}).

4.18 gives the losses for laboratory tests of the Andersen Mark III, University of Washington, and Brink impactors. The losses are size dependent and can contribute significant errors in particle size distribution and mass concentration measurements.

e. Collection Substrate Material

Most impactors have collection plates that are too heavy to permit accurate measurements of the mass of the particles collected. Weighing accuracy can be improved by covering the plate with a lightweight collection substrate made of aluminum foil, Teflon, glass fiber filter material, or other suitable lightweight materials, depending on the particular application. Some

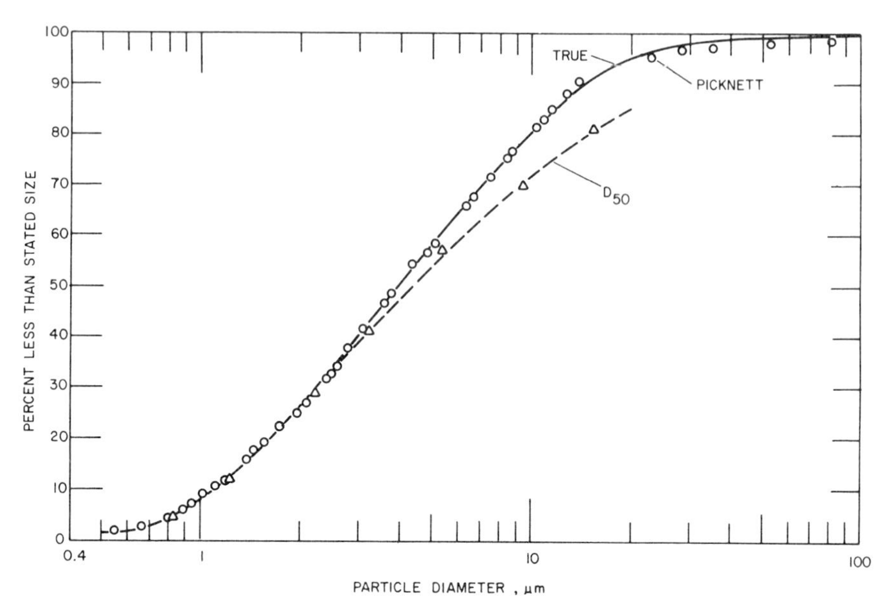

Figure 4.17. Comparison of the D_{50} and Picknett methods for cascade impactor data reduction.

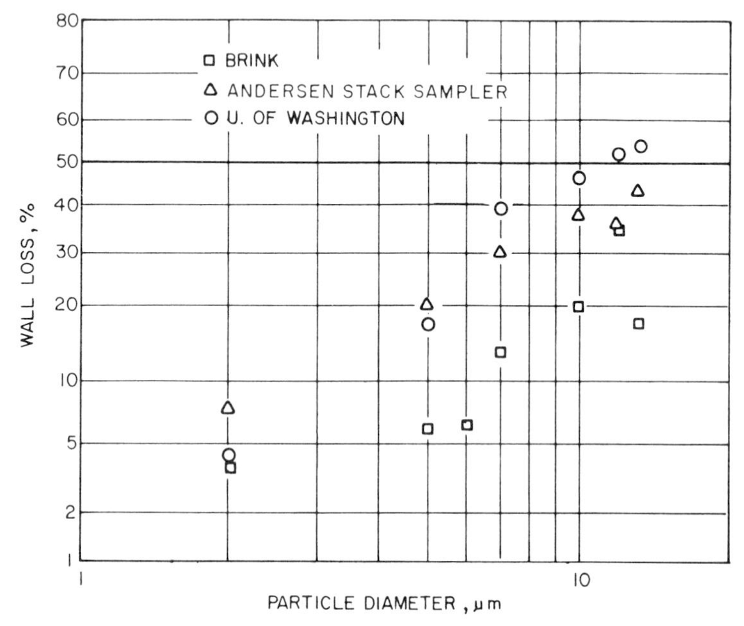

Figure 4.18. Wall losses measured during laboratory tests of the Brink, Andersen, and University of Washington impactors.

manufacturers now furnish lightweight inserts to be placed over the collection stages. With such arrangements it is possible to make accurate determinations of the masses collected by the various stages without risking reentrainment due to overloading.

The use of greased foils or lightweight glass fiber filter mats as collection substrates tends to alleviate the problem of weighing accuracy, but problems due to substrate weight changes are introduced. Although normal substrate preparation includes baking and desiccation before the initial weighing, it is frequently found that weight losses occur when sampling clean air. With careful handling, weight losses for glass fiber substrates can be kept below 0.2 mg.[28] This loss is attributed to loss of fibers that stick to seals within the impactor and to "superdrying" when sampling hot, dry air. Weight losses of 0.2 mg are small compared to normal stage catches when sampling particulate, thus are within a tolerable range for sampling errors.

When clean, hot air is sampled with greased substrates, more severe weight losses can occur. Some of the weight lost on upper impactor stages reappears as weight gained on the backup filters. This is interpreted as an indication that grease flows or is blown off the collection surfaces.

Anomalous substrate and filter weight *gains* have been a source of very large errors in the sampling of industrial flue gases. These weight gains, due to gas phase reactions, seem to occur at many sites tested and often have been larger than normal particulate catches. Ten different types of filter medium have been tested for susceptibility to this interfering mechanism at three industrial sites: the outlet of a hot side ESP on a coal-fired boiler, the outlet of a hot side ESP on a cement kiln, and the outlet of a cold side ESP on a coal-fired boiler. The results of these tests can be summarized as follows:[28]

- The pH of the filters varied widely from batch to batch before testing.
- There was a definite correlation between high initial pH and weight gains upon testing.
- The pH decreased during testing.
- A large fraction of the weight gain in every case was found to be the result of sulfate formation on the filter medium.

It is presumed the sulfate was formed by the reaction of sulfur dioxide with basic sites on the surface of the glass fibers. This phenomenon was known to occur in ambient sampling,[34] but it had been neglected or ignored in stack sampling.

Two approaches to avoiding the problems of anomalous substrate and filter weight gains are possible. In the first, substrates are preconditioned by long exposure to the flue gas, neutralizing most of the reaction sites on the filters before they are used. In the second, a search can be made for substrate materials that do not react with the flue gas. Although preliminary, the results of tests of preconditioned materials indicate that the preconditioning reduces the magnitude of anomalous weight gains by approximately a factor of 10. Also, Teflon, Whatman GF/D and GF/A, and Reeve Angel 934AH are filter media that have shown little weight change when exposed to flue gas. These materials are not currently being used in cascade impactors, and their particulate retention properties could differ significantly from those of standard substrates, perhaps altering the impactor calibrations to some extent.

f. Electrostatic Effect

Electrostatic forces due to charges that may exist on the flue gas particulate are a potential source of error for most sampling techniques other than light scattering. The magnitudes of these forces are difficult to estimate for impactors because of the complicated geometry. Smith et al. have reported results from a limited series of experiments performed to quantify this effect.[28]

The first two stages of a Brink cascade impactor were employed with a removable polonium α-source attached to the nozzle of the impactor. A flow rate of 0.1 $ft^3\ min^{-1}$ was maintained through the impactor. Impactor inlet and outlet particle concentrations were measured using a Climet Particle Analyzer (Optical Single-Particle Counter). The aerosol exiting the impactor was charge neutralized with a second polonium α-source to minimize sample line and instrumental losses due to electrostatic forces between the impactor and particle counter.

Three different configurations of the impactor were tested: (1) with glass fiber filter substrates, (2) with bare metal plates, and (3) extractive sampling with a 1 m, 6 mm ID copper probe. Each condition was tested with and without the charge neutralizer upstream of the impactor nozzle.

The results of this investigation are presented in Figure 4.19. The data are normalized to the concentration measured with the charge neutralizer in place. Curves are shown for the probe alone, for the impactor with glass fiber substrates (no probe), and for the impactor operated with bare metal collection plates (no probe). Losses in the probe and with bare substrates are seen to be rather large. For the tests made using glass fiber substrates, there was no appreciable difference in the concentrations measured with and without the charge neutralizer.

This study shows that losses due to electrostatic forces can be large, and possibly such losses have introduced serious errors into impactor data reported from field tests. The losses would be more severe for extractive

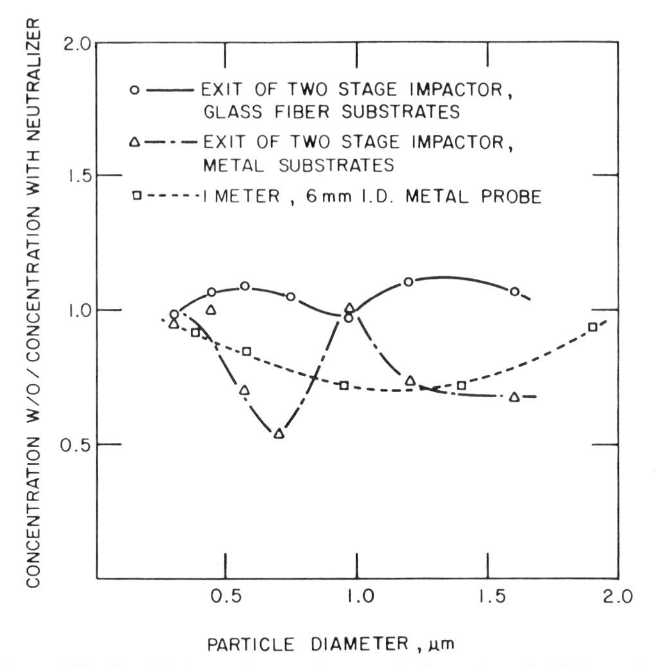

Figure 4.19. Particle losses due to electrostatic forces in a Brink impactor.

sampling through a probe. To our knowledge, no one has used charge neutralizers in conjunction with cascade impactors for field testing.

g. Reentrainment due to Improper Operation

When impactors are operated at flow rates higher than some critical value, particle bounce can lead to incorrect sizing of noncohesive particles. Errors are especially severe for the lower impactor stages, where high jet velocities often cause noticeable scouring. Cohen and Montan[30] proposed a set of design limits for cascade impactors based on experimental and theoretical impactor work before 1967 and taking into account some manufacturing difficulties. These limits were as follows: (1) the Reynolds number in the jet should lie between 100 and 3200, (2) the jet velocity should be at least 10 times the settling velocity of the particles, and (3) jet velocities should not exceed 10^4 cm sec^{-1}.

The importance of maintaining reasonable jet velocities is illustrated in Figure 4.20, presenting deposition patterns for a hard-dry ammonium fluorescein aerosol in an experimental impactor. The particles were all 2.8 μm in diameter and the stage D_{50}'s were 1.80, 0.83, 1.8, and 0.38 μm for Figures

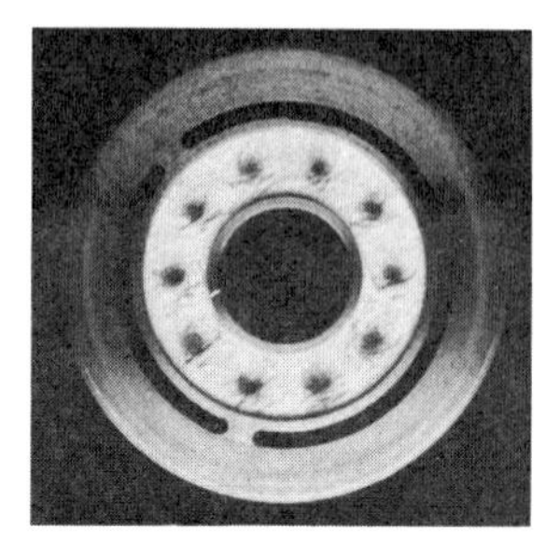

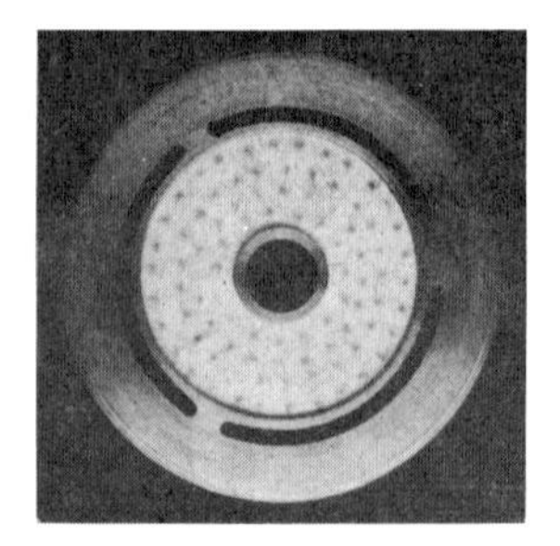

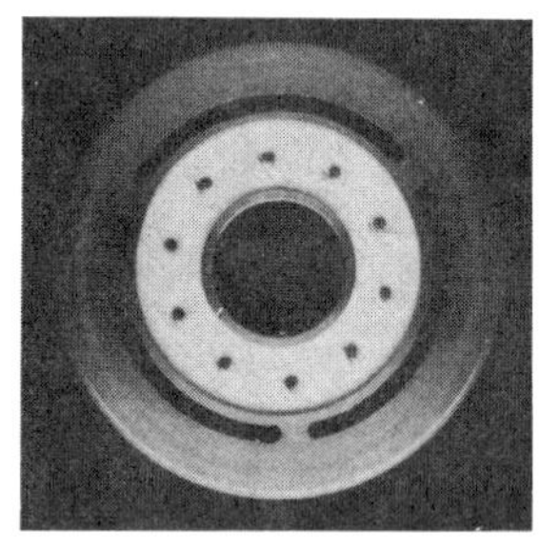

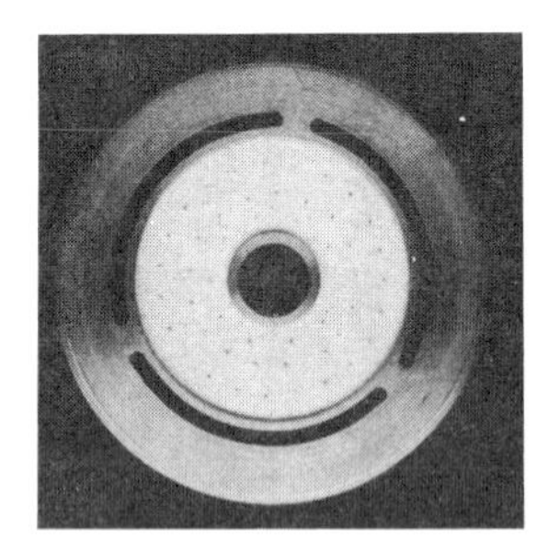

Figure 4.20. Particulate deposition patterns for different flow rates.

4.20*a* to *d*. The deposition patterns formed at 11.4 and 45.1 m sec^{-1} jet velocity show definite evidence of scouring, whereas those obtained at the lower velocities are compact and well defined. Also, as noted in the figure captions, the collection efficiencies of the stages deteriorated significantly and the wall losses increased for the higher jet velocities.

Figure 4.21 graphs the relationship among the jet diameter, the number of jets per stage, the jet velocity, and the stage cut point. A large number of very small holes is required to achieve small cut points at jet velocities below 10^4 cm sec^{-1} with reasonably large total volumetric flow rates.

*h. Isokinetic Sampling and Accuracy**

It is difficult to always sample isokinetically when using cascade impactors because the flow rate must remain fixed during any single test and because the

* See also Chapter 3.

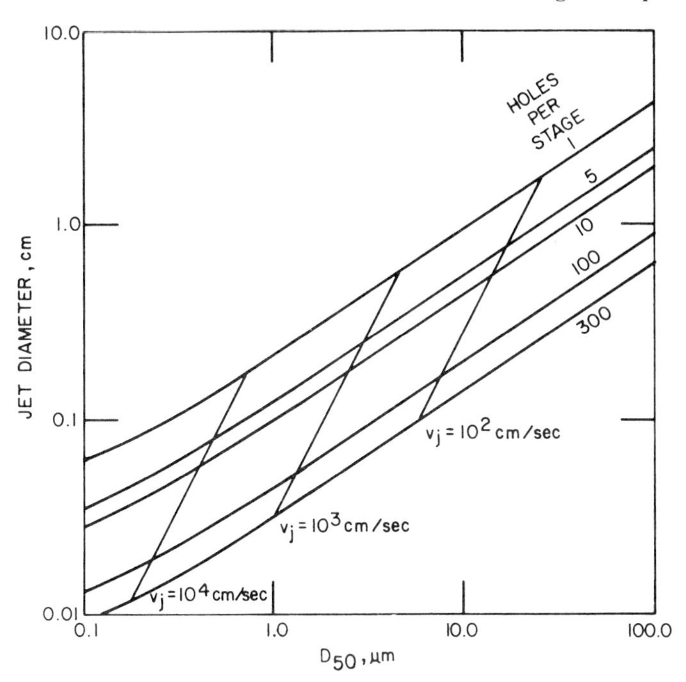

Figure 4.21. Approximate relationship among jet diameter, number of jets per stage, jet velocity, and stage cut point for circular jet impactors.

selection of practical nozzle sizes is limited. It has been general practice to avoid the use of sampling nozzles smaller than 2 mm in diameter, even if smaller nozzles are required for isokineticity.[35]

Fuchs[36] and Davies[37] have reviewed theories and experiments designed to study errors due to anisokinetic sampling in measured aerosol concentrations. According to Fuchs, the following equation is applicable:

$$\frac{C_i}{C_e} = K + \frac{V_e}{V_i}(1 - K) \tag{24}$$

where C_i = the measured concentration
C_e = the true concentration
V_i = the sampling velocity
V_e = the velocity of the gas
K = an empirically determined constant

Here K is zero for very large Stokes number Stk and 1 when Stk is 0. Values tabulated by Fuchs from experimental data are given in Table 4.5.

Table 4.5. Values of K Calculated by Fuchs From Experimental Data

Stk	K	Stk	K
0	1	0.6	0.39
0.05	0.87	0.8	0.32
0.1	0.77	1	0.26
0.2	0.64	2	0.17
0.3	0.55	3	0.13
0.4	0.48	4	0.10
0.5	0.43	5	0.08

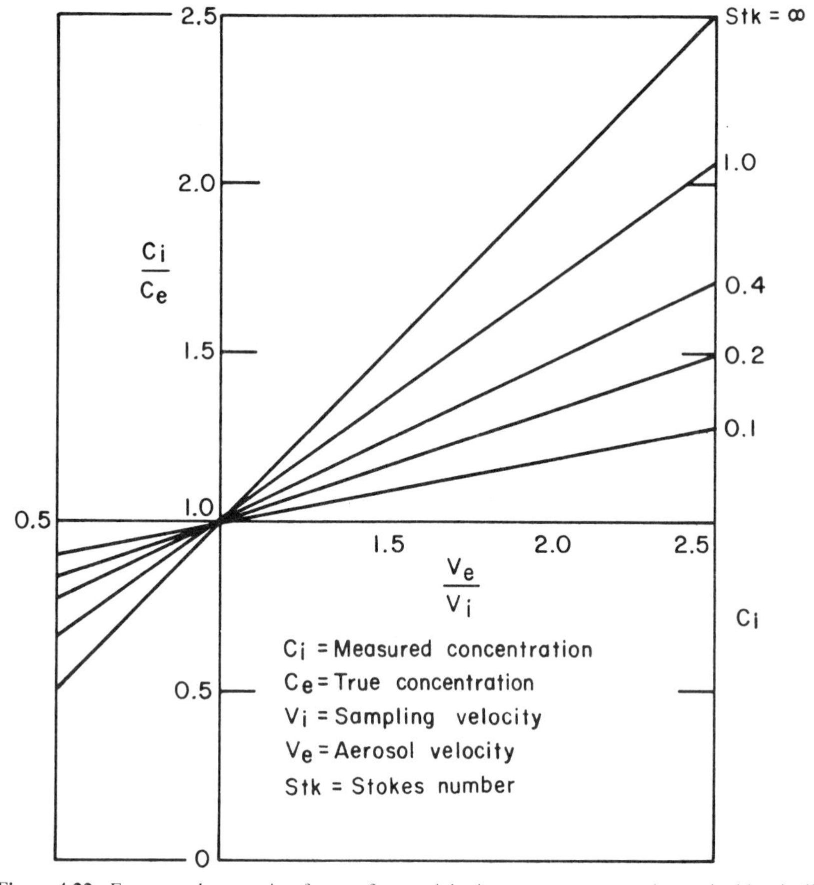

Figure 4.22. Errors and correction factors for particle size measurements taken anisokinetically.

The parameter ψ, previously defined in eq. 21, is half the Stokes' number, Stk. In this case D_j, which appears in the definition of $\sqrt{\psi}$, is the diameter of the sampling nozzle. Figure 4.22 shows correction factors calculated using eq. 24, for aerosols sampled at various degrees of anisokineticity. Clearly this is an effect that must be considered in the interpretation of sampling data.

In addition to the relative concentration with respect to size, cascade impactors give a measure of the total concentration of particulate in the volume of aerosol sampled. It is of interest to compare these results with those obtained simultaneously using standard methods for emission testing of either the Environmental Protection Agency (EPA) or the American Society

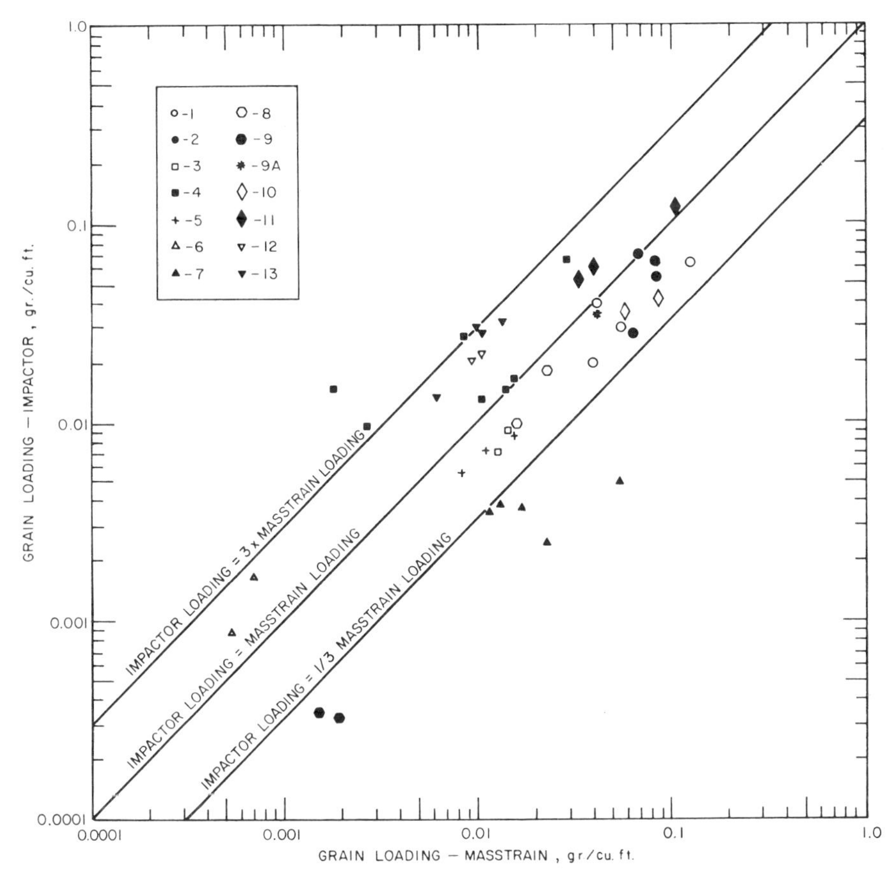

Figure 4.23. Grain loading by conventional (ASME, EPA) mass train versus grain loading by cascade impactor for 13 sampling locations (1 gr/ft^3 = 2.3 g m^{-3}).

Table 4.6. Sources for Data in Figure 4.25

1.	Coal-fired power boiler, 315°F, low sulfur
2.	Refuse incinerator, 410°F
3.	Coal-fired power boiler, 300°F, low sulfur
4.	Coal-fired power boiler, 315°F, high sulfur
5.	Coal-fired power boiler, 300°F, low sulfur
6.	Open hearth furnance, 200°F
7.	Coal-fired power boiler, 620°F, low sulfur
8.	Electric arc smelting, 180°F
9.	Outlet, aluminum reduction potlines, 120°F
9a.	Inlet, aluminum reduction potlines, 120°F
10.	Asphalt plant, 170°F
11.	Coal-fired power boiler, 315°F, moderate sulfur
12.	Coal-fired power boiler, 320°F, high sulfur
13.	Cement kiln, 550°F

of Mechanical Engineers (ASME). Results from tests at 14 sites are reproduced in Figure 4.23, where both cascade impactors and mass trains were used. Table 4.6 identifies the sources tested. Clearly better agreement would be desirable. It is certain that these errors are not due to departures from isokinetic sampling alone, however. Substrate reactions, weighing errors, sample handling errors, poor sample locations, and other factors probably contributed. In addition, it is likely that problems with the mass trains were responsible for a significant portion of the scatter.

i. Present Research on Cascade Impactors

It is possible to extend the sizing capability of a cascade impactor to submicron particles by operating the device at pressures of about 0.01 to 0.1 atm. If all parameters except the pressure are held constant, the cut point of a stage, or D_{50}, is inversely proportional to the Cunningham correction factor:

$$D_{50} \propto \frac{1}{\sqrt{C}} \tag{25}$$

However C increases rapidly with decreasing pressure, which means that D_{50}'s of 0.02 μm or less can be obtained.[38,39] Pilat[40,41] has designed and tested such an impactor for use in source testing.

The EPA is supporting the development of a low pressure cascade virtual impactor (impaction into a dead air space) with beta gauge readouts by the Environmental Research Corporation.[42] An evaluation of this system is not yet available.

Automatic readout mechanisms that have been used with cascade impactors for real-time data analysis are piezoelectric microbalances[43–45] and beta gauges.[45] Beta gauges seem to be the most practical because they do not suffer from overloading and electrical contact difficulties, and they need not be exposed directly to the hot gas stream. GCA Corporation has completed a prototype impactor with beta gauge readouts for the EPA and is presently developing a miniaturized version for in-stack use.

j. Summary

Cascade impactors are the most useful means presently available for measuring particle size distributions in industrial flue gases. This technique is currently limited, however, to the size range larger than 0.3 μm in diameter, and the samples collected are frequently too small for wet chemical analysis. Impactors require skill and experience for proper use, thus poor calibration and misapplication can result in inaccurate estimates of particle size distributions. Recent research has pointed out some of the problem areas; however it has been demonstrated that impactors can be used effectively and accurately for fine particle measurement if reasonable precautions are taken in obtaining and interpreting the data.

Ongoing research is devoted to the optimization of basic impactor performance, handling ease, and the automation of sampling and data reduction. There is no clear alternative to impactors in cases where the main objective is to make rapid, detailed studies of particle size distributions from 0.3 to 10 μm diameter in an industrial environment.

4.2.2.2. Cyclone Sampling Systems

a. General Discussion

Cyclones have been used less than impactors for making particle size distribution measurements because they are bulky and give less resolution. When larger samples are required, however, or when sampling times with impactors may be undesirably short, cyclones are better suited for testing than impactors. Cyclones are also frequently used as precollectors in impactor systems to remove large particles that might overload the upper stages.

Chang[46] developed a system of parallel cyclones that separates particles into four size fractions. This system is too large for in-stream sampling; thus it employs a probe for sample extraction. Although the system is impractical for stack sampling, the discussions of cyclone design and calibration included in Chang's report are a good starting point for the design of small cyclones. Figure 4.24 schematizes a much simpler series cyclone system that was described by Rusanov[47] and is used in the USSR for obtaining

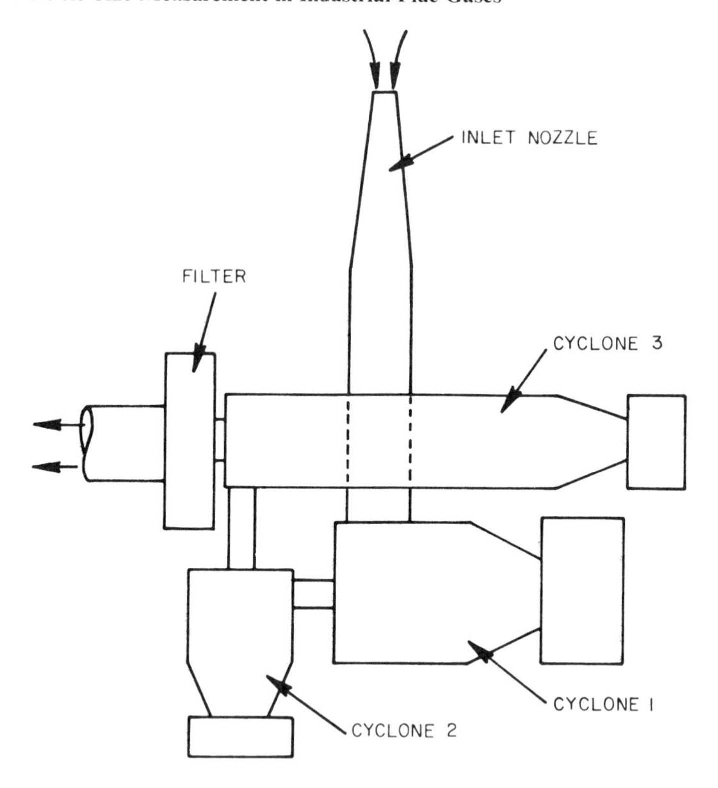

Figure 4.24. Series cyclone used in the USSR for sizing flue gas aerosol particles.[47]

particle size information. This device is operated in stack, but because of the rather large dimensions an 8 in. port is required for entry. Smith et al.[28] have developed and tested a series cyclone system that is designed to operate at a 1 ft^3/min^{-1} flow rate and is compact enough to fit through a 6 in. port. Complete calibration and preliminary performance testing of this system have been done. Series cyclone systems have adequate resolution for many purposes, and they should be much less susceptible to operator error than impactors, as well as free from gas-substrate interferences. The main advantage of such systems, however, is the ability to collect large sized samples for subsequent analysis.

b. Cyclone Design

In addition to Chang's work, several other papers are useful in developing cyclone designs.[48–51] Figure 4.25 displays the nomenclature that is used in

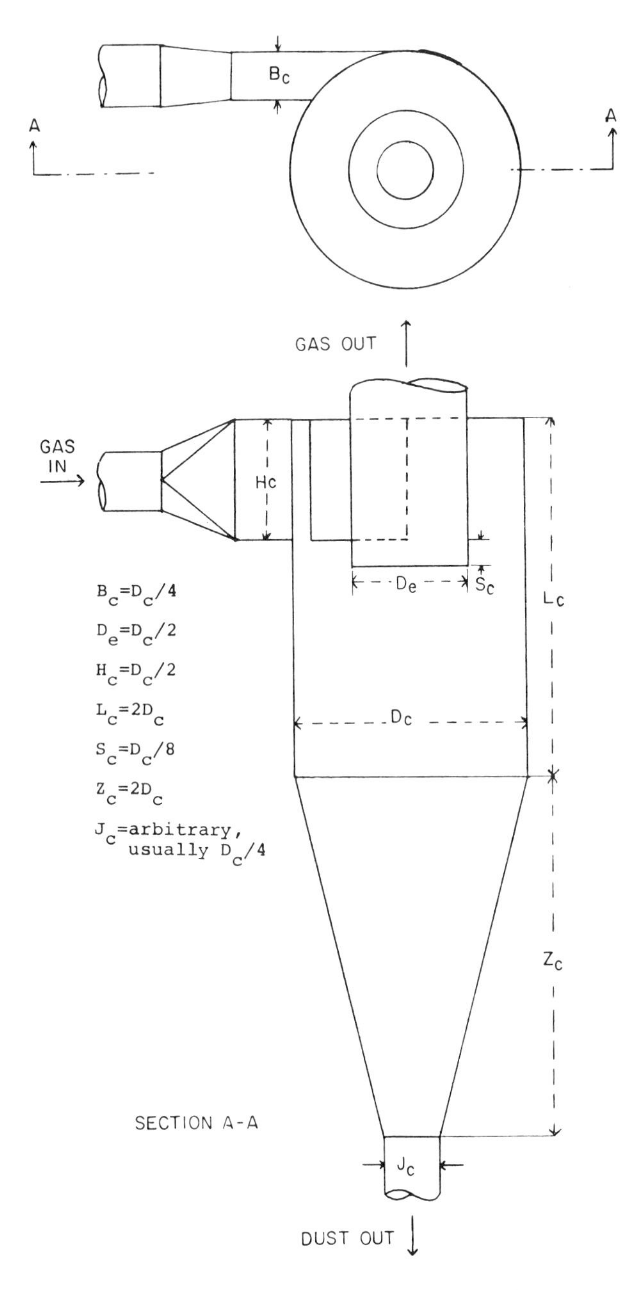

Figure 4.25. Generalized cyclone design for the application of Lapple's equation. From *Engineering Manual* by Perry & Perry. Copyright 1959. Used with permission of McGraw-Hill Book Company.

Lapple's equation to calculate the D_{50}, or cut point, for a cyclone of arbitrary size. Lapple's equation[52] is

$$D_{50} = \left[\frac{9\mu B_c}{2\pi N_c V_c(\rho_p - \rho)} \right]^{1/2} \tag{26}$$

where D_{50} = cyclone D_{50} (μm)
μ = gas viscosity (poise)
ρ_p = density of particle (g cm^{-3})
ρ = gas density (g cm^{-3})
N_c = number of turns made by gas stream in the cyclone body and cone (usually between 1 and 5)
V_c = inlet air velocity (cm sec^{-1})
B_c = width of cyclone inlet (cm)

The square root relationship between D_{50} and V_c can be used to predict the cyclone performance over a wide range of flow rates if the cyclone has been calibrated at a known flow rate and if N_c is not sensitive to the flow rate. In this case, Lapple's equation can be written as follows:

$$D_{50}(1) = \frac{C'}{\sqrt{V_1}} \tag{27}$$

where $D_{50}(1)$ = the cyclone cut point at flow rate 1
C' = the cyclone calibration constant
V_1 = aerosol flow rate 1

Then the relationship between $D_{50}(1)$ and $D_{50}(2)$, the cut points at the two different flow rates, is

$$D_{50}(2) = D_{50}(1) \left(\frac{V_1}{V_2} \right)^{1/2} \tag{28}$$

The results of calibration tests on several cyclones (Figure 4.26) indicate that this relationship is qualitatively valid over reasonable ranges of flow rates.

There are three approaches for arriving at final designs for cyclones having the desired cut points: (1) Figure 4.25 and Lapple's equation can be used, (2) previously calibrated prototypes can be used at various flow rates, and (3) dimensions can be selected by extrapolation for cyclones that are between sizes or are related to the sizes of the prototypes tested previously.

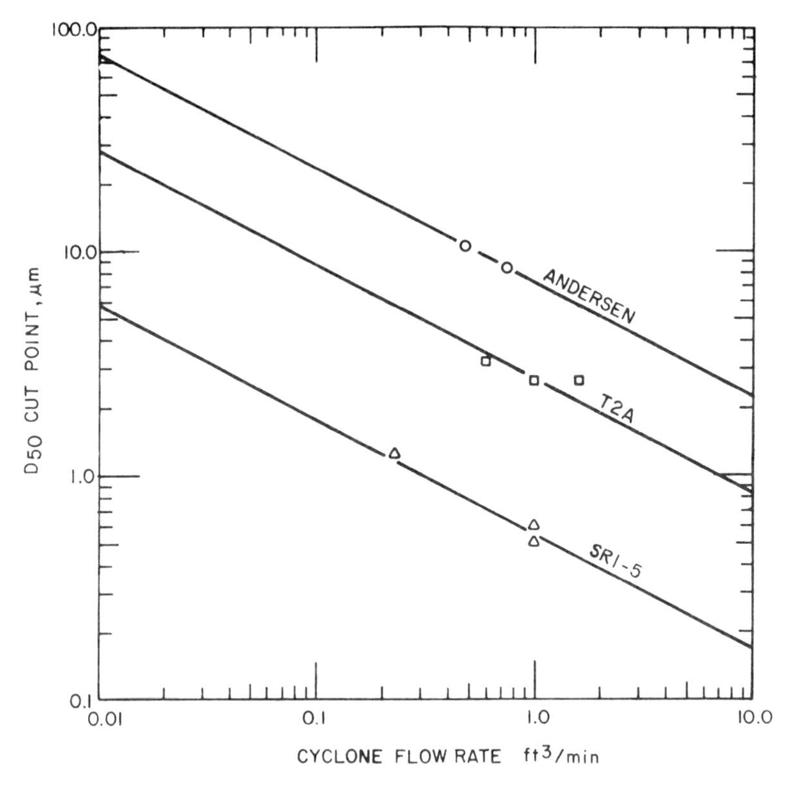

Figure 4.26. D_{50} cut point versus cyclone flow rate for three calibrated cyclones (1 ft^3/min = 1.7 m^3 hr^{-1}).

c. *Series Cyclone Sampling System*

A series cyclone system has been developed by Smith et al. to satisfy the specific objectives of achieving longer sampling times in high grain loading situations and to collect gram quantities of size-fractionated particulate for chemical analysis. Figure 4.27 presents calibration curves for the three-stage series cyclone that was developed for use in this arrangement.

Figure 4.28 contains micrographs of fly ash retained by the various cyclones in sampling the effluent from a coal-fired boiler, revealing that the series cyclones are very effective in classifying the fly ash with respect to size.

In conjunction with this calibration, a detailed study was done to determine the ultimate location of particles on the surfaces of the cyclones. These surfaces are the nozzle, body, cup, cap and outlet, and backup filter. Some results of this study are given in Figure 4.29. Only the large particles ultimately reach the collection cup.

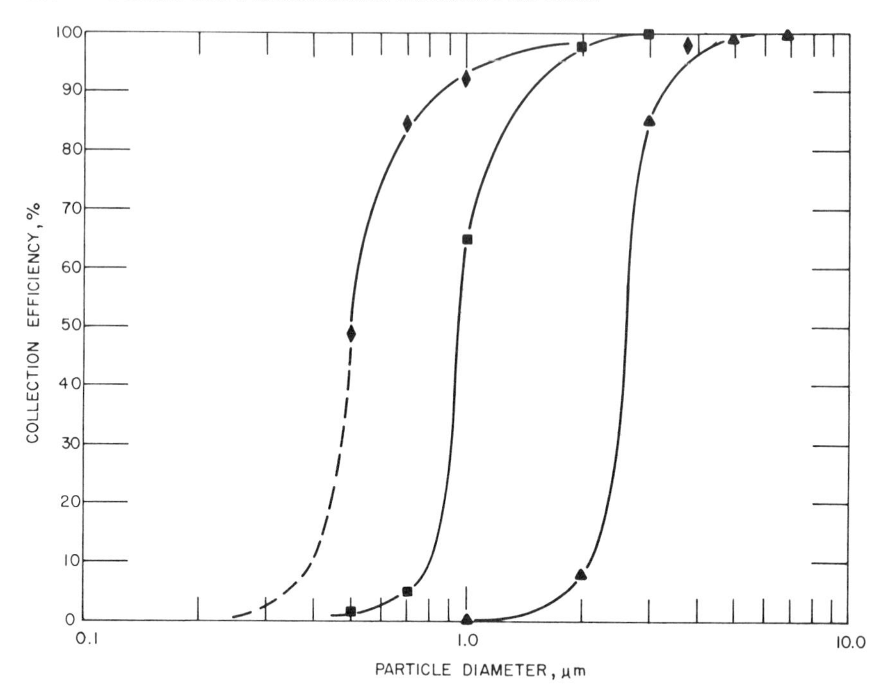

Figure 4.27. Calibration curves for the 1 ft^3/min (1.7 m^3 hr^{-1}) three-stage series cyclone. The particle density is 1.35 g cm^{-3}.

Figure 4.30 illustrates the configuration of the prototype 1 ft^3/min^{-1} (1.7 m^3 hr^{-1}) flow rate series cyclone system. A "stacked" arrangement was used to join the cyclones to minimize the diameter required for sampling ports. The system can be used with 6 in. ports.

4.2.3. Scattered Light Intensity Techniques

Figure 4.31 shows representative scattered light intensity patterns calculated from the Mie theory. All these patterns are rotationally symmetric about the vertical axis. Figure 4.31*a* is descriptive of Rayleigh (or dipole) scattering, where the particle diameter is less than one wavelength in magnitude. Figures 4.31*b* and 4.31*c* are scattering diagrams for intermediate and large values of the scattering coefficient α' ($\alpha' = \pi D_p/\lambda'$, where D_p = particle diameter and λ' = radiation wavelength). For a given index of refraction, the shape of the patterns depends only on the value of α', that, is on the relative values of D_p and λ', whereas the intensity of the scattered radiation depends

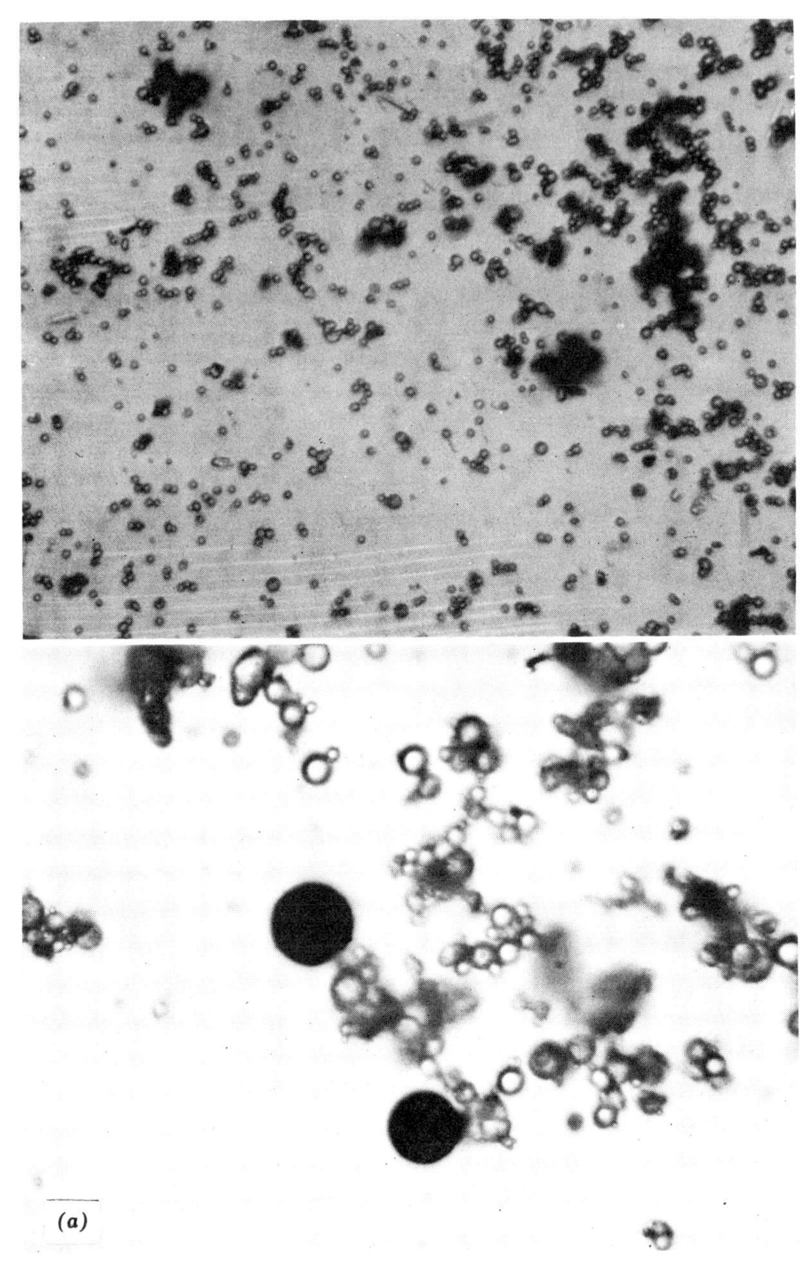

Figure 4.28. (*a*) Photomicrographs of cyclone catches acquired at a coal-fired steam plant.

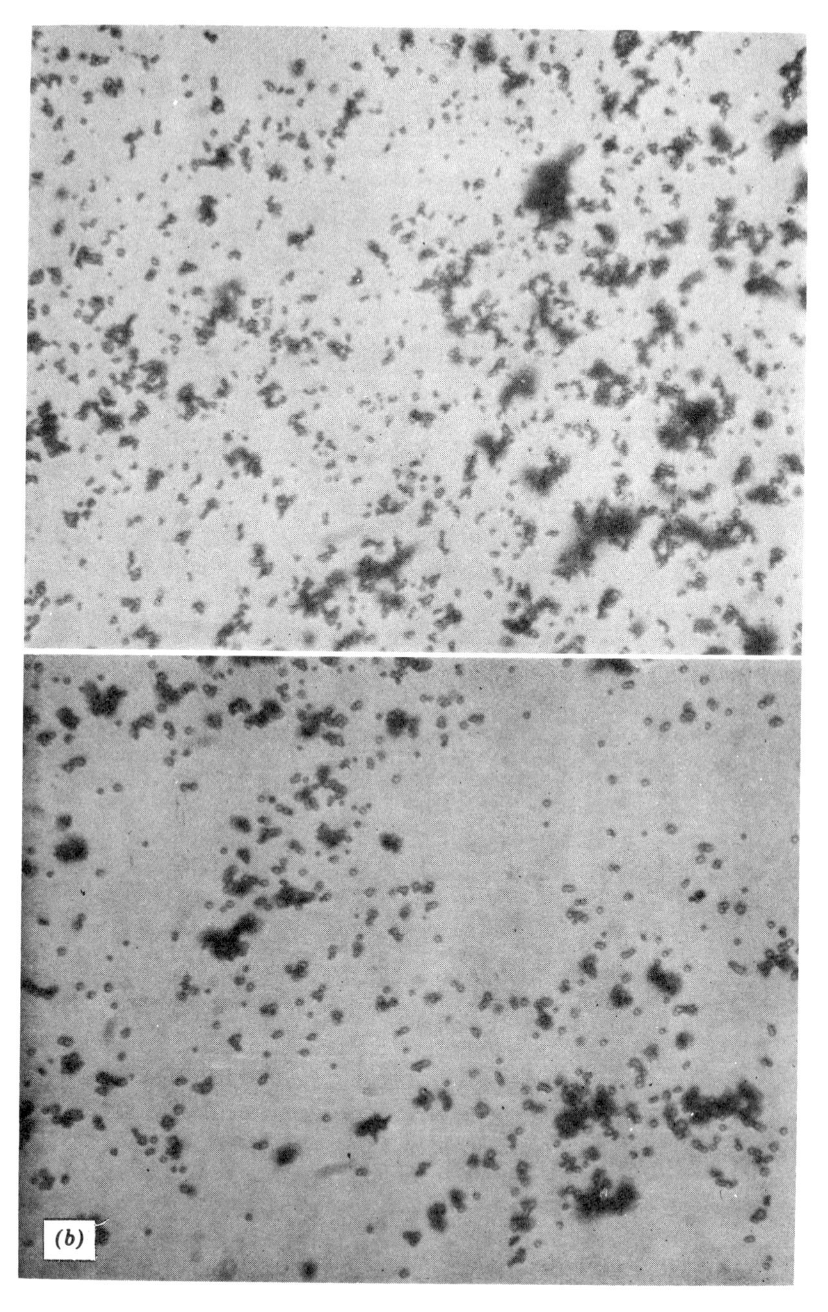

Figure 4.28. (*b*) Photomicrographs of cyclone and backup filter catches acquired at a coal-fired steam plant.

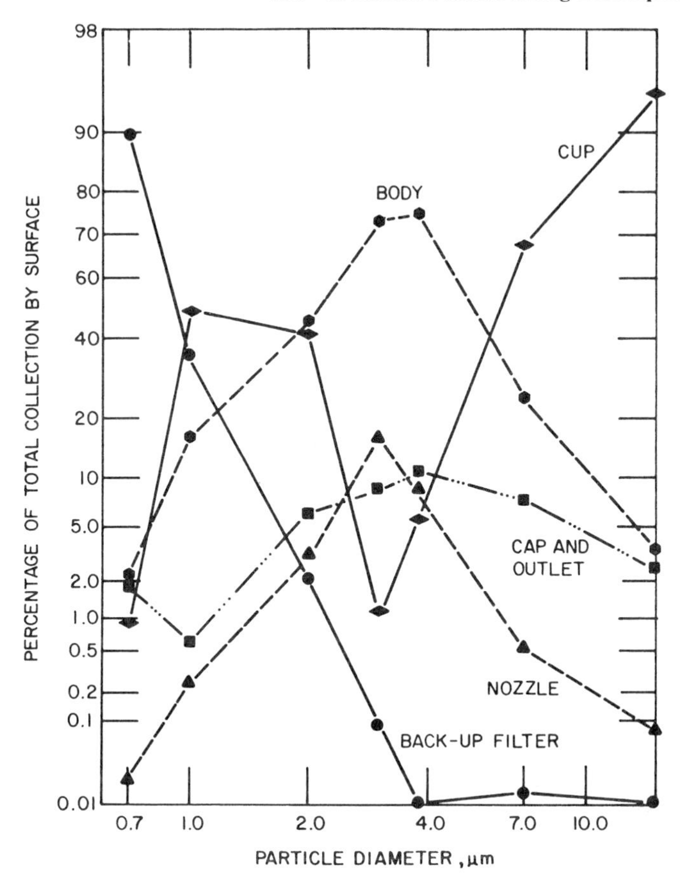

Figure 4.29. Percentage of total cyclone catch found on the cyclone's nozzle, body, cup, cap, and backup filter versus particle size.

on the particle size. The pattern changes progressively from the dipole pattern to a pattern for large particles, which can be calculated from classical diffraction, refraction, and reflection. Details of the patterns that are important in applying light scattering to particle size analysis are the positions of the maxima and minima, positions at which the intensity is relatively independent of the index of refraction, and the relative intensities of the plane-polarized components at various angles.

Procedures applicable to the measurement of particle sizes and concentration by light scattering have been reviewed by Hodkinson,[53] and more recently by Berglund.[54] In addition, Berglund compared the response

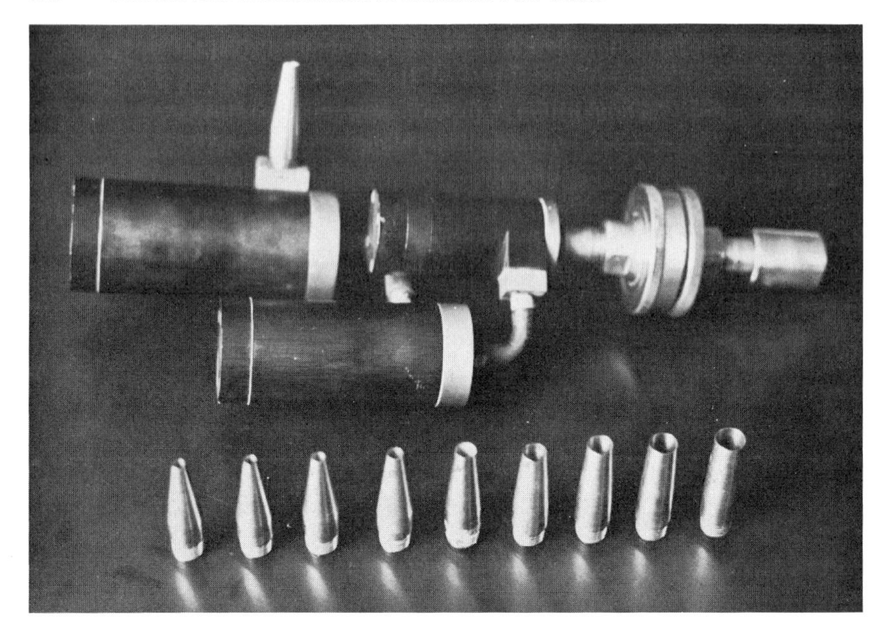

Figure 4.30. Series cyclone showing the three cyclones and Gelman backup filter.

of three commercially available optical particle counters with theory, using particles with different indices of refraction.

The basic operating principle for an optical particle counter is illustrated in Figure 4.32. Light is scattered by individual particles as they pass through a small viewing volume, the intensity of the scattered light being measured by a photodetector. The sizes of the particles determine the amplitude of the scattered light pulses. The rate at which the pulses occur is related to the particle concentration. Thus a counter of this type gives both size and concentration information. The occurrence of more than one particle in the viewing volume is interpreted by the counter as a larger single particle. To avoid errors arising from this effect, dilution to about 300 particles cm^{-3} is generally necessary. Errors in counting rate also occur as a result of electronics dead time and because of statistical effects resulting from the presence of high concentrations of subcountable ($D_p < 0.3$ μm) particles in the sample gas stream.[55] The intensity of the scattered light depends on the viewing angle, particle index of refraction, particle optical absorptivity, and shape, in addition to the particle size. The schematic in Figure 4.32 is for a system that utilizes "integrated near forward" scattering. Different viewing angles might be chosen to optimize some aspect of the counter performance. For example,

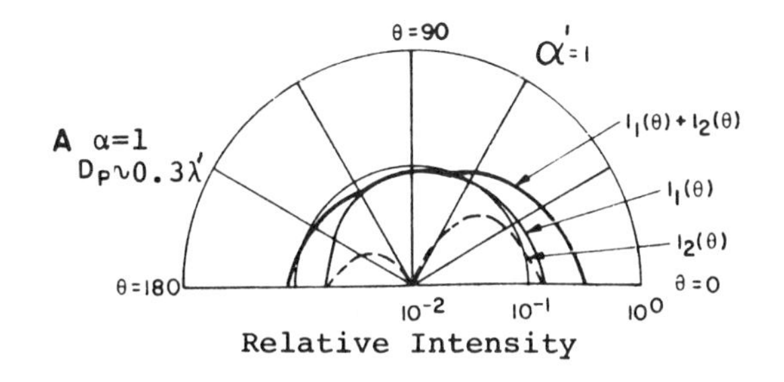

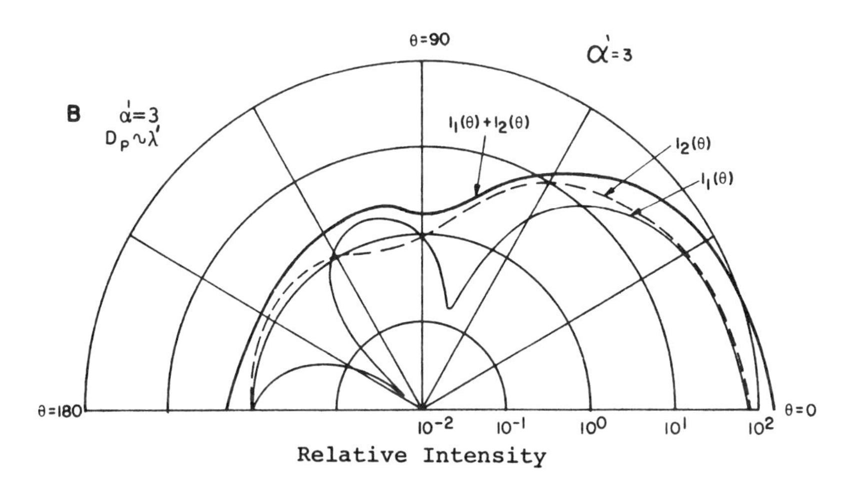

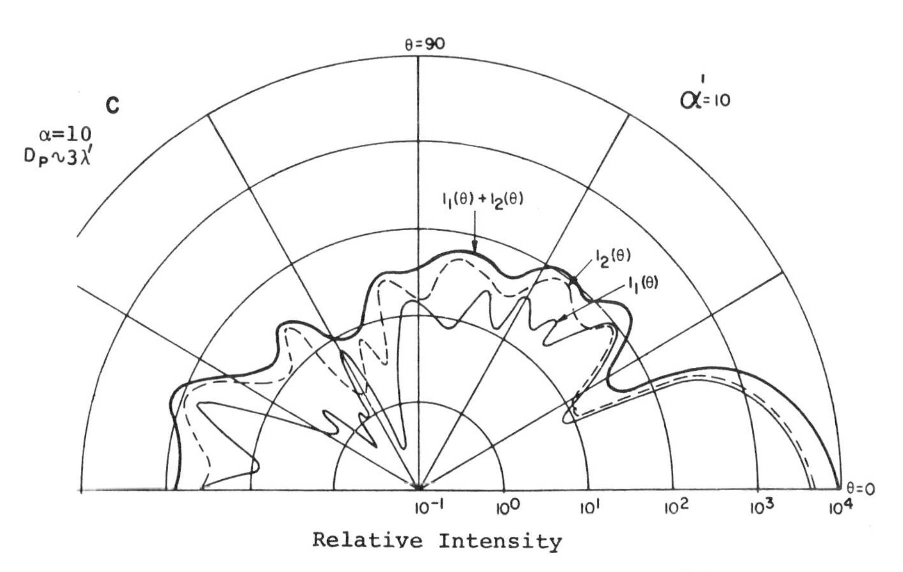

Figure 4.31. Light scattering diagrams for spherical particles.

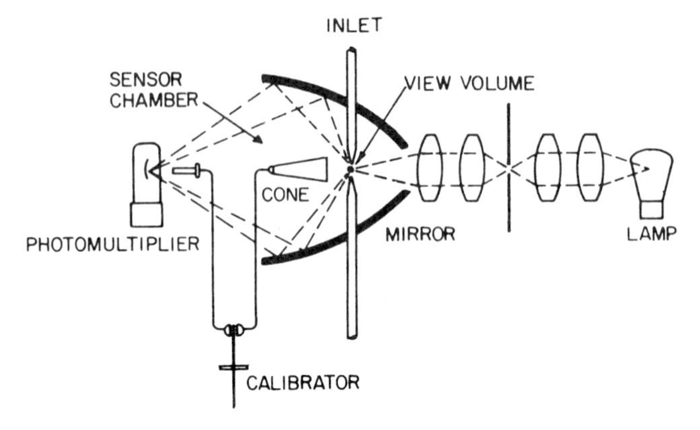

Figure 4.32. Operating principle for an optical particle counter. Courtesy of Climet Instruments Company.

near forward scattering minimizes the affect of variations in the indicated particle size with index of refraction, but for this geometry, there is a severe loss of resolution for particle sizes near 1 μm. Right angle or 90° scattering smooths out the response curve, but the intensity is more dependent on the particle index of refraction. Available geometries are as follows:

Bausch & Lomb 40-1	Near forward scattering
Royco 220	Right angle
Royco 245, 225	Near forward
Climet CI-201, 208	Integrated near forward

Current research is aimed at extending the concentration limits of optical particle counters, minimizing problems due to the variation in the particulate index of refraction, and improving the technique of using these devices for stack testing.

Gravatt[56] has developed a device that measures the *ratio* of the scattered light intensities at forward scattering angles of 5 and 10°. He states that the response of this instrument is independent of the particle index of refraction and that the device can be used effectively at particle concentrations up to 10^4 cm^{-3}.

Shofner et al.[57] have developed a laser-illuminated system that measures light scattered into two cones, similar in principle to Gravatt's device. This device is reported to sample aerosols *in situ* where concentrations are as high as 10^6 particles cm^{-3}.[57] Particle size information is obtained over any single decade range in sizes for sizes larger than about 0.2 μm in diameter.

Single particle counters with a diluter have been used in conjunction with impactors for making particulate control device fractional efficiency measurements. In general, agreement is within about a factor of 2 in concentration for a given size between the inertial and optical instruments.

4.2.4. Ultrafine Particle Sizing Techniques

Two physical properties of ultrafine particles (diameter < 0.5 μm) are size dependent and can be predicted with sufficient accuracy under controlled conditions to be used to measure particle size. These are particle diffusivity and electrical mobility. Although ultrafine particle size distribution measurements are still in a developmental stage, instruments are available that can be used for this purpose, and some field measurements have been made. Practical limitations on the lower size limit for this type of measurement are the loss of particles by diffusion in the sampling lines, and instrumentation. These losses are excessive for particle sizes below about 0.01 μm, where the samples are extracted from a duct and diluted to concentrations within the capability of the sensing devices.

4.2.4.1. Diffusional Sizing

Fuchs[58] has reviewed diffusion battery sizing work up until 1956, and Sinclair,[59,60] Breslin et al.,[61] Twomey,[62] and Sansone and Weyel[63] have reported more recent work, both experimental and theoretical.

Diffusion batteries consist of a number of long, narrow, parallel channels, a cluster of small bore tubes, or a series of screens. A typical parallel channel diffusion battery and the aerosol penetration characteristics of this geometry at two flow rates appear in Figure 4.33*a* and 4.33*b*, respectively. The parallel plate geometry is convenient because of ease of fabrication and the availability of suitable materials, and also because sedimentation can be ignored if the slots are vertical, while additional information can be gained through settling, if the slots are horizontal. Sinclair[60] and Breslin et al.[61] report success with more compact, tube-type and screen-type arrangements in laboratory studies.

Although the screen-type diffusion battery must be calibrated empirically, it offers convenience in cleaning and operation, as well as compact size. Figure 4.34 shows Sinclair's geometry. This battery is 21 cm long, approximately 4 cm in diameter, and weighs 0.9 kg.

Variations in the length and number of channels (tubes, or screens) and in the aerosol flow rate are used as means of measuring the number of particles in a selected size range. As the aerosol moves in streamline flow through the channels, the particles diffuse to the walls at a predictable rate, depending on the particle size and the diffusion battery geometry. It is assumed that every particle that reaches the battery wall will adhere; therefore only a fraction

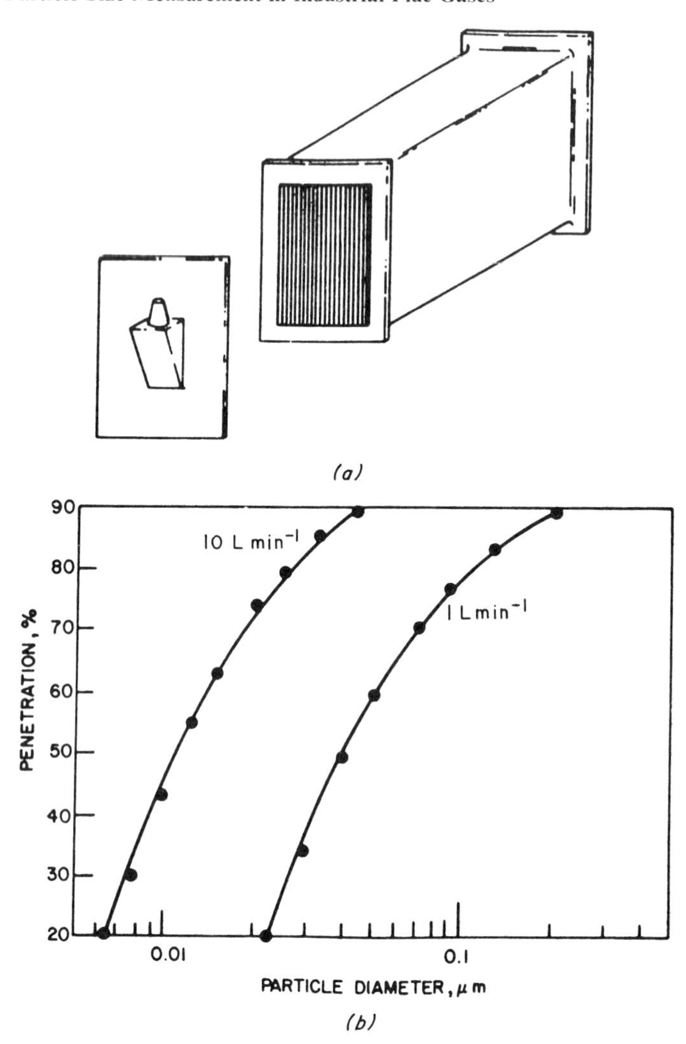

Figure 4.33. (*a*) Parallel plate diffusion battery. (*b*) Penetration curves for monodisperse aerosols (12 channels, 0.1 × 10 × 48 cm).

of the influent particles will appear at the effluent of a battery. It is necessary only to measure the total number concentration of particles at the inlet and outlet to the diffusion battery under a number of conditions to calculate the particle size distribution.

The mathematical expression for the penetration (n/n_0) of a rectangular slot or parallel plate diffusion battery by a monodisperse aerosol was given

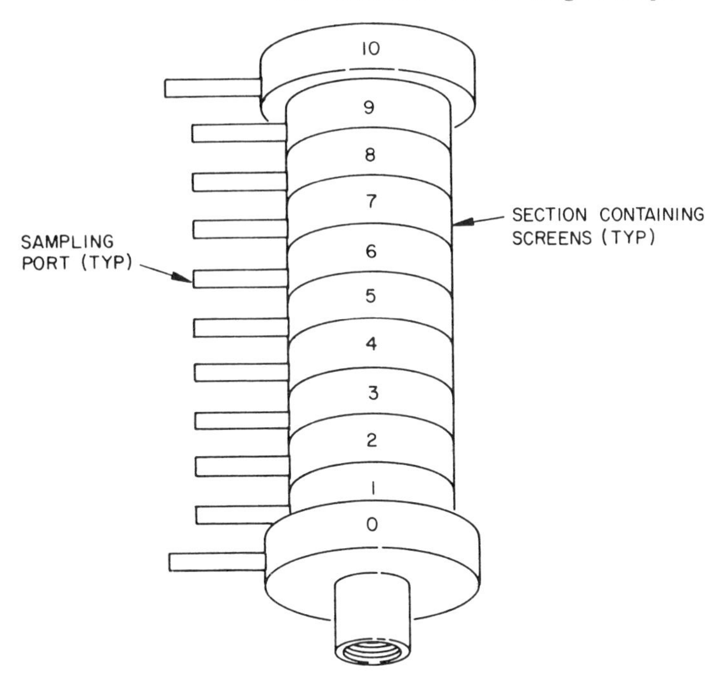

Figure 4.34. Screen-type diffusion battery. The battery is 21 cm long, 4 cm diameter, and contains 55 stainless steel screens, 635 mesh.[60]

in series form by Gormley and Kennedy.[64] The coefficients were calculated and tabulated by Twomey[62] using a computer. The equation is

$$\frac{n}{n_0} = 0.91e^{-xm\mathscr{D}} + 0.053e^{-11.37xm\mathscr{D}} + 0.015e^{-33.06xm\mathscr{D}} + 0.0068e^{-66.06xm\mathscr{D}} + 0.0037e^{-110.4xm\mathscr{D}} + \cdots \quad (29)$$

where $x = 3.77\ hl/aQ$
h = height of each rectangular slot (cm)
l = length of each rectangular slot (cm)
$2a$ = width of each rectangular slot (cm)
Q = volumetric flow rate ($\mathrm{cm^3\ sec^{-1}}$)
n = outlet concentration/($\mathrm{cm^{-3}}$)
n_0 = inlet concentration/($\mathrm{cm^{-3}}$)
m = number of rectangular channels,
$\mathscr{D} = kTb$ = diffusion coefficient of particles ($\mathrm{cm^2\ sec^{-1}}$),
k = Boltzmann constant (1.38×10^{-16} erg $\mathrm{K^{-1}}$),
T = temperature (°K)
b = particle mobility (cm $\mathrm{dyne\text{-}sec^{-1}}$).

The mobility of a particle is defined as the velocity for a unit force. For particles having diameter smaller than approximately 1 μm, the slip correction factor must also be included. The corrected mobility[65–70] is

$$b = \left[1 + A\frac{\lambda}{r} + B\frac{\lambda}{r}\exp\left(-C\frac{r}{\lambda}\right)\right]\frac{1}{6\pi\mu r} \tag{30}$$

where r = particle radius
λ = mean free path of the gas molecules
μ = gas viscosity

and A, B, and C are empirically determined constants. For air at standard temperature and pressure, $A \sim 1.2$, $B \sim 0.4$, $C \sim 0.9$, and $\lambda = 0.1$ μm. If A

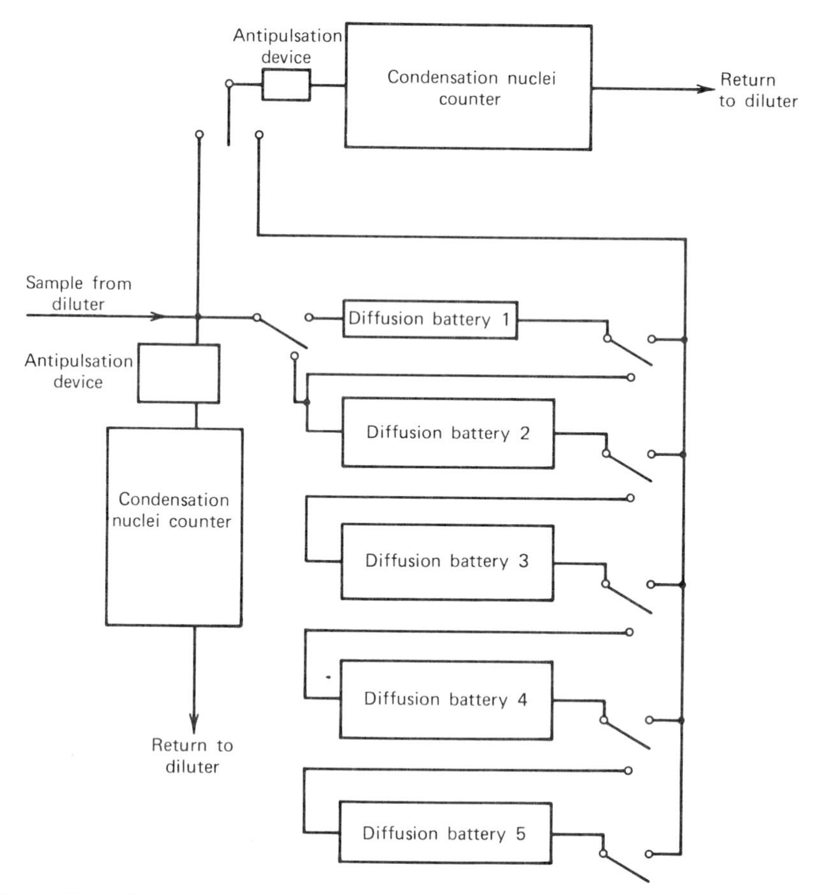

Figure 4.35. Diffusion battery and condensation nuclie counter layout for fine particle sizing.

is set equal to unity and B to zero, the numerator of eq. 30 is the Cunningham correction factor. When the Stokes diameter is used to describe particle size, the penetration of diffusion batteries is virtually independent of physical properties of the individual aerosol particles.

Fuchs et al.[71] presented a technique for calculating the particle size distribution from raw data, assuming that the distribution is log normal. The technique suggested by Sinclair[59] does not include this restriction. In Sinclair's technique a nomograph is prepared using the penetration for each diffusion battery geometry and flow rate and a large number of monodisperse particle

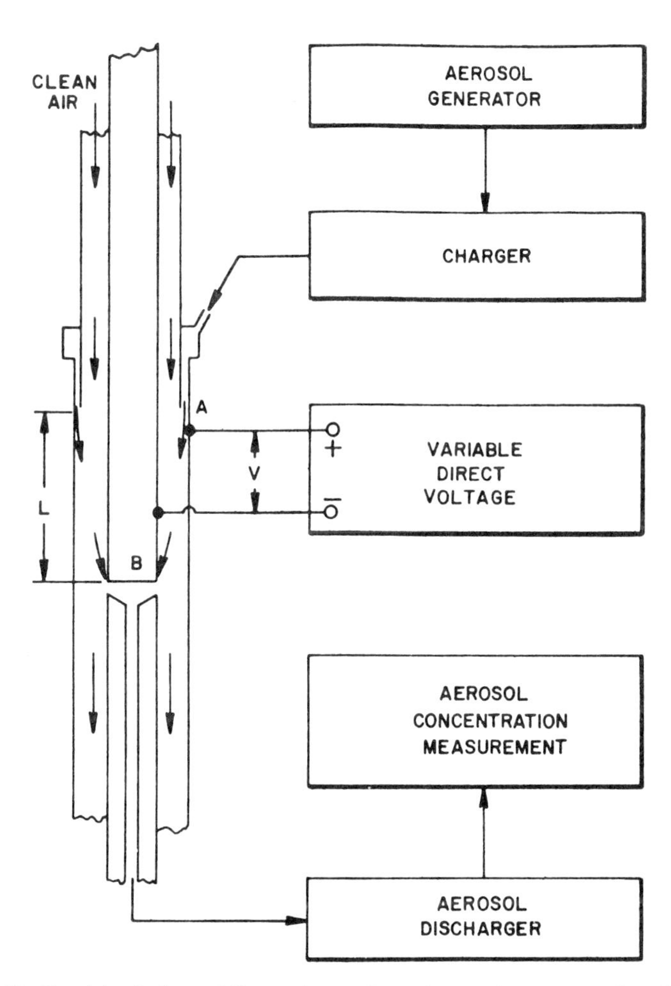

Figure 4.36. Coaxial cylinder mobility analyzer. Charged aerosol enters at A. Sampling groove is at B in inner cylinder, which is adjusted axially. After Hewitt.[72]

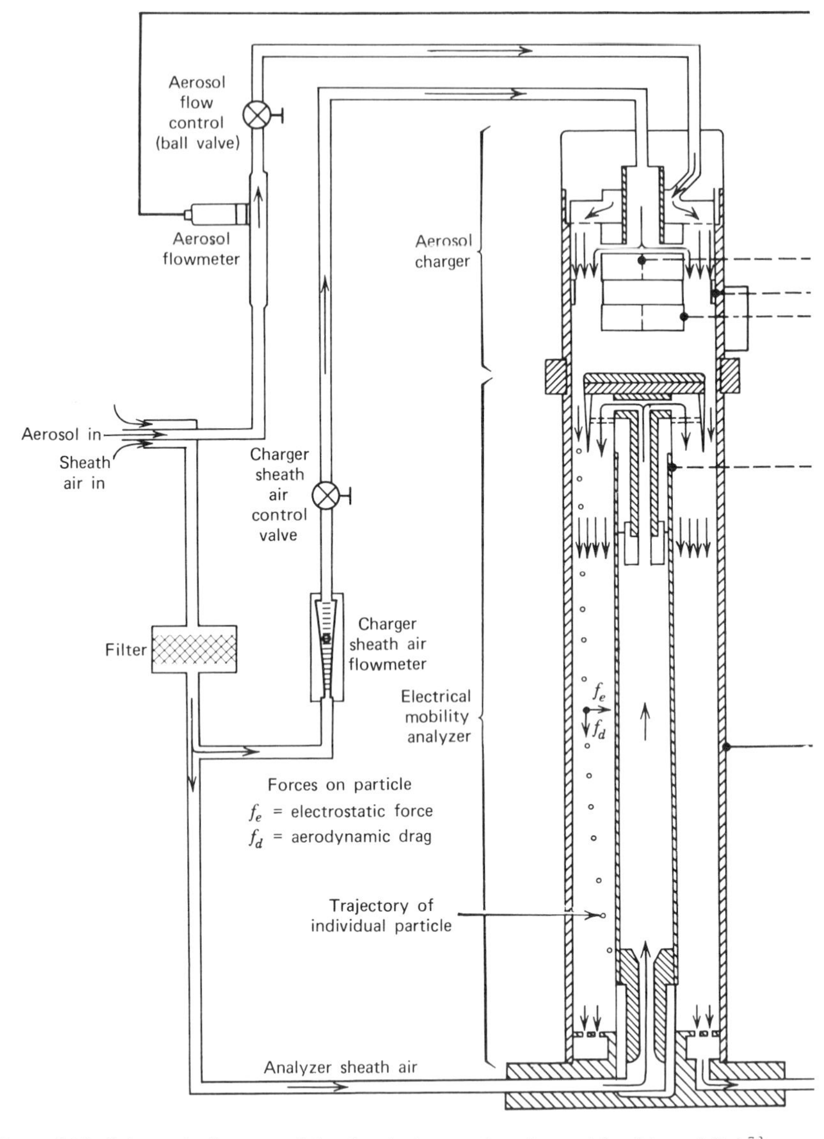

Figure 4.37. Schematic diagram of the electrical aerosol analyzer. After Liu and Pui.[73]

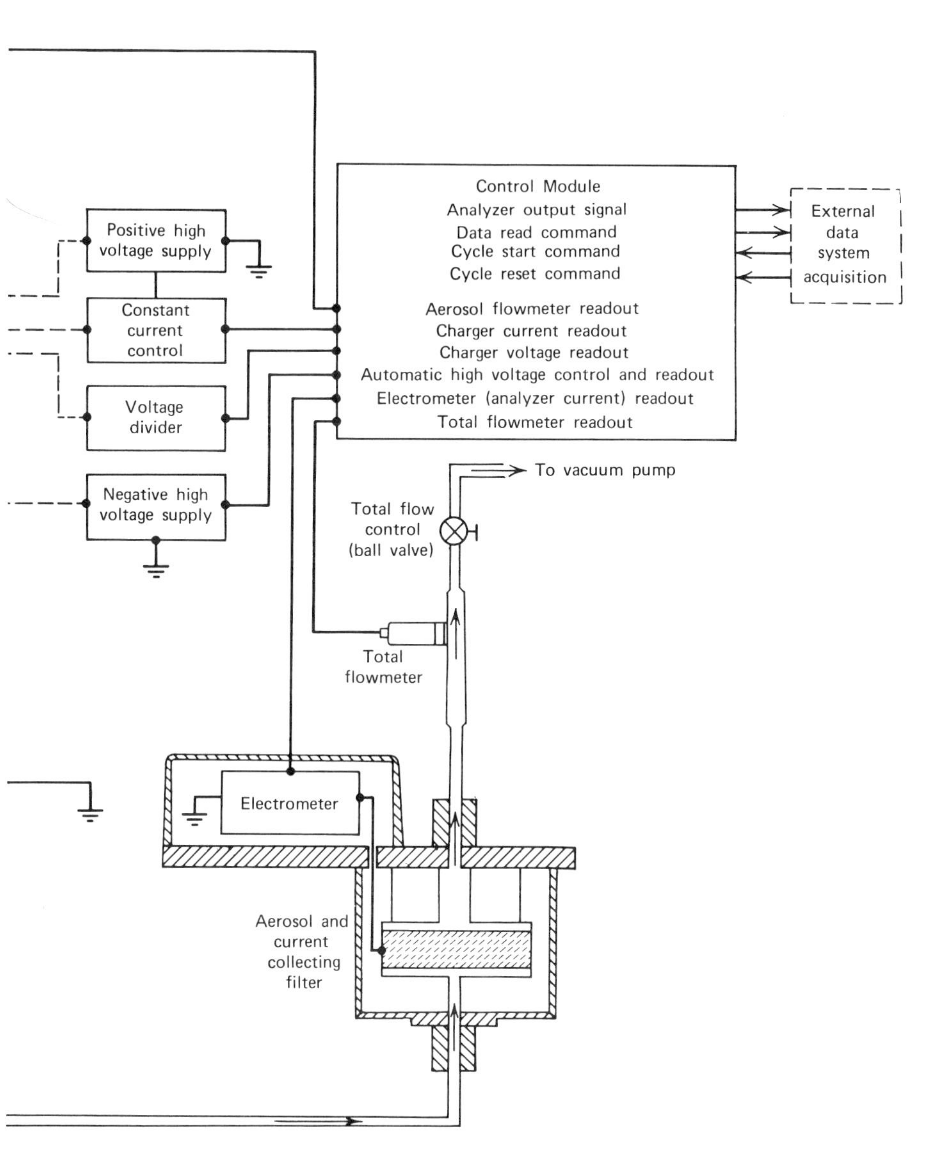

Figure 4.37. *Continued.*

sizes. Comparing this nomograph with experimental penetrations, one calculates the particle size distribution using a "graphical stripping" process. Alternatively, it is sometimes convenient to use a "D_{50}" technique like that previously described in the discussion of impactor reduction.

Figure 4.35 is a block diagram of a diffusion sizing system for field testing.

Diffusional measurements are less dependent on the aerosol parameters than the other techniques discussed and perhaps have a theoretically firmer basis.

Disadvantages of this technique are the bulk of the diffusional batteries (although advanced technology may alleviate this problem), the long time required to measure a size distribution, and problems with sample conditioning when condensable vapors are present.

4.2.4.2. Electrical Particle Counters

The most complete set of experiments performed in the determination of the relationship between particle size and charge was reported by Hewitt[72] in 1957. This work confirmed theoretical predictions that there exists a unique charging rate for each particle size if the charging region is homogeneous with respect to space charge density and electric field.

In the course of his work, Hewitt developed a mobility analyzer (Figure 4.36). Charged particles enter through the narrow annular passage at *A* and experience a radial force toward the central cylinder because of the applied field. By moving the sampling groove *B* axially, or by varying the magnitude of the applied field, the mobility of the charged particles can be measured. In Hewitt's experimental work, the particle size was known, and the mobility determined the charge. If, however, the particle charge were known, the mobility would determine the particle size. This concept has been used by Whitby et al.[73] at the University of Minnesota to develop a series of electrical aerosol analyzers (EAA). A rugged version of the field test unit* based on the earlier University of Minnesota designs is now commercially available (Figure 4.37).

The reliance on electrical mobility as a measure of particle size is sound in principle. If calibration studies and field tests show this technique to be accurate and reliable, it could become the most popular method for making ultrafine particle size measurements. The EAA has the distinct advantage of very rapid data acquisition compared to diffusion batteries and condensation nuclei counters (2 min as opposed to 2 hr for a single size distribution analysis.)

* Model 3030 (EAA), Thermo-Systems, Inc., 2500 North Cleveland Ave., St. Paul, Minnesota 55113.

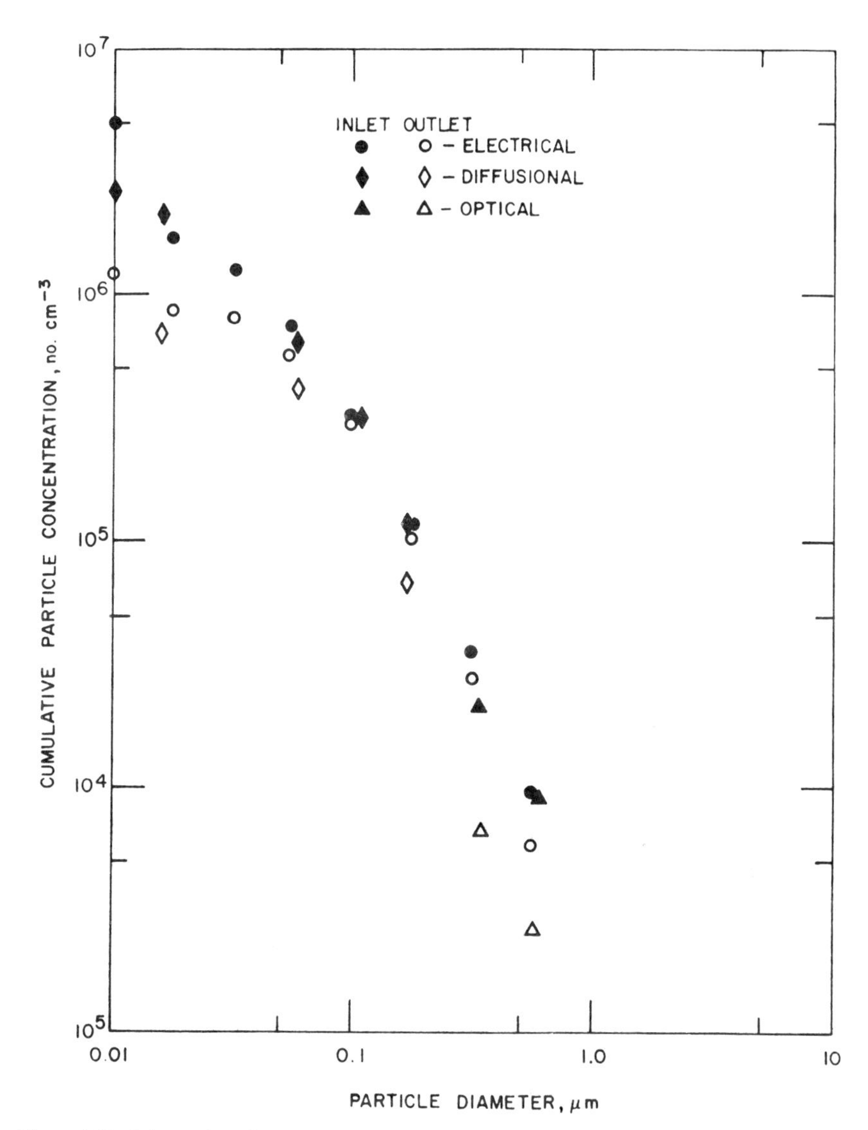

Figure 4.38. Inlet and outlet distributions as obtained with optical, diffusional, and electrical techniques.

Disadvantages of this type of measurement system are difficulties in predicting with sufficient accuracy the particle charge and the fraction of the particles bearing a charge, and the requirement for sample dilution when making particle size distribution measurements in flue gases.

Figure 4.38 is a graph of particle size distributions measured simultaneously in the field with diffusional and electrical apparatus. The agreement between the methods is good for this particular test.

4.3. OFF-LINE SIZE ANALYSIS

4.3.1. General Discussion

The techniques previously discussed for the measurement of particulate size distributions are far more useful than any laboratory methods in control device evaluation. The size distribution in a flue gas aerosol is a dynamic function that is continually changing by means of agglomeration, settling, and condensation. It is impossible to redisperse the particulate exactly as it was in the gas or to calculate the original size distribution from laboratory data. Such attempts to measure the size distribution invariably lead to results different from those obtained in stream, and careful interpretation is required to avoid confusion. For example, microscopic examination of stage catches on upper impactor stages with cut points near 10 μm often reveal only particles having diameters much less than 10 μm. This could lead to the interpretation that cascade impactors are not effective as instruments for measuring size distributions. Thorough calibration studies, however, have clearly demonstrated that impactors do behave according to theory; thus the proper interpretation of such observations is that the smaller particles on the upper impactor stages existed as parts of agglomerates having much larger equivalent diameters.

In spite of these limitations, laboratory investigations are the only current methods for studying particle shape, degree of agglomeration, surface features and other physical properties of interest such as density and compostion.

Laboratory methods for the measurements of particle size can be put into four categories based on optical and electron microscopy, electrical conductivity, inertia, or sedimentation. Cadle[4] has written an excellent monograph that includes discussions of most of the methods and instruments developed before 1955. This section includes more recently developed techniques not covered by Cadle.

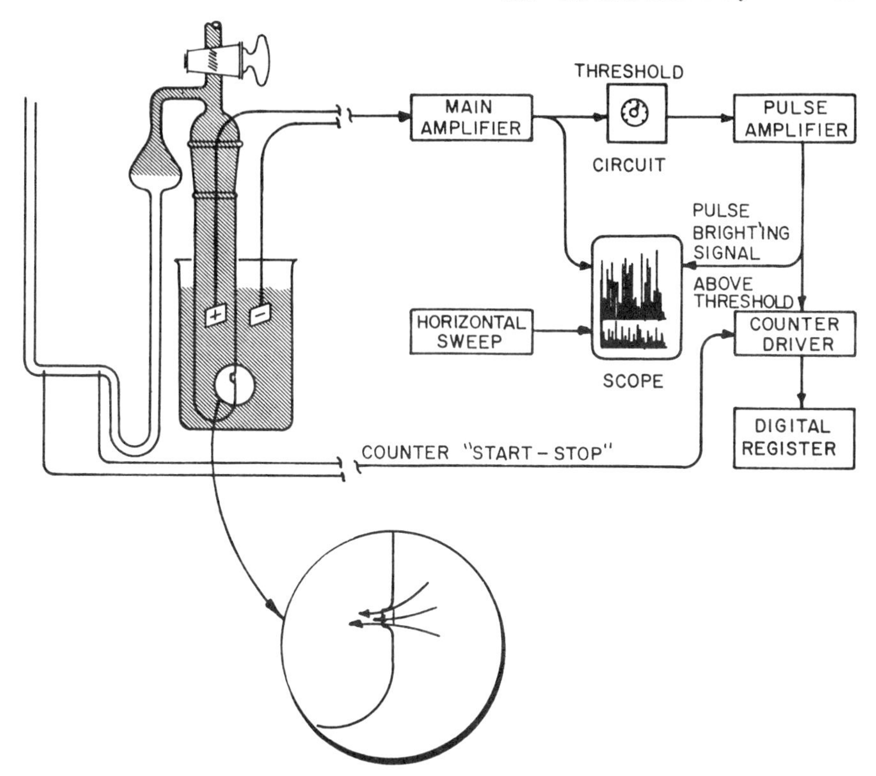

Figure 4.39. Operating principle of the Coulter counter. Courtesy of Coulter Electronics.

4.3.2. Electrical Conductivity

A very convenient technique for measuring the size distribution of powders that can be suspended in an electrolyte dispersing medium is conductivity modulation. A commercially available device, the Coulter Counter,* is illustrated in Figures 4.39 and 4.40. Figure 4.39 gives the operating principle of the Coulter Counter. Particles suspended in an electrolyte are forced through a small aperture in which an electric current has been established. Each particle displaces electrolyte in the aperture, providing an electrical pulse that is proportional to the particle electrolyte interface volume. A special pulse height analyzer is included with the system, allowing convenient data acquisition. A bibliography related to the operation of the Coulter Counter has been compiled by the manufacturer.

* Available from Coulter Electronics, Inc., 590 West 20th Street, Hialeah, Florida 33010.

Figure 4.40. The Coulter Counter. Photograph. Courtesy of Coulter Electronics.

A disadvantage of this device is the limited size span that can be covered with any one orifice. The range is limited at the fine end by rapidly declining resolution as the particle volume becomes small compared to the orifice dimensions, and on the large end by the physical size of the orifice itself. A secondary problem is obtaining a suitable carrier liquid having the required conductivity, in which the particle can be dispersed without dissolving.

4.3.3. Microscopy

Cadle[4] has discussed methods of illumination, the selection of optics, sample preparation, and counting techniques used in obtaining the best aerosol characterization by means of microscopy. The major technological innovations since Cadle's discussion are the development of the scanning electron microscope and computerized systems that can scan microscopic samples and do the statistical analysis very rapidly. Scanning electron microscopes (SEM) are now familiar to most researchers and have the definite advantage of easier sample preparation and much improved depth of field as compared to transmission-type electron microscopes. These devices are unparalleled for convenient studies of surface features, particle shape, agglomeration, and semi-quantitative compositional analysis.

Computerized scanning devices have been developed for use with optical microscopes, scanning or transmission electron microscopes, and even photographs, to obtain size and shape information on several bases.

Figure 4.41 shows the Bausch & Lomb "Omnicon" image analysis system. Analyses of images in various forms of particulate samples can be done automatically or manually in terms of area, projected length, perimeter, volume, and other dimensions.

A number of fully equipped commercial laboratories specialize in aerosol analyses. It is now possible to have complete physical and chemical characterizations of dusts done rapidly, conveniently, and relatively inexpensively.

A useful adjunct for microscopic examination of airborne particulate pollutants is the McCrone *Particle Atlas*,[74] which includes techniques for microscopic study of pollutants and an extensively illustrated atlas of the shapes, sizes, and other physical properties of typical particulate pollutants and means for identifying them.

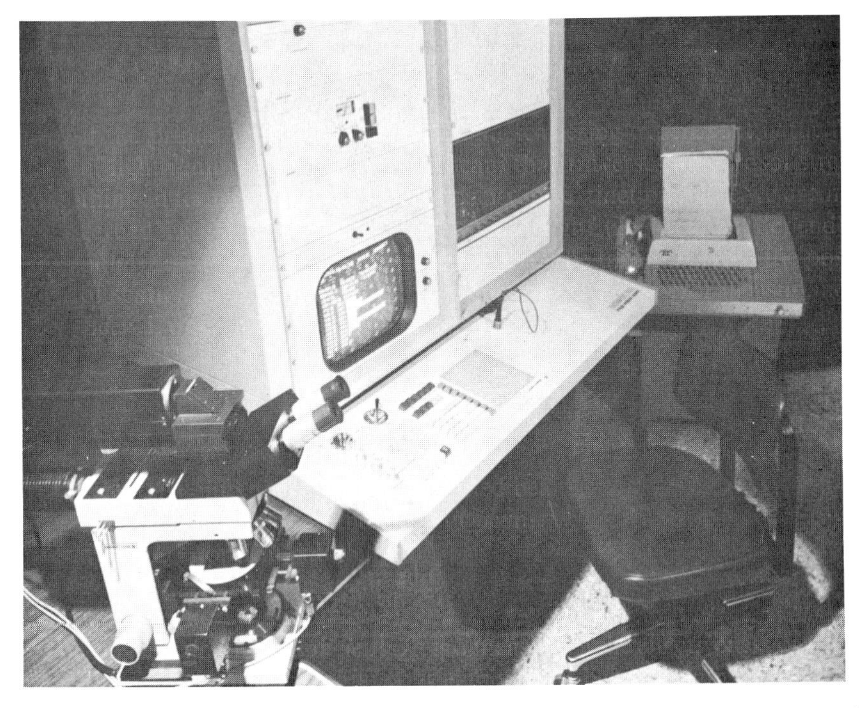

Figure 4.41. The Bausch & Lomb "Omnicon" image analysis system. Photograph courtesy of Bausch and Lomb.

4.3.4. Centrifugal Separation

The process of separating particles according to the Stokes diameter can be accomplished more quickly if a strong centripetal acceleration is applied. The commercial centrifugal particle classifier (Bahco) in Figure 4.42 has been accepted by the ASME[75] and is routinely used to measure the size distribution of powders. Figure 4.43 illustrates its principle of operation. The entire assembly, with the exception of components 1 to 6, which make up the powder feeding assembly, rotates about the vertical axis. Clean air enters past the throttle nut 13, past the symmetrical disks 11, through the sifting chamber 10, radially toward the axis, and outward through the fan vanes 8. Particles enter the device through the feed nozzle 6 and enter the airstream through the rotary duct opening 9. Large particles are carried outward by the centrifugal force and are collected in the catch basin 12. Small particles are carried with the airstream and emerge through the fan vanes 8, where they impact on the inner wall of the rotor casting 7. Thus the classification results in two size fractions, one with diameters smaller than the cutoff size.

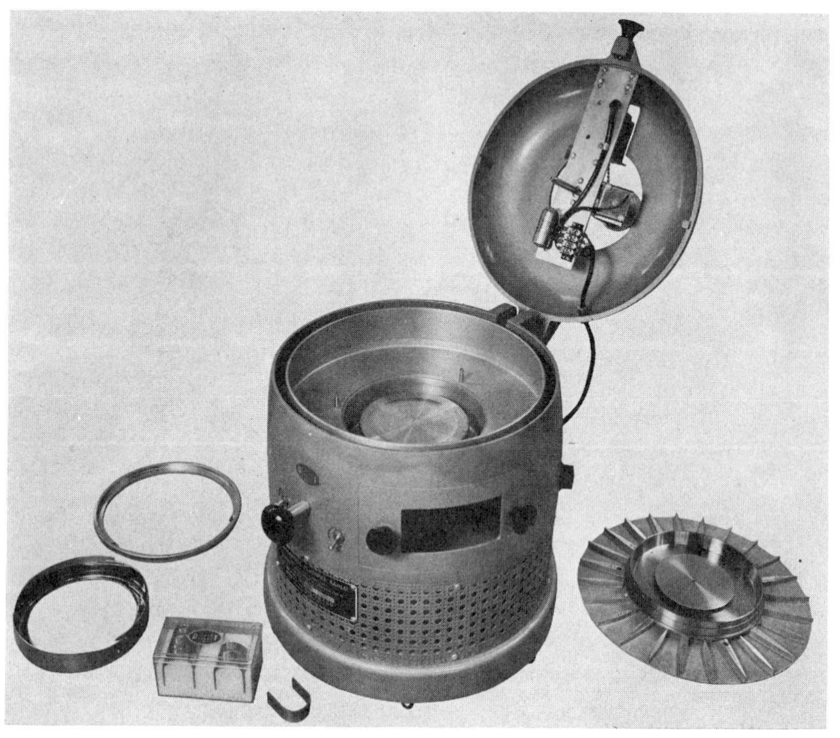

Figure 4.42. The Bahco particle classifier. Photograph courtesy of the Harry W. Dietert Co.

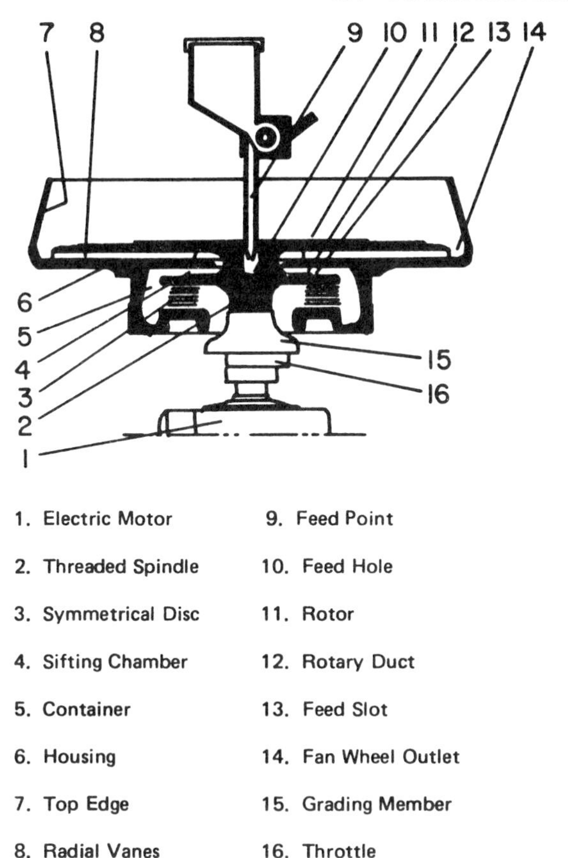

Figure 4.43. Operating principle for the Bahco particle classifier.

The cutoff size can be varied from about 2 to 50 μm by changing the throttle nut position (13) by means of the throttle spacer 14. This changes the amount of air passing through the device and thus altering the magnitude of the aerodynamic drag force acting on the particles. To measure the size distribution, the operator adjusts the machine for the smallest cutoff, runs an analysis, weighs the large fraction, then reruns this fraction with a larger cutoff. This process is repeated until the entire sample is classified.

Spiral centrifuges have been developed by Goetz et al.[76–78] and by Stöber and Flachsbart.[79,80] With these devices, particle size distributions can be measured from a few hundreds of a micrometer to approximately 2 μm in diameter. The major elements of the Goetz and Stöber centrifuge spectrometers appear in Figures 4.44 and 4.45. The basic element of such a device is

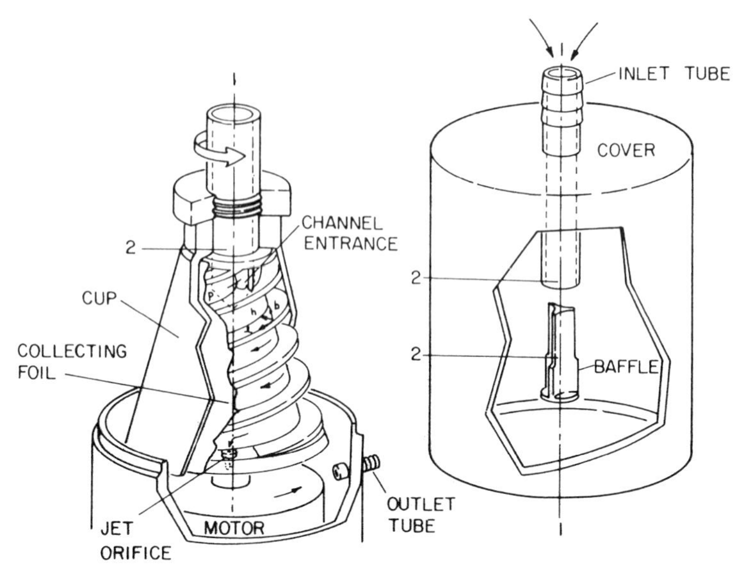

Figure 4.44. Cutaway sketch of the Goetz and Stöber centrifuge spectrometer. After Gerber.[78]

the spiral channel, which may be rotated at high speed to achieve large radial accelerations. The particulate is deposited on a flexible foil that is subsequently removed for microanalysis. If the time of residence in the channel is sufficiently long for all the aerosol particles to reach the foil, the deposit will contain a complete size spectrum. The larger particles are deposited in regions of small radius and the small particle in regions of large radius along the helical path. The locus of the deposition of each size is dependent on the

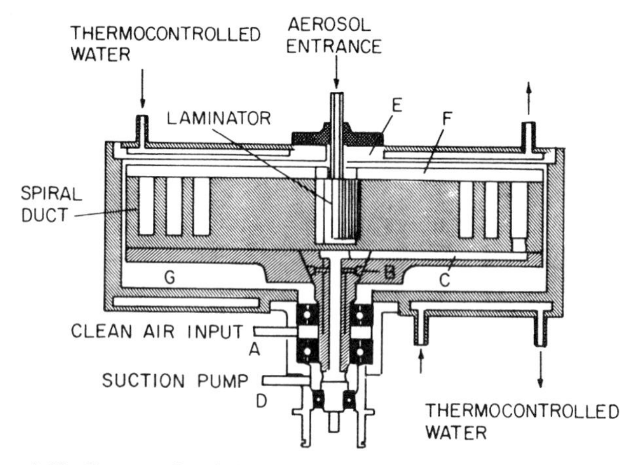

Figure 4.45. Cross-sectional sketch of the Stöber centrifuge. After Stöber.[79]

geometry of the channel, the flow rate, the angular velocity of the spiral, and various idiosyncrasies of the individual type of system used. The size distribution can be determined from the deposits collected on the foil by microscopy, by photometry, and perhaps in some cases gravimetrically. A cylindrical centrifuge capable of 0.5% size resolution over the range of aerodynamic diameters from 0.1 to 1.0 μm has been reported by Preining et al.[81] The Goetz and Stöber centrifuge instruments were manufactured commercially for a time, but are now available only on special order.

4.3.5. Sedimentation and Elutriation

The mobility of a particle was given in eq. 28 and is defined as the velocity for a unit force; that is,

$$b = \frac{C}{3\pi\mu D_p} \tag{31}$$

Thus the velocity for any net applied force F is

$$v = \left(\frac{C}{3\pi\mu D_p}\right)F \tag{32}$$

If only gravitational forces are present, the terminal settling velocity of a small particle diameter D_p in the viscous flow (Stokes) regime is (expressing F as Mg or $g\pi D_p{}^3\rho_p/6$):

$$V_s = \frac{\rho_p C D_p{}^2 g}{18\mu} \tag{33}$$

In the discussion above

M = particle mass
ρ_p = particle density
C = slip correction factor
μ = viscosity of air
D_p = particle Stokes diameter
g = acceleration of gravity

Figure 4.46 shows the settling velocity of particles with a density of 2.0 g cm^{-3} in air. The relationship between the Stokes diameter of a particle and settling velocity can be exploited in a number of ways to obtain particle size distributions.

If the aerosol that contains the particles of interest is introduced into a chamber, which is then sealed to form a quiescent zone, the particles will immediately begin to settle to the bottom with velocities as in Figure 4.46.

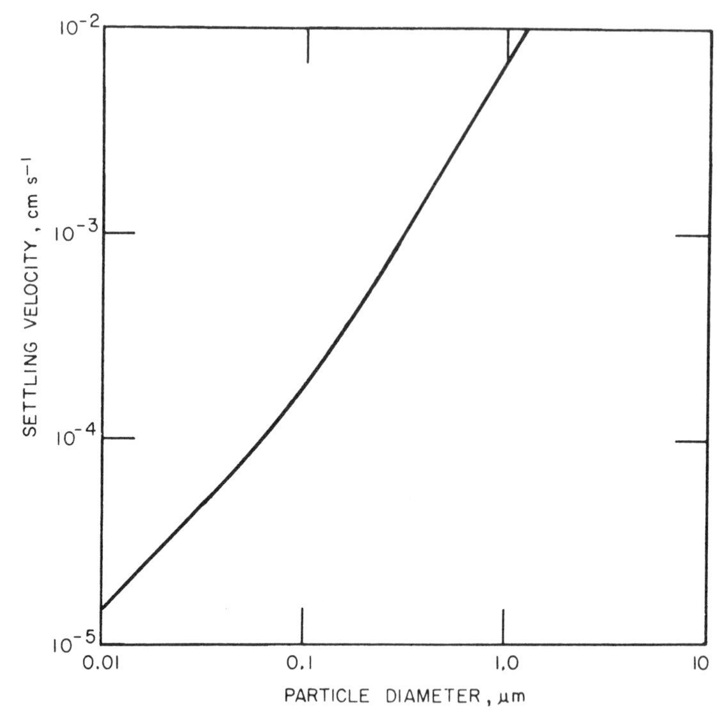

Figure 4.46. Settling velocity in air for spherical particles with density of 2.0 g cm^{-3}.

By measuring the rate at which the aerosol concentration changes at various levels, or the rate at which mass accumulates on the bottom, it is possible to calculate a particle size distribution.

Odén[82] has developed a technique for calculating the size distribution from accumulated weight versus time curves. This is done by use of the equation

$$w(t) = w + t\left(\frac{dw}{dt}\right) \tag{34}$$

where $w(t)$ = total mass accumulated by settling at time t
t = elapsed time
w = fraction of the total mass consisting of particles with settling times less than t
dw/dt = rate at which mass is accumulating by settling at time t

The relationship among the three terms of eq. 34 is plotted in Figure 4.47. By drawing a number of tangents along the $w(t)$ versus t curve, several

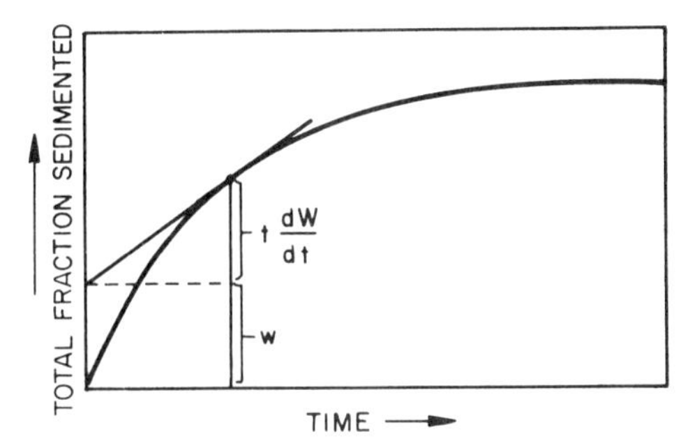

Figure 4.47. Intercept method of determining $t(dW/dt)$ and w. After Odén.[82]

points are obtained for w which are used to construct a cumulative particle size distribution.

Sedimentation is particularly well suited for automated readout of the $w(t)$ versus t information which could be done by gravimetric means. Cahn Instrument Company* has available a settling chamber attachment for their electronic microbalance. Vibrating crystal microbalance sensors could also be utilized to obtain the data.

If the air in the aerosol chamber is not stagnant but moves upward, particles with settling velocities equal to or less than the air velocity will have a net velocity up, and particles that have settling velocities greater than the air velocity will move down. This is the principle of "elutriation," which is used frequently to measure the size distribution of dusts. Figure 4.48 shows the Roller particle size analyzer, which is frequently used for this purpose and is available commercially.† It consists of an air inlet, a U-shaped Pyrex glass vessel that holds the powder sample, oscillator connections for the glass vessel, four stainless steel chambers corresponding to different cut points at a single flow rate, and the filter sample collector. The main advantage of this type of instrument over sedimentation devices is that the sample is actually collected according to size. Disadvantages are the coarseness of the cut points, the time required to accomplish the classification, and lower size limit, which is about 5 μm.

Instrumentation for sedimentation sizing has been developed by Chabay[83] using optical heterodyne spectroscopy. In this technique a continuous wave laser is used in an interferometric mode to determine Doppler shifts resulting from the sedimentation velocity of the particles. Doppler shifts as small as a few parts in 10^{14} are measurable permitting diameter measurements to be

* Cahn Instrument Company, 7500 Jefferson St., Paramount, California 90723.

† The Roller particle size analyzer is available from the American Standard Instrument Co., Inc., Silver Springs, Maryland.

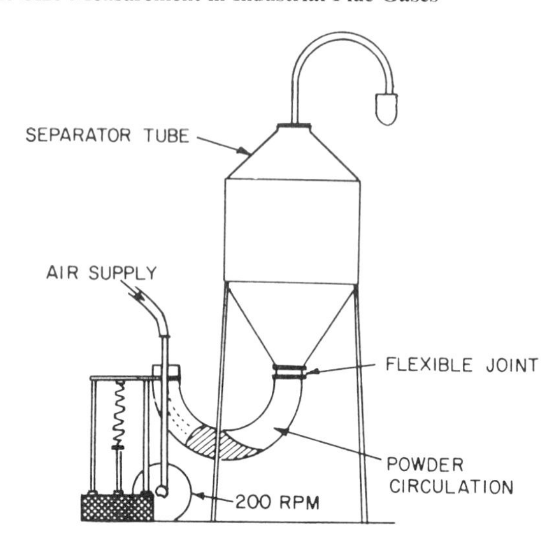

Figure 4.48. The Roller elutriator. After Allen.[3]

obtained for particles as small as about 0.5 μm for settling in air. However the data obtained require knowledge of the Mie scattering coefficients for conversion of measured scattered intensity versus Doppler shift (photocurrent power spectrum) to particle concentration versus diameter. Under some conditions with the use of a suitable stilling chamber, this technique might be capable of on-line operation, but such an application has not been attempted to date.

4.4. ELECTRONIC AND ELECTROOPTICAL *IN SITU* TECHNIQUES

4.4.1. General Discussion

The ideal particle sizing system is one wherein data are obtained from a representative sample with no perturbation of the aerosol. In most tests it is likely that the only practical means of accomplishing this is through measurements of extinction or scattering of electromagnetic radiation. This section discusses methods that show promise for applications in flue gas sampling. These systems or methods are in various stages of development ranging from theoretical proposals to commercial prototypes. None, however, is currently in use on a routine basis to obtain aerosol particle size information for industrial aerosols.

Several *in situ* techniques based on optical methods are available or are in developmental stages. Many of these methods are amenable to real-time or near-real-time, on-line data analysis, making them potentially useful for continuous monitoring purposes. *In situ* electrooptical measurement techniques are particularly useful in sizing liquid aerosols because of their noncontact operation, which alleviates difficulties arising from possible evaporation or condensation of liquid particulates during measurements made with other devices, such as cascade impactors.

4.4.2. Nonimaging Optical Methods

Most of the methods to be discussed in this section are based on the scattering and/or absorption of electromagnetic waves by particles; thus a brief review of these phenomena precedes the descriptions of specific optical techniques. Applications of these to monitoring are discussed in Chapter 2.

4.4.2.1. Light Extinction by Spherical Particles

If the primary measurement in a light scattering experiment is the fraction of light removed from an initial beam that traverses an aerosol stream, we think in terms of transmittance, opacity, or extinction. Here we are not concerned with the nature of the scattered light (i.e., angular dependence of polarization or intensity); in fact, care must be taken to avoid the possibility of scattered light confusing the results. For this type of measurement, the Beer–Lambert law applies:

$$\frac{I}{I_0} = \exp(-Q_{ex} L) \tag{35}$$

where I = intensity of the transmitted beam
I_0 = intensity of the incident beam
L = optical path length
Q_{ex} = extinction coefficient

The extinction coefficient Q_{ex} is a measure of the effectiveness of the aerosol in decreasing the amplitude of the radiation, per unit path length, and is the sum of two parts. Mathematically, $Q_{ex} = Q_{ab} + Q_{sc}$, where Q_{ab} is due to absorption and Q_{sc} is due to scattering; Q_{ex} can also be expressed in terms of the projected cross-sectional area of the particles. When this is done, the Beer–Lambert law becomes

$$\frac{I}{I_0} = \exp\left(-\sum_i n_i a_i E_i L \right) \tag{36}$$

where n_i = number of particles in ith size band
$a_i = \pi d^2/4$, the cross-sectional area of particle in the ith size band
E_i = scattering efficiency or particle extinction coefficient of particles in the ith size band

Figure 4.49 shows E as a function of the dimensionless parameter α, where $\alpha = \pi D_p/\lambda'$. For convenience in interpretation, a horizontal scale is also given in terms of particle radii for $\lambda' = 0.52\ \mu$m, which is typical of the response in the visible spectrum. Some commonly observed characteristics of light scattering and extinction can be explained in terms of this graph. For example, ultrafine particles and molecules are effective scatterers only at short wavelengths. Therefore a medium containing only fine particles will appear blue when viewed obliquely with respect to an illumination source of white light (scattered light) and red or orange when viewed with backlighting (transmitted light in a dense medium or with a long optical path). On the other hand, when $D_p \gg \lambda'$, the extinction coefficient is insensitive to wavelength and all frequencies of visible light are scattered with equal intensity. In this region, Q_{ex} is directly proportional to the total projected

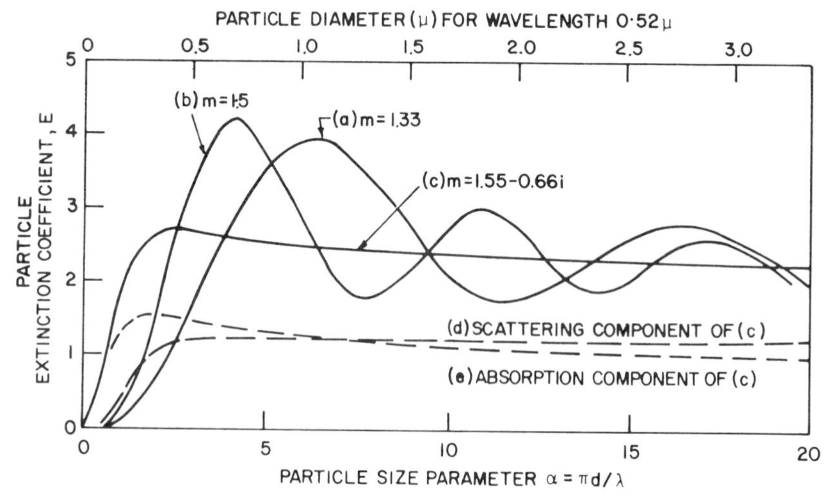

Figure 4.49. Particle extinction coefficient E calculated from Mie theory. After Hodkinson.[53]

cross-sectional area of the dust or particulate. The oscillations in E with respect to D_p can be interpreted as interference between rays that pass through a sphere and rays that do not. For absorbing spheres, polydisperse aerosols, or white light, these oscillations are strongly damped.

4.4.2.2. *Measurement of Optical Transmittance: Opacity*

According to Figure 4.49, for α greater than about 3 or 4, the Beer–Lambert law can be written as follows:

$$\frac{I}{I_0} = \exp\left(-2 \sum_i n_i a_i L \right). \tag{37}$$

The summation $\sum_i n_i a_i$ is equal to A_p, the total projected cross-sectional area of the particulate that intercepts the light beam. Thus a single measurement of opacity, transmittance, or extinction can give an accurate measure of the total projected area of the dust present in the volume illuminated by the light source.

The relationship between opacity and mass has been studied by several workers. The basic problem is to relate the total cross-sectional area of the particulate to the total volume, or mass. This can be done analytically only if the shape of particle size distribution is known. If done empirically, a change in the particle size distribution can change the calibration of the system. Pilat and Ensor[84] have developed a theoretical expression for the relationship between opacity and particulate mass concentration. The particulate mass concentration can be calculated from opacity using the known particle size distribution and index of refraction, or values that are assumed to be typical of the installation being tested. The authors assume a log-normal distribution and show that for an index of refraction equal to 1.33, the relationship between opacity and mass is insensitive to geometric mass mean radius over a range from about 0.1 to 1.5 μm for very polydisperse distributions ($\sigma_g > 4$). This indicates that opacity might provide a good measure of the particulate emitted under some conditions or from special sources. These criteria, however, are too stringent to apply in general.

4.4.2.3. *Measurement of Optical Transmittance: Multiple Discrete Wavelengths*

The preceding paragraphs illustrated how a single measurement of optical transmittance can be used to give an accurate estimate of the total projected area of the illuminated aerosol. This type of measurement yields the product

$N\bar{A}$ (or A_p), where N is the total number of particles and $\bar{A}$ is the mean projected area. If either N or $\bar{A}$ can be determined by some other method, the optical transmittance data can be converted to average particle size or particulate mass. Sem et al.[85] have reviewed attempts prior to 1971 to accomplish this by means of multiwavelength transmittance. These authors expressed reservations regarding the applicability of multiwavelength transmissometers because of the difficulty of accomplishing data reduction and the technical problems associated with extending the upper size limit beyond 2 μm. Grassl,[86] however, has developed and tested a transmissometer that employs wavelengths up to 10 μm. His calculations require a computer and consist of assuming a particle size distribution and computing extinction coefficients as functions of wavelength. Comparison of these calculated values with experiment, and repeated iteration, eventually lead to a particle size distribution for which the calculated extinction coefficients agree with those which are measured at seven wavelengths.

A more basic but also more complicated technique was applied by Yamamoto and Tanaka[87] to experimental data. The technique consists of solving simultaneously several differential equations of the form

$$Q_{\text{ex}}(\lambda') = \int_0^\infty \pi r^2 E(\alpha, m) n(r)\, dr \tag{38}$$

for a number of wavelengths; Q_{ex} is measured experimentally, and the $E(\alpha, m)$ are calculated from the Mie theory. This technique, if practical, would yield more detailed size distribution.

Gunther[88] has proposed a hybrid system, where $N\bar{A}$ is measured from transmittance and $\bar{A}$ is determined from the angle at which the first minimum occurs in the scattered light intensity pattern.

A rather simple method for measuring the mean volume-surface diameter of a polydisperse aerosol is the first maximum method. Dobbins and Jizmagian[89] have shown that a curve similar to c in Figure 4.49 exists for polydisperse aerosols. By varying the wavelength, a value of α can be found for which the extinction coefficient is a maximum. This value of α is then used to calculate the mean volume-surface diameter.

Dobbins and Jizmagian[90] also demonstrated a different technique whereby the ratio of turbidities (I_0/I) measured at two well separated wavelengths was used to measure D_{vs}, and thus the mass of several laboratory aerosols. Good agreement was found although a very limited range of D_{vs} was used.

This technique may be useful for industrial applications, although only an average size is obtained, because the volume-surface diameter can be related directly to the mass concentration. The ratio of turbidities method is outlined below.

It can be shown that it is possible to apply the Beer–Lambert law to polydisperse aerosols if the following relations are used:

$$\frac{I}{I_0} = \exp\left[-\frac{3}{2}\left(\frac{\bar{E}}{D_{vs}}\right)C_v L\right] \tag{39}$$

where $\bar{E}$ = mean scattering coefficient
D_{vs} = mean volume-surface diameter $(\sum_i n_i D_i^3 / \sum_i n_i D_i^2)$
C_v = volume concentration, or volume of particulate per unit volume of aerosol
L = path length traversed by the beam

It follows that if $\bar{E}$ and D_{vs} can be determined from extinction measurements, C_v can be calculated.

The suggested general procedure is to measure I/I_0, the transmittance, at two well separated wavelengths. If this is done, and the logarithm of the inverse transmittance (optical density) values is taken, one finds

$$\ln\left(\frac{I_0}{I}\right)_1 = \frac{3}{2}\left(\frac{\bar{E}_1}{D_{vs}}\right)C_v L \quad \text{and} \quad \ln\left(\frac{I_0}{I}\right)_2 = \frac{3}{2}\left(\frac{\bar{E}_2}{D_{vs}}\right)C_v L$$

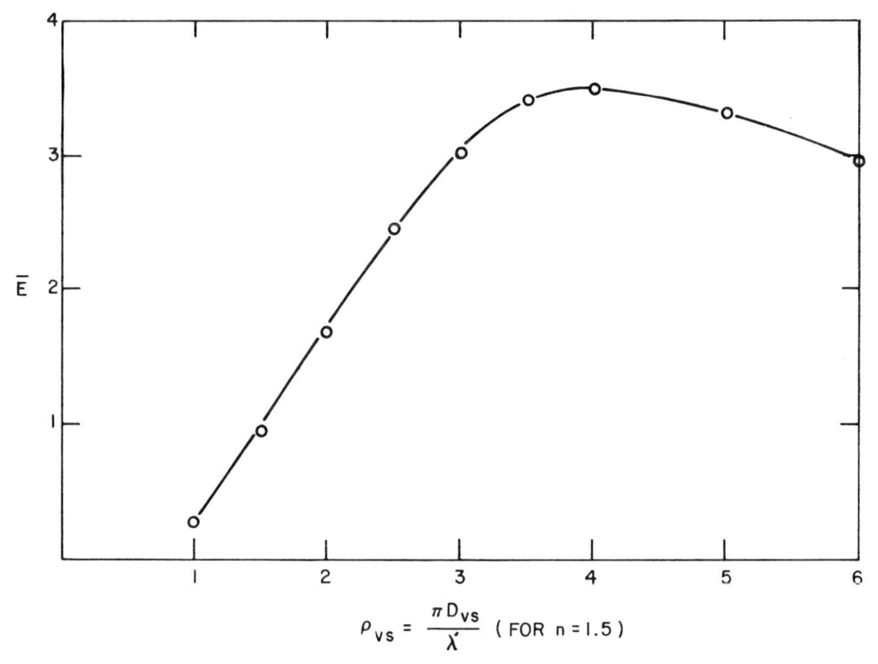

Figure 4.50 Mean extinction coefficient as a function of the phase shift parameter $\rho_{vs}[= 2(n - 1)\pi D_{vs}/\lambda']$.

where the subscripts 1 and 2 refer to the wavelengths used. In this argument $\lambda_2' > \lambda_1'$. The ratio of these values is the quantity of interest.

$$\frac{\ln(I_0/I)_2}{\ln(I_0/I)_1} = \frac{\bar{E}_2}{\bar{E}_1} \tag{40}$$

The measurement of two values for $\bar{E}$ is required to remove ambiguities that exist in a single measurement because $\bar{E}$ is a double-valued function of D_{vs} for some regions in the size range of interest. Figure 4.50 plots $\bar{E}$ versus the phase shift parameter ρ_{vs} for a particle index of refraction of 1.5. The wavelengths to be used in the measurement scheme are theoretically dictated by the size range of interest and the requirement that one of the $\bar{E}$ values always be located on the initial positive slope, as in Figure 4.50.

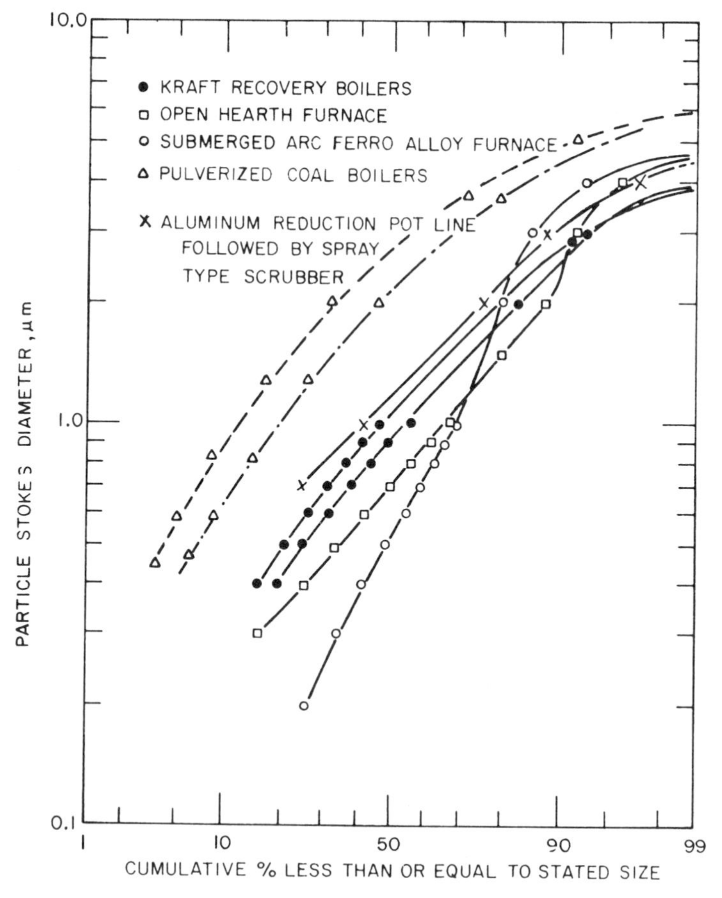

Figure 4.51 Particle size distributions from typical pollution sources (on a mass basis).

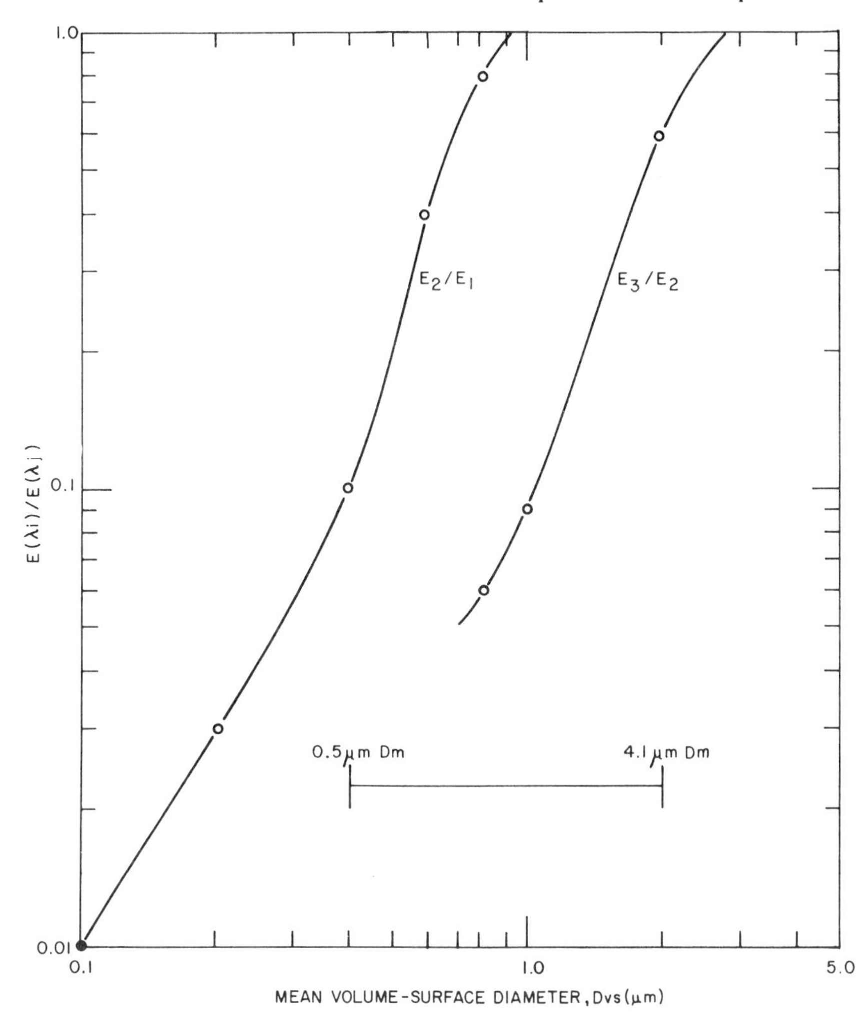

Figure 4.52. Ratio of the extinction coefficients measured at two wavelengths plotted versus the particulate mean volume-surface diameter: $\lambda_3' = 3.39$ μm, $\lambda_2' = 1.06$ μm, and $\lambda_1' = 0.4$ μm.

Experimentally, the wavelengths must be chosen to avoid interference due to molecular absorption by gases and water vapor. In addition, care must be taken to avoid the Fraunhofer absorption lines that occur in the visible spectrum. For purposes of discussion, the values of λ' are chosen to be 0.4, 1.06, and 3.39 μm. These wavelengths are available in commercial lasers.

Figure 4.51 gives particle size distributions for emissions from a number of pollution sources. These curves reveal that the system must be responsive to particles having diameters between <1 and about 10 μm. For some sources, the mass concentration changes by an order of magnitude in less than 5 min. The response time of the measurement system should be short, compared to these source fluctuations.

Figure 4.51 also indicates that mass mean diameters from about 0.5 to about 4 or 5 μm are typical of mass emissions from various sources. Although there is no simple relationship between the mass mean diameter D_m and D_{vs}, in the cases shown, the range of D_m corresponds to a range in D_{vs} from about 0.4 to approximately 2.1 μm.

Figure 4.52 demonstrates that no combination of two of the three wavelengths can cover the entire size range of interest. In fact, this is true of any combination of two wavelengths. Consequently, for the expected range in particle size distribution, more than two wavelengths must be used. The size range of interest can easily be spanned, with excellent resolution, by using the three wavelengths mentioned previously. In the case of coal-fired boilers (Figure 4.51), wavelengths of 3.39 and 1.06 μm would be appropriate. For the other sources where the mass mean diameter is much smaller, the 1.06 and 0.4 μm combination of wavelengths would be more suitable.

4.4.2.4. Scattered Light Intensity Measurements

Hodkinson[53] suggested a method for minimizing the effects of particle index of refraction in sizing measurements from a study of the Fraunhofer diffraction formulation at small forward scattering angles. The basic concept of this method involves the measurement of the intensity of light scattered by a single particle at two small scattering angles and the calculation of the ratio of these two intensities.

Figure 4.53 plots the scattered light intensity versus scattering angle for two spherical particles with equal diameters. One particle is a glassy non-absorbing sphere with index of refraction equal to 1.55. The other is absorbing, such as carbon, and has an index of refraction of $1.96 - 0.66i$. At small angles, the intensity of the scattered radiation is approximately the same for both spheres, although at large angles there is a difference of orders of magnitude.

Shofner et al.[57] and Gravatt[56] have developed prototype systems for particle sizing based on the intensity ratio concept of Hodkinson. Figure 4.54 is a schematic of Shofner's system, which is called the "PILLS IV." The intensities of the scattered light pulses at the angles θ_1 and θ_2 are normalized to the reference pulse at $\theta = 0$ for synchronization and to account for fluctuations in intensity of the laser source. The optics and sensors are kept clean and cool by the use of a purge air system.

The laser used in this system is a semiconductor junction diode ($\lambda' = 0.9$ μm). The useful size range for particle sizing is from 0.2 to 3.0 μm diameter.

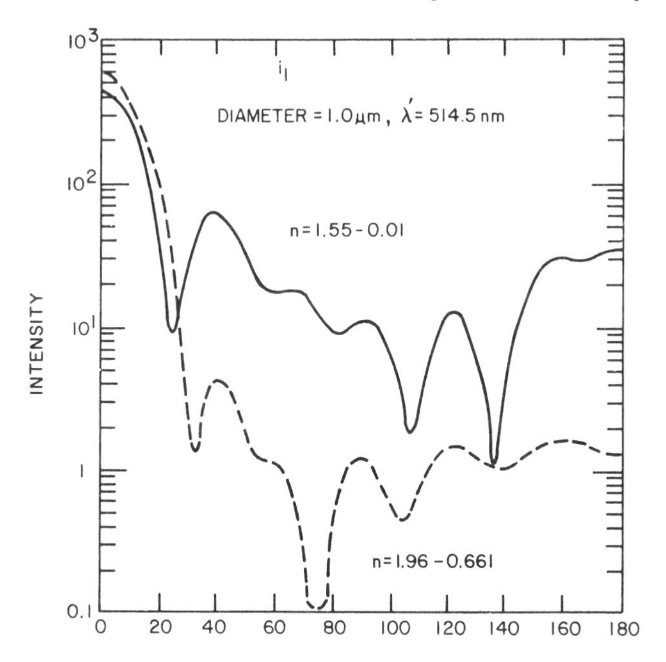

Figure 4.53. Scattered intensity as function of scatttering angle in the plane of polarization. n, index of refraction; λ', wavelength of incident radiation. After Gravatt.[56]

Shofner states that the view volume of his system is approximately 2×10^{-7} cm^3. The upper concentration limit for single particle counters is determined by the requirement that the probability of more than one particle appearing in the view volume of a given time be much less than unity. For Shofner's system this would set the concentration limit at approximately 10^6 particles cm^{-3}, a value much higher than for conventional single particle counters.

Kuykendal[91] has used the PILLS IV prototype system simultaneously with several inertial sizing devices to measure particle sizes. Results from these tests appear in Figure 4.55. It is clear that the PILLS IV does not agree well with impactor data for this source, and further calibration may be required.

An optical particle sizing device developed by R. G. Knollenberg[92] for atmospheric application might be applicable to industrial emission measurements. In some respects the device is similar to a conventional near forward scattering optical single particle counter, except that the sensing zone is contained within an open cavity laser as illustrated in Figure 4.56. This configuration results in very high illumination levels, permitting the detection of particles smaller than those sized by most light scattering instruments,

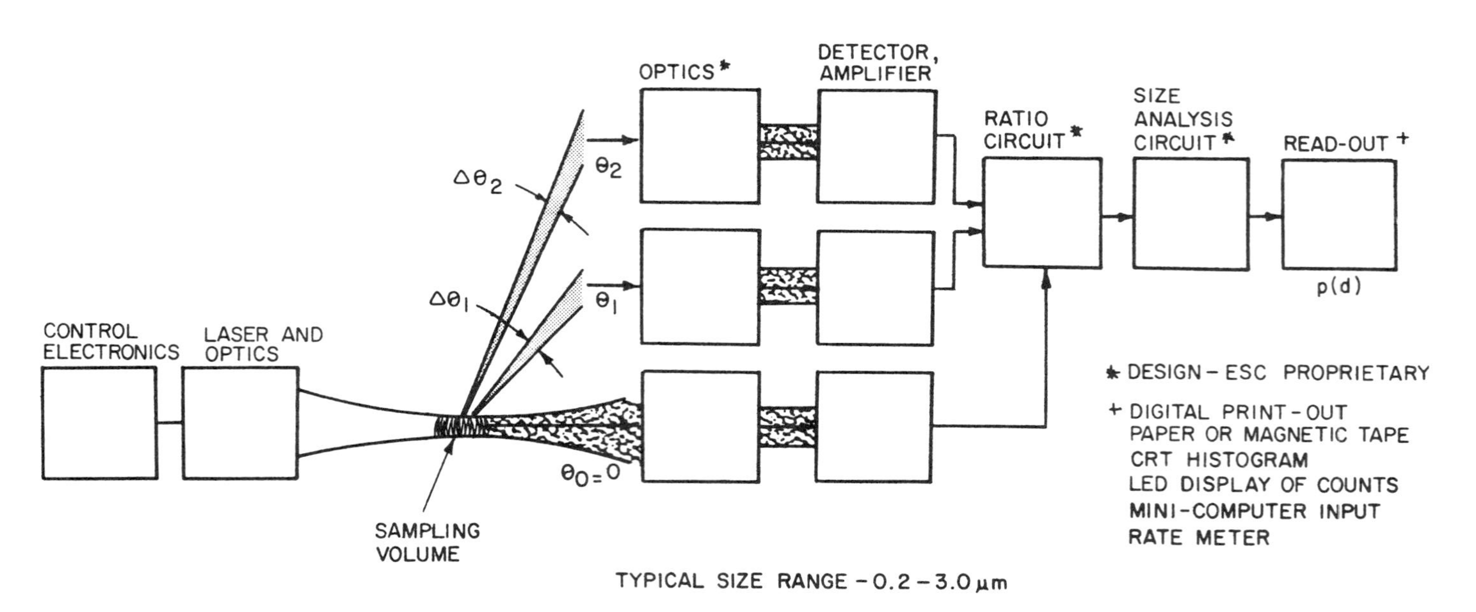

Figure 4.54. The PILLS IV optical particle counter. After Shofner et al.[57]

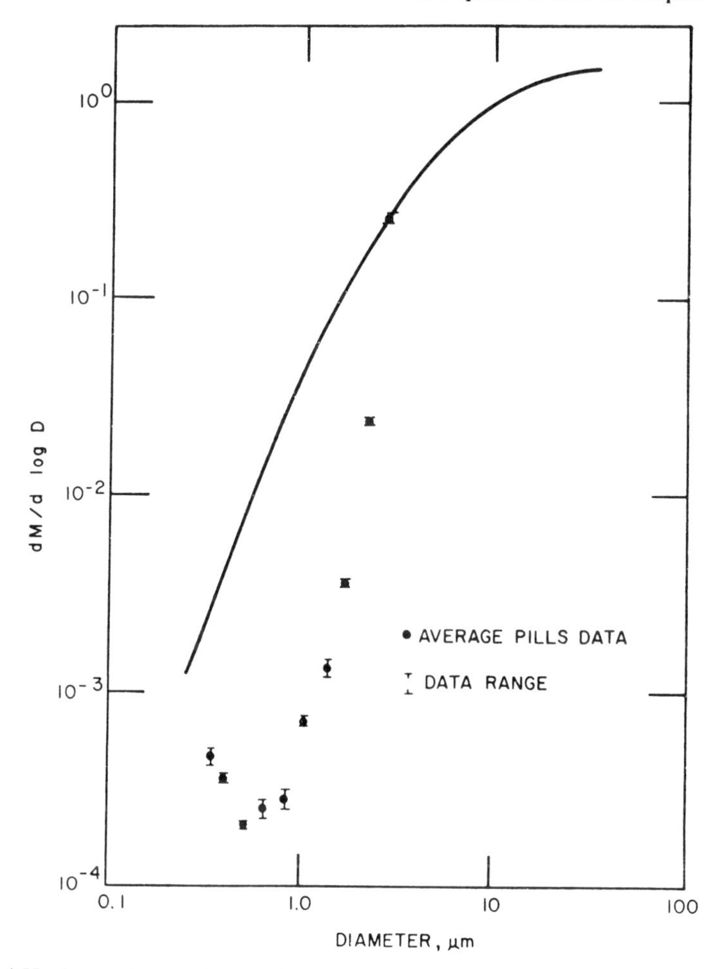

Figure 4.55. Comparison of size distributions measured by impactors and the PILLS IV. After Kuykendal.[91]

and it has been used in vacuum applications for sizing particles in the range from 0.1 to 30 μm. The method also has potential for the measurement of particle velocities (across the beam) through an optical heterodyne effect. The particle imaging and detection system of the device permits the rejection of signals from particles outside the nominal view volume so that the sample gas stream flows in a fairly unrestricted manner through the relatively large open cavity. However substantial modifications in terms of cooling and purge air for the optical components would be required before the device could be used in stack.

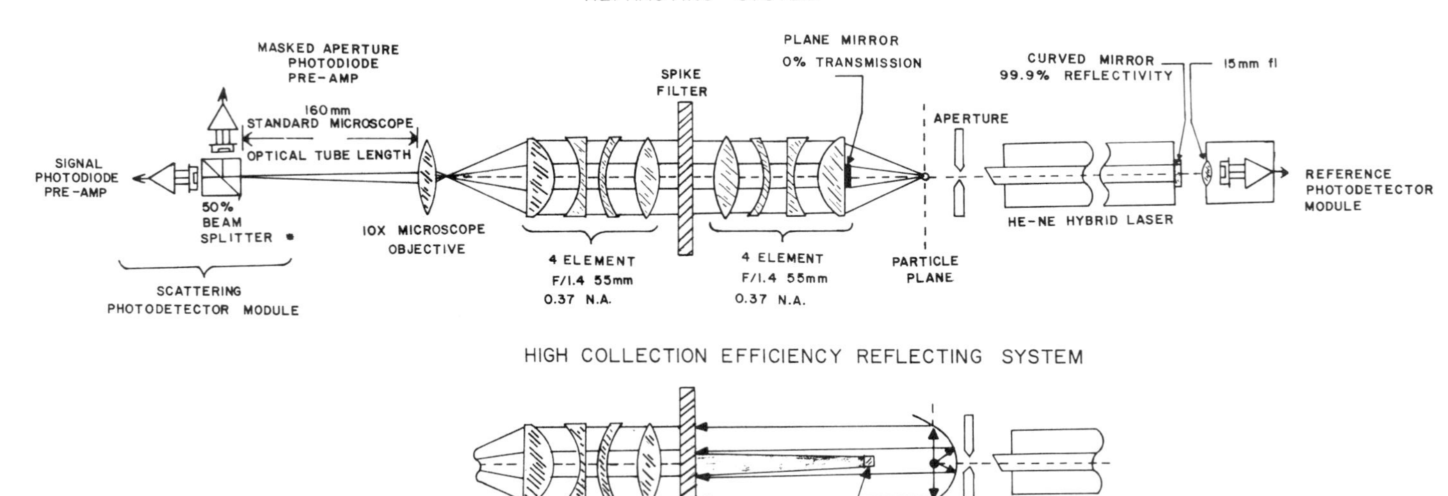

Figure 4.56. Schematic of open cavity laser active scattering particlc counter. After Knollenberg.[92]

Optical Fourier transform systems for obtaining particle size distributions in the 5 to 100 μm diameter range have been described by Cornillault[93] and McSweeny.[94] In this technique a moderately large diameter collimated beam of spatially filtered coherent light is used to produce a diffraction pattern from all particles in a known volume of space. The diffraction pattern (Figure 4.57*a*) is imaged on a detector array having circular symmetry, which permits a determination of the radial distribution of the intensity of light in the superimposed diffraction patterns of the randomly distributed particles in the view volume. Figure 4.57*c* is a conceptual diagram of the system. A numerical inversion process, which can be adequately achieved by matrix multiplication of the intensity data with an inversion matrix, provides the

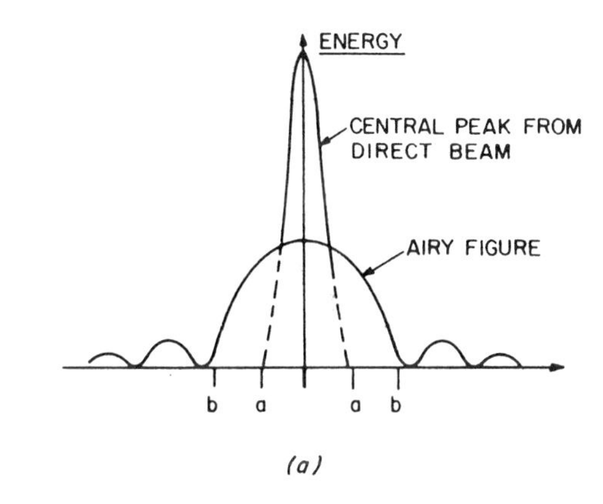

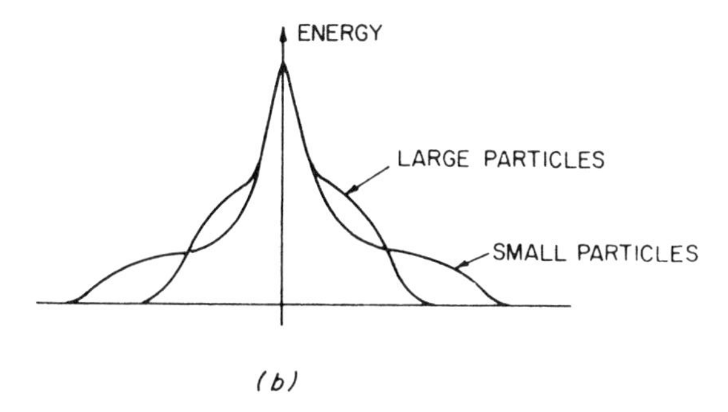

Figure 4.57. (*a*) Diffraction pattern due to identical spherical particles.[93] (*b*) Evolution of diffraction pattern with the size of particles.[93]

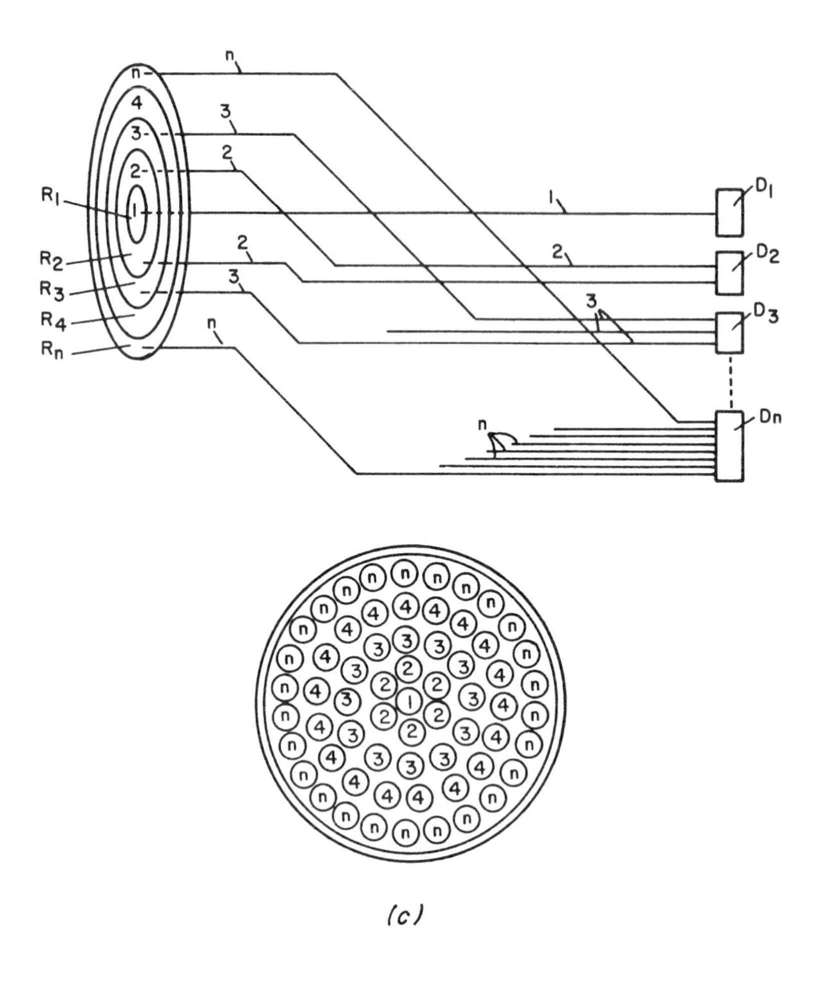

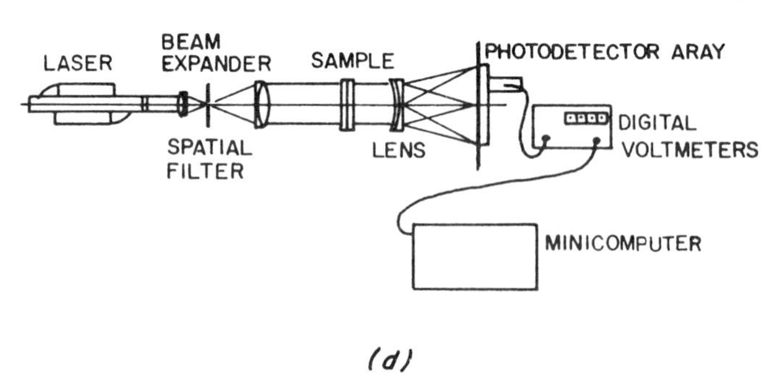

Figure 4.57. *Continued.* (*c*) Diagram of a fiber-optic array.[94] (*d*) Particle size analyzer.[93] By permission from CILAS.

required size distribution. The inversion process can be carried out in real time using a minicomputer. With the proper selection of measurement points in the diffraction pattern, the size interval covered by the technique can be extended to particles both larger and smaller than those in the previously mentioned 5 to 100 μm range. To date the method has been applied only to measurements of particles in liquid suspensions, but current efforts will extend the technique to measurements at atmospheric conditions.[94]

Imaging systems, either of a direct type or of a type using reconstructed images from holograms, have not been widely used for size distribution analysis in flue gases but have served routinely for work with liquid aerosols —particularly to size aerosols produced by spray nozzles of various types. In both direct imaging systems and holographic systems a short light pulse of high intensity illuminates the particles. The pulse durations from available illuminators are short enough to effectively eliminate blur due to particle motion for velocities up to 300 m sec^{-1}.

Flash television particle counters providing real-time size distributions have been described by Hotham[95] using pulsed ultraviolet laser illumination and by Simmons and Lapera[96] using xenon flash tubes for illuminators. The reported range for size distribution determinations for the latter device is 0.3 to 10,000 μm. Figure 4.58 is a conceptual illustration of the system described by Hotham. Image size analysis can be performed instantaneously on a basis of image height, length, perimeter, or projected area. Dynamic

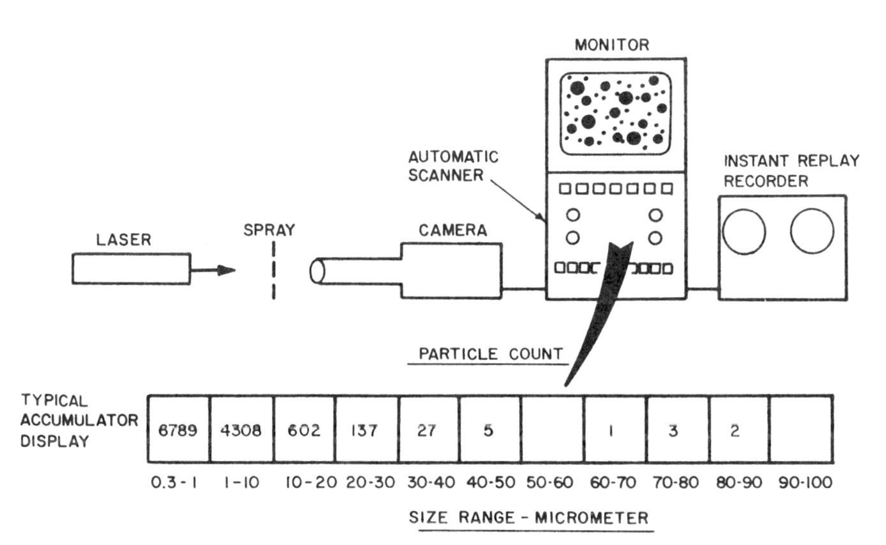

Figure 4.58. Flash television particle counter using pulsed UV laser illuminator. After Hotham.[95]

processes and particle motion can be readily observed and studied by use of a video tape recorder. The view volume in systems of this type is defined electronically in width and height to exclude particles that are only partially within the field of view; focus detection circuits define the depth of the view volume and exclude out-of-focus particles. Because of cost and the practical difficulties involved in the use of such a system in a flue gas environment, such applications of these systems will probably be limited to special research projects.

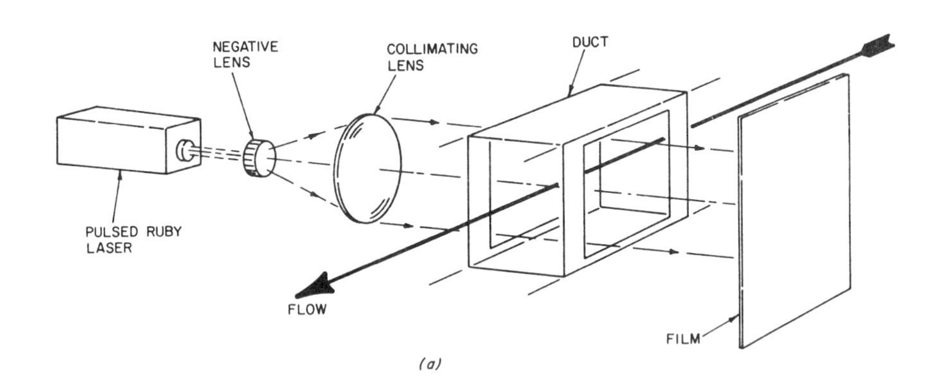

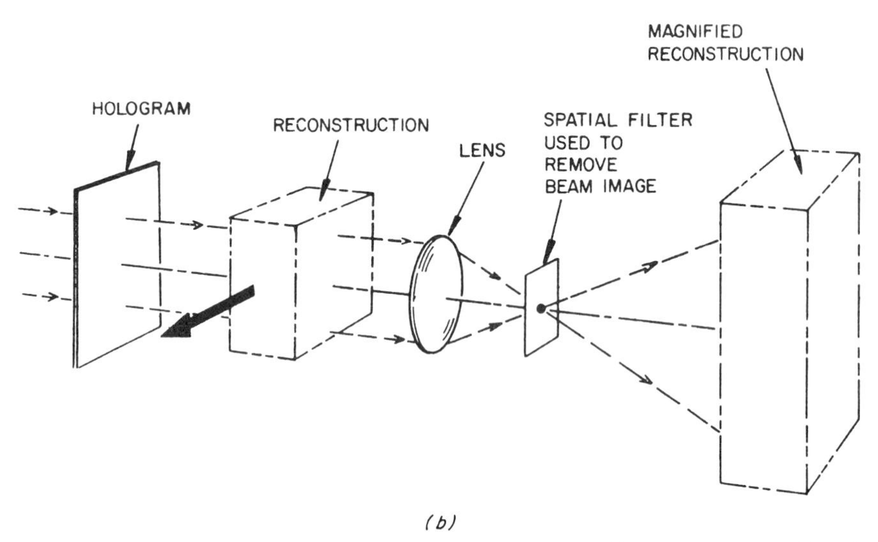

Figure 4.59. (*a*) Single-beam holographic system. After Allen et al.[97] (*b*) System to magnify and spatially filter the reconstructed image. After Allen et al.[97]

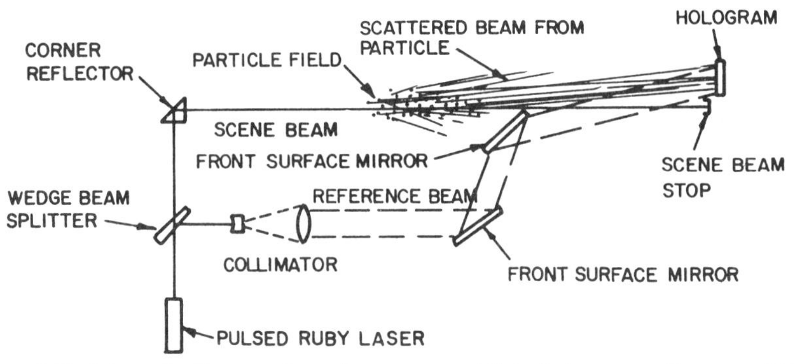

Figure 4.60. Schematic diagram of two-beam scattered light holocamera. Courtesy of B. J. Matthews, TRW, Inc.[98]

Holography as a technique for investigating aerosols has several advantages over most of the methods previously described. The aerosol is not disturbed by the measurement process, a large depth of field is possible and, as in the flash television method, the particles can be effectively "stopped" for examination at speeds up to a few hundred meters per second. Typical system resolution limits, however, result in a lower limit in sensitivity for particle sizing of about 5 μm. By double-pulsing the laser illuminator, one can obtain holograms that permit the determination of particle velocities in three dimensions. Allen et al.[97] have described the use of a single-beam holographic system on a bench scale system for obtaining velocity profiles together with size distributions and particle shapes for particles larger than 5 μm (Figure 4.59). Matthews and Kemp[98] have discussed the use of a two-beam holographic system for determining the spatial distribution of limestone particulate injected into an operating 140 MW pulverized coal fired steam boiler, 24 ft wide. Figures 4.60 and 4.61 represent the two-beam holocamera utilized and the test setup around the boiler. The systems described by both Matthews and Kemp, and Allen et al. utilize low angle forward scattered light from a pulsed ruby laser. By using pulsed ultraviolet laser illumination some gain probably can be achieved in resolving smaller particles. Image Analyzing Computers, Inc., Monsey, New York, offers an automatic analyzer for reading out and analyzing aerosol data from holograms, making it possible to eliminate manual analysis. The analyzer was designed in collaboration with the Meteorological Office and the Chemical Defense Establishment at Porton Down, England. When used with holograms obtained with a pulsed ruby laser, the analyzer provides information on the size, shape, and location of all particles having diameters larger than a few micrometers in a sample volume up to one liter in size.

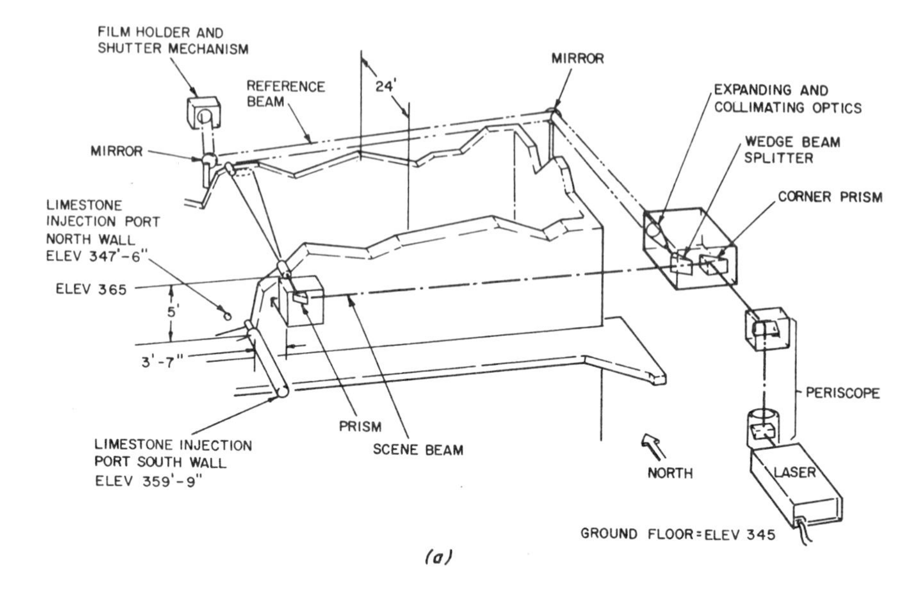

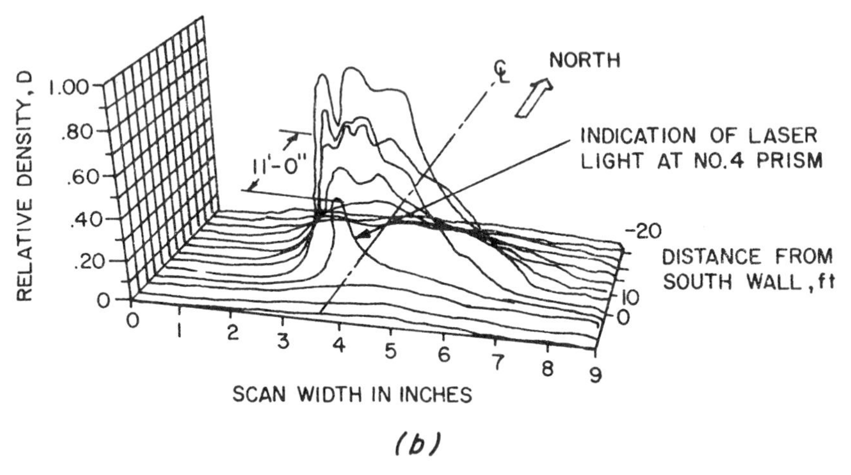

Figure 4.61. (*a*) Perspective view of two-beam holocamera installed on boiler.[98] (*b*) Densitometer traces from a mapping of a hologram into a composite perspective.[98]

An electronic (nonoptical) instrument has been developed by Medecki and Magnus[99] of KLD Associates, Inc., Huntington, New York, for sizing liquid droplets, especially in scrubbers. The instrument operates by inertial deposition of 1 to 600 μm spray droplets on a 5 μm diameter by 1 mm long platinum sensing element of the type used in hot-wire anemometry. Droplets smaller than 1 μm can be measured with a change in sensor geometry. The sensing element is electrically heated to a predetermined temperature. Impinging particles cool the sensing element, resulting in changes in resistance that are related to the sizes of the impinging droplets. The commercially available version of the device provides concentration outputs in six selectable size channels. Size calibrations for the channels are for water droplets; yet in principle, the application of the method is not limited to water. Because the device is essentially a modification of a hot-wire anemometer, it could also be used to measure flow velocity and temperature, permitting impingement rates to be converted to aerosol concentrations.

The foregoing discussion indicates that with further research and development, particle sizing *in situ* can be done. If this capability is developed, aerosol sampling in industrial flue gases can be performed in a more convenient and economical manner than is now possible.

SYMBOLS

a	half-width of rectangular slot
A	empirical constant
a_i	cross-sectional area of particle in ith size band
A_p	total projected cross-sectional area of particles
b	particle mobility
B	empirical constant
B_c	width of cyclone inlet (cm)
C'	cyclone calibration constant
C	slip (Cunningham) correction factor (defined quantitatively in eq. 23)
C	empirical constant (eq. 30)
C_e	true concentration
C_i	measured concentration
C_v	volume concentration
D	Stokes diameter
D_{aer}	aerodynamic diameter
D_g	geometric mean diameter
D_{gm}	geometric mean diameter of mass
D_{gs}	geometric mean surface area diameter

D_{gvs}	geometric mean volume surface area diameter
D_n	number mean diameter
D_j	diameter characteristic of the *j*th interval (eqs. 2 to 7)
D_j	jet diameter (cm) (eq. 21)
D_m	volume or mass mean diameter
D_p	particle diameter (cm)
D_s	surface diameter
D_{vs}	surface–volume mean diameter (Sauter diameter)
$\mathscr{D}$	particle diffusivity
E_i	scattering efficiency in *i*th size band
f_j	relative mass, surface area, or number of particles in interval
F	applied force
g	acceleration of gravity
h	height of rectangular slot (cm)
I	light intensity of transmitted beam
I_0	light intensity of incident beam
j	size interval
k	Boltzmann constant
K	empirical constant
l	length of rectangular slot (cm)
L	optical path length
m	number of rectangular channels
M	mass of particle
M_j	cumulative mass below size D_j
M_t	mass contained between sizes D_t and D_{t-i}
n	outlet concentration
n_0	inlet concentration
n/n_0	penetration
n_i	number of particles in *i*th size band
N	number of intervals
N_c	number of turns made by gas stream in cyclone
N_j	number of holes per stage
P_j	air pressure at jet stage (atm)
P_0	air pressure at impactor inlet (atm)
Q	aerosol flow rate (l min^{-1}) volumetric flow rate ($\text{cm}^3\ \text{sec}^{-1}$) eq. 29.
Q_{ex}	extinction coefficient
r	particle radius
Stk	Stokes No. (dimensionless)
t	time
T	absolute temperature
V_c	inlet air velocity (cm sec^{-1})
V_e	gas velocity

V_s particle settling velocity
V_i aerosol flow rate i ($i = 1, 2, \ldots$)
V_i sampling velocity (eq. 24)
w fraction of mass settling in time t
x $3.77\ hl/aQ$
α relative standard deviation
α' scattering coefficient
λ mean free path of gas molecules
λ' radiation wavelength
μ viscosity of gas
ρ gas density (g cm^{-3})
ρ_p particle density (g cm^{-3})
σ standard deviation
ψ inertial impaction parameter (dimensionless)

REFERENCES

1. Green, H. D., and W. R. Lane, *Particulate Clouds: Dusts, Smokes, and Mists*, 2nd ed., Van Nostrand, New York, 1964.
2. Davies, C. N., *Aerosol Science*, Academic Press, London, 1966.
3. Allen, T., *Particle Size Measurement*, Wiley, New York, 1974.
4. Cadle, R. D., *Particle Size Determination*, Wiley-Interscience, New York, 1955.
5. Irani, R. R. and C. F. Callis, *Particle Size: Measurement, Interpretation, and Application*, Wiley, New York, 1963.
6. Mitchell, R. I. and J. M. Pilcher, *Ind. Eng. Chem.*, 51 (9), 1039 (1959).
7. Berglund, R. N. and B. Y. H. Liu, *Environ. Sci. Technol.*, 7 (2), 147 (1973).
8. Kolmogrov, A. N., *C. R. (Dokl.) Acad. Sci. USSR*, 31 (1941).
9. Drinker, P., *J. Hyg. Toxicol.*, 7, 305 (1925).
10. Austin, J. B., *Ind. Eng. Chem.* (Anal. Ed.), 11, 334 (1939).
11. Hatch, T. and S. Choate, *J. Franklin Inst.* 207, 369 (1933).
12. Epstein, B., *J. Franklin Inst.*, 244, 471 (1947); *J. Appl. Phys.*, 19, 140 (1948); *Ind. Eng. Chem.*, 40, 2289 (1948).
13. Kottler, F., *J. Franklin Inst.*, 250, 339 (1950); 250, 419 (1950).
14. Phelps, G. W. and S. G. Maguire, *J. Am. Ceram. Soc.*, 40, 403 (1957).
15. Menis, O., H. P. House, and C. M. Boyd, ORNL, Report No. 2345, Chemistry—General, Oak Ridge, Tenn., Atomic Energy Commission unclassified report.
16. Ames, D. P., R. R. Irani, and C. F. Callis, *J. Phys. Chem.*, 63, 531 (1959).
17. Ranz, W. E. and J. B. Wong, *Ind. End. Chem.*, 44, 1371 (1952).
18. Mercer, T. T. and H. Y. Chow, *J. Colloid Interface Sci.*, 27, 75 (1968).
19. Mercer, T. T. and R. G. Stafford, *Ann. Occup. Hyg.*, 12, 41 (1969).
20. Yuu, S. and K. Iinoya, *Kagaku Kōgaku*, 34 (4), 427 (1970).
21. Davies, C. N. and M. Aylward, *Proc. Phys. Soc.*, 64, 889 (1951).
22. Marple, V. A., Doctoral thesis, University of Minnesota, 1970.

23. Rao, A. K., Doctoral thesis, University of Minnesota, 1975.
24. Brink, J. A., Jr., *Ind. Eng. Chem.*, 50 (4), 645 (1958).
25. Pilat, M. J., D. S. Ensor, and J. C. Bosch, *Atmos. Environ.*, 4, 671 (1970).
26. Bird, A. N., J. D. McCain, and D. B. Harris, Particulate Sizing Techniques for Control Device Evaluation, Paper No. 73-282, presented at the 66th Annual Meeting of the Air Pollution Control Association, Chicago, 1973.
27. Smith, W. B., K. M. Cushing, and J. D. McCain, *Particulate Sizing Techniques for Control Device Evaluation*, U.S. Environmental Protection Agency Report EPA-650/2-74-102, July 1974.
28. Smith W. B., K. M. Cushing, G. E. Lacey, and J. D. McCain, *Particulate Sizing Techniques for Control Device Evaluation*, EPA-650/2-74-102a, August 1975.
29. Cushing, K. M., J. D. McCain, and W. B. Smith, Final Report on EPA Contract No. 68-02-0273, to be published.
30. Cohen, J. J. and D. N. Montan, *Am. Ind. Hyg. Assoc. J.*, 28 (March–April) 95-104 (1967).
31. Picknett, R. G., *J. Aerosol Sci.*, 3, 185 (1972).
32. Lundgren, D. A., *J. Air Pollut. Control Assoc.*, 16, 225 (1967).
33. Gussman, R. A., A. M. Sacca, and N. M. McMahon, *J. Air Pollut. Control Assoc.*, 23, 778 (1973).
34. Forrest, J., and L. Newman, *Atmos. Environ.*, 7, 671 (1973).
35. Harris, D. B., Tentative Procedures for Particle Sizing in Process Streams—Cascade Impactors. Issued by IERC-EPA, Research Triangle Park, N.C.
36. Fuchs, N. A., *Atmos. Environ.*, 9, 697 (1975).
37. Davies, C. N., *Brit. J. Appl. Phys.*, Ser. A, 1, 921 (1968).
38. McFarland, A. R., and R. B. Husar, *Development of a Multi-Stage Inertial Impactor*, University of Minnesota Particle Technology Laboratory Publication 120, 1967.
39. Parker, G. W., and H. Buchholz, *Size Classification of Sub-Micron Particles by a Low-Pressure Cascade Impactor*, ORNL 4226 UC-80-Reactor Technology, 1968.
40. Pilat, M. J., Submicron Particle Sampling with Cascade Impactor, Paper No. 73-282, presented at the 66th Annual Meeting of the Air Pollution Control Association, Marquette Mich., 1973.
41. Pilat, M. J., G. M. Fioretti, and E. B. Powell, Sizing of 0.02 to 20 Micron Diameter Particles Emitted from Coal-Fired Power Boiler with Cascade Impactors, presented at APCA–PNWIS meeting, 1975.
42. Knapp, K., private communication, 1974.
43. Carpenter, T. E. and D. L. Brenchley, *Am. Ind. Hyg. Assoc. J.*, 33, 503 (1972).
44. Chuan, R. K., *J. Aerosol. Sci.*, 1, 111 (1970).
45. Sem, G. J., J. A. Borgos, J. G. Olin, J. P. Pilney, B. Y. H. Liu, N. Barsic, K. T. Whitby, and F. D. Dorman, *State of the Art: 1971. Instrumentation for Measurement of Particulate Emissions from Combustion Sources*, Vol. IV, *Particulate Mass—Summary Report.* NTIS Publication PB 202668.
46. Chang, H., *Am. Ind. Hygiene Assoc. J.*, 75 (9), 538 (1974).
47. Rusanov, A. A., "Determination of the Basic Properties of Dusts and Gases," in *Ochistka Dymovykh Gazov v Promshlennoi, Energetike*, A. A. Rusanov, I. I. Urbackh, and A. P. Anastasiadi, Energiya, Moscow, 1969, pp. 405–440.
48. Muschelknautz, E., *Staub-Reinhalt. Luft*, 30 (5), 1–12 (1970).

49. Blachman, M. and M. Lippman, *Am. Ind. Hyg. Assoc. J.*, 35, 311 (1974).
50. Leith, D., and D. Mehta, *Atmos. Environ.* 7, 527 (1973).
51. First, M. W., Doctoral thesis, Harvard University, 1950.
52. Perry, J. H. and R. H. Perry, *Engineering Manual*, McGraw-Hill, New York, 1959, pp. 5–61.
53. Hodkinson, J. R., "The Optical Measurement of Aerosols," in *Aerosol Science*, C. N. Davies, Ed., Academic Press, New York, 1966, pp. 287–357.
54. Berglund, R. N., Doctoral thesis, University of Minnesota, 1972.
55. Whitby, K. T., and B. Y. H. Liu, *J. Colloid Interface Sci.*, 25, 537 (1967).
56. Gravatt, C. C., "Light Scattering Methods for the Characterization of Particulate Matter in Real Time," in *Aerosol Measurements*, NBS Special Publication 412, 1974, p. 21.
57. Shofner, F. M., G. Kreikebaum, H. W. Schmitt, and B. E. Barnhart, "*In Situ*, Continuous Measurement of Particle Size Distribution and Mass Concentration Using Electro-Optical Instrumentation," presented at the Fifth Annual Industrial Air Pollution Control Conference, Knoxville, April 1975.
58. Fuchs, N. A., *The Mechanics of Aerosols*, Macmillan, New York, 1964, pp. 204–212.
59. Sinclair, D., *Am. Ind. Hyg. Assoc. J.*, 33 (11), 729 (1972).
60. Sinclair, D., *Am. Ind. Hyg. Assoc. J.*, 36, 39 (January 1975).
61. Breslin, A. J., S. F. Guggenheim, and A. C. George, *Staub* (English transl.), 31 (8), 1–5 (1971).
62. Twomey, S., *J. Franklin Inst.*, 275, 121 (1963).
63. Sansone, E. B., and D. A. Weyel, *J. Aerosol Sci.*, 2, 413 (1971).
64. Gormley, P. G. and M. Kennedy, *Proc. Roy. Irish Acad.*, 52A (1949).
65. Zebel, G., "Coagulation of Aerosols," in *Aerosol Science*, C. N. Davies, Ed., Academic Press, New York, 1966, p. 33.
66. Knudsen, M. and S. Weber, *Ann. Phys.*, 36, 982 (1911).
67. Millikan, R. A., *Phys. Rev.*, 22, 1 (1923).
68. Cunningham, E., *Proc. Roy. Soc.*, A83, 357 (1910).
69. Whitby, K. T. and B. Y. H. Liu, *J. Colloid Interface Sci.*, 25, 27 (1967).
70. Cadle, R. D., *Particle Size Determination*, Wiley-Interscience, New York, 1955, p. 291.
71. Fuchs, N. A., I. B. Stechkina, and V. I. Starasselskii, *Brit. J. Appl. Phys.*, 13, 280 (1962).
72. Hewitt, G. W., "The Charging of Small Particles for Electrostatic Precipitation," Paper No. 73-283, presented at the AIEE Winter General Meeting, New York, 1957.
73. Liu, B. Y. H., K. T. Whitby, and D. Y. H. Pui, A Portable Electrical Aerosol Analyzer for Size Distribution Measurement of Submicron Aerosols, Paper No. 73-283, presented at the 66th Annual Meeting of the Air Pollution Control Association, 1973.
74. *The Particle Atlas*, W. C. McCrone and J. G. Delly, Eds. Ann Arbor Science Publishers, Ann Arbor, Mich., 1975.
75. Crandall, W. A., "Development of Standards for Determining Properties of Fine Particulate Matter," presented at the ASME Winter Meeting, New York, November 29–December 4, 1964.
76. Goetz, A., H. J. R. Stevenson, and O. Preining, *J. Air. Pollut. Control Assoc.*, 10 (5), 378 (1960).
77. Goetz, A. and T. Kallal, *J. Air Pollut. Control Assoc.*, 12 (10), 479–486 (1962).
78. Gerber, H. E., *Atmos. Environ.*, 5, 1009 (1971).

79. Stöber, W. and H. Flachsbart, *Environ. Sci. Technol.*, 3 (12), 1280 (1969).
80. Kops, J., L. Hermans, and J. F. Van De Vate, *J. Aerosol Sci.*, 5, 379 (1974).
81. Preining, O., M. Abed-Navandi, and A. Berner, "The Cylindrical Aerosol Centrifuge," presented at the Symposium on Fine Particles, Minneapolis, May 28–30, 1975.
82. Odén, S., *Proc. Roy. Soc., Edinburgh*, 36, 219 (1916).
83. Chabay, I., Rapid Measurement of Droplet Size Distributions by Optical Heterodyne Spectroscopy, NBS Special Publication 412, October 1974.
84. Pilat, M. J. and D. S. Ensor, *Atmos. Environ.*, 4, 163–173 (1970).
85. Sem, G. J., *State of the Art: 1971 Instrumentation for Measurement of Particulate Emissions from Combustion Sources*, Vol. II, NTIS PB-202-666, 1971.
86. Grassl, H., *Appl. Opt.*, 10, 11, 2534–2538 (1971).
87. Yamamoto, G. and M. Tanaka, *Appl. Opt.*, 8 (2), 447–454 (1969).
88. Gunther, R., *Staub-Reinhaltung Luft*, 33 (9), 345–352 (1973).
89. Dobbins, R. A. and G. S. Jizmagian, *J. Opt. Soc. Am.*, 56 (10), 1345–1350 (1966).
90. Dobbins, R. A. and G. S. Jizmagian, *J. Opt. Soc. Am.*, 56 (10), 1351–1354 (1966).
91. Kuykendal, W. B. and C. H. Gooding, "New Techniques for Particle Size Measurements," presented at the Workshop on Sampling, Analysis, and Monitoring of Stack Emissions, October 2–3, 1975.
92. Knollenberg, R. G., "Open Cavity Laser 'Active' Scattering Particle Spectrometry from 0.05 to 5 Microns," in *Fine Particles, Aerosol Generation, Measurements, Sampling, and Analysis*, Benjamin Y. H. Liu, Ed., Academic Press, New York, pp. 669–696.
93. Cornillault, J., *Appl., Opt.*, 11 (2), 265–268 (1972).
94. McSweeney, A., "An Optical Transform Technique for Measuring the Size Distribution of Particles in Fluids," in *Aerosol Measurements*, NBS Special Publication 412, 1974.
95. Hotham, G. A., "Sizing Aerosols in Real Time by Pulsing UV Laser Machine," in *Aerosol Measurements*, NBS Publication 412, 1974.
96. Simmons, H. C. and D. J. Lapera, "A High-Speed Spray Analyzer for Gas Turbine Fuel Nozzles," presented at ASME Gas Turbine Conference, Cleveland, March 12, 1969.
97. Allen, J. B., D. M. Meadows, R. F. Tanner, and L. M. Boggs, "Velocity of Particulate in Laminar and Turbulent Gas Flow by Holographic Techniques," Lockheed–Georgia Company Report on Contract EHSD-71-34, 1971.
98. Matthews, B. J. and R. F. Kemp, "Holography of Light Scattered by Particulate in a Large Steam Boiler," presented at the 63rd Annual Meeting of the AIChE, Chicago, November–December, 1970.
99. Medecki, H., and D. E. Magnus, "Liquid Aerosol Detection and Measurement," Paper No. 75-24.1, presented at the APCA Meeting, Boston, June 1975.

5

COLLECTION AND ANALYSIS OF TRACE ELEMENTS IN THE ATMOSPHERE

D. F. S. Natusch and C. F. Bauer

Department of Chemistry,
Colorado State University,
Fort Collins, Colorado

A. Loh

Eli Lilly and Company
Greenfield, Indiana

5.1. INTRODUCTION

Consideration of the periodic table shows that the elements that are most toxic to living organisms also have extremely low availability because of their low abundance or their geochemical stability. Indeed, it can be argued that contemporary life has evolved in an environment that is essentially free from elements such as antimony, arsenic, beryllium, cadmium, lead, mercury, silver, tellurium, and thallium. and that as a consequence, living organisms have developed little or no tolerance to these elements.

During the past 100 years, however, man's burgeoning needs for energy and for raw materials to feed a metal-based technology have mobilized large quantities of normally unavailable trace elements in the environment. This sudden availability of trace elements, coupled with their toxicity to living organisms, constitutes the problem of trace element (primarily trace metal) pollution.

Although trace elements are mobilized mostly in the lithosphere and the hydrosphere, significant amounts also reach the atmosphere. There is now convincing evidence that the concentrations of many elements are substantially elevated over background levels both locally[1–6] and in the global atmosphere.[7–12] At present this enrichment is of concern mainly from a toxicological standpoint;[13,14] however it is also possible that particulate trace elements participate in chemical reactions with gases.[15–18]

To date, determinations of trace elements in the atmosphere have been concerned for the most part with local regions of high population density and with emissions from specific sources. The most common type of measurement is of volume concentrations ($\mu g\ m^{-3}$) of individual elemental mass, often as a function of particle size. Such measurements are useful for establishing source emission rates and the amounts of trace elements that can be inhaled or are available for interaction with other atmospheric constituents. Consequently, emphasis is placed on these measurements in discussing source characteristics, sample collection, and elemental analysis in the following sections.

Determination of volume concentrations is undoubtedly the most common and easiest type of measurement to make, but it does not provide much insight into the physicochemical characteristics of airborne trace elements or into their potential environmental behavior. To obtain such information, it is often desirable to establish specific concentrations (micrograms per gram) of particulate trace elements, variations of specific concentrations with particle size, the actual chemical compounds in which a trace element exists, and the manner in which an element is distributed within an airborne particle or collection of particles. Extension of information in these areas, it would appear, will constitute the most significant advance in our future understanding of the environmental impact of trace elements in the atmosphere. The extent to which such information is currently available and the means whereby it can be obtained are discussed in the following sections.

5.2. SOURCE CHARACTERISTICS

Most metallic elements are emitted to the atmosphere in association with solid particles, although certain volatile species are emitted as vapors. In general, essentially all anthropogenic sources of particles are also significant sources of trace elements that are usually present in much higher specific concentrations (micrograms per gram) than in natural crustal dusts. Thus although anthropogenic particles make only a small contribution to the total global aerosol mass, as indicated in Table 5.1,[19,20] their trace element contributions are considerable. Some idea of these contributions[21–27] can be obtained by comparing the representative specific concentrations in Table 5.2 with the particle mass emissions listed for each source in Table 5.1.

The data in Table 5.2 indicate that specific concentrations of trace elements in particles from a single source type can vary considerably. Even greater variations are found for elemental volume concentrations that depend on air flow rates and dilution effects in or near the source. The data also imply that specific concentrations (though not necessarily volume concentrations) of trace elements emitted from a single source are independent of particle size. This is not true. Indeed, recent studies have shown that many elements (including arsenic, cadmium, chromium, lead, nickel, selenium, thallium, vanadium, and zinc) exhibit increasing specific concentrations with decreasing size in particles derived from several high temperature combustion sources.[28–33] It has been postulated[28,34] that the observed elemental size dependences are due to volatilization of certain elements during combustion, followed by condensation or adsorption onto the surfaces of coentrained particles as the temperature decreases. Since small particles have greater

Table 5.1. Mass Emissions of Particles from Main Natural and Anthropogenic Sources[20]

	Natural		Anthropogenic	
Source	$\times 10^6$ tons year^{-1}	%[a]	$\times 10^6$ tons year^{-1}	%
Primary particle production				
Coal fly ash			39.7	1.4
Iron and steel industry emissions			9.9	0.34
Nonfossil fuels (wood, mill wastes)			8.8	0.31
Petroleum combustion			2.2	0.08
Incineration			4.4	0.15
Cement manufacture			7.7	0.27
Agricultural emission			11.0	0.38
Miscellaneous			17.6	0.61
Sea salt	1102	38		
Soil dust	220	7.7		
Volcanic particles	4.4	0.15		
Forest fires	3.3	0.12		
Subtotal	1329.7	46.3	101	3.5
Gas-to-particle conversion				
Sulfate from H_2S	225	7.8		
Sulfate from SO_2			162	5.6
Nitrate from NO_x	476	16.6	33	1.1
Ammonium from NH_3	297	10.3		
Organic aerosol from terpenes, hydrocarbons, etc.	220	7.7	30	1.1
Subtotal	1218	42.4	225	7.8
Total	2548	88.7	326	11.3

[a] Percentage of total: 2874 $\times$ 10^6 tons year^{-1}.

specific surface areas than large particles, this volatilization-deposition mechanism leads to higher specific concentrations of deposited trace elements in small particles.

From an environmental standpoint, the observed increase in specific concentration of some trace elements with decreasing particle size is of considerable significance. Thus it is the small, mainly submicrometer size particles that pass most readily through sampling and control devices, have the longest atmospheric lifetimes, and reach the lungs when inhaled.[13,20,32,36] Furthermore, it has been established, in support of the

Table 5.2. Range of Specific Concentrations of Elements in Source Materials[21–27]

Element	Crustal Dust[a]	Sea Salt[a]	Coal-Fired Power Plant Ash[a]	Oil-Fired Power Plant Ash[a]	Municipal Incinerator Ash[a]	Cement Manufacturing Ash[a]	Automotive Exhaust Particulates[a]	Iron and Steel Foundry Emissions[a]
Al	8.6%	4.6–5.5	1–10%	100–5000	1–10%			10–1000
As			10–500	30	10–500			
Ba	538	1.4	100–1000	0.05–1%	100–5000			
Be	2.4		1–10		1–10	1–10		
Br		0.19%			50–2000		7.9%	
Ca	1.25%	1.16%	1–5%	10–1000	1–10%			
Cd			10–100		10–5000	10–1000		
Cl		55%			0.5–20%		6.8%	
Co	17		10–100	90	10–1000			
Cr	98		10–1000	66	100–5000	100–1000		10–100
Cu	48		10–1000	50–2000	100–5000	0.01–1%		10–100
Fe	4.8%	0.5–5	1–50%	1–10%	0.1–10%		0.4%	0.1–10%
Hg	0.37		0.1–1					
K	1.68%	1.1%	0.5–5%	1000				
Mg	1.62%	3.7%	0.1–1%	500–5000	10–1000			100–1000
Mn	550	0.025–0.25	100–1000	1–100	50–5000	100–1000		10–1000
Mo			10–100					
Na	0.9%	30.6%	0.5–5%	0.2–5%	1–10%			
Ni	57		10–1000		10–1000	100–1000		10–1000
Pb	168	0.12–0.14	100–5000	200–2000	1–10%	100–1000	40%	10–1000
Sb			1–100	5	10–1000	100–1000		
Se	15		10–100	5	1–10			
Si	24%	1.4–94	10–50%	0.1–1%				
Sn	11		1–10					
Sr						0.01–1%		
Ti	0.51%		0.1–2%		0.2–2%			
V	100	0.009	50–5000	0.01–20%	10–100	10–100		
Zn	281	0.14–0.40	100–10,000	200–3500	1–10%		0.14%	0.1–1%

[a] Concentrations are micrograms per gram unless noted as percentage.

above-mentioned mechanism, that particles derived from coal combustion, municipal incineration, open hearth furnaces, and automobile exhausts have much higher (~ 10 to 50 times) concentrations of trace elements on the particle surfaces than in their interior. This means that the actual trace elemental concentrations in contact with the external environment (e.g., lung fluids) are much greater than is indicated by conventional bulk analyses.

Although the identities of trace elements emitted from different sources have been established and the general characteristics of these elements are known, there is considerable variation in the actual concentrations and characteristics of trace elements emitted from specific sources. It is appro-

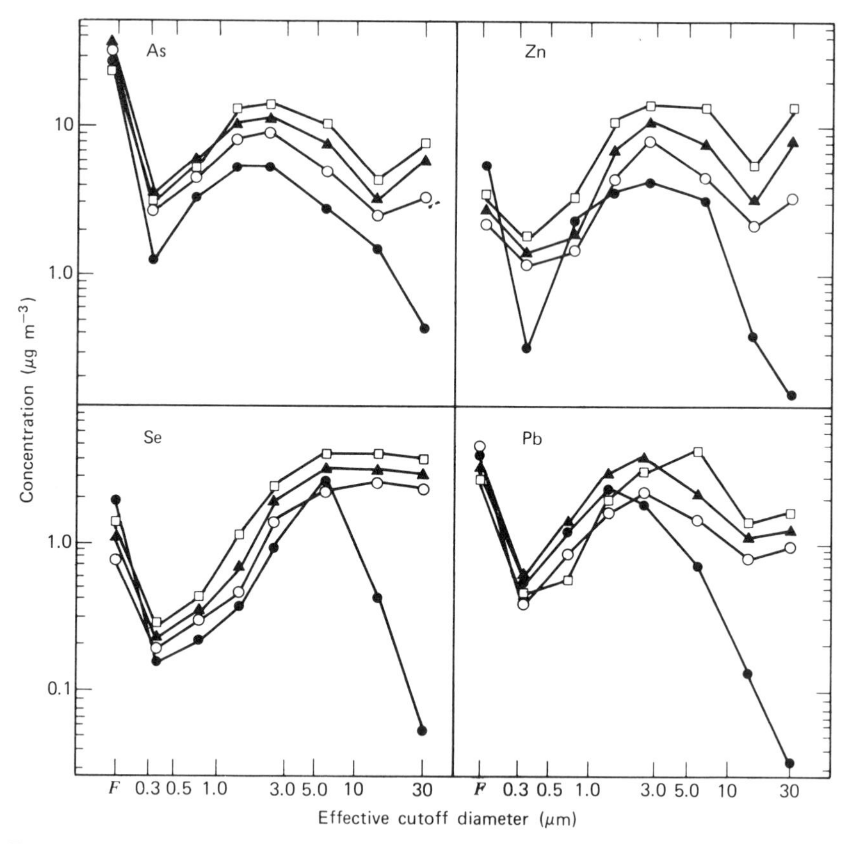

Figure 5.1. Representative size dependences of arsenic, lead, selenium, and zinc in the stack emissions from a coal-fired power plant; *F* refers to the final filter, and the four sets of results indicate the variability between sampling runs.[24]

Table 5.3. Example of the Distributions[a] of Antimony and Potassium with Respect to Physical Size, Density, and Ferromagnetism in a Bulk Sample of Coal Fly Ash[37]

		Density (g cm^{-3})			
		<2.1	2.1–2.5	2.5–2.9	>2.9
Potassium (%)					
Particle size (μm)					
Nonmagnetic	<20	2.69	2.34	2.22	1.73
	20–44	2.63	2.28	1.33	1.09
	44–74	1.89	1.63	1.05	0.45
	> 74	1.79	1.48	1.06	0.13
Magnetic	<20	—[b]	—	0.76	0.70
	20–44	—	1.92	1.48	0.73
	44–74	1.78	1.60	1.27	0.85
	>74	1.37	1.62	1.49	0.83
Antimony (μg g^{-1})					
Particle size (μm)					
Nonmagnetic	<20	24	30	28	19
	20–44	7	11	20	23
	44–74	4	7	17	14
	>74	5	8	15	16
Magnetic	<20	—	—	38	33
	20–44	—	29	38	21
	44–74	11	22	33	—
	>74	13	—	—	—

[a] Determinations by instrumental neutron activation analysis.
[b] No meaningful data.

priate, therefore, to summarize the information presently available for each of the major anthropogenic particulate sources.

In the case of fly ash from coal combustion in electric power generating plants, essentially all the so-called toxic trace elements are represented and some typical size distributions are depicted in Figure 5.1. Measurements of specific concentrations of trace elements in coal fly ash show that they exhibit dependences on physical size, particle density, and ferromagnetic character such as those indicated in Table 5.3.[37,38] These dependences are often obscured when aerodynamic sizing (which depends both on particle density and on physical size) is employed, and size dependences presented in

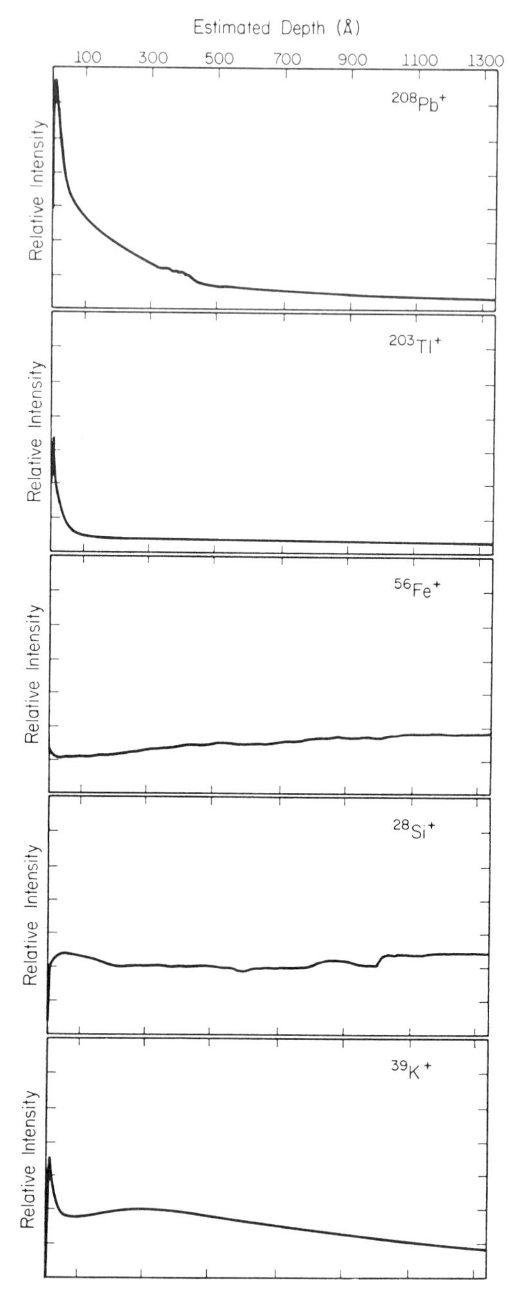

Figure 5.2. Relative concentrations of lead, thallium, iron, silicon, and potassium as a function of depth in individual fly ash particles as determined by ion microprobe mass spectrometry.[36]

terms of specific concentration are different from those presented in terms of volume concentration.

There is now substantial evidence indicating the concentrations of the volatilizable trace elements beryllium, carbon, chromium, lead, lithium, manganese, phosphorus, potassium, sodium, sulfur, thallium, vanadium, and zinc present within 10 to Å of the fly ash particle surface are at least an order of magnitude greater than in the particle interior (Figure 5.2), and it is reasonable to expect the same behavior for arsenic, antimony, cadmium, and selenium.[36] The actual chemical forms of trace elements emitted from coal combustion are reliably known only for bromine, mercury, and selenium, which all occur as the element.[39] It is generally supposed, however, that most trace elements in fly ash are present as oxides or sulfates.

Only a few trace elements are present at significant specific concentrations in fly ash from oil burning (Table 5.2). However this source is noteworthy for the relatively high specific concentrations of vanadium emitted in small particles.[26] Indeed vanadium, whose size distribution is compared with that of calcium in Figure 5.3, is often employed as a characteristic tracer element for oil fly ash.[41–44]

Particles emitted from the high temperature processing of iron and steel have, of course, a high iron content. In addition, however, surprisingly high specific concentrations of a number of other trace elements are present, with the volatilizable elements antimony, arsenic, bromine, chlorine, gallium,

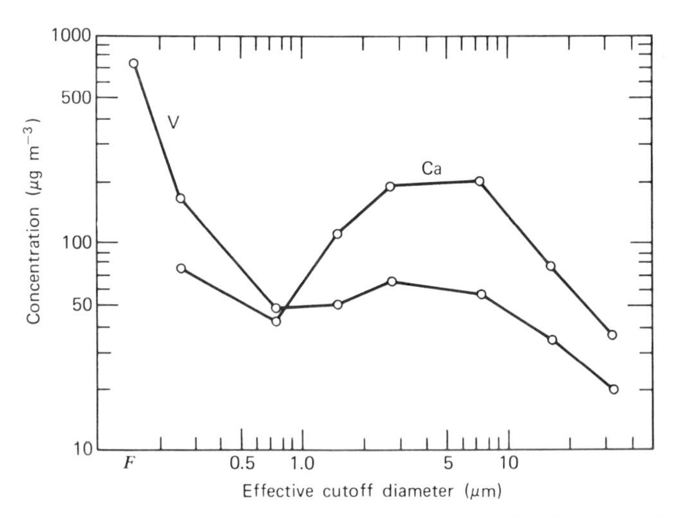

Figure 5.3. Representative size dependences of calcium and vanadium in stack emissions from an oil-fired power plant; *F* refers to the final filter.[24]

selenium, and zinc being most concentrated in particles having aerodynamic diameters less than about 3 μm.[45]

Only a few measurements of trace elements present in stack emissions from nonferrous metal smelters are available. As expected, however, specific concentrations are high.[21,46] Interestingly these data indicate much lower specific concentrations in lead smelter dusts than in those derived from nickel and zinc smelting. This may be a consequence of the lower temperatures employed in lead smelting.

Cement manufacturing is a significant anthropogenic source of atmospheric particles (Table 5.1) and one of the more ubiquitous. Specific concentrations of trace elements in stack particulates are high, and the data suggest that cement manufacture may be an important source of cadmium, antimony, and possibly beryllium.[21] There appears to be no information about the particle size distribution of these elements as emitted, although it is reasonable to suppose that the high temperatures involved may result in volatilizable elements being most concentrated in small particles.

Incineration of solid wastes in centralized municipal and industrial installations contributes a relatively small amount of particulate matter to the atmosphere (Table 5.1), but a large number of trace elements are present at high specific concentrations (Table 5.2). Particular attention is drawn to antimony, cadmium and lead. Incinerator fly ash is noteworthy in a number of respects: the particles are almost completely soluble in water, the volatile trace elements are most concentrated in small particles, and there is little variation in either the specific or volume concentrations over long operating periods (~ 6 months) and between different incinerators.[24,25] It has also been demonstrated that the efficiency of collection of volatile elements such as cadmium and lead by spray scrubbers is very much poorer than for nonvolatile elements[32] and that some elements predominate on the surface of incinerator fly ash particles.[36] This suggests that volatilizable elements may exhibit an inverse dependence of specific concentration on particle size as occurs for coal fly ash.

Studies of trace elements emitted from mobile sources have been primarily directed toward automobiles, and it has been shown that barium, bromine, chlorine, cobalt, lead, and zinc are definitely derived from automobile and truck traffic.[24,46] In addition, there is some evidence to suggest that antimony, aluminum, cadmium, calcium, chromium, and iron may also be derived from this source.[24]

It is now well established[47,48] that bromine, chlorine, and lead are emitted in large amounts from the exhausts of automobiles using leaded gasoline. Approximately 70% of these elements are present as $PbBrCl$ in iron-containing particles whose mass median diameter is about 10 μm.[49,50] These particles are normally deposited on or close to roadways and can be uniquely identified because they contain ferromagnetic Fe_3O_4.[37,49] Approximately 25% of the lead is emitted in very small (~ 0.5 μm) particles that contain PbBrCl, $NH_4Cl \cdot 2PbBrCl$, and $2NH_4Cl \cdot PbBrCl$.[37,47,51,52] In addition, some evidence suggests that approximately 5% of the lead is present as unburned tetraethyl or tetramethyl lead vapors, which pass through particulate sampling devices.[53–55]

We are not certain why particulate bromine, chlorine, and lead from automobile exhausts are bimodally distributed with respect to particle size. It has been suggested,[33] however, that condensation of PbBrCl onto iron-containing nuclei derived either from the intake air or from corrosion of the exhaust system results in rapid particle growth, whereas the higher energy required for self-nucleation[20] produces small particles.

It is apparent from this brief summary that methods for determination of trace elements in source emissions should be capable of quantitatively collecting and analyzing very small particles, of determining both the specific and volume concentrations of a wide variety of elements as a function of particle size, and of determining elements present as gases or vapors. It would also be desirable to determine surface concentrations of trace elements in particles and the chemical forms in which these elements exist.

5.3. ATMOSPHERIC DISTRIBUTION

Measurements of trace elements present in the atmosphere have been reported for a number of localities, most of them urban.[1–12,56–69] It is unwise to draw generalized conclusions from the existing data, which have been obtained using different sampling and analytical methods and are often unique to a particular locality and sampling period; however the following characteristics are recognizable.

1. Individual trace element concentrations vary widely with location and range from a few nanograms (or even picograms) per cubic meter in remote areas to several micrograms per cubic meter in polluted urban areas (Table 5.4).
2. Essentially all trace elements are enriched in urban aerosols when compared with specific concentrations in crustal dusts.[1] There is also substantial evidence indicating that the greatest enrichment factors are exhibited by the most readily volatilizable elements (Table 5.5).
3. Particle size distributions of individual trace elements are fairly constant in urban aerosols that approach stability. In general, volatilizable elements have somewhat smaller mass median diameters than those occurring in a refractory form.

Table 5.4. Representative Volume Concentrations of Elements from Remote, Rural, and Urban Areas

Element	Quillayute[a]	Hawaii[b]	Niles[c]	Northwestern Indiana[d]	New York[e]	Boston[f]
Ag			<1	<0.5–5		
Al		0.5–48	1200	1380–3100		1320–1630
As	<0.05		4.6	2–12		
Br	0.53–6.35		38	26–300		224
Ca	4.7–210		1000	1400–7000	66–184	
Cd					4–12	2
Ce			0.82	1.4–13		3.0–4.1
Cl	760–8500					590–2300
Co		<1	0.95	0.47–2.6		1.3
Cr	<0.22	<1	9.5	6.2–113	5–19	6.8
Cu	1.0	0.2–12	280	25–4000	47–80	
Fe	3.6–14.2	1–45	1900	1420–13,000	1260–1940	1370–1480
Ga			0.9	0.25–3.5		
Hg			1.8	0.8–4.9		
K	18–111		720	730–1860	45–458	
La			0.82	0.9–5.9		1.8
Mg			500	530–2700	17–31	
Mn	<0.18	0.02–0.6	62	63–390	24–35	30
Na		850–14,400	170	160–500	391–1990	1480–2340
Ni	<0.01–0.16	<1			31–328	90
Pb	0.58–0.83	0.3–13			1140–2000	1900
Sb			6.0	2.2–31		7.4–12
Sc			1.2	0.92–3.1		0.25–0.36
Se			2.5	0.8–44		1.4–4.4
Th			0.27	0.17–1.3		0.11–0.28
Ti	0.98–1.3		120	120–280	64–82	40
V		0.04–0.74	5.0	4.0–18.1		860–980
Zn	0.8–5.3	<20	160	100–1550	280–400	260–340

[a] Quillayute AFB, Washington, remote area.[61]
[b] Oahu, Hawaii, remote area.[8]
[c] Niles, Michigan, rural.[3]
[d] Northwestern Indiana, urban.[3]
[e] New York City, urban.[62]
[f] Boston, urban.[24,25]

The most extensive data on particle size distributions of trace elements in urban atmospheres have been reported by Lee et al.[2] These data were obtained in a number of major cities in the United States by the National Air Surveillance Network (NASN) and show quarterly averages in terms of both specific and volume concentrations of a number of trace elements (Figures 5.4 and 5.5). The availability of these types of data is of considerable

Table 5.5. Concentrations and Enrichment Factors of Elements in the Aerosol over Ghent, Belgium[1]

Element	Volume Concentration Range ($ng\ m^{-3}$)	Average Specific Concentration ($\mu g\ g^{-1}$)	Enrichment Factor[a]
Ag	<0.2–10	6.1	580
Al	510–2800	15,000	1.2
As	5–130	178	660
Au	<0.01–0.21	0.57	950
Ba	<30–150	570	8.9
Br	43–1800	2250	6000
Ca	1100–11,000	39,700	7.3
Cd	<15–260	490	16,000
Ce	0.55–14	16	1.8
Cl	1300–15,000	71,600	3700
Co	0.47–12	28	7.4
Cr	3.7–40	143	9.5
Cs	0.22–2.4	7.4	16
Cu	13–215	665	80
F	<70–1000	2940	31
Fe	780–13,000	36,100	4.8
Ga	0.6–13	28	12
Hf	0.04–0.5	1.9	4.2
Hg	0.1–15	14.7	1200
I	3.7–13	87	1200
In	0.07–11	2.8	190
K	430–2300	13,600	3.5
La	0.6–8.3	28	6.2
Mg	<150–1500	5380	1.7
Mn	40–850	1940	14
Na	890–10,000	39,800	9.3
Ni	<10–1400	354	31
Pb	180–3400	9150	4700
Sb	2–470	108	3600
Sc	0.12–1.3	4.6	1.4
Se	0.95–120	43	5700
Si	1300–8700	41,700	1
Sm	0.05–1.0	1.35	1.5
Th	0.06–1.3	2.45	2.3
Ti	<30–290	1170	1.8
V	11–73	465	23
Zn	160–2000	7740	740

[a] Enrichment factor $= \dfrac{[\text{conc}(X)/\text{conc}(St)]\ \text{aerosol}}{\text{conc}(X)/\text{conc}(St)\ \text{crust}}$.

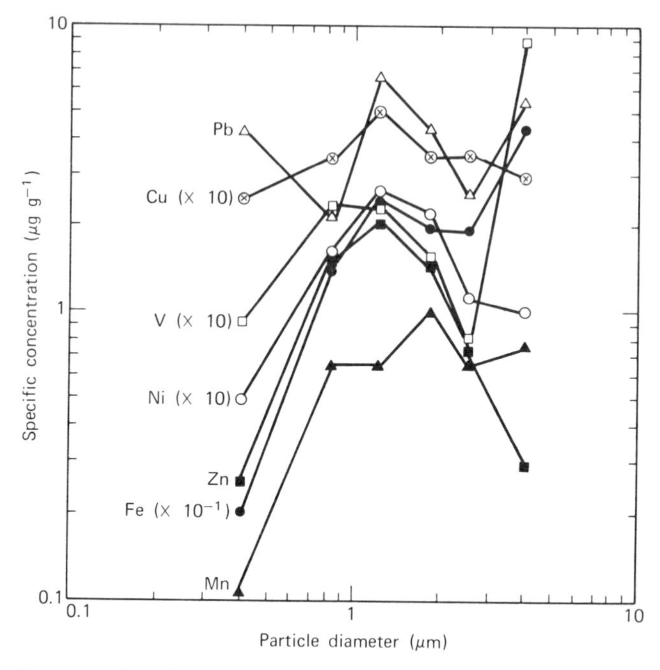

Figure 5.4. Representative size distributions of the *specific* concentrations of copper, iron, manganese, nickel, lead, vanadium, and zinc in a typical urban aerosol.[2]

practical importance, since specific sources of particulate trace elements can be identified by comparing atmospheric and source distribution data for elements that are typical of the source in question.[41–44,63–69]

Current practice in reporting atmospheric size distribution data is to fit them to a tractable mathematical distribution function—most commonly a log-normal distribution. However most cascade impactors do not have sufficient resolution to define particle size distributions precisely, and recent results have indicated[70] that actual atmospheric distributions are far from log-normal. Indeed some elements, notably lead, exhibit a very definite bimodal distribution in urban aerosols. Since the atmospheric residence time and inhalation characteristics of a particle depend on its aerodynamic size,[13] it is important to be able to define elemental size distributions precisely in assessing environmental and toxicological impact.

Very little information is available about the levels of trace metals that exist in the atmosphere as gases or vapors. The most extensive measurements have been made for mercury, which predominates as elemental vapor at ambient levels in the range of 4 to 8 ng m^{-3}. It has been shown, however,

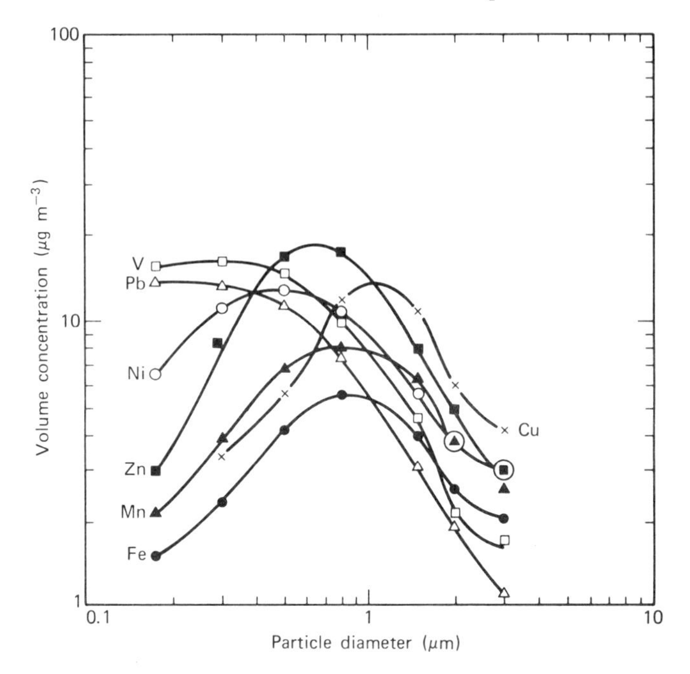

Figure 5.5. Representative size distributions of the *volume* concentrations of copper, iron, manganese, nickel, lead, vanadium, and zinc in a typical urban aerosol.[2]

that significant atmospheric concentrations (0.5 to 1.5 ng m^{-3}) of methyl mercury species exist, especially in the vicinity of sewage plants and other installations where bacterial methylation of inorganic mercury can take place.[71]

The sampling and analytical requirements for determining trace elements present in the atmosphere are essentially similar to those for source emissions. The main difference, and it is an important one, is that trace elemental levels in the atmosphere are generally very much lower than those in or near emitting sources. Collection of a given amount of atmospheric material therefore requires longer sampling times or higher air flow rates than is the case of source material. There are several consequences of this.

1. Very sensitive analytical procedures are generally needed to determine trace elements present in collected atmospheric samples.
2. Determination of elemental volume concentrations is much easier than that of specific concentrations, since the former does not require measurement of the weight of collected material.

3. Determination of realistic atmospheric particle size distributions is frequently difficult because of the need for prolonged sampling periods during which changes can occur in meteorological, thus sampling conditions. In addition, detailed physicochemical characterization of trace elements in urban aerosols is difficult because of the small amounts and considerable heterogeneity of the material available.

5.4. AEROCHEMISTRY

In principle, knowledge of the atmospheric chemistry of trace elements is important in choosing suitable analytical and sampling methodology, since the chemical and physical forms of an element may undergo considerable modification in the atmosphere. An obvious example involves vapor-to-particle transformations by condensation or absorption. In practice, however, very little is known about physico-chemical transformations of trace elements in the atmosphere.

It has been demonstrated that a number of metals such as iron and manganese can promote the oxidation of sulfur dioxide to sulfate,[15,16,72] and the processes by which particle growth can occur have been established.[20,73–75] However neither of these considerations is likely to influence the choice of sample collection methodology. Probably of greater significance are the apparent photochemical breakdowns of PbBrCl to release bromine (probably as vapor) and of methylmercuric vapor to form elemental mercury.[52,76,77] Photochemical dissociation of tetraethyl and tetramethyl lead is also thought to occur,[77] but the nature of the final products is not known.

5.5. SAMPLE COLLECTION

The great majority of atmospheric samplings for trace elements involve collection of particulate material, since only a few elements are present as vapors. In both cases material must be physically collected, often in substantial quantities, although methods are available for determining trace metal vapors (e.g., Hg) at ambient levels in a flowing airstream without preconcentration.

The amount of particulate material that must be collected is determined by the detection limits and configuration (i.e., single or multielemental) of the subsequent analytical technique and by the need to obtain accurate weights or to separate the sample into components of different size, density, or composition. In this regard attention is again drawn to the distinction between specific concentration measurements, which require an accurate determination of the weight of particulate material collected, and volume concentration measurements, which require only a knowledge of the volume of air from which the particles (or vapors) are removed.

As indicated in the previous sections, particle size is an important parameter in determining the environmental behavior of associated trace elements. It is also an important parameter in determining particulate collection efficiency. The most commonly used measure of airborne particle size is aerodynamic diameter, which is the diameter of a spherical particle of unit density whose aerodynamic behavior in an airstream is the same as that of the particle in question. The relationship between aerodynamic diameter D_{aer} and physical diameter D_r (as determined by Zeiss particle size analysis) is

$$D_{aer}^2 = D_r^2 \frac{C(D_r)\rho_p}{C(D_{aer})\rho_{aer}}$$

where $C(D_{aer})$, $C(D_r)$, and ρ_{aer}, ρ_p are Cunningham slip correction factors and particle densities for the aerodynamic equivalent and real particles, respectively.[20] It will be noted that ρ_{aer} is equal to 1 g cm^{-3} by definition.

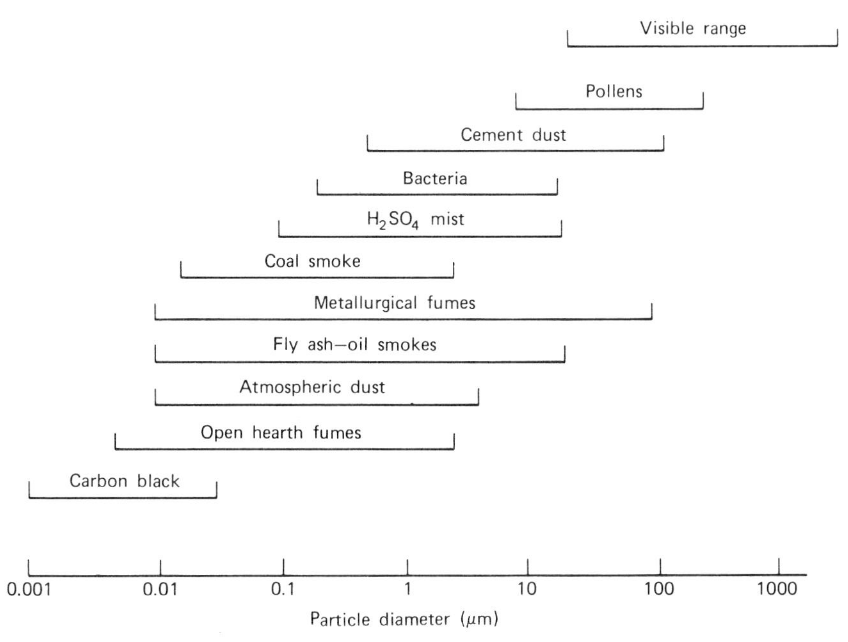

Figure 5.6. Physical particle size ranges for common particle types that contribute to aerosols.

The particle collection characteristics of most common sampling devices are determined by equivalent aerodynamic diameter rather than physical diameter; this is not always the case, however, and a clear distinction between the two quantities should be maintained. The range of physical particle sizes encountered for several atmospheric particulates is indicated in Figure 5.6.

5.5.1. Filtration

The most commonly used particulate sampling methods employ the principle of filtration, whereby particles are quantitatively removed from an airstream flowing through a dense material containing submicrometer size pores.[78] Any material that passes through such a filter is considered, rather arbitrarily, to be nonparticulate.

It is widely believed that a particle will be collected by a filter only if its physical size is too great to permit it to pass through the filter pores (direct interception) or if its inertia is sufficiently great for it to impact on the solid structural material of the filter (inertial deposition). However particle collection is achieved by a combination of five distinct processes, namely, direct interception, inertial deposition, diffusional deposition, electrical attraction, and gravitational attraction. Only in the first of these processes must the particle be larger than the filter pores. The other four all involve migration of a particle, which would otherwise pass through a filter pore, from a stream line to the filter surface under the action of inertial, diffusional, gravitational, or electrostatic forces.

The cooperative effect of these five collection processes gives rise to an important dependence of collection efficiency on particle size in most common filter materials. Thus larger particles are efficiently removed from the airstream by inertial deposition on filter fibers, where they are retained by weak gravitational, electrostatic, and adhesive forces. Similarly, very small particles, whose motion is governed by diffusion, can move rapidly to and be retained efficiently by the filter microstructure. Particles of intermediate size, however, are unable to leave the airstream under the influence of either inertial or diffusional forces; thus they pass through the filter. As a result, essentially all filters exhibit a "window" within which size range particles are inefficiently collected.[54]

This phenomenon has been treated theoretically by several authors[79,83] and can be expressed quantitatively by the equation

$$E = \varepsilon_i + \varepsilon_D + 0.15\,\varepsilon_R - \varepsilon_i\varepsilon_D - 0.15\,\varepsilon_i\varepsilon_R$$

where E is the efficiency of particle collection and ε_i, ε_D, ε_R are partial collection efficiencies describing impaction, diffusion, and direct interception, respectively. The partial efficiency of impaction is given[79] by

$$\varepsilon_i = \frac{2\varepsilon_i'}{1+\xi} - \frac{\varepsilon_i'^2}{(1+\xi)^2}$$

where $\varepsilon_i' = 2\text{Stk}\sqrt{\xi} + 2\text{Stk}^2\xi \exp\left[\frac{-1}{\text{Stk}\sqrt{\xi}}\right] - 2\text{Stk}^2\xi$

$$\xi = \frac{\sqrt{\epsilon}}{1-\sqrt{\epsilon}}$$

$$\text{Stk} = \frac{mu}{6\pi\mu r R_0}$$

$$m = \tfrac{4}{3}\pi r^3 \rho_p$$

In these relations ϵ and R_0 are the filter porosity and initial pore radius, respectively, r is the physical radius of the particle whose density is ρ_p, u is the face velocity of the airstream at the filter, and μ is the viscosity of the air.

The partial efficiency of diffusion is given by

$$\varepsilon_D = 2.56{N_D}^{2/3} - 1.2N_D - 0.0177{N_D}^{4/3}$$

for $N_D < 0.01$ and by a sum of several exponential terms in N_D for $N_D > 0.01$.[79] The quantity N_D is a coefficient of diffusive collection such that

$$N_D = \frac{L\mathscr{D}\epsilon}{{R_0}^2 u}$$

where $\mathscr{D}$ is the particle diffusivity and L is pore length. The other symbols have the same meaning as previously.

The partial efficiency of direct interception is given by

$$\varepsilon_R = \frac{r}{R_0}\left(2 - \frac{r}{R_0}\right)$$

These relations show that particle collection efficiency by filtration is a complicated function of the face velocity of the airstream, the nature and composition of the filter, the size of the particles, and their composition. Consequently, collection efficiencies vary considerably for different filter and particle types. Generalized behavior, however, can be represented as in Figures 5.7 and 5.8, which illustrate the dependence of collection efficiency on particle size as a function of flow rate and pore size, and in Figure 5.9, which shows the interdependence on particle size and flow rate.

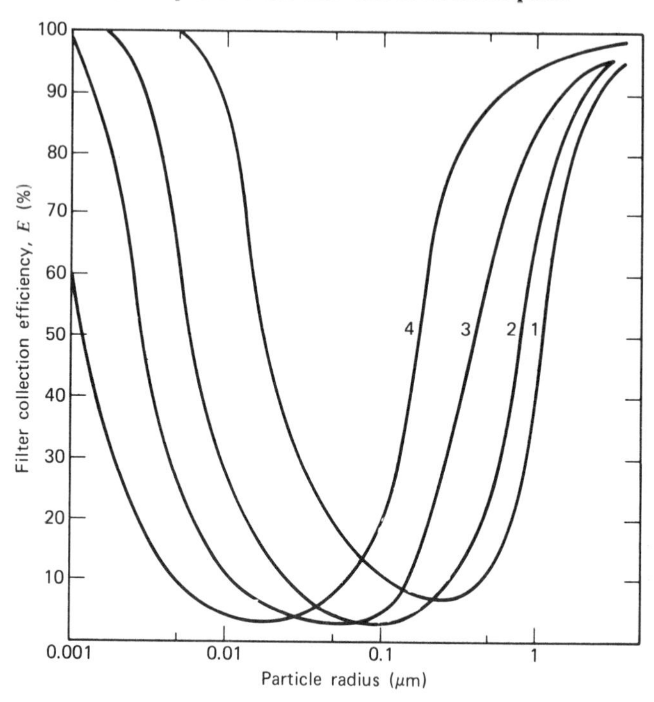

Figure 5.7. Computed dependence[79] of collection efficiency on particle size r and face velocity u for a filter having porosity $\epsilon = 0.05$, pore radius $R_0 = 4.0\ \mu m$, and particle density $\rho_p = 2.1$ g/cm^{-3}; curve 1, $u = 0.1$ cm sec^{-1}; curve 2, $u = 1.0$ cm sec^{-1}; curve 3, $u = 5$ cm sec^{-1}; curve 4, $u = 25$ cm sec^{-1}.

Since significant mass fractions of some trace elements (e.g., lead) are associated with particles whose sizes fall within the range of common filter windows, it is apparent that considerable care should be exercised in assigning quantitative values to collected elemental mass. As indicated in Figures 5.7 and 5.8, however, better quantitation can be achieved by employing high flow rates and filters having small pore sizes. Collection efficiencies are also improved by increasing pore length. In fact the actual process of particle collection somewhat improves filter performance,[79] since pores tend to become clogged as filter loadings increase, thereby decreasing the width of the filter window (Figure 5.8).

Other factors that lead to collection inefficiency in filtration are particle reentrainment and vaporization. Although both these processes have been the subject of considerable speculation, we are not aware of any data describing the extent to which they may occur in practical sampling situations. Both loss processes are promoted by high air flow rates, and it has been

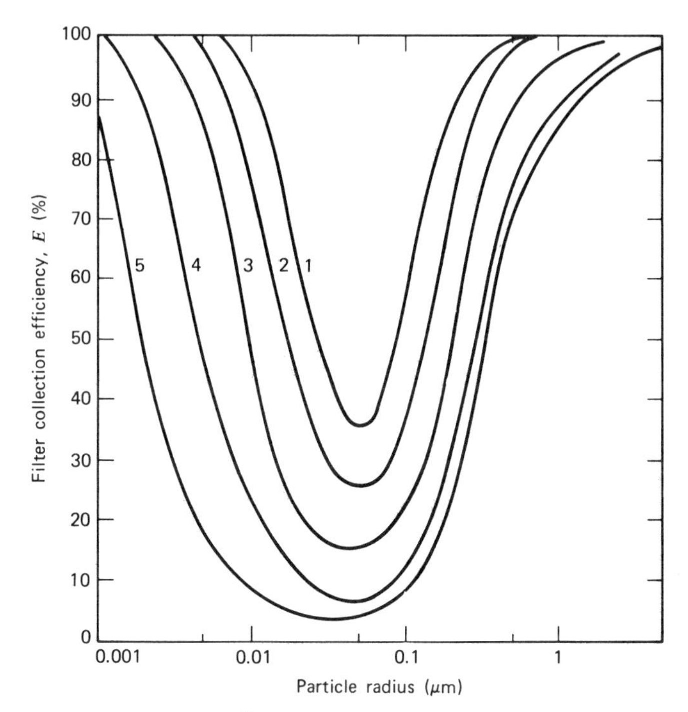

Figure 5.8. Computed dependence[79] of collection efficiency on particle size r and pore size R_0, for a filter operating at a face velocity $u = 7.5$ cm sec^{-1} for collection of particles of density $\rho_p = 2.1$ g cm^{-3}; curve 1, $R_0 = 0.4$ μm; curve 2, $R_0 = 0.5$ μm; curve 3, $R_0 = 1.0$ μm; curve 4, $R_0 = 2.5$ μm; curve 5, $R_0 = 4.0$ μm.

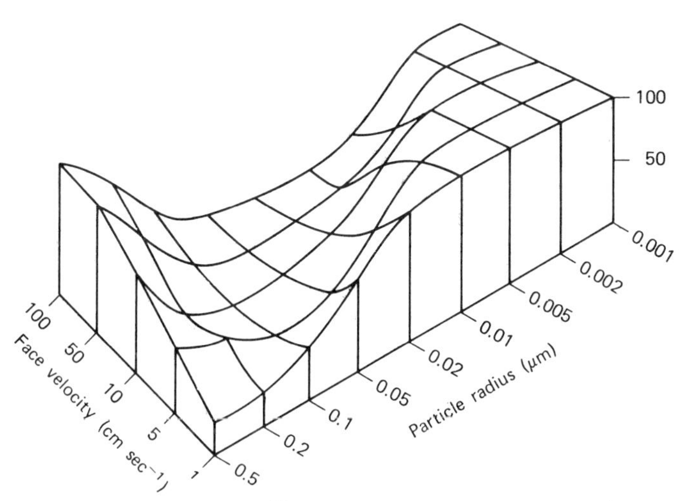

Figure 5.9. Computed interdependence[84] of collection efficiency E, face velocity u, and particle radius r for a filter having pore size $R_0 = 0.4$ μm, with particle density $\rho_p = 2.1$ g cm^{-3}.

Table 5.6. Elemental Impurity Levels in Common Air Filter Materials (ng cm $^{-2}$)

Element	Materials												
	a	*b*	*c*	*d*	*e*	*f*	*g*	*h*	*i*	*j*	*k*	*l*	*m*
Ag	< 2	2	3	< 4		< 3	< 1						
Al	20	12	200	20	10	15	10	60	740				
Ba	< 500	< 100	< 100	< 100	< 100	< 100	< 100						
Br	1000	5	20	4	3	< 5	< 2	6	4				
Ca	300	140	3800	670	250	500	370	570	1250	30	1.1		
Ce	< 1	< 0.5	< 0.3	< 0.5	< 1	< 0.5	< 0.3						
Cl	27,000	100	300	1000	1000	1700	1000	1800	600	1.1	669		
Co	0.2	0.1	0.8	0.2	< 1	0.4	0.1						
Cr	2	3	12	15	14	20	15					80	60
Cu	320	< 4	90	20	40	85	60	25	30	1.2	0.8	20	20
Fe	85	40	300	40	< 300	80	40			1	0.3	4000	300
Hg	1	0.5	3	< 0.4	< 1	< 1	0.5						
K	8	15	200	130	100	120	100		11				
La	< 0.1	< 0.2	< 0.3	< 0.1	< 0.2	< 0.5	< 0.2						
Mg	< 1500	< 80	2400	< 300	< 200	400	200						
Mn	2	0.5	80	7	2	2.5	2	6	2			400	30
Na	90	150	700	600	330	520	400	1800	2200				
Ni	< 25	< 10	60	< 8	< 50	14	< 20			0.2	0.3	< 80	100

Pb										0.6	0.6	800	200
S	±30,000		6000			4800				1	13		
Sb	1	0.15	0.5	0.5	3	0.4	1					30	
Sc	<0.01	<0.05	0.2	<0.05	<0.01	<0.01	<0.05						
Si										34	61	7×10^6	13,000
Ti	70	10	50	15	5	10	<10					800	200
V	<0.6	<0.03	0.5	<0.06	0.09	<0.2	<0.05	<0.05	<0.05	0.2	0.05	30	
Zn	515	<25	30	25	20	10	7			30	0.2	1.6×10^5	10

[a] Delbag, polystyrene.[87]
[b] Whatman No. 41, cellulose-organic binder.[87]
[c] C. H. Dexter & Sons, cellulose.[87]
[d] Millipore, cellulose ester, 0.45 μm pore size, 25 mm diameter.[87]
[e] Millipore, cellulose ester, 0.45 μm pore size, 47 mm diameter.[87]
[f] Millipore, cellulose ester, 0.8 μm pore size, 25 mm diameter.[87]
[g] Millipore, cellulose ester, 0.8 μm pore size, 47 mm diameter.[87]
[h] Millipore "Celotate," cellulose acetate, 0.5 μm pore size, 47 mm diameter.[87]
[i] Gelman "Metricel," cellulose triacetate, 0.45 μm pore size, 47 mm diameter.[87]
[j] Pallflex, E70.[87]
[k] Acrapor, AN800.[87]
[l] Glass fiber.[86]
[m] Silver membrane.[86]

pointed out[13,29] that species such as Hg, SeO_2, and As_4O_6 have sufficient vapor pressures at ambient temperatures for substantial losses to occur.

Several types of filter material have been used and are currently commercially available for collecting atmospheric particulates. They can be classified into five categories in terms of the substrate material, namely, glass fiber, paper, organic membranes, metal membranes, and graphite. The advantages and disadvantages of each are discussed in terms of their physical properties, their hygroscopicity, their inertness to extracting solvents, their intrinsic trace element content, and their relative utility for specific applications.

Probably the simplest and cheapest filter is a dense cellulose paper such as Whatman No. 41 analytical filter paper. This material consists of closely interwoven cellulose fibers that confer sufficient mechanical stability to enable use of the paper in the form of 20.3 × 25.4 cm sheets for conventional high volume sampling. Paper does, however, have a broad distribution of pore sizes and can vary considerably in density, thus collection inefficiencies are observed for particle sizes below about 0.3 μm when sampling is conducted for short periods.[85] This problem is somewhat overcome by pore

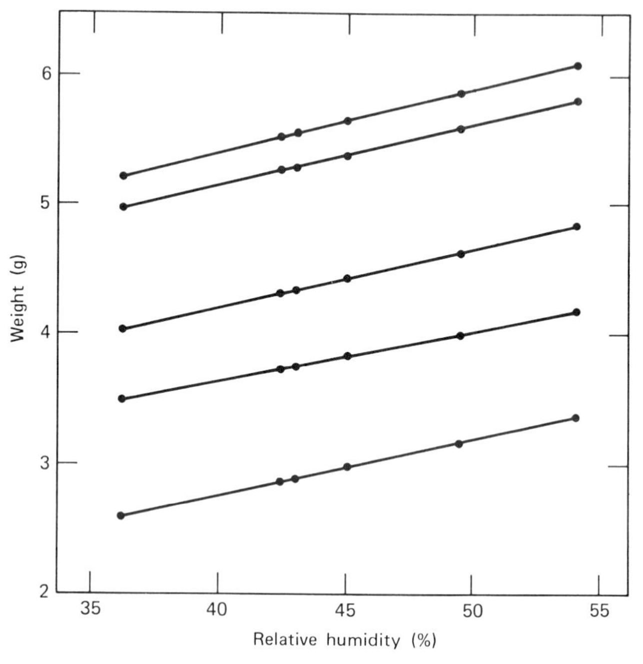

Figure 5.10. Dependence of filter weight on relative humidity for equilibrated Whatman No. 41 filter paper.[85]

clogging effects when 24 hr sampling periods are employed. A significant operational disadvantage of filter paper is its tendency to sorb water; this makes careful equilibration to standard relative humidity necessary if accurate weights of collected material are required. The extent of weight change with relative humidity is depicted for several Whatman No. 41 filters in Figure 5.10. Paper filters are suitably inert to enable the use of wet ashing techniques; and although several elemental impurities have been reported (Table 5.6), all represent relatively major constituents of airborne particles, rendering contributions of these impurities essentially negligible.

Glass fiber filters are the most widely used materials for collection of airborne particles. Despite mechanical properties inferior to those of cellulose paper, they are suitable for use in a high volume sampling configuration. Furthermore, glass fiber is inert to extracting acids (other than hydrofluoric acid) and to organic solvents. For accurate determination of collected sample weights, filters must be equilibrated in an atmosphere at constant relative humidity; however weight changes are closely reversible with humidity. Unfortunately glass fiber filters contain substantial amounts of several trace elemental impurities, and large amounts of sample must be collected to render the filter impurity contribution negligible.[54,86,87] For example, the approximately 500 cm^2 face area of a high volume filter may contain as much as 64 μg of lead, although the actual impurity levels can vary substantially even within a single production batch (Table 5.7).

Glass and quartz fiber filters that have been subjected to high temperatures and washed to remove ashed inorganic impurities are reputed to have much lower impurity levels than conventional filters. In particular, high temperature ashing followed by a phosphoric acid wash removes so-called alkaline sites,[88] which are responsible for conversion of sulfide and sulfite to sulfate.

Table 5.7. Variability of Several Elemental Impurity Levels in Air Filters from the Same Production Batch[54]

Filter Type	Impurity Element	Number of Filters Analyzed	Average μg Impurity Element per Filter	Standard Deviation (μg)
Glass fiber (8 × 10 in)	Pb	7	39	20
Glass fiber (47 mm)	Pb	3	0.8	0.6
Millipore (47 mm)	Pb	4	0.08	0.05
Nucleopore (47 mm)	Al	3	7.8	1.5
	Ca	3	55	4
	Mg	3	3.8	1.5

This is important if determination of individual sulfur species in aerosols is required. Furthermore many commercially available glass fiber filters contain substantial amounts of several organic impurities. These do not interfere with the determination of inorganic species, but they do make it impossible to use some filters for collection of particles that are to be analyzed for both organic and inorganic constituents.

Several forms of organic membrane filter have come into use for particle collection during the last few years. The most common are Millipore, Nucleopore, Pallflex, and Acrapor, which are composed of cellulose acetate or nitrocellulose, and Fluoropor and Millex, which are Teflon filters. With the exception of Millipore filters, which because of high cost have limited use as high volume (20.4 × 25.4 cm) filters, all are employed as 47 mm diameter disks for low volume sampling. In general organic membrane filters have high flow resistance, thus must be used in conjunction with a relatively pressure-insensitive pump. Some face velocity data representative of several different filter types are presented in Table 5.8. Cellulose membrane filters are mechanically fragile and are best suited to analyses that can be performed directly on the filter. They are readily soluble in most common organic solvents for ease of removal of collected particles but tend to disintegrate if subjected to neutron bombardment as required for neutron activation analysis. Millipore filters are slightly hygroscopic[89] and should be equilibrated at constant humidity for precise weight determination; however

Table 5.8. Some Typical Face Velocities and Areas Employed for Several Common Air Filters[87]

Filter Type	Face Area and Dimensions	Face Velocity ($l\ min^{-1}\ cm^{-2}$)
Delbag, polystyrene	400 cm^2 (20 × 25 cm)	4.5
	9.62 cm^2 (47 mm diameter)	6.5
	3.68 cm^2 (25 mm diameter)	13
Whatman No. 41	400 cm^2 (20 × 25 cm)	4.5
	9.62 cm^2 (47 mm diameter)	6.5
	3.68 cm^2 (25 mm diameter)	13
Millipore, 0.45 μm pore size	9.62 cm^2 (47 mm diameter)	2.6
	3.68 cm^2 (25 mm diameter)	4.3
Millipore, 0.8 μm pore size	9.62 cm^2 (47 mm diameter)	4.8
	3.68 cm^2 (25 mm diameter)	7.3
Millipore "Celotate," 0.5 μm pore size	9.62 cm^2 (47 mm diameter)	2.6
Gelman "Metricel," 0.45 μm pore size	9.62 cm^2 (47 mm diameter)	2.6
Glass fiber	400 cm^2 (20 × 25 cm)	1.4

equilibration is rapid being complete within 200 sec for 47 mm diameter filters. Inorganic impurity levels in Millipore filters are generally substantially lower than those in glass fiber filters, although Pallflex and Acrapor exhibit impurity levels that are unacceptably high[54,90] for determination of many collected particulate elements (Table 5.6). Little information is available on the characteristics and impurity levels of Teflon filters, developed only recently, although it should be noted that their inertness to both acids and organic solvents permits multiple use.

Graphite filters, which consist of spectroscopically pure graphite, have been employed in two basic configurations. The first is as a 47 mm diameter disk, and the second is as a partially hollow rod such as is conventionally employed as the lower electrode in dc arc atomic emission spectrometry.[54,91] Both configurations have substantial flow resistance. The main advantages of graphite filters are their very low level of background impurities, their efficient particle collection capabilities, and their capacity to be employed directly as electrodes for dc arc atomic emission spectrometry or, with suitable change in configuration, for spark source mass spectrometry. If required, graphite filters can also be dissolved in perchloric acid.

Table 5.9. Typical Collection Efficiencies of Several Common Air Filters for Specific Elements at 1 cm sec^{-1} Face Velocity[54]

Filter Type	Run Number	Element	Total Collection Contained on the Prefilter (%)
0.8 μm Millipore	1	Al	99.6
		Ca	99.7
		Mg	98.7
		Pb	94.6[a]
0.8 μm Nucleopore	1	Al	99.5
		Ca	99.7
		Mg	98.7
		Pb	94.3[a]
	2	Al	99.8
		Ca	99.8
		Mg	99.5
		Pb	94.4[a]
~ 0.4 μm Glass fiber		Pb	94.2[a]
Paper (Whatman No. 42)		Pb	96.3[a]

[a] Mean value of four replicate samplers run simultaneously under the same conditions.

Metal membranes have also been applied to airborne particulate sampling on a limited basis.[86] The most common metal is silver; apart from their insensitivity to humidity and ease of use as spark source mass spectroscopic electrodes, however, silver membranes have no advantage over other filter types. As noted from Table 5.6, impurity levels can be substantial in commercially available silver membranes.

In practice the choice of a filter material and configuration depend on the specific sampling situation and the analytical technique(s) employed for determination of the collected trace elements. The main considerations should be the level of impurities in the filter compared with levels of each element collected, the collection efficiency for elements of concern, and the ease of manipulation of the filter in preparation for subsequent analysis. Studies have been conducted[54] demonstrating that several representative elements in urban aerosols, notably lead, are collected with efficiencies in

Table 5.10. Efficiency of Collection[a] of Airborne Lead by Individual Units of a Series Sampling Train[54]

Run Number	Distance from Street (m)	Collection Unit	Atmospheric Concentration Indicated ($\mu g\ m^{-3}$)	Filter Efficiency of Total Collected (%)
1	2	0.45 μm Millipore	0.37	33.6
		Graphite cup	0.21	22.9
		HNO_3 scrubber	0.48	43.5
		Total	1.11	
1	12	0.45 μm Millipore	0.28	43.9
		Graphite cup	0.15	23.8
		HNO_3 scrubber	0.21	32.3
		Total	0.64	
2	2	0.45 μm Millipore	0.33	53.5
		Graphite cup	0.16	25.0
		HNO_3 scrubber	0.13	21.5
		Total	0.62	
2	12	0.45 μm Millipore	0.26	42.7
		Graphite cup	0.32	51.8
		HNO_3 scrubber	0.03	5.5
		Total	0.60	

[a] Samples were collected simultaneously in each run.

excess of 94% by most common filters (Table 5.9). This may well be taken as evidence that the so-called window effect exhibited by filters can reasonably be disregarded. However care should be taken in extrapolating these results to other sampling situations. For example, sampling for lead close to a roadway requires collection of both very small lead-containing particles and vapor phase lead. By deploying a nominal 0.4 μm Millipore filter in series with a graphite cup filter and a nitric acid scrubber, it has been shown (Table 5.10) that a substantial amount of lead passes through both the Millipore and the graphite filters.[54] Furthermore, the proportions change substantially with distance from the roadway—possibly because of particle growth.

5.5.2. Particle Size Discrimination

In discussing the physicochemical characteristics of emitted particulates and of composite aerosols, it was frequently remarked that many elements show a dependence of specific and volume concentrations on particle size. This phenomenon is of considerable importance,[13,35,92] and often there is a need to determine the size distributions of particulate trace elements.

Several methods have been employed for determining the size distributions of airborne particle mass or number, but only a few of these can be used for the determination of *elemental* size distributions. This is because the particles must be physically collected in sufficient quantities for subsequent elemental analysis of separate size fractions. The principles of inertial separation, electrostatic precipitation, and filtration have been used to ensure that this requirement is met.

Undoubtedly the most common method employed for simultaneously collecting and size differentiating an aerosol *in situ* in cascade impaction. In cascade impactors, such as those depicted in Figure 5.11, a stream of air containing particles of different sizes is constrained to make increasingly sharper changes in direction at increasingly higher flow velocities as it passes through the impactor. A particle whose inertia is too great to enable it to follow the airstream will impact on a collection surface; thus successive impaction stages will collect successively smaller particles. In most cascade impactors the last stage is followed by a backup filter (usually glass fiber) for collecting the particles that are sufficiently small to pass right through the impactor.

The theory of cascade impaction has been treated extensively by several authors[93–100] and is not discussed in detail here. It has been shown, however, that the efficiency of collection of a spherical particle of physical radius r is related to a single dimensionless parameter, ψ (inertial impaction parameter),

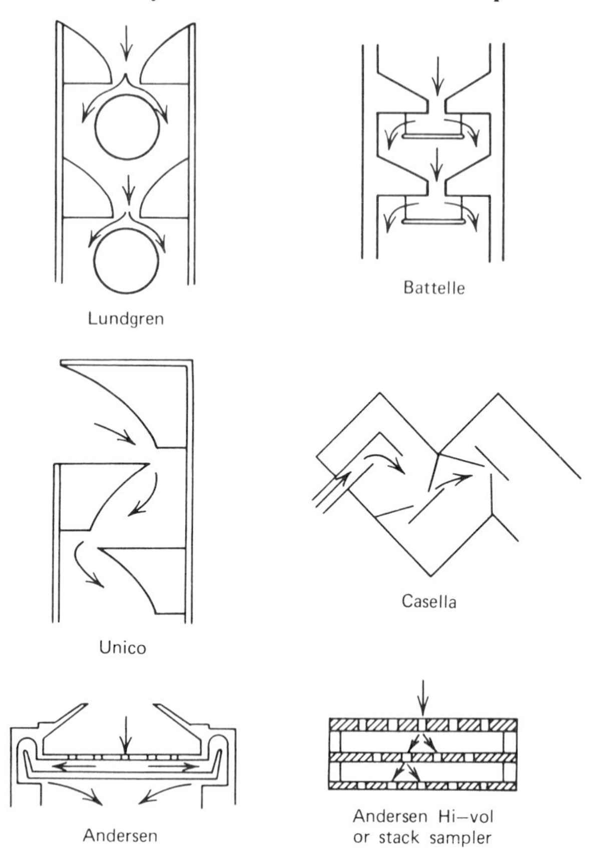

Figure 5.11. Basic design and flow configuration schematics of several common cascade impactors.[37]

whose value is determined for each impaction stage by the geometry and working conditions. Thus[93,95,99]

$$\psi = \frac{4r^2\rho_p FC}{18\mu L D_j^2} \tag{1}$$

where D_j and L are the width and length of the impactor slit, ρ_p is the particle density and C its Cunningham slip correction factor, and μ and F are the viscosity and flow rate of air passing through the impactor, respectively.

For practical purposes the operating parameters of cascade impactors are usually expressed in terms of effective cutoff diameter (ECD), which is the aerodynamic particle diameter for which a given impactor stage exhibits 50% collection efficiency. It is convenient, therefore, to write eq. 1 in the form

$$\psi = \frac{F}{18\mu D_j}(ECD)^2 \tag{2}$$

where $ECD = (\rho_p C)^{1/2} D_{aer}$

and D_{aer} is the equivalent aerodynamic diameter of a particle. Most cascade impactors are designed so that there is a fairly constant ratio R between ECD values for successive stages where

$$R = \frac{(ECD)_i}{(ECD)_{i+1}}.$$

Values of R and ECD values for typical operating conditions are presented for some representative cascade impactors in Table 5.11.

Equations 1 and 2 show that the size differentiating ability of a cascade impactor depends on its geometry, its operating conditions, and the nature of the particles being separated. As far as impactor geometry and operation is concerned, all designs are more or less subject to the following shortcomings:

Bounce-off. Particles can bounce off a collection surface, proceeding to and being collected by a later stage. Such bounce-off tends to distort a distribution toward small particle sizes.

Reentrainment. Deposited particles can be reentrained into the airstream, proceeding to a later stage as in the case of bounce-off.

Wall Losses. Particles can stick to the impactor walls instead of depositing on an impaction surface.

Cross Sensitivity. Since the collection efficiency of an impaction stage does not change abruptly from 0 to 100% at a given particle diameter, the particles having aerodynamic diameters close to ECD values are usually distributed between two or more impaction stages.

Discreteness. Cascade impactors contain only a limited number of stages (usually four to nine), and errors can result from interpretation of the discrete mass data in terms of a continuous distribution.

The significance of these performance characteristics has been considered by several authors.[94,95,100–105] It has been concluded that mass median diameters of aerosol constituents determined using cascade impactors are generally within 20% of the true value; however determinations of mass or concentration present in particles smaller (or larger) than a given aerodynamic size are subject to extremely large errors. Indeed, where the fraction

Table 5.11. Operational Characteristics of Several Common Cascade Impactors

Type	Number of Stages	Flow Rate ($m^3\ hr^{-1}$)	Typical ECD[a] (μm)	Stage Constant, R	Reference
Casella	4	1	12.2, 3.7, 1.5, 0.45	3.0–4.0	105
Scientific Advances	6	0.75	16, 8, 4, 2, 1, 0.5	2.0	4, 104
Lundgren	4	0.85–8.5	17, 5.2, 1.7, 0.5	3.0–3.3	103
Andersen Sampler	6	1.7–3.4	4.6, 2.75, 1.65, 1.0, 0.5	1.65–2.0	111
NASN	5	8.5–10	3.19, 2.17, 1.49, 0.96, 0.61	1.47–1.57	112
Andersen Stack Sampler	7	2	11.3, 7.3, 4.7, 3.3, 2.06, 1.06, 0.65	1.41	29
Andersen Hi-Vol	4	34	7.0, 3.3, 2.0, 1.1	2.0	102
ERC–EPA Classifier-Analyzer	5	17	10, 3, 1, 0.3, 0.07	3.0–4.3	116
University of Washington Mark III	7	0.5–2.5	24, 10.5, 4, 2, 1.05, 0.54, 0.26	1.9–2.5	115

[a] Effective cutoff diameters vary with flow rate according to eq. 2.

lies in the wings of a particle distribution, measurements can be in error by several orders of magnitude. In this regard it should be noted that the emerging practice of separating airborne particles into only two size fractions, above and below 5 μm (the approximate upper limit of pulmonary deposition), could result in substantial errors when most of the mass of an element is concentrated in one fraction.

It is generally agreed[100,103,106,107] that wall losses and particle bounce-off are the two performance factors that contribute most to distortion of particle size distributions determined by cascade impaction. Wall losses can be essentially eliminated by using an impactor with a high impaction surface area compared with the wall area (e.g., the Andersen Hi-Vol sampling head). However bounce-off can also constitute a significant problem whose extent is determined by the "stickiness" of the impaction surface and the velocity with which the particles strike the surface.[103,108,109] This dependence is illustrated in Figure 5.12.

Cascade impaction surfaces most commonly consist of a bare metal plate, glass, aluminum foil, or a glass fiber filter.[2,106,107] In addition, surfaces such as organic membranes, paper, Teflon, and silicone oil have been used. The actual choice of an impaction surface material generally involves a tradeoff between attainment of an authentic size distribution and ease of operation. Thus bounce-off is minimized if glass fiber or silicone oil is employed;[37,109,110] however glass fiber contains significant levels of many trace elements (Table 5.6) and must be equilibrated to constant humidity before weighing. Also, oil is messy and susceptible to loss, making it difficult to determine particulate mass, thus specific concentrations of trace elements.

The final factor that controls the size differentiating ability of cascade impactors is the nature of the particles themselves.[99] Thus certain particles, such as those which constitute fly ash, are hollow[38,40] and exhibit a high tendency to bounce. Particles derived from sea spray and gas-to-particle conversion are often highly hygroscopic and tend to grow rapidly at high relative humidity,[73–75] and a few gaseous species, notably sulfur dioxide, are capable of reacting and crystallizing on impactor collection surfaces.[37,111]

All these factors tend to obliterate the finer details of aerosol size distributions as determined by cascade impactors, and there is considerable evidence that the apparent log-normal distributions of total and elemental mass in urban and remote aerosols are largely due to impactor artifacts and the averaging effect of long sampling times.[20,70,100] Nonetheless, several of the cascade impactors currently available are capable of size discrimination that provides a useful approximation to actual atmospheric aerosol size distributions.

The actual choice of a cascade impactor for determination of elemental size distributions in an atmospheric aerosol is the subject of considerable

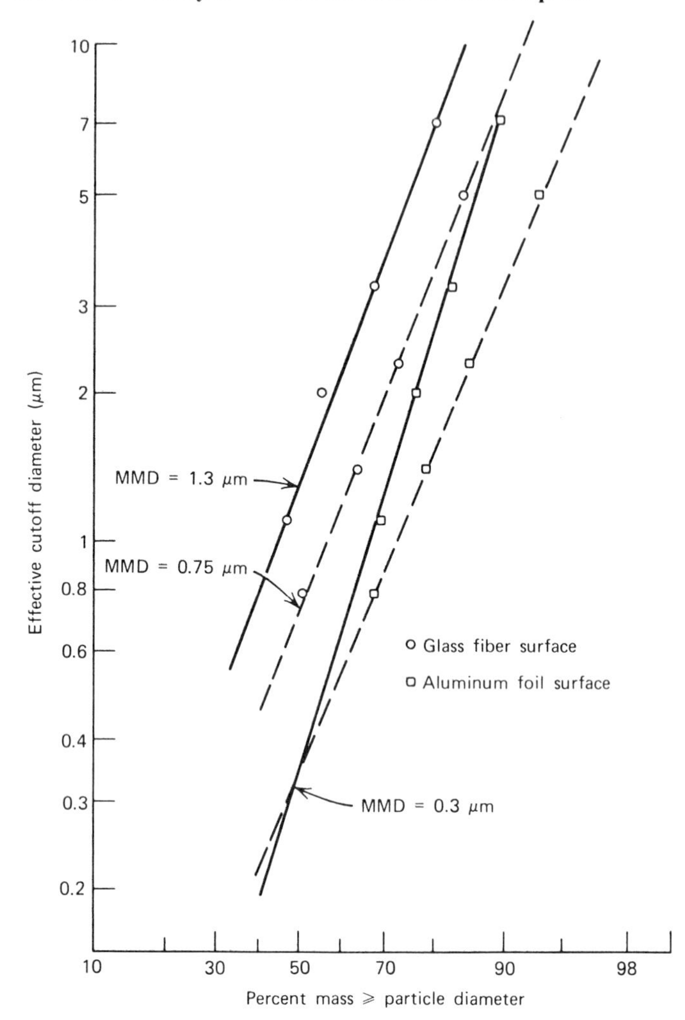

Figure 5.12. Log-probability plots of the size dependences of a typical aerosol mass determined simultaneously using Andersen Hi-Vol cascade impactors with different flow rates (solid lines, 20 $ft^3\ min^{-1}$; broken lines, 40 $ft^3\ min^{-1}$) and impaction surfaces.[37]

debate. This is because few truly comparative data are available[24] and different impactors may be appropriate to different sampling and analytical situations.

Probably the most important criterion in calculating elemental size distributions is the need to collect sufficient material to enable precise determination of individual elements in each size fraction. When airborne particles are being collected at their source, this requirement is not usually difficult to meet. Indeed, overloading of impactor plates can easily occur. For atmospheric sampling, however, it is necessary to employ a cascade impactor that operates at a high flow rate, to sample for an extended period, or to employ an analytical method capable of achieving low limits of detection. If long-term integrated size distributions are required, extended sampling periods do not constitute a problem; if 24 hr samples are called for, however, as is often the case, it may be necessary to collect several hundred milligrams of sample. The most readily available cascade impactor for this purpose is the Andersen high volume sampling head, which operates at a flow rate of 10 l sec^{-1} (20 ft^3 min^{-1}) in conjunction with a standard Hi-Vol sampler. This impactor has four impaction stages (Table 5.11), has negligible wall losses, and, for minimum bounce-off, is best employed with glass fiber impaction surfaces. Other similar devices have also been developed, and so-called low volume cascade impactors have been adapted to enable operation at high flow rates.[2,112,113] The high volume devices are generally suitable for use in conjunction with such analytical methods as conventional flame atomic absorption spectrometry, which have low, but not very low, elemental detection limits.

Where long sampling times (e.g., several days) are acceptable, or highly sensitive analytical methods are available, a number of cascade impactors are suitable. Foremost among these are the standard Andersen eight-stage impactor, and the so-called Battelle, Scientific Advances, and Casella impactors. The choice should be based on performance characteristics, which appear best for the Scientific Advances model,[4,24] although the Andersen impactor has been most extensively employed.[2,112] As a general rule a single-jet impactor appears to achieve better size discrimination than a multijet impactor.[24] If time resolution is required in conjunction with size discrimination, this can be achieved with the cascade impactor developed by Lundgren[103,114] in which the impaction surfaces are slowly revolving drums; this impactor must be employed in conjunction with a highly sensitive analytical technique, however, since the subdivided sample sizes are small.

Other applications of the principle of cascade impaction involve extension to aerodynamic particle sizes below 0.1 μm, which has been achieved by operating the latter stages of an impactor at low pressure,[115] and source sampling at elevated temperatures. In the latter case, the most readily available device is the Andersen stack sampler, an eight-stage cascade impactor capable of insertion into the stack of an emission source such as a coal-fired power plant.[102]

An important modification of cascade impaction is the replacement of the impaction surface regions by orifices through which particles can pass and be collected in a region of low or zero air flow.[116–119] This approach, which is referred to as virtual impaction, has the following advantages: particle bounce-off and reentrainment are minimized, substantial amounts of sample can be collected at high flow rates without encountering loading problems, and the collected particles are subjected to much less mechanical shock than occurs by impaction. A particle size discriminating device of this type has been developed by the U.S. Environmental Protection Agency (the ERC/EPA Classifier-Analyzer) and is designed to achieve separation of particles in the size range 10 to 0.07 μm ECD. Published data[116] indicate that cross-sensitivity between successive stages is unimpressive compared with conventional cascade impactors,[100] but this shortcoming may be expected to be improved as the basic design is refined.

The principle of inertial separation of airborne particles is also employed in cyclones that are used with a range of flow rates to collect particles of sequentially smaller size.[119–123] In a cyclone the air flow rate F required for a particle of equivalent spherical radius r to be collected in the cyclone settling chamber of diameter D is[123]

$$F = \frac{\pi g r^2(\rho_p - \rho)D^2}{18\mu} \tag{3}$$

where g is the gravitational constant, ρ_p is the particle density, and ρ and μ are the density and viscosity of air. As is apparent from eq. 3, the particle sizes collected by a cyclone can be adjusted by varying either D or F, thereby enabling considerable flexibility of performance.

Cyclones are not easily applicable to the collection of particles having aerodynamic diameters below about 5 μm. However this size coincides with the onset of small particle deposition in the pulmonary region of the human respiratory system,[13] and the EPA is currently exploring the use of a single cyclone followed by a backup filter to separate so-called respirable and nonrespirable particles. Cyclones have been used for many years to separate bulk particulate samples in the laboratory[123] and in some sampling trains for stack gas particulates. This approach is useful for the collection and separation of samples of source particulates previously sampled from those collected by emission control devices such as precipitators and bag filters or by central heating and air conditioning filters.

Airborne particles can also be separated into discrete size fractions for subsequent elemental analysis by electrostatic precipitation or successive filtration. Small electrostatic precipitators capable of field operation have been designed and exhibit good separation characteristics for particles having aerodynamic diameters in the range of 0.01 to 20 μm. Particles in the airstream are charged and are attracted to a collection surface bearing the opposite charge. The velocity with which they move to the collection surface depends on their inertia, on the charge they carry, and on the precipitator voltage. Size discrimination is achieved because both particle parameters depend on particle size.[35] Portable electrostatic precipitators are not widely available and are rarely used for atmospheric sampling. Indeed, the choice of such equipment for this purpose should be made with caution, since the ability of a particle to hold an electrostatic charge is strongly dependent on particle resistivity,[124,125] which may vary greatly for particles of the same aerodynamic size in a composite aerosol.

The use of several filters (e.g., Millipore or Nucleopore) in series, each having successively smaller pore diameters, is an inexpensive but not very effective means of size discrimination, since particles considerably smaller than the physical pore size are collected quite efficiently by processes other than direct interception. Also, collection characteristics change with time as filter loading increases. On the other hand, separation of particles by sieving bulk samples in the laboratory is an effective method, applicable to particles having physical diameters as small as a few micrometers. For best reproducibility, sonic agitation or an "air jet" sieve is recommended.

5.5.3. Gases and Vapors

Strictly speaking, discussion of methods for the collection of trace elements present in the atmosphere as gases or vapors should include consideration of species containing elements such as sulfur, nitrogen, and the halogens. However collection of these species is a topic in itself, and only gases and vapors containing metallic and metalloid elements are covered here.

Most metals are associated with particles in the atmosphere; exceptions include elemental mercury and mercury alkyls, tetraethyl and tetramethyl lead, arsenic and selenium oxides, elemental selenium, arsine and arsenic alkyls, and postulated species such as nickel carbonyl.[126] Recently it has been shown[54] that small amounts of several elements, including cadmium, lead, and iron, can pass through apparently efficient filters, although it is currently fashionable to consider the uncollected fraction as being nonparticulate. Such assumptions should be treated with extreme caution in view of the demonstrated filter inefficiencies. It should also be recognized that most nonparticulate metals and metalloids exist in the atmosphere as vapors rather than true gases and may be subject to condensation or crystallization during the process of collection.

Collection of vapors from the atmosphere is normally by absorption into solution, adsorption onto a solid, or condensation to a liquid or solid in a cold trap. In some cases the collection process is an integral part of the subsequent analytical evaluation. With the exception of those used for mercury and arsenic species, collection methods for metal- and metalloid-containing vapors have not been fully evaluated with respect to their ability to achieve quantitative sample collection and stabilization. Methods are therefore presented in terms of their general characteristics and the types of compound to which they are applicable.

Absorption of vapors into solution can take place either physically or chemically. Physical absorption obeys Henry's law; thus the collection efficiency theoretically cannot reach 100%. As a result, physical absorption is rarely employed for collection of vapors, even though soluble species can be collected with good efficiency by using several absorption vessels in series. Superior collection efficiencies and capacities are achieved when a chemical reaction occurs between the gas or vapor being collected and some species in the collecting solution. For example, promotion of ionization, acid–base behavior, complex formation, or formal chemical reaction can result in conversion of the species of interest to a form that is both highly soluble and stable in solution. Examples of this are the use of acidic potassium permanganate solutions for the collection and oxidation of mercury species to Hg^{2+} (References 127–131) and the use of iodine monochloride solution for collection of tetraalkyl lead[132] and mercury species.[133] In the last case it has been shown that iodine monochloride solutions are really suitable only for the quantitative collection of elemental mercury vapor at levels encountered in source emissions. Inefficient collection occurs for atmospheric mercury levels.[134]

The efficiency of collection of vapors by absorbing solutions depends very much on the pertinent gas–liquid mass transfer characteristics. These depend, in turn, on the bubbler design and flow rates involved.[135] Some of the more efficient designs include that of Wartburg et al.,[136] the "petticoat" bubbler, and the Greenburg–Smith impinger,[137,138] which can be operated at flow rates in the range 10 to 20 l min^{-1}. Midget impingers,[138] Dreschel bottles,[136] and packed columns[139] are useful when low flow rates are acceptable. In general, chemically reactive solutions have better collection and stabilization characteristics than physical absorbants, but many of the chemicals involved contain substantial contaminant levels of an element of interest. For example, detection limits for mercury are usually determined by the extent of contamination of potassium permanganate and iodine monochloride where these compounds are used. In addition, solutions are subject to loss by evaporation, collected species often tend to adsorb on container walls, and manipulation of glassware and tubing in the field is inconvenient. For these reasons it is better to avoid absorbing solutions when possible.

Adsorption of gases or vapors onto a solid surface is a convenient and popular method of collection. As in the case of absorption, adsorption can be either physical or chemical. Physical adsorption is generally reversible, either by heating the adsorbant or by extracting it in a Soxhlet or other suitable extractor. The most common physical adsorbants are activated charcoal (for mercury species[140] and lead alkyls)[141] and adsorbent resins such as Tenax (for organometallic vapors). Chemical adsorption is also widely used—most valuably where a known specific reaction takes place between the vapor species and the solid. The best known reactions of this type involve amalgamation of elemental mercury vapor with gold[142,143] or silver[143–147] and the reaction of alkyl lead with iodine crystals.[148,149] Amalgamation with gold and silver has been shown to be a highly efficient method for the collection of elemental mercury;[143] however collection of lead alkyls by iodine crystals is inefficient[149] and should be avoided. Several chemical adsorbants have recently been employed with success for specific collection of individual mercury and arsenic compounds.[143,150] Since these methods form integral parts of procedures for the speciation of airborne mercury and arsenic, they are discussed in a later section.

Condensation by cold trapping at liquid nitrogen or solid carbon dioxide temperatures is a very simple means of collecting atmospheric vapors. Since the vapor pressure of the species in question must exceed its saturation vapor pressure for condensation to occur, however, condensation is not necessarily achieved simply by lowering the temperature when low concentrations of vapors are present. Furthermore, the associated condensation of copious amounts of water vapor frequently prohibits the use of cold traps.

5.5.4. Other Methods

Several additional methods are occasionally encountered for the collection of atmospheric trace elements—mainly in particulate form. These include solution impingers, settling buckets, and wind deposition devices. Although not of primary importance for the quantitative determination of airborne trace elements, these methods do provide alternative information that may be appropriate to certain studies.

Solution impingers consist of a drawn-out tube through which air containing particles flows into a solution, where the particles are trapped.[138,151] There is little to recommend such a method of collection save that the particles can be readily isolated by evaporation, and both particulate and vapor species can be collected together with suitable choice of solution. For this purpose an acidic oxidizing solution is best employed. On the other hand, collecting solutions can be operated only at low flow rates (<20 $l\ min^{-1}$), are subject to evaporation, and suffer from dependence of collection efficiency on particle size similar to that exhibited by filters.[152,153] Consequently applications are limited to sampling in the following situations: high atmospheric particulate loadings exist, simultaneous collection of particles and gases is required, or alternative equipment is not available.

Dustfall buckets, as the name implies, are used to collect atmospheric particulate matter that reaches the ground as a result of various sedimentation processes,[20] including gravitational settling, washout, and rainout. Several designs are available; however the most common consists of a cylindrical polypropylene container 15 cm in diameter and 23.5 cm long. The bucket is partially filled with water, which wets collected particles, thus preventing reentrainment. To obtain enough material for subsequent analysis, dustfall buckets usually have to be exposed for prolonged periods (several weeks), and it is common practice to locate them on the roof of a high building to prevent ground level contamination. The main problem encountered when dustfall buckets are located at ground level (e.g., for studies of automobile particulate settling) is interference by man.

Though not a generally recognized method for measuring sedimentation, collection of precipitated snow provides a simple and convenient means of obtaining both a time and an area resolved sample. This approach has been employed with considerable success for determining sedimentation rates of lead in polar regions[154] and of a variety of elements from power plant plumes.[84]

Wind deposition devices are used to collect particulate matter that reaches a surface by advection.[20] The surface, which consists of Teflon tape, double-sided Scotch tape, or a nylon mesh,[155] is mounted vertically to minimize deposition by sedimentation. Depending on the type of information required, the surface can be positioned in a fixed orientation or can be mounted on a rotating base that maintains the collector at right angles to the wind direction. Clearly calibration of such devices is largely subjective; however the relative amounts and compositions of particulate matter collected by a given surface can provide useful information about wind deposition.

5.6. ANALYSIS

Nowadays there is a disquietingly widespread belief that once an environmental sample has been collected, its analysis is largely routine and trivial. In the case of airborne particulate matter, for example, attention is focused on the few elements of interest, yet these elements usually constitute only a fraction of the mass present, and many other elements and compounds that can influence the analytical result are present. Furthermore, too little attention is paid to the environmentally significant questions whose answers are being sought, with the consequence that the numbers resulting from an analysis are considered to be an end in themselves.

Such attitudes have resulted in the generation and publication of large quantities of analytical data that are inaccurate, imprecise, and only tangentially pertinent to environmental questions. There are several reasons for this. The first, and most important, is failure to recognize that an analytical evaluation step is only one part of an integral process involving delineation of the question(s) to be asked, experimental design, sample collection, preservation of sample integrity, pretreatment for analysis, analysis, and data interpretation. Poor performance of any one step or inappropriate interfacing between steps will result in inaccurate final answers. The second widespread problem involves lack of recognition of the importance of interferences that can affect several stages of the overall process. Such interferences are particularly important in determining trace elements in environmental samples because of the variability and complexity of matrix compositions encountered and the large number of elements present. These remarks are well illustrated by the results of several interlaboratory comparison studies of fly ash and other airborne particulates: frighteningly large discrepancies were revealed between results obtained in different laboratories and by different techniques.[156,157] In the following sections, therefore, attention is drawn to the appropriate choice of methodology for a given analytical problem, to suitable interfacing of individual stages of the analytical process, and to recognition of sources of interference.

The types of information sought by analysis of airborne particles or vapors can be subdivided into three categories.

Bulk Analysis. This involves determination of either the mass or the concentration of one or more elements present in a collected sample. These values can then be converted to volume concentrations (micrograms per cubic meter) or specific concentrations (micrograms per gram) by relationship to the conditions of collection as discussed earlier.

Elemental Distributions. It is also necessary to determine how elemental mass or concentration is distributed in the atmosphere, in an aerosol or group of particles, or in an individual particle. Determination of atmospheric distribution relies primarily on the siting of collection devices to provide maximum definition of elemental mass or concentration with respect to region, height, wind direction, or other meteorological conditions. Distribution within an aerosol or group of particles involves subdivision of the sample with respect to size (either aerodynamic or physical), density, ferromagnetic character, matrix composition, resistivity, or some other appropriate physical parameter. This is accomplished either during collection, as in the case of size, or during pretreatment of the collected bulk sample for analysis. Distribution within a single particle involves determination of the nature of heterogeneous distribution (e.g., surface association, crystal formation) and is generally accomplished by deployment of an appropriate analytical technique.

Speciation. The actual chemical compound(s) in which elements exist must be found. Approaches to date have involved both specific instrumentation and analytical sequences encompassing sample collection, pretreatment, and analysis.

It is apparent from these remarks that the choice of analytical methodology and instrumentation will depend greatly on the type of information sought. In many cases several alternative analytical approaches can be employed, and it is well worthwhile to consider the various options and tradeoffs involved.

Probably the foremost consideration in choosing an analytical method is its detection limit for the element(s) in question. The masses of most environmentally interesting elements present in atmospheric samples (and also many source samples) are generally small (1 to 500 μg), thereby limiting the choice of analytical methodology. Indeed, some samples (e.g., those from individual size fractions, low volume collection, or remote atmospheric aerosols) contain such small amounts of each element that only one or two highly sophisticated techniques can be employed. In this regard the choices of sample collection and analytical techniques are interdeterminate—a restriction that must be borne in mind when choosing each during the project design stage.

A second consideration is the choice between analytical techniques capable of sequentially determining individual elements and those suitable for simultaneous multielemental analysis. In principle, multielemental techniques for the analysis of airborne particles is preferable, since generally several elements are of interest. However analytical conditions employed in multielemental determinations are rarely optimum for any one element, thus there must be a compromise between having a wide range of accessible elements and determining any one with optimum precision and accuracy. Furthermore, no so-called multielemental technique is capable of determining all elements in an airborne particulate sample, both because of limitations intrinsic to the method and because of interferences from other species present. In practice, therefore, it is usually necessary to employ a multielemental technique in conjunction with one or more single element techniques, that is, to focus on the elements that can be determined by available methodology.

Relatively little recognition is given to the distinction between situations that require precise, accurate, or semiquantitative data in choosing an analytical technique. In the vast majority of cases requests are made for highly accurate data, yet these are frequently not necessary (even though they may be desirable). For example, investigators often want to make an initial survey to establish the identity and approximate concentrations of elements present. For this purpose a semiquantitative multielemental survey technique can be employed. High precision is desirable when comparative data are required among similar samples to establish elemental concentration dependences (e.g., on particle size, density, atmospheric region), although absolute accuracy is not necessary in such cases. It is, however, necessary for establishing elemental emission rates from sources and for compiling elemental inventories and mass balances. As pointed out in later sections, different analytical methods can provide semiquantitative, precise, or accurate results with a minimum of operator time and effort, and such characteristics should be carefully considered in choosing methods most appropriate to a given investigation.

A major consideration in selecting an analytical method for particulate material is whether the sample should be analyzed directly in its solid form, or digested and analyzed as a solution. Direct analysis of the solid retains sample integrity, avoids time-consuming and messy digestions, and usually requires minimal sample manipulation. On the other hand, calibration is usually difficult, and matrix interferences are often considerable. Analysis of solutions is amenable to easy and precise calibration, and matrix effects usually can be avoided; however sample manipulation is considerable, and sample integrity is in jeopardy through loss of volatile or adsorbing elements, and through changes in chemical species. The last consideration prohibits the use of digestion in the identification of inorganic compounds present in the original sample. Overall, if suitable calibration can be achieved and if matrix effects are small, direct analysis of solids is preferable.

The presence of a large number of elements in airborne particles gives rise to a number of interferences that may be either chemical or intrinsic to a particular analytical method. The occurrence of such interferences should be recognized in choosing an analytical technique, although not all elements will be affected.

Finally, although not strictly a scientific consideration, the cost and time required for analyses by different methods should be taken into account. In many cases these factors may be those limiting both the choice of an analytical method and the project design. Usually the multielemental techniques are considerably more costly and time-consuming; they are frequently less expensive per element than single elemental analyses, however.

5.6.1. Sample Pretreatment

Collected airborne particulate and gaseous samples are often in an unsuitable form for introduction to an analytical technique. In such cases some pretreatment is necessary, the most common being the dissolution of particulate matter.

As discussed in a previous section, airborne particles are usually collected on a surface by filtration, impaction, sedimentation, or advective deposition. Quantitative removal of solid particles from collection surfaces is normally impractical except in the case of stainless steel, glass, or aluminum foil surfaces used in cascade impactors. From these the collected material can be removed with a soft brush or by washing, followed by evaporation of the wash solvent. Also, organic membrane filters can be dissolved by addition of alcohol or acetone. Direct analysis by techniques such as instrumental neutron activation analysis or X-ray fluoresence spectrometry, however, can be accomplished without separating solid particulate matter from the collection surface, provided background impurity levels in the surface material are sufficiently low (Table 5.6).

For most analytical techniques, however, it is necessary either to destroy the collection surface material or to remove the collected material by wet oxidation, dry ashing, or low temperature ashing. In general, wet oxidation procedures require only simple apparatus and are not subject to significant losses of elements by volatilization or adsorption onto container surfaces. Considerable manipulation is necessary, however, and there is high risk of contamination of samples, both by contact with glassware and from impurities present in the chemicals used.[158] Dry ashing, on the other hand, requires few or no reagents, limited manipulation, and little operator attention. Losses of elements by volatilization, retention, and formation of analytically intractable species can be severe, however. Such losses are substantially reduced if low temperature ashing techniques are employed.

Depending on the elements to be analyzed and the type of collection surface with which they are associated, wet oxidation may be variously required to achieve extraction of elements from essentially intact particles, complete solution of particles, or destruction of the collection surface material. In most cases rigorous digestion procedures are required to solubilize refractory species contained in airborne particles, although in the case of lead, it has been shown[158] that digestion with 2*M* HCl will quantitatively solubilize lead without dissolving the particles or glass fiber filters. ASTM Method D2681-68T specifies the use of a mixture of nitric, sulfuric, and perchloric acids[148] and recommends continuation of digestion until sulfur trioxide (SO_3) fumes are observed. This mixture is usually adequate for complete digestion, although problems are often encountered with the coprecipitation of some elements (e.g., lead) with insoluble species such as calcium sulfate. Since the presence of sulfate can also introduce interferences in subsequent analysis by atomic absorption, a nitric–perchloric acid digestion is often preferred.[54]

Before choosing a wet oxidation procedure from among a number of alternatives,[54,158,160] it is well to recognize the functions of the various acids. Thus oxidation of the particulate matrix and of filter materials such as paper is achieved by nitric and sulfuric acids. When carbon is present, either in the particles or in the filter material, the presence of perchloric acid ensures its solubilization. Many airborne particles, notably fly ash, have silicate matrices whose solution requires hydrofluoric acid. Since this acid will also dissolve glass fiber filters and chemical glassware, it should be used with care, and nonglass containers should be employed. For complete digestion of particulate matter having a high silicate content, the use of a mixture of aqua regia and hydrofluoric acid in a Teflon-lined stainless steel digestion bomb is extremely effective.[29] Digestion is carried out at 110°C for 2 hr, and after cooling, boric acid is added to neutralize excess hydrofluoric acid as boron trifluoride. Dissolution is better than 96% efficient for all trace elements of interest.

Probably the major problem associated with wet oxidation methods is the presence of impurities in the reagents employed. For example, lead blanks in A.R. grade acids can be as high as several micrograms per milliliter,[158] and so it is essential to run reagent blanks routinely. For trace level analyses, acids should be redistilled from quartz or exhaustively electrolyzed. Significant blank levels may also originate from glassware, which should be scrupulously cleaned with hot nitric or hydrochloric acid immediately before use.[158]

Dry ashing, usually in a muffle furnace at 450 to 550°C, is widely employed to destroy particle matrices and to remove associated organic matter such as collection surface material. Three processes are involved.

1. Evaporation of physically and chemically bound water.
2. Evaporation of volatile materials and progressive thermal destruction.
3. Oxidation of the nonvolatile residue until all organic matter is converted to inorganic.

The resulting ash can then be either analyzed directly or dissolved in acid (e.g., HCl) for solution analyses.

It is now well established that dry ashing at 450 to 550°C results in significant losses of some elements (Table 5.12). This effect is probably due both to volatilization of the metal salts and to formation of insoluble metal silicates by reaction with borosilicate or silica containers and filter materials.[54,86,158,161] These losses can be minimized by gradually raising the temperature of the muffle furnace to a maximum of 450°C over a period of 4 hr. This prevents flash burning and allows conversion of volatile halides to less volatile oxides at low temperatures.[54] The addition of fluxing materials such as sulfuric acid or other "ashing acids" also reduces losses; however

Table 5.12. Recovery of Trace Elements from Typical Sample Ashing Procedures[86]

	Recovery (%)		
Metal	Low Temperature Ashing	Dry Ashing Muffle Furnance (550°C)	Wet Ashing
As	63		
Ba	97	99	81
Cd	92	53	
Co	96	97	
Cr	112	100	
Cu	98	92	83
Mn	99	107	
Mo	98	116	
Ni	97	99	
Pb	101	46	98
Sb	99	46	
Sn	95	87	
Ti	95	92	
Zn	96	39	

such additions involve the risk of increasing blank levels. It should be remarked that losses of lead from airborne particulates collected on filters can be extremely variable and may depend on the collection and storage history of the filters. It has been suggested[158] that this is due to the slow conversion of volatile lead halides to less volatile forms during storage.

Low temperature ashing is a variation of conventional dry ashing in which a plasma of radiofrequency-excited oxygen is formed around the sample at a pressure of approximately 1 torr.[162–166] Slow burning takes place at temperatures in the range 50 to 250°C (typically below 150°C), thus minimizing losses by volatilization (Table 5.12). Particles collected on conventional high volume filters can be ashed in one hour by this technique. However there are substantial losses of elements such as arsenic and selenium and of volatile metal halides.[166] Low temperature ashing has been applied extensively to airborne particulate samples with considerable success.[165] Its main disadvantages are the complexity of the equipment and the long ashing times (~ 24 hr) required for highly refractory matrices.

When studies of the distribution of elements within a group of particles are required, it is sometimes necessary to fractionate a bulk sample before analysis. Since considerable amounts of material (~ 1 g) are usually necessary, fractionation is usually practical only when particles can be collected from a precipitator or bag filter associated with an emission source. Such samples can be subdivided according to physical size by sequential sieving and according to aerodynamic size by cyclonic separation.[123] Both methods enable separation of particles of nominal spherical diameter greater than about 5 μm. Fractionation of particles according to density is conveniently accomplished by means of a float-sink technique[167] employing successively more dense liquids. A suitable density range is spanned by employing chloroform, bromoform, and diiodomethane, or mixtures of these.[38,168] An aqueous solution containing thallium formate and thallium malonate provides the highest density (4.3 $g\,ml^{-1}$), although the possibility of dissolving some elements or compounds should be recognized. It also should be noted that because many particles derived from high temperature combustion sources are hollow or highly voided,[38,40] the observed density is not representative of the particle matrix unless the particles are crushed. Substantial mass fractions of particles derived from blast furnaces, automobiles, and the combustion of coal are ferromagnetic, and separation of such fractions is sometimes required. This can be accomplished by repeated contact between the particles and a powerful magnet (preferably through a smooth surface such as glass or a Millipore filter). Separation is best achieved if concurrent sonic agitation is employed.[38]

Pretreatment of samples collected from the atmosphere as vapors frequently forms part of the actual analysis step. In the case of vapors that have been adsorbed onto solid materials such as activated carbon or an adsorbant resin, however, removal must be accomplished before analysis. Where the sample has to be transferred to a solution, either thermal evolution with subsequent solution trapping or solvent extraction in a Soxhlet apparatus is appropriate. The latter method is preferable in the case of thermally sensitive species such as nickel carbonyl.

5.6.2. Analytical Methods

Most of the analytical methods employed for the determination of elements present in atmospheric samples are spectroscopic or utilize spectroscopic detection. Almost all respond only to the presence of an *element* and provide little or no information about the chemical compound in which the element exists.[169–171] Indeed, we have a recognizable paradox in that methods capable of trace level detection indicate only the presence of elements, whereas methods such as X-ray diffraction, capable of distinguishing compounds, are extremely insensitive. The extent to which this paradox can be resolved is considered for each method in the following sections.

5.6.2.1. *Spark Source Mass Spectrometry*

Of all the methods available for elemental analysis of airborne particles, spark source mass spectrometry (SSMS) most nearly approaches the goal of being able to determine all environmentally interesting elements simultaneously. In conventional spark source mass spectrometry the sample is presented in the form of two normally identical electrodes between which a high voltage radiofrequency spark discharge is established. Ions characteristic of the electrode material are sputtered from the electrode surface and are accelerated into a double-focusing mass spectrometer for mass analysis.[171,172] Photographic detection is usually preferred for multielemental analysis because it averages out fluctuations in the electrical discharge and records all ions passing through the spectrometer. Electrical detection provides considerably faster readout, but either the spectrum must be scanned across the detector, with consequent loss of sensitivity, or only a small number of ions may be recorded simultaneously, which sacrifices the multielemental capability.

Airborne particles can be analyzed directly as the solid, after ashing, or in solution. Solids are homogeneously mixed with spectroscopic graphite or silver powder and pressed into an electrode approximately 2 mm × 2 mm in cross section and about 20 mm long. Solutions are evaporated to dryness with the powder. Direct analysis of particles on a filter is not currently practical unless silver membrane or graphite filters are employed. Both these materials can be readily fashioned into suitable electrodes.

Spark source mass spectrometry is capable of detecting essentially all elements having atomic numbers greater than that of lithium. An attractive characteristic of the method is that it detects all elements with quite similar sensitivity (usually within a factor of 3). Detection limits depend on the detection system and spectrometer resolution employed but are generally in the range 10^{-11} to 10^{-12} g (Figure 5.13) using photographic detection and a resolution of around 10,000. This corresponds to a specific concentration of 0.1 to 1.0 $\mu g\ g^{-1}$ in the original sample for direct analysis of solid particles. Actual detection limits for different elements depend on elemental vaporization and ionization characteristics, the electrode composition, and the extent to which the sample can be preconcentrated by ashing. Some representative values are presented in Table 5.13.

Spark source mass spectrometry is relatively free from chemical interferences, although some interelemental ions are formed—probably during the sputtering process. However there can be severe spectral interferences, resulting from the formation of multiply charged ions whose mass to charge ratio is close to that of a single charged species of interest (e.g., $^{9}Be^{+}$ and $^{27}Al^{3+}$), from formation of molecular ions, and from carbon-containing ions derived either from the graphite electrodes or from organic species

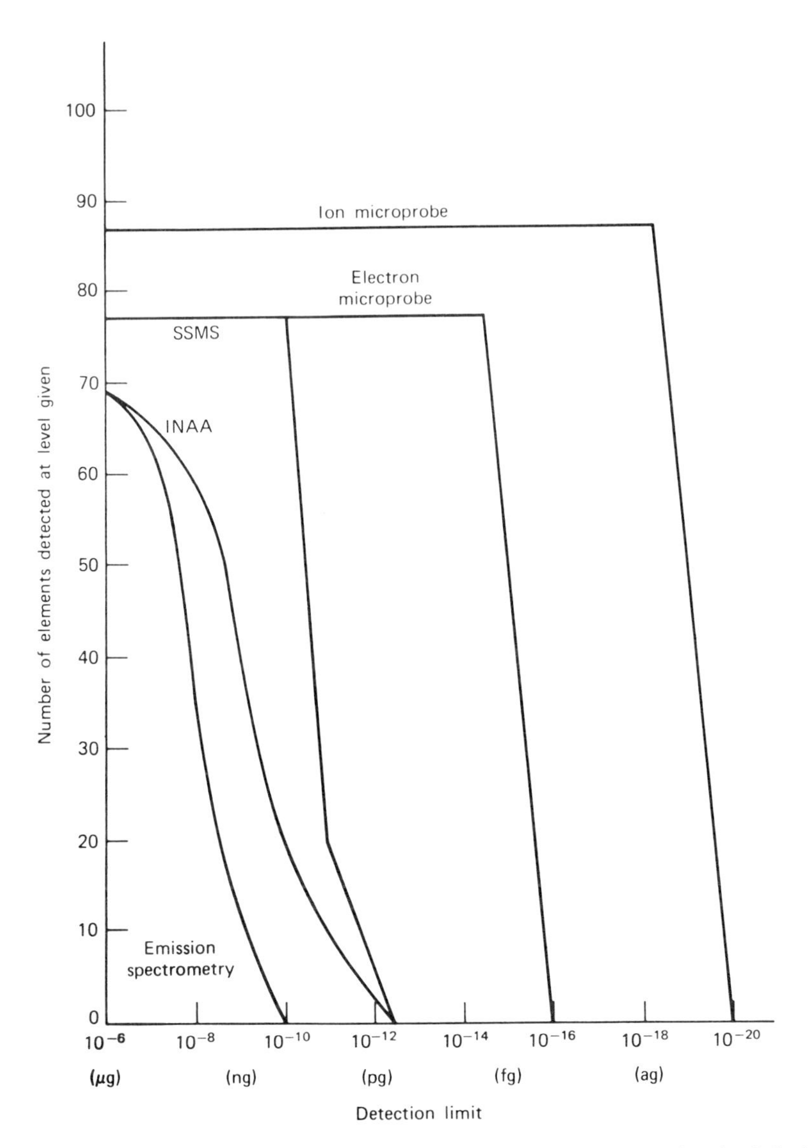

Figure 5.13. Comparison of the number of elements that can be detected at given levels by ion microprobe mass spectrometry, electron microprobe X-ray spectrometry, instrumental thermal neutron activation analysis, spark source mass spectrometry, and dc arc emission spectrometry.

Table 5.13. Typical Elemental Detection Limits in Airborne Particulate Samples for Spark Source Mass Spectrometry (SSMS) and Instrumental Neutron and Photon Activation Analysis (INAA and IPAA)

	Detection Limits (ng m^{-3})			
Element	INAA[a]	INAA[b]	IPAA[c]	SSMS[d]
Ag		1		0.02
Al	1	8		0.002
As		4	0.2	0.006
Au		0.1		0.02
Ba	0.1	0.1		0.02
Be				0.0008
Bi				0.02
Br	2	2.5	30	0.01
Ca	1000	200	30	0.003
Cd				0.03
Ce	0.01	0.25	0.4	0.01
Cl	200	100	0.4	0.004
Co	0.001	0.025		0.005
Cr		0.25	4.5	0.005
Cs				0.01
Cu	20	5		0.008
Eu	0.01	0.01		0.02
Fe	1	0.02		0.005
Ga				0.009
Hf	0.01			0.04
Hg		0.1		0.06
I		20	0.17	0.01
In		0.04		0.01
K		7.5		0.003
La	0.05	0.2		0.01
Li				0.0006
Lu	0.01			0.01
Mg		600		0.003
Mn	1	0.6		0.005
Na	1	40	2	0.002
Ni	10	20	0.05	0.007
Pb			12	0.03
Rb			0.2	0.02
S		5000		0.003
Sb	0.01	1	0.3	0.02
Sc	0.02	0.004		0.004
Se	0.02	0.1		0.01

Table 5.13. (*Continued*)

Element	Detection Limits (ng m^{-3})			
	INAA[a]	INAA[b]	IPAA[c]	SSMS[d]
Si			2.5	0.003
Sm	0.5	0.005		0.05
Sn				0.03
Sr				0.009
Th	0.01	0.04		0.02
Ti		40	0.9	0.005
Tl				0.02
V	0.5	2		0.004
W		0.5		0.05
Y			0.4	0.007
Yb	0.01			0.05
Zn	0.02	1	3	0.01

[a] 10 to 25 m^3 of air passed through Delbag polystyrene or Millipore (0.5 μm pore size) filters; neutron flux of 2×10^{13} n sec^{-1} cm^{-2}; 26 cm^3 Ge(Li) γ-ray detector.[177]
[b] 5 to 80 m^3 of air passed through polystyrene filters; neutron flux of 10^{12} to 10^{13} n sec^{-1} cm^{-2}; 30 cm^3 Ge(Li) detector.[179]
[c] 1000 m^3 of air passed through polystyrene filters; bombardment with 50 μA of 35 to 40 MeV electrons; 50 cm^3 Ge(Li) detector.[183,184]
[d] Adopted from absolute (nanogram) detection limits[175] by assuming a 10 m^3 sampling volume. Values are optimum for each element and are not necessarily representative of air particulate applications.

in the sample. Airborne particles generally contain such small amounts of organic material that the last type of interference is not a problem; but if carbon-containing ions are anticipated, samples must be ashed before analysis. In principle spectral interferences can be overcome by employing a spectrometer having sufficiently high resolution to separate analyte and interfering ions.[29] Experience has shown that a resolution of around 10,000 is suitable for airborne particle analysis, although problems are sometimes encountered in determining elements such as beryllium that experience interferences.

The precision and accuracy attainable depend very much on the way an analysis is performed. The most common procedure is to obtain a quantitative estimate of one element by another technique (e.g., atomic absorption spectrometry) and to assume that spectral intensities are proportional to the amount of each element present. This internal standard method provides analyses that are usually accurate to within a factor of 3 when account is taken of the abundances of each isotope of an element. Such semiquantitative results are usually sufficient for initial surveys of the elements present and of their approximate amounts. Where similar sample types are involved, as is the case with airborne particles, the same calibration can reasonably be applied to all samples. More precise calibration is achieved by establishing so-called sensitivity factors for each element of interest by standard addition. This can be done either by solution doping or by adding solid material to the sample.[173] Concentrations C for a given element X are then calculated from the expression[29]

$$C_X = C_{St} \frac{I_X}{I_{St}} \left[\frac{M_X}{M_{St}} \right]^2 \frac{\phi_{St}}{\phi_X} \cdot k$$

where I is peak intensity of the ion beam, ϕ is isotopic abundance, M is mass, and k is the sensitivity factor for X, relative to these values for a standard element St. This expression assumes that the photographic line width of an ion signal is proportional to $M^{1/2}$ [174] Such calibration is time-consuming and requires that at least one element be determined precisely in each sample. Precisions of $\pm 30\%$ are generally claimed for analyses performed in this way, although our experience with airborne particles has indicated that the precision is in the range ± 10 to 20% relative standard deviation.[29]

The highest precision is obtained using the method of isotope dilution, which involves standard addition of one or more isotopes of each element being calibrated. This procedure results in excellent precision,[156,173,175] but it is extremely time-consuming and hardly applicable to routine analysis of airborne particles.

As indicated earlier, spark source mass spectrometry is best suited to semiquantitative survey analyses or analyses of a number of similar samples in which essentially all the environmentally interesting elements must be determined. For this purpose it is probably the best technique currently available. The presence of several isotopes of most elements and multiply charged ions can be a significant aid in positive identification in such surveys. No useful information about the chemical species in the sample can be obtained. The main disadvantages of the technique are the high cost (thus low availability) of the instrumentation and the long time required to perform an analysis, much of it spent in extracting numerical data from photographic plates. Improvements in detection systems and ion sources (e.g., hollow cathode sources), however, may enhance significantly the application of mass spectrometry to analysis of airborne particles.[176]

5.6.2.2. *Activation Analysis*

Activation analysis is a general term encompassing several distinct analytical methods. Those that have been applied to the analysis of airborne particles include both thermal and fast (14.3 MeV) neutron activation analysis and photon activation analysis.[87,169,170,174,177–185] The most useful and widely used technique is that of instrumental thermal neutron activation analysis (INAA), which enables simultaneous determination of some 70 elements (Figure 5.13) with high precision, sensitivity, and accuracy.[186]

In INAA the sample, which can be either a solid or a liquid, is bombarded with thermal (slow) neutrons in a nuclear reactor. Many elements are capable of capturing a neutron and being converted to a radioactive isotope of one higher mass unit. For example,

$$^{23}_{11}\mathrm{Na} + {}^{1}_{0}n \longrightarrow {}^{24}_{11}\mathrm{Na}$$

For most of the radioactive isotopes thus formed, the decay process involves emission of a γ-ray whose energy is characteristic of the emitting isotope. This enables qualitative identification of the isotope involved. Quantitative determination is achieved by relating the number of γ-rays of each energy emitted, to the weight of each isotope present. γ-Ray energy discrimination is achieved by employing a lithium-drifted germanium detector, Ge(Li), in conjunction with a 4096 channel multichannel analyzer. The resulting γ-ray spectrum is processed by computer to determine the area under each peak and to correct for background.[177] Other detector-analyzer configurations can be employed (e.g., an intrinsic germanium detector), but the one just described is the most common.

INAA has one very distinct advantage over other methods, namely, results are extremely insensitive to variations in sample matrices because of the high penetrating power of both neutrons and γ-rays. Consequently airborne particulates can be analyzed directly on a filter or other collection surface, provided background impurity levels of the elements of interest are sufficiently low.[87] Samples are sealed in clean polyethylene vials for irradiation, and several vials are normally irradiated simultaneously. To circumvent spectral interferences between long- and short-lived isotopes, it is usually necessary to employ two irradiations: a short irradiation ($\sim$ 5 min), after which the sample carrier (rabbit) is delivered to the counting system within a few seconds, and a long irradiation (up to 9 hr), after which the sample is allowed to "cool" for several days to enable the short-lived isotopes to decay. It is frequently necessary to count samples that have been irradiated for long periods twice. This minimizes spectral interferences from isotopes having very short half-lives.

In general INAA can be employed to determine a minimum of 20 and up to

35 elements of interest in airborne particles. The elements that can be determined after short irradiations include aluminum, bromine, calcium, chlorine, copper, indium, magnesium, manganese, sodium, sulfur, titanium, and vanadium. Long irradiations enable determination of antimony, arsenic, cerium, chromium, cobalt, europium, gallium, gold, iron, lanthanum, mercury, potassium, samarium, scandium, selenium, silver, tungsten, and zinc.[177] For determination of some elements (e.g., vanadium) that have very shoft lifetimes, it is necessary to employ specialized equipment to enable samples to be counted immediately after irradiation.[26]

The extreme sensitivity attainable with radioactive counting techniques provides excellent detection limits[7,177] for many elements of interest in airborne particles (Table 5.13). Unfortunately it is sometimes believed that this high sensitivity is a general property of the technique. It is not. In fact sensitivity depends on the neutron capture cross section of the isotope being activated, its radioactive lifetime, and its fractional abundance. These quantities vary greatly among the elements to the extent that some elements (e.g., sulfur) exhibit poor sensitivities and can be determined by INAA only when present at high concentrations in airborne particles. Furthermore, several elements of real interest, including beryllium, cadmium, and lead, do not emit γ-rays under the normal conditions of thermal neutron activation, thus are not accessible by INAA. It should also be pointed out that sensitivity is proportional to the neutron flux reaching the sample. For most commercially available reactors this is of the order of 10^{12} to 10^{13} neutrons sec^{-1} cm^{-2}, with the highest fluxes occurring in the reactor core, which is not always accessible. However several published detection limits were obtained using fluxes in the range 10^{13} to 10^{14} neutrons sec^{-1} cm^{-2}, or even higher, to which most neutron activation analysts do not have access.[177,183]

Calibration of INAA is relatively straightforward due to the lack of matrix effects associated with the technique and the extreme linearity of the dependence of γ-ray counting on the amount of an element present. Consequently a so-called working curve calibration is usually employed in which a standard solution containing known amounts of all analyzable elements is sealed in a polyethylene vial and irradiated alongside the sample.[177] Provided both standards and samples are irradiated with the same neutron flux, excellent precision and accuracy can be attained. Dams et al.[87] have shown that the precision can be as good as the counting statistics (2 to 3%), but reproducibility between apparently identical samples is only about 10%. This discrepancy is attributed to sample variations. The accuracy of INAA has also been shown to be consistently good for analysis of airborne particles in interlaboratory comparisons.[156,157]

Activation by fast (14.3 MeV) neutrons and by 35 to 40 MeV electrons has also been employed in the analysis of airborne particles. Although not widely available, these techniques have the advantage of being able to detect several elements in addition to those normally accessible by INAA. Fast neutron fluxes in the range 10^8 to 10^9 neutrons sec^{-1} cm^{-2} are usually derived from a tritium source available on many university campuses. Neutron capture cross sections for the fast neutrons supplied by such sources are superior to those for thermal neutrons for several elements, including fluorine, oxygen, and silicon. Irradiation with bremsstrahlung from 35 to 40 MeV electrons, such as can be achieved with a linear accelerator,[183,184] results in the induction of (γ, n), (γ, p) and similar reactions, whose products emit γ-rays having energies and intensities that can be determined in the same way as in INAA. This technique of photon irradiation (IPAA) enables determination of antimony, arsenic, bromine, calcium, cerium, chlorine, chromium, iodine, nickel, sodium, titanium, zinc, and zirconium. Several other elements, including iron, rubidium, selenium, and yttrium, are marginally observable. Although these techniques are generally less sensitive than INAA, they do enable determination of elements such as arsenic, fluorine, iodine, lead, nickel, oxygen, silicon and titanium, which are often difficult or impossible to measure by INAA.[183,184]

Activation analysis, particularly INAA, is probably the best available technique for the precise determination of a large number of elements in airborne particulate samples. Its undoubted sensitivity makes it especially useful for analyzing small amounts of material (a milligram or less) such as obtained from individual stages of low volume cascade impactors or from remote atmospheres.[7] The ability to analyze particles without removing them from collection surface materials is a major advantage. In addition, since the method is nondestructive, samples that have been analyzed by any of the activation techniques can later be subjected to analysis by other, possibly destructive, means. It is sometimes argued that the multielemental capability of INAA is overrated in that approximately half the measurable elements (the rare earths) are of no environmental interest and several environmentally important elements are not measurable. However it can also be argued that rare earth elements are potentially useful as tracers and provide useful analogues of several particle matrix elements. Similarly, although no information about chemical species is provided by INAA, its ability to determine isotopes precisely makes it potentially useful for crude isotope tracing studies.

The major disadvantages of activation analyses are, of course, the scarcity of facilities, the cost, and the long time ($\sim$ several weeks) required for analyses. The choice of INAA for analysis of airborne particles essentially implies ease of access to a suitable nuclear reactor and counting equipment (which may cost between $25,000 and $30,000). Given such access, however, the cost of analysis is quite low, whether considered on a per element or per sample basis.

5.6.2.3. X-Ray Spectrometry

X-Ray spectrometry encompasses a number of analytical techniques, many of which are finding increasing application to the analysis of airborne particles.[169,170,188–204] Unfortunately, different forms of X-ray spectrometry have distinct analytical characteristics and equipment requirements, and failure to distinguish between them has led to considerable confusion and misunderstanding. This section points out the important distinctions between the X-ray spectrometric methods and indicates how the performance characteristics of each make it more or less applicable to the analysis of airborne particles.

All forms of X-ray spectrometry involve the bombardment of a sample, which can be a solid, a liquid, or a gas, with radiation of sufficient energy to eject inner shell electrons from the atoms in the sample. Ejected electrons are immediately replaced by electrons from outer atomic shells, with simultaneous emission of X-rays whose energies are characteristic of the electron transitions (thus the elements) responsible for their emission. Since several electron transitions can occur for a given element, several characteristic X-ray emission wavelengths are available for identification of a sample element. Quantitative analysis is achieved by determining the number of X-rays of each energy emitted.[189–193] Variations of this basic process lie primarily in two areas, the form of the bombarding radiation, and the method of detection and energy discrimination of the emitted X-rays.

Two types of X-ray detection system are employed in X-ray spectrometry. The first, which has been used analytically for many years, is wavelength dispersive detection,[190–192] in which emitted X-rays are collimated and diffracted by a so-called analyzing crystal. Here the angle of diffraction θ is related to the wavelength λ of an X-ray, according to Bragg's law:

$$n\lambda = 2d \sin \theta$$

where n is a small integer and d is the interplanar atomic spacing of the analyzing crystal used for diffraction. X-Rays having different wavelengths are thus diffracted at different angles. Detection is achieved by means of a conventional Geiger, scintillation, or proportional counter, which can be positioned so that only X-rays diffracted at a given angle are detected, or scanned through all angles to record an X-ray spectrum. The latter mode provides lower sensitivity per unit analysis time, since less time is spent in counting X-rays in any given angular range. The second type of detection system is relatively new and employs a lithium-drifted silicon detector, Si(Li), in conjunction with a 1024-channel multichannel analyzer.[190,193] All X-rays entering this detector are sorted according to their energies and stored as digital signals in different channels of the multichannel analyzer. Consequently a complete X-ray spectrum is recorded.

Both detection systems provide the same X-ray spectral information, since wavelength and energy are directly related ($E = hc/\lambda$), and both have certain advantages for the analysis of airborne particles. Thus wavelength dispersive detection exhibits excellent resolution (0.5 eV), thereby maximizing signal to noise ratios for the narrow spectral peaks observed. However the transmission efficiency of wavelength dispersive spectrometers is low, and analyses require lengthy counting periods at each wavelength of interest to achieve adequate detection. Indeed, analysis by wavelength dispersive X-ray spectrometry is hardly a true multielemental technique in that each element must be determined separately. In contrast to this, energy dispersive detection records the complete X-ray emission spectrum of a sample during a continuous analysis period, but with a resolution that is considerably inferior (150 to 160 eV) to that of wavelength dispersion. Both systems are subject to spectral interferences, though these are of different origin. In the case of wavelength dispersion, second and higher order ($n > 1$) diffraction lines occasionally can produce interference with an analytical line of interest. Spectral overlap is frequently a problem with energy dispersive detection because of the relatively poor resolution of the Si(Li) detector.

The greatest variations among X-ray spectrometric methods are due to the different types of exciting radiation that can be employed. These can be conveniently divided into four categories:

1. Monochromatic X-rays.
2. Polychromatic X-rays.
3. Charged particles such as electrons, protons, or α particles.
4. γ-Rays.

Only when X-rays are used for excitation is the much-abused term "X-ray *fluorescence* spectrometry" applicable.

Excitation of X-ray emission requires that the exciting radiation possess energy greater than a threshold value, which increases approximately linearly with atomic number Z. Furthermore, the efficiency of excitation, consequently the sensitivity of detection, falls off steeply as the energy of the exciting radiation increases above this absorption threshold or "edge."[192,193] Consequently maximum sensitivity is achieved for a given element when the exciting radiation has an energy just above that of the absorption edge. This is an important consideration when monochromatic X-rays are used for excitation, since only a few elements will be excited with near maximum efficiency. To cover a wide range of elements with monochromatic X-ray excitation, therefore, it is usually necessary to employ several X-ray tubes, or a series of secondary targets.[193] To achieve sufficient secondary target emission intensity, a high intensity primary X-ray tube must be used.

The term "monochromatic," as applied to exciting X-radiation is a misnomer, since X-ray tubes actually emit continuous radiation or bremsstrahlung upon which are superimposed lines characteristic of the tube target material. This continuous radiation ensures some excitation of all elements but produces substantial background because of scattering from low Z elements such as aluminum, carbon, and silicon present at high levels in airborne particles or filter matrices.[188,193] The use of secondary targets greatly reduces this background. In general, excitation with monochromatic X-rays derived from secondary targets provides the best detection limits for trace elements in airborne particles.[193,194] A potentially important advance in this area involves the recent development of so-called pyrolyzed graphite monochromators, whose high X-ray transmission characteristics enable their use with conventional low intensity X-ray tubes.

Irradiation with polychromatic X-rays has the advantage of providing more or less uniform excitation for all elements. However the X-ray background due to scattering is normally too great to enable analysis of trace elements in airborne particles.

Irradiation with γ-rays furnishes an attractive means of excitation, since energies are sufficient to excite heavy elements and background is low because radiation is monoenergetic.[193,195] Sensitivities are improved substantially by employing secondary targets so that the sample is actually irradiated with monochromatic X-rays derived therefrom. γ-Rays are generally derived from a radioactive isotope whose choice and activity level is limited by safety and half-life considerations. The most useful isotopic source currently available is ^{241}Am, which emits γ-rays of energy 60 keV and 26 keV with a half-life of 458 years. There is little need to apply decay corrections for ^{241}Am, although these are important for isotopes such as ^{109}Cd (453 days), ^{57}Co (270 days), ^{153}Gd (242 days). Isotopic γ-ray sources have the advantages of simplicity, reliability, and low background stimulation, and they can be applied readily to the analysis of airborne particles. In general, however, detection limits are somewhat inferior to those attainable with monochromatic X-rays.[195]

Excitation of X-ray emission by charged particle bombardment primarily involves the use of electrons and protons. Electron bombardment suffers because stimulation of bremsstrahlung is considerable, leading to high scattering backgrounds in the X-ray spectra. Consequently little use has been made of electrons in the analysis of airborne particles.[190,193] An important specialized application, however, is in the electron microprobe or X-ray analyzer associated with an electron microscope, where excitation is achieved with the electron beam. Monoenergetic proton excitation does have some advantage, however, in that bremsstrahlung is quite low and good detection limits have been demonstrated[193,196] for analysis of airborne particles (Table 5.14). A significant disadvantage is that access to a van de Graaff generator is required.

It is apparent from the foregoing remarks that X-ray spectrometry can be performed in a variety of configurations and that sensitivities for trace elements vary considerably. Therefore it is essential to establish which instrumentation is involved before assessing the utility of X-ray spectrometry as a whole. The detection limits presented in Table 5.14 are for monochromatic X-ray excitation, thus they may not be universally attainable.[194]

X-Ray spectrometric analysis is a true multielemental technique applicable to all elements having atomic numbers greater than that of potassium. To analyze for elements below argon, however, it is necessary to exclude air from the system, either by evacuation or by purging with helium, since there are substantial spectral interferences from oxygen and, to a lesser extent, from argon.

Some information on the oxidation states and binding characteristics of elements in the first two rows of the periodic table can be obtained from X-ray spectrometry, since inner electron energy levels are somewhat influenced by the valence shell. The high resolution of wavelength dispersive X-ray spectrometry is required to observe the small chemical shifts involved, but distinction can be made, for example, between S^{2-}, S, $SO_3{}^{2-}$, and $SO_4{}^{2-}$. In principle this speciation capability can be extended to elements having atomic numbers up to approximately 50.

X-ray spectrometry has the major advantage that airborne particles can be analyzed nondestructively on a collection surface such as a filter, provided background impurities are sufficiently low.[171,174] Its major limitation lies in the difficulty associated with achieving quantitative calibrations. This is because the X-ray penetration and escape depths are only the order of a few micrometers and vary considerably with the energy of the X-rays involved and the nature of the substrate. Thus a "soft" X-ray such as sulfur K_α (2.5 keV) is emitted only from close to the sample surface, whereas higher energy X-rays, such as lead L_α (10.5 keV), are derived from considerably greater depths. Consequently it must be recognized that X-ray spectrometry is biased toward surface sampling, a fact that must be taken into account when samples exhibit surface predominance of elements of interest.[29]

The importance of matrix effects in determining emitted X-ray intensities makes it necessary to have standards as similar in composition to samples

Table 5.14. Typical Elemental Detection Limits (ng m^{-3}) in Airborne Particulate Samples for Several X-ray Spectrometric Methods[a]

		Elements											
Detection	Excitation	Al	Au	Br	Ca	Cu	Fe	K	Pb	S	Se	V	Zn
Wavelength dispersion	X-Ray tube												
	Cr	110		120	3.1	51	47	0.9	310	16	260	17	55
	Rh	26		67	9.1	15	9.5		83	4	47	1	16
	W			51		13	11			16	31	9	
Energy dispersion	X-Ray tube												
	Mo				110		38		35		15	50	38
	W				10		12		60		35	11	52
	W tube–Ni foil				44		38	7	35	180	26	28	35
	Secondary fluorescer												
	Mn				18							16	
	Cu				20		14					22	
	Ag				97				60		22	120	57
	Cr–Zr				27		11				14		68
	Isotope												
	^{55}Fe; 7 mCi				35							57	
	^{109}Cd; 70 mCi				1900		680		224		132	680	440
	Particle												
	Proton		1.6	0.31		0.64		0.31					
	Proton[b]	0.01	0.1	0.06	0.01	0.01	0.01	0.01	0.1	0.01	0.04	0.01	0.02
	Alpha		6.4	3.1				0.31					

[a] Determined from synthetic multielement standards deposited on Millipore or Whatman No. 41 filter paper; adapted from nanogram per square centimeter values[295] by assuming passage of 50 m^3 of air through a 47 mm diameter Millipore filter; all determinations based on 100 sec counting time except ^{55}Fe with 2000 sec.

[b] Adapted from absolute (nanogram) detection limits[296] by assuming a 50 m^3 sampling volume; 10 μC of charge accumulated, 10 min irradiation time, 3.7 MeV protons. Limits are not necessarily representative of air particulate applications.

as possible. This can be achieved by standard addition, provided the standard can be distributed analogously to the analyte within the sample. In the case of particulate material, such identity is usually difficult to achieve. Calibration by internal standardization achieves matrix similarity, but considerable errors may result when analyte and standard elements have significantly different X-ray emission energies. In practice, however, it is more common to prepare a set of working standards in different matrices[204] and to use these for calibration with suitable absorption corrections.

Two sample configurations are normally employed.

1. So-called thin samples, in which it is assumed that X-rays can penetrate and escape from all parts of the sample. To satisfy this approximation, particles must be ground to a few micrometers in size and must be presented in extremely thin layers.[167,194] Detection limits are poor for such samples.
2. "Thick" samples, in which X-rays only penetrate to and escape from depths of a few micrometers. Corrections for the extent of absorption of X-rays must then be applied, or it must be assumed that the X-ray penetration and escape depths are the same for the sample and the standard.[193,194,205] This assumption is generally reasonable for solution samples; however scattering from the light elements that comprise the solvent usually inhibits trace analysis.[193]

Regardless of whether corrections are made for absorption of X-rays, it is desirable to have high depth uniformity between samples and standards. This can be achieved by pressing the particles into a pellet or disk with 1 to 5% of a binding material (pressures $> 20{,}000$ psi are usually required for airborne particles) or by fusing them at high temperature ($\sim 800°C$) before pressing.[167] Such treatment, however, generally requires that particles be separated from a collection surface, and volatile elements may be lost during fusion. Most frequently, therefore, particles are analyzed directly on a filter and absorption corrections are applied.[194]

X-Ray spectrometry is capable of high precision (~ 2 to 5%) and reproducibility for analysis of airborne particles. The problems associated with calibration, however, make good accuracy difficult to obtain unless considerable care is taken to match samples and standards or to apply matrix absorption corrections.[169,194] So for occasional use the technique is best suited to problems requiring good relative, rather than absolute, accuracy. The cost of equipment is moderately high ($20,000 to $80,000), but with on-line computerized baseline stripping and data manipulation capabilities, analysis times are attractively short (5 to 30 min). Despite its ability to detect most elements of interest in airborne particles, however,

X-ray spectrometry is highly susceptible to misuse, and detection limits can be optimized for only a few elements in any one analysis.

5.6.2.4. *Atomic Absorption Spectrometry*

Since its presentation as an analytical tool in 1955,[206–208] atomic absorption spectrometry (AAS) has developed almost explosively. It is not surprising, therefore, that it is currently the technique most widely employed for elemental analysis of airborne particles.[65,158,165] There are several reasons for this, and the foremost are its low cost, wide availability, simplicity and speed of operation, accuracy, and adequate detection capabilities for essentially all elements of environmental interest (Table 5.15). However the technique is sample destructive, and individual elements must be determined sequentially. Thus although in principle some 60 elements can be determined, airborne particulate samples are often not large enough to enable determination of more than half a dozen unless precautions are taken to conserve sample wherever possible.

Atomic absorption spectrometry is the atomic counterpart of conventional molecular absorption spectrophotometry. Thus a light beam containing wavelength components that can be absorbed by the sample atoms is passed through an absorption "cell" containing atomic vapor. The beam intensity is attenuated according to Beer's law. The main instrumental differences between molecular and atomic absorption spectrometry are that in the latter case the light beam is derived from a hollow cathode lamp, which emits only the wavelengths characteristic of the element being determined, and that the so-called cell consists of a flame (or other thermal source) capable of acting as a reservoir for atoms.

Several types of atom reservoir are in current use. The most common is a flame supported by hydrogen and air, hydrogen and oxygen, acetylene and air, acetylene and nitrous oxide, or some other fuel–oxidant mixture. Each flame type has distinct temperature and oxidation-reduction characteristics, and different flames may be necessary for determining different elements.[209] Normally the flame is extended along the axis of the light beam to provide a long absorption path. Samples, which must be in the form of solutions, are aspirated into the flame, where evaporation of solvent and analyte occur, and the analyte is further thermally dissociated into atoms in their ground electronic energy states. Since a high steady state concentration of ground state atoms is desirable, flame temperatures must be sufficient to achieve atom production but not so hot that extensive thermal excitation is caused to electronic energy levels above the ground state.

Significantly improved elemental detection limits can be achieved if the sample is rapidly converted to atoms in a transient, rather than a steady

Table 5.15. Typical Elemental Detection Limits in Airborne Particle Samples for Flame and Flameless Atomic Absorption Spectrometry

	Detection Limits (ng m^{-3})				
Element	Flame[a]	Flame[b]	Carbon Furnace[c]	Carbon Rod[d]	Tantalum Strip[e]
Al	400	400	0.27	0.03	0.15
As	20				0.15
Ba	20				0.005
Be		0.04			0.001
Bi		2			0.02
Ca	0.02	30	0.53	0.01	0.0005
Cd	0.2	0.3	0.007	0.0005	0.001
Co		0.8		0.005	0.15
Cr	2	0.3		0.005	0.01
Cu	1	3	0.16	0.005	0.005
Fe	10	70	0.14	0.003	0.05
K		40			0.0005
Mg		10	0.012	0.021	0.00005
Mn	1	0.3	0.14	0.0005	0.0035
Na		30			0.0005
Ni	4	2	0.036	0.01	0.1
Pb	2	7	0.028	0.005	0.005
Si		800			5
Sn				0.05	0.05
Sr		3			0.005
Ti		70			0.5
Tl	10				0.035
V	10	6			0.2
Zn	0.2	30	0.02	0.001	0.0005

[a] 2000 m^3 of air was passed through glass fiber filters; samples were pretreated by low temperature ashing.[165]
[b] 2000 m^3 of air was passed through polystyrene filters; samples were solubilized by a combination of dry and wet ashing.[297]
[c] Adapted from absolute (nanogram) detection limits[298] by assuming a 2000 m^3 sample volume, sample uptake into 10 ml of solution, and 10 μl of solution applied to the atomizer.
[d] Extrapolated from 0.6 to 2000 m^3 of air passed through Nucleopore filters; samples were pretreated by wet ashing and ultrasonic agitation; 5 μl of solution was applied to the atomizer.[299]
[e] Adapted from absolute (nanogram) detection limits[300] by assuming a 2000 m^3 sample volume, sample uptake into 10 ml of solution, and 10 μl of solution applied to the atomizer.

state, flow configuration (Table 5.15). This can be achieved by placing either a solution or solid sample in a container such as a Delves cup or tantalum boat[210,212] in the flame. Good detection limits can also be obtained using a graphite furnace or carbon rod (Table 5.15) in which the solution or solid sample is electrically heated to produce high transient concentrations of atoms in a nearly static atmosphere.[213–220] Electrically heated tantalum strips have also been developed for producing a transient atom reservoir, but these appear to offer little advantage over the graphite furnace or carbon rod.

Despite repeated statements in the early literature on atomic absorption spectrometry that the technique was completely free from interferences,[221] very significant interferences do occur. Indeed, failure to recognize their existence can lead to substantial errors. Chemical reactions, such as the formation of alkaline earth phosphates and lead silicate,[222] can occur for many elements and can result in reduction of the free atom population available for absorption of radiation. Such negative errors generally can be eliminated by addition of a suppressing agent such as a lanthanum salt.[209] Before performing an analysis by atomic absorption spectrometry, therefore, inexperienced analysts are strongly encouraged to consult recent literature[171] to obtain information about probable chemical interferences and their avoidance for each element being determined.

Spectral interferences, however, are potentially more serious than chemical interferences, especially when working close to elemental detection limits with the chemically complex solutions obtained from digestion of airborne particles. Spectral interferences can be due to reduction of transmitted radiation both by scattering from particles present in the atom reservoir and by molecular absorption at the analytical wavelength. Occasionally more than one element has an absorption line at the same wavelength, although this is unusual. Light scattering is greater at short wavelengths than at long, is greater in the case of carbon rod and graphite furnace systems than with flames, and is enhanced by the presence of high sulfate levels in solution.[158] Consequently the use of concentrated sulfuric acid (which can result in formation of insoluble sulfate particles) in digestion mixtures is not recommended. Similarly, the wavelength region between 200 and 250 nm is degraded by absorption of radiation by nitric oxide derived from nitric acid.

Such spectral interferences can be circumvented by employing background correction techniques. These involve determination of the absorbance at a wavelength very close to, but distinct from, the analytical wavelength and are *absolutely essential* if accurate results are to be obtained. Background correction is most frequently accomplished by monitoring a hollow cathode emission line which is not absorbed, or, more elegantly, by employing a continuum source such as a hydrogen or deuterium arc lamp in conjunction with the hollow cathode source.[223] Both elemental and background absorptions are measured with the hollow cathode source, but only background is measured with the continuum source. The net elemental absorbance is obtained from the difference between the two measured absorbances. Instruments can now be purchased or modified to obtain the correction automatically in a single measurement.[224,225] In complex samples such as airborne particulate digests, however, care should be taken to ensure that background corrections are not made at wavelengths where some other element present absorbs.

The precision attainable with conventional flame atomic absorption spectrometry is 1 to 5% but is only 5 to 15% when a carbon rod or graphite furnace is used. This is due to the greater matrix effects and required background corrections encountered with these devices and dictates careful work and cautious interpretation of results.[158] Accuracies attainable with flame instruments are the order of 5% with careful calibration. For airborne particulate samples, calibration by standard addition usually gives the best results, provided the response curve is linear, this procedure is time-consuming, however, and is not generally used for routine work. It should be stressed that calibration by standard addition *does not* compensate for background absorption as some investigators[226] seem to assume. For routine elemental analysis of airborne particles by atomic absorption spectrometry, the method of Thompson et al.[165] is recommended.

Although most trace elements present in airborne particles can be determined by atomic absorption spectrometry, several (including antimony, arsenic, and mercury) require the application of special methodology. In the case of antimony and arsenic this is because the useful analytical absorption wavelengths are in the ultraviolet region close to the air cutoff, because interferences from the sample matrix are substantial, and because refractory oxide formation occurs when aqueous samples are introduced into a flame. These problems are overcome by forming volatile metal hydrides by reaction of the sample with sodium borohydride. The hydride is then introduced into a flame or a heated tube. Hydride formation can be accomplished for antimony, arsenic, bismuth and selenium[227] and has the advantages of preconcentrating the elements of interest, separating them as a gas from the sample matrix and, in some cases, enabling the use of weak absorption lines well removed from the short wavelength, ultraviolet region. This approach is still being developed,[171] and care should be taken in its application.

Analysis for mercury by conventional flame atomic absorption spectrometry is impractical because the high volatility of elemental mercury makes it difficult to maintain a sufficient concentration of unexcited atomic mercury in a flame. Since atomic mercury vapor is stable at ambient temperatures, however, it is possible to determine its absorption directly in a so-called cold vapor system.[128] Both particulate and vapor phase mercury collected from the atmosphere can be presented for analysis either in solution or adsorbed on to gold, silver, or graphite.[128,133,142–147,228–230] Where adsorption is employed, particles are normally retained (though probably not quantitatively) by the packing material in the adsorption tube. Analysis then requires generation of either a steady state or a transient concentration of atomic mercury vapor in a flow-through absorption cell. Mercury present in digested sample solutions is reduced to the elemental state and then swept out of solution by sparging with a stream of air in which a steady state concentration of mercury vapor is attained. Reduction can be accomplished in several ways; the most common[128,133,228–230] involves oxidation of all forms of collected mercury to Hg(II) with acid permanganate/persulfate followed by reduction to Hg^{0} with stannous chloride. This procedure, though widely used, is limited by the level of mercury contamination of the several reagents employed.[229] Detection limits as low as 2 ng can readily be accomplished by employing sodium borohydride as the reducing agent, however.[231] Where mercury vapor is collected together with particulate mercury using an adsorption tube, the vapor can be generated directly for analysis by heating the collection tube to approximately 650°C.[143,144,147] This produces a high transient concentration of mercury vapor, thus increasing the sensitivity of detection ($\sim$ 1 ng) over that obtainable from generation of a lower steady state concentration. However transient vaporization of adsorbed mercury may be accompanied by a significant interference from covolatilized water, which also absorbs radiation at the analytical wavelength (2536.7 Å). It is necessary, therefore, to employ background correction or to remove water vapor before evolution of the adsorbed mercury.[144,147]

Atomic absorption spectrometry is undoubtedly an extremely attractive technique for performing elemental analyses of airborne particles, one that is widely available to most investigators because of its relatively low instrumental cost ($4000 to $16,000). To some extent the speed and precision of analyses are instrument (and frequently cost) dependent. For example, some desirable features of more expensive units include the following:[158]

1. Dual beam optics, which enable electronic compensation for source instability.
2. Dual channels, to enable simultaneous determination of two elements.
3. Signal integration for increased precision.

4. Background compensation systems, to correct for any nonatomic absorption.
5. Automatic sample changing and data processing.

Such features do not, however, overcome the undoubted disadvantage of having to analyze for each element sequentially, and the associated problem of sample shortage. Consequently, when a large number of elements must be determined, it is desirable to conserve sample by dilution for the elements present in substantial amounts, or to achieve maximum sensitivity by using transient techniques. The other major disadvantage of atomic absorption spectrometry is the considerable sample preparation required.

Insofar as atomic absorption spectrometry, by definition, detects only atoms, it exhibits no intrinsic capability for distinguishing chemical species. However when transient generation of atoms is employed, as with a graphite furnace or carbon rod atomizer, different compounds containing the same element (e.g., $PbCl_2$, PbS, PbO_2) are often evolved at different rates, with the result that more than one lead absorption peak is frequently observed. The peaks are not sufficiently resolved in time to enable quantitative determination of individual species, but their observation does provide qualitative evidence of the presence of more than one compound. In such cases peak integration is necessary to achieve determination of total elemental concentration.

5.6.2.5. *Atomic Emission Spectrometry*

Several forms of atomic emission spectrometry have been or are being used for the elemental analysis of airborne particles or vapors. These include arc–spark emission, plasma emission, atomic fluorescence spectrometry, and flame photometry.[171] With the exception of atomic fluorescence spectrometry, all operate on the same basic principle.

The sample, which can be either a solid or a solution, is subjected to a high temperature generated by an electrical discharge, a flame, or a sustained plasma. The solvent, if present, and the analyte are vaporized and dissociated into atoms, which are then thermally excited to an electronic energy level higher than the ground state. The wavelength of radiation from excited atoms returning to the ground state is characteristic of the emitting atom, and the intensity is proportional to the number of emitting atoms of each type. Emitted radiation can be isolated with a conventional monochromator and detected by a photomultiplier or dispersed across a photographic plate.

Before 1960 arc–spark emission spectrometry was the most useful and versatile technique available for trace element analysis and was employed in much of the early work on airborne particulates.[232,233] Its use, however, has diminished greatly (possibly too greatly) with the advent of atomic absorption spectrometry, and modern methods of spark source mass spectrometry, X-ray spectrometry, and instrumental neutron activation analysis. For qualitative or semiquantitative analysis, the sample, which can be the particles themselves or the solid residue from digestion, is placed in a graphite electrode cup and burned in a dc arc. Solutions are best analyzed by spark excitation of residue after evaporation on the wetted surface of a rotating disk electrode.[234–236] The resulting spectral emission is dispersed across a photographic plate that integrates the light emitted at each wavelength over the burning period. More rapid and convenient though somewhat less sensitive detection is achieved by so-called direct reading instruments, in which slits and photomultipliers are arrayed to isolate and simultaneously record the spectral emission from several predetermined elements.[237,238] Quantitative analysis by arc–spark emission spectrometry is usually feasible only in the hands of a skilled operator; an ac arc or spark discharge may be employed to enhance precision.[237,238]

Arc–spark emission spectrometry is applicable to some 60 elements (Figure 5.13), many of which may be at the microgram per gram level in a few milligrams of sample (Table 5.16). Because of the expense and complexity of the equipment, determination of a single element is generally impractical given the alternative methods available, although emission spectrometry is one of the few techniques capable of determining beryllium. Multielemental analysis using photographic detection does, however, provide one of the most readily available methods for establishing the approximate concentrations of most trace metals and some nonmetals present. Visual estimation of line intensities on a photographic plate can rapidly establish concentrations to within a factor of 2 or 3, and precisions of the order of 10 to 20% can be achieved by optical densitometry.

Calibration can be obtained by working curve, internal standard, standard addition, or isotope dilution methods as described for spark source mass spectrometry. It should be noted, however, that the intensity of a spectral line depends on both the physical and the chemical makeup of the sample, thus internal standardization, which is necessary to overcome problems of source instability, must employ elements that are both chemically and spectrally similar to the analytes. Matrix effects can also be reduced by addition of a radiation buffer whose function is to desensitize the influence of different matrix materials on the elemental excitation efficiency. Spectral interferences are generally not of major importance in emission spectrometry, although use of a dc arc discharge in air results in the appearance of band emission because of the formation of cyanogen in the discharge. This can be reduced by employing an inert atmosphere, but quantitative work requires the careful use of spectral background correction.

Table 5.16. Typical Elemental Detection Limits in Airborne Particulate Samples for Arc, Spark, and Inductively Coupled Plasma Emission Spectrometry

	Detection Limits ($ng\ m^{-3}$)			
Element	DC Arc[a]	Plasma[b]	Copper Spark[c]	Graphite Spark[c]
Al	17	2	2	0.3
As	[60][c]	40	50	10
B	[3]	5		
Be	1.7	0.5	0.02	
Cd	[10]	2	10	2
Co	17	3	4	0.5
Cr	8	1	2	0.1
Cu	[2]	1	10	0.05
Fe	[0.3]	5	8	0.3
Hg	8	200	50	1
Mg	8	0.7	1	
Mn	8	0.7	0.3	0.03
Ni	8	6	0.1	0.1
Pb	50	8	3	
Sb	[10]	200	50	1
Se	[1000]	30		
Ti	17	3	1	0.3
V	34	6	0.8	0.1
Zn	50	2	20	1

[a] Adapted from absolute (nanogram) detection limits[91] by assuming air sampled for 60 min at 1 l min^{-1} with a graphite cup filter—0.06 m^3.

[b] Adapted from solution values (micrograms per milliliter)[240] by assuming a 10 m^3 sampling volume and sample uptake into 10 ml of solution. Values are optimal for each element and are not necessarily representative of air particulate applications.

[c] Adapted from absolute (nanogram) detection limits[175] by assuming a 10 m^3 sampling volume. Values are optimal for each element and are not necessarily representative of air particulate applications.

Conventional emission spectrometry utilizing an electrical discharge cannot be applied directly to the analysis of airborne particulates that have been collected on filters. Direct analysis, however, can be achieved by actually collecting the particles, and possibly also vapors, in the graphite cup electrode. This procedure has been developed and described by Seeley and Skogerboe[91] and offers an attractive method for multielemental analysis of airborne particles by laboratories possessing capabilities in conventional emission spectrometry.

The main advantages of electrical discharge emission spectrometry are its wide availability and the comparative ease of performing rapid multielemental survey analyses. Quantification is difficult and time-consuming, however, and detection limits are often borderline for several elements of interest in airborne particles.

The sensitivity limitation of arc–spark emission spectrometry can be largely overcome if excitation is achieved in a plasma (Table 5.16). Several types of plasma emission source can be used, with the plasma being sustained by an inductively coupled radiofrequency, a microwave field, or a dc discharge.[171,239] The source most fully developed for analytical purposes is the inductively coupled radiofrequency plasma (ICP), which operates at about 30 MHz, and a generator output power of 1.0 to 1.5 kW.[240,241] The plasma is sustained in a stream of argon or helium flowing in a toroidal configuration, and samples are introduced as solutions that enter the plasma after nebulization. Solid samples cannot be analyzed directly, thus in this respect the pretreatment requirements are the same as those for atomic absorption spectrometry.

Some 60 elements are accessible by ICP–emission spectrometry, and the high excitation efficiency of this source provides detection limits that are generally significantly lower than those normally attainable by atomic absorption spectrometry or conventional dc arc emission spectrometry (Table 5.16). Furthermore, analysis of solution samples affords ease of calibration, although the intrinsic precision of the method is realized only in a simultaneous multielemental mode when detection is achieved by direct reading or when quantitative densitometry is employed following photographic detection. Despite some early claims that ICP–emission spectrometry was effectively free from matrix interferences, such interferences do exist for many elements, notably the alkali and alkaline earth metals.[239–242] Therefore it may be necessary to employ standard addition techniques or to add scavenging species for quantitative work. It is, of course, necessary to correct for spectral background and to recognize that the spectra contain a number of strong emission lines because of the presence of excited argon species.

At the present time ICP–emission spectrometry is potentially a very attractive technique for the multielemental analysis of airborne particles. Its main advantages over, for example, atomic absorption spectrometry are its improved detection limits and its capability for simultaneous multielemental analysis, thus sample conservation. Its main disadvantages are the cost and complexity of equipment and the requirement for a skilled operator. These drawbacks should be reduced as the technique is further developed and refined. In the meantime caution should be exercised in the use of this new technique whose idiosyncrasies are not fully known.

One application of plasma emission spectrometry for the determination of atmospheric mercury is worthy of mention. In this technique elemental mercury vapor, together with several other vapor species containing mercury, is adsorbed by formation of an amalgam onto gold or silver wire, chips, or coated glass beads. All adsorbed mercury is then removed as elemental mercury vapor into a flowing helium stream by heating to 500 to 600°C, and it enters a dc plasma sustained between two platinum or tungsten electrodes. Atomic mercury emission from the plasma is monitored using a conventional monochromator–photomultiplier combination.[71,143] Mercury detection limits with this system can be as low as 0.02 ng, thereby enabling determination of total airborne mercury in samples of a few liters of ambient air. This type of detection system is relatively free from interferences, and good results can be obtained even without background correction.

The third emission spectroscopic technique is that of flame photometry in which sample vaporization, atomization, and excitation are achieved by a flame. In common with all atomic emission techniques, flame photometry is intrinsically capable of simultaneous multielemental determination; however the classical mode of operation is to analyze for one element at a time using a conventional monochromator–photomultiplier combination.[171] The monochromator also can be used to scan the atomic emission spectrum when sample sizes and elemental concentrations are sufficiently large. Flame photometry offers detection limits superior to those of conventional flame atomic absorption spectrometry for several elements (Table 5.17). It is perhaps, useful to think of the two techniques as being complementary and having similar sample requirements, calibration methods, precision, and accuracy. In general, flame photometry is most useful for determining elements whose analytical wavelengths are greater than 400 nm.

Flame photometry requires that a high proportion of the atoms in the flame be excited to emission, whereas atomic absorption demands a high ground state atomic population. In both cases, however, detection limits are primarily determined by the efficiency of the atomization process. Consequently it is reasonable practice to employ similar flame compositions

Table 5.17. Typical Elemental Detection Limits in Airborne Particulate Samples for Atomic Absorption Spectrometry, Flame Photometry, Atomic Fluorescence Spectrometry, and Differential Pulse–Anodic Stripping Voltammetry (DP–ASV)

	Detection Limits (ng m^{-3})			
Element	Atomic Absorption[a,b]	Flame Photometry[b]	Atomic Fluorescence[b]	DP–ASV[c]
Al	0.15	0.025	0.025	
As	0.5	250	0.5	
Be	0.01	0.5	0.05	
Bi	0.25	10	0.25	
Ca	0.005	0.0005	0.000005	
Cd	0.005	4	0.00005	0.005
Co	0.025	0.15	0.025	
Cr	0.015	0.02	0.02	
Cs		0.04[d]		
Cu	0.01	0.05	0.005	0.02
Fe	0.025	0.15	0.04	
Hg	2.5	200	0.1	
Li		0.3[d]		
Mg	0.0005	0.025	0.005	
Mn	0.01	0.025	0.01	
Na	0.01	0.0005		
Ni	0.025	0.1	0.015	
Pb	0.05	0.5	0.05	0.025
Rb		0.5[d]		
Se	0.5	500	0.2	
Si	0.5	25		
Sn	0.1	1.5	0.25	
Ti	0.45	1.0	0.5	
Tl				
V	0.1	0.05	0.35	
Zn	0.01	250	0.0001	0.02

[a] Adapted from solution values (micrograms per milliliter)[240] by assuming a 2000 m^3 sampling volume and sample uptake into 10 ml of solution. Values are optimal for each element and are not necessarily representative of air particulate applications.
[b] Compare these optimal values with those of real samples in Table 5.15.
[c] Adapted from absolute (nanogram) detection limits[248] by assuming a 2000 m^3 sampling volume.
[d] 2000 m^3 of air was passed through polystyrene filters; samples were solubilized by a combination of dry and wet ashing.[298]

for flame photometry as for atomic absorption, except that a total consumption burner is required for flame photometry to concentrate the emitting atoms in a small volume.

It is commonly held that interferences are greater in flame photometry than in atomic absorption. This is not intrinsically correct for chemical interferences, but several of the elements (e.g., alkali and alkaline earth metals) that are most accessible by flame photometry are subject to substantial interferences from ions such as phosphate and aluminum in solution. The simplest method of reducing these interferences is to employ a hot flame, such as nitrous oxide or oxyacetylene, which is capable of vaporizing the partially refractory species whose formation is responsible for interference. Alternatively, lanthanum ions can be added to the sample solution to tie up the phosphate, and ammonium ions can be added to depress the influence of aluminum (the mechanism by which such depression occurs is not fully understood). Spectral interferences are generally greater in flame photometry than in atomic absorption, and it is necessary to employ background correction with particular care to ensure that the region of background determination is free from atomic emissions due to elements other than the analyte.

Flame photometry has the advantages of low cost, simple instrumentation, ease of calibration, and speed of analysis. Its complementary relationship with atomic absorption makes combined use of the two techniques attractive, although the combination of sample size requirements and sequential analysis of elements frequently limits its utility for determining a large number of elements in airborne particles.

Atomic fluorescence spectrometry, though correctly classified as an atomic emission technique, differs from those described previously in that a hollow cathode lamp is used to excite atoms to emission. Solution samples are entrained into a conventional flame whose function is to establish a reservoir of ground state atoms. Specific atomic excitation is achieved by radiation from the hollow cathode lamp, and the resulting emission is viewed at right angles to the incident radiation.[243] The advantages of this approach over those of atomic absorption and flame photometry are the ability to use a cool, low background flame, plus enhanced spectral specificity, high excitation efficiency, and improved detection limits for many elements of interest (Table 5.17). Atomic fluorescence spectrometry combines many of the advantages of atomic absorption and flame photometry, although it is still subject to the chemical interferences intrinsic to flame techniques. In practice, however, atomic fluorescence spectrometry has not been widely employed for determination of trace elements in airborne particules, probably because of its relatively recent advent. However the technique should be included with atomic absorption and flame photometry to provide a suite of relatively low cost methods for the precise determination of trace elements with maximum capability for sample conservation.

Overall, atomic emission spectrometry encompasses several techniques that have undoubted applicability to the multielemental analysis of airborne particles. Both semiquantitative survey analyses and precise determination of individual elements can be achieved as appropriate. None of these methods, however, provides information about chemical species, and none is capable of analyzing particulate material without removing it from the collection system. The one exception involves direct collection of particles in a graphite electrode cup for subsequent analysis by arc–spark emission spectrometry.[91] A special advantage of emission spectroscopy in general is that the analytical quantity of interest is the absolute intensity of a spectral emission line, as opposed to the small difference in intensity between incident and transmitted radiation as required for absorption measurements. This means that in principle at least, and all other things being equal, emission spectrometry affords improved sensitivity and precision over absorption spectrometry.

5.6.2.6. *Electrochemical Methods*

A variety of electrochemical methods have been used for determining either individual elements or groups of elements present in airborne particles. These involve potentiometry using ion selective electrodes and both polarography and stripping voltammetry in their dc pulse, and differential pulse modes.[165,244–249] By comparison with techniques such as atomic absorption spectrometry, electrochemical methods have few significant advantages for airborne particulate analysis; however individual techniques may offer real advantages in terms of low cost, simplicity, sensitive detection, and nondestructive analysis. As a result, such techniques will find continued use on a limited scale, and it is appropriate to point out some of their characteristics here.

Despite the wide variety of configurations in which electrochemical analyses can be performed, several general characteristics are recognizable. Thus all methods require the sample to be in solution, thereby enabling ease of calibration; all involve either oxidation or reduction of the analyte species being determined, and all determine the presence of an *ionic* form of the analyte. In the last respect electrochemical analyses are species selective. Furthermore, all methods respond to activity rather than concentration—a property that may be of considerable environmental significance in natural water systems but is hardly important for digested airborne particles.

Ion selective electrodes have found limited use for the determination of bromine, cadmium, calcium, chlorine, copper, fluorine, iodine, lead, silver, and sodium in airborne particles.[244] These elements can be detected at solution concentrations in the range 10^{-5} to 10^{-8} M. In view of the simplicity and low cost of ion selective electrodes, it is unfortunate that their use must be strongly discouraged in complex matrices such as digestates. This is because the electrodes are selective rather than specific, and they respond to several ions likely to be present in addition to those for which they are selective. The one possible exception to this rule is the fluoride selective electrode, which responds only to fluoride and hydroxyl ions. Consequently its use in acid solutions is acceptable.[250] Otherwise, results obtained for digestates of airborne particles should be treated with skepticism.

Polarographic methods can be employed for determining several trace elements including arsenic, bismuth, cadmium, cobalt, copper, lead, manganese, nickel, selenium, and zinc.[251–253] Detection limits are best in the pulse and differential pulse modes of operation and are generally in the range 10^{-6} to 10^{-8} M using a dropping mercury electrode.[245,251] Despite the relatively high ionic strengths of digested samples, it is usually necessary to add a supporting electrolyte such as sodium fluoride, calcium chloride, or potassium chloride. Pulse polarography has the advantage over nonpulse methods of enhanced resolution of individual waves, this feature is of considerable importance in multielemental analysis. Even so, it may be necessary to vary the nature of the supporting electrolyte or to add a specific complexing ligand, to modify the relative positions of polarographic waves from different ions, thus improving discrimination between overlapping waves.[245] In general polarography has little to offer over anodic stripping voltammetry except for a few elements such as arsenic and manganese.

Anodic stripping voltammetry is really the only electroanalytical method that can lay claim to inclusion in the repertoire of techniques for determining trace elements in airborne particles. This method employs a hanging mercury drop or thin mercury film electrode[251,254] and involves a preelectrolysis period to concentrate the analyte ions as an amalgam in this electrode. The potential is then reversed, and preconcentrated elements are sequentially stripped from the electrode by means of a time-dependent potential sweep.[255] Detection limits in the range 10^{-7} to 10^{-10} M are attainable, depending on the preelectrolysis time employed and the electrode system chosen. In general the hanging mercury drop electrode provides good stability and routine relative precisions of 5 to 10%; however the sensitivity attainable for a given preelectrolysis time is somewhat less than that for thin film electrodes.[256,257] Essentially the same elements can be analyzed as by polarography.

Pulse anodic stripping voltammetry has generally been found to be somewhat more reliable than the dc mode for routine analyses, although both are eminently usable. Yet care should be taken in comparing analyses of solutions having widely different resistances, since both the peak position and peak current depend on solution resistance.[258] For solutions of digested particles, however, the ionic strength is usually sufficiently great to make the effects of solution resistance small, and it is not generally necessary to add a supporting electrolyte. As in the case of polarography, overlapping waves can be separated to some extent by addition of complexing species. For example, thallium can be analyzed in solutions containing lead concentrations about 500 times higher by addition of EDTA to shift the lead wave cathodically.[259]

Anodic stripping voltammetry and polarography have limited application to multielemental analysis of airborne particles. They do, however, enable extremely sensitive and precise determination of several elements of interest for a very low equipment cost (about $3000). Consequently they are best employed when a small number of appropriate elements must be determined precisely, or when sample conservation is of primary importance. The use of anodic stripping voltammetry in conjunction with atomic absorption and flame emission spectrometry to extend the range of elements accessible will probably constitute the most common application.[248]

5.6.2.7. *Microscopic and Surface Methods*

The great majority of analytical studies of trace elements contained in airborne particles are designed to determine the bulk concentration of one or more elements present in a composite sample. However analysis of individual particles and determination of the distribution of a given element within a single particle can provide considerable insight into the probable environmental behavior of that particle.[36,260,261] Several techniques are available for performing such analyses, including electron microprobe X-ray spectrometry, Auger electron spectrometry, electron spectroscopy for chemical analysis (ESCA), and ion-microprobe mass spectrometry, and these are currently being applied to studies of airborne particles.[171]

Before discussing the capabilities of individual techniques, it is important to be aware of several general characteristics of microscopic and surface analyses. First, the elemental analysis of individual airborne particles provides information that cannot be correlated easily with that derived from bulk analyses. This is because of the considerable heterogeneity between particles in any given sample. Thus in gaining the advantage of specificity by analyzing a single particle, one loses statistical knowledge of the sample composition as a whole. In principle it is possible to analyze many particles to obtain a statistical distribution of elemental composition; however this is usually too time-consuming to be practical. Second, all the techniques available for microscopic and surface analysis are sensitive only to material close to the particle surface. Since the dependence of signal intensity on depth beneath the surface is poorly defined, and since individual elements may be nonuniformly distributed in a particle, truly quantitative analyses are not generally possible. Third, the sensitivities and detection limits of microscopic and surface analysis techniques must be considered in a different light from those associated with bulk analyses. Thus most techniques are extremely sensitive in terms of the actual elemental mass being determined, but the mass contained in the analytical volume is very small; this means that detection limits expressed in micrograms per gram are usually poor. Consequently only major and minor elements are routinely accessible. An example of the comparative sensitivities and numbers of elements accessible by ion-microprobe mass spectrometry, electron microprobe X-ray spectrometry, instrumental neutron activation analysis, spark source mass spectrometry and dc arc atomic emission spectrometry is provided for illustration in Figure 5.13. Finally, all the techniques discussed require the sample to be maintained under high vacuum, and each may produce considerable local heating. This effect may result in the loss of volatile material and changes in sample structure during analysis.

The instrument most widely available for microscopic analysis is the electron microprobe. Closely similar in operational principle to a scanning electron microscope, the microprobe excites characteristic X-ray emission from elements in the sample by bombardment with a beam of electrons approximately 3 μm in diameter. The emitted X-rays are analyzed by one or more wavelength dispersive analyses or by an energy dispersive analyzer.[262–264] Essentially the same capability is available with a scanning electron microscope equipped with an energy dispersive X-ray analyzer, by increasing the electron beam intensity, except that the specially designed electron microprobe is generally more sensitive and is afforded better X-ray spectral resolving power by virtue of its wavelength dispersive analyzers.

Airborne particulate samples must be made to conduct electrons by coating them with a thin vapor-deposited layer of carbon or gold and are mounted under vacuum. Analyses can be performed either by maintaining the electron beam in a fixed position or by rastering (scanning) it over the area of interest, which is observed using the scanning electron microscopic capability. In the latter case it is possible to obtain a map of the distribution of each element in the analytical field. This can be useful for determining whether an element of interest is uniformly distributed and, if not, with which other elements it is associated. Sensitivities are such that elements present at concentrations below about 1% are not detectable in particles having equivalent spherical diameters smaller than about 1 μm. Exceptions to these general rules do occur, however, notably when an element is highly concentrated on a particle surface.[36,265,266]

As discussed earlier for X-ray emission spectrometry, this type of analysis is partially surface selective in that X-ray escape depths are of the order of

1 to 10 μm depending on the nature of the solid matrix. It is therefore possible to obtain qualitative information about the surface predominance of an element by varying the energy of the incident electron beam or by etching into part of the surface using a stream of positive argon ions. In the latter case we have obtained useful results by comparing the intensity ratios of X-rays emitted from the etched and unetched sides of a single particle.[36] Apart from the difficulty of obtaining comparable particulate standards, quantitative analysis of particles is complicated because X-ray intensities depend on the takeoff angle from the surface. Recently, however, it has been suggested[266] that correction factors can be calculated and applied by approximating particle shapes to mathematically tractable geometrical structures. Precisions of 5 to 10% are claimed, although it is doubtful whether this procedure is worthwhile for routine analysis of individual particles. Without such corrections, the relative amounts of each element present in the analytical volume can be approximated to a useful degree, but care should be taken not to overinterpret such data.

The ion microprobe is at present probably the ultimate instrument for performing sensitive microscopic analyses (Figure 5.13). In this instrument a beam of ions (typically negative oxygen ions) is focused onto the sample surface and produces a sputtering or emission of secondary ions characteristic of the sample surface material. The secondary ions then enter a mass spectrometer, where they are mass analyzed and detected either electrically or photographically as in a spark source mass spectrometer.[267,268] Detection limits depend on the intensity and cross-sectional area (> 5 μm) of the primary ion beam but are of the order of 100 $\mu g\ g^{-1}$ on a bulk basis. Consequently this instrument can detect many of the trace elements of interest in airborne particles. Its main utility, however, is in determining depth profiles of elements beneath a particle surface. Extreme depth resolution is provided because secondary ions are essentially derived from a monomolecular layer of material at the particle surface, thus observation of signal intensity as a function of time supplies a qualitative profile of elemental concentration.[36] Quantitative analysis of particles is impractical now because of the diffi culties associated with obtaining suitable standards and imperfect understanding of the sputtering process.[267]

Auger electron spectrometry and ESCA are related techniques in that both determine the energy and intensity of electrons ejected from the inner electron energy levels of atoms. In Auger spectrometry the sample is bombarded with an electron beam, which may be focused as a microprobe, and inner shell electrons are ejected. Electrons from higher energy levels fill the vacancy, and excess energy is removed either by emission of a characteristic X-ray or emission of a secondary electron (the Auger electron) from an outer energy level. Since X-ray and Auger electron emission are competitive processes, Auger electron spectrometry is primarily applicable to elements for which X-ray spectrometry is least sensitive, that is, elements in the first two rows of the periodic table.[269] In ESCA the sample is irradiated with a beam of X-rays (these cannot be focused to the same degree as electrons or ions), which produce photoionization and ejection of electrons from inner electron levels.[270] In both Auger spectrometry and ESCA the energies of emitted electrons are characteristic of the atoms from which they were derived, thereby enabling elemental identification following energy discrimination and detection. The main utility of these techniques, however, lies in the fact that electron escape depths are only of the order of 5 to 20 Å, so that only a thin surface layer of material is being analyzed.[36]

Neither Auger spectrometry nor ESCA has been widely applied to the analysis of airborne particles;[25,36,271] however they do afford useful capabilities for surface and regional analyses, and their growing availability may well result in more extensive use. Both techniques are comparable to the electron microprobe in sensitivity (Figure 5.13), being capable of detecting concentrations in the range 0.1 to 1.0% on a bulk basis. Both are difficult to calibrate because of difficulties in obtaining standards,[272,273] although precisions of 30% are claimed. In addition, both techniques require exposure of fresh surface (usually by intermittent etching with a positive argon ion stream, a technique for which most Auger spectrometers are equipped), to be able to perform depth profiling studies.[36] Also, both techniques experience considerable operational problems as a result of sample charging when insulating materials (e.g., fly ash particles) are being studied. Auger spectrometry can be employed either in a microprobe configuration or otherwise, but ESCA is incapable of analyzing small fields. Auger spectrometry is particularly useful for studying elements of low atomic number (unlike electron microprobe X-ray spectrometry), and it offers potential for characterizing matrix surfaces in single particles.

ESCA is perhaps best known for its intrinsic ability to provide information about elemental oxidation states and binding electron energies because the energies of ejected electrons are significantly influenced by the valence shell. Unfortunately relating electron binding energies to chemical species is not at all straightforward, although it is a relatively simple matter to define the oxidation states of elements present in the analytical volume.

Although microscopic and surface analytical techniques are growing rapidly in popularity, their use is likely to remain limited because of the high cost (about $100,000) of the equipment required. On the other hand, some insight into the surface predominance and availability of trace elements associated with particles can be obtained simply by determining the rates of dissolution of each element in different solvents.[38] Such an approach can hardly be classified as surface analysis, but it does provide a means by which almost any laboratory can obtain a qualitative indication of the probable physicochemical distribution of an element in an airborne particulate sample.

5.6.2.8. *Speciation Methods*

Despite the number of analytical methods suitable for determining the concentrations of elements present in atmospheric samples, very few methods are available for the identification and determination of the chemical compounds in which these elements exist. As a result, many environmental scientists tend to think in terms of the presence of the elements themselves, and losing sight of the fact that the environmental behavior of an element is determined primarily by the chemical form in which it is present. As mentioned earlier, a paradox exists in the analytical chemistry of elements and of inorganic species; thus several methods are applicable to the determination of individual compounds present in substantial amounts, but determination of other than the compound elements at trace levels is not generally possible. One important reason for this is that most of the inorganic species of interest lose their identity when removed from their solid or vapor state into solution; thus the field of solution chemistry is scarcely available.

When high concentrations or large amounts of a particular species are available, several general approaches to speciation can be employed. For solids, the most useful are X-ray powder diffraction and infrared spectrometry. These techniques have enabled identification (Figure 5.14) of the minerals α-quartz (SiO_2), hematite (Fe_2O_3), magnetite (Fe_3O_4), mullite ($3Al_2O_3 \cdot 2SiO_2$), and gypsum ($CaSO_4 \cdot 2H_2O$) in fly ash particles,[37,274] and of sulfate, nitrate, silicate, ammonium, and carbonate groups in airborne particles.[275] Concentrations must be greater than about 0.1 to 1.0% in the solid, although somewhat lower levels may be accessible by Fourier transform infrared spectrometry.[276] For vapor phase species concentrations must generally be in the parts per million range by volume, at which levels, in principle, infrared and ultraviolet–visible spectrometry can be employed, provided the species of interest has sufficiently high absorption coefficients. For the latter method it is probably desirable to employ derivative spectrometry.

It is apparent that the above-mentioned approaches to speciation are applicable mainly to the determination of matrix compounds. Determination of chemical species present at trace levels, however, is currently an open field. Methods that have been investigated to date, or are potentially useful, can be subdivided into those which may be generally applicable to compounds of many different elements and those which have been specifically designed for compounds of a single element.

One of the most successful general approaches to trace metal speciation

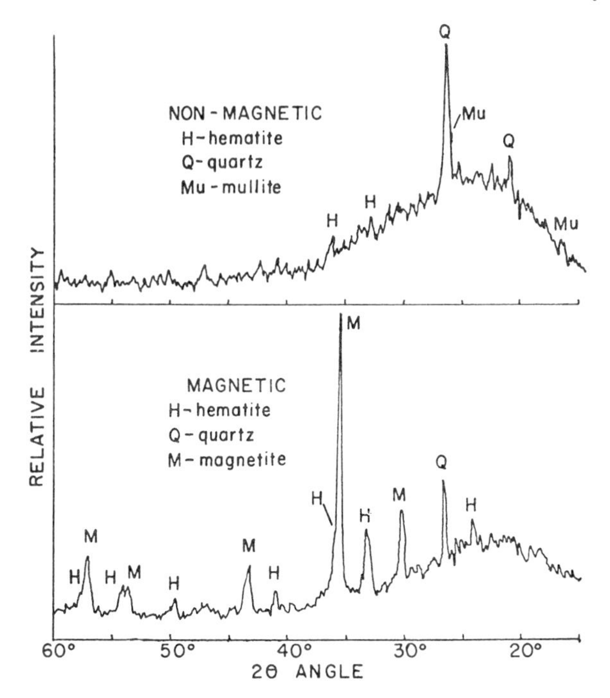

Figure 5.14. X-Ray powder diffraction patterns of magnetic and nonmagnetic fractions of coal fly ash showing the presence of polycrystalline mullite (Mu), α-quartz (Q), hematite (H), and magnetite (M).

in particulate matter involves isolation of particles containing high levels of the species of interest, followed by determination of X-ray powder diffraction patterns. This technique requires that some particles have high concentrations (several percent) of the species of interest in polycrystalline form. Isolation can be accomplished by sequential fractionation of the bulk sample in terms of particle size, density, ferromagnetism, bubble flotation characteristics, and morphology, as described in the earlier section on sample pretreatment. The approach has been successful in identifying the species Pb, PbO_2, $PbSO_4$, and $PbO \cdot PbSO_4$ in urban dust particles.[168] Quantification is not possible because of imperfect fractionation of individual compounds. A second approach that holds considerable promise for trace level speciation is gas evolution analysis. In this technique the sample is progressively heated in a stream of inert gas, and evolved vapors are swept into a flame or plasma for determination of the elements present by emission spectrometry. Alternatively, the vapor can be subjected to mass analysis in a mass spectrometer.[277–280] Speciation is achieved by determining which elements are evolved simultaneously and over which temperature range, although the latter indicator may depend markedly on the sample matrix composition and on the distribution of compounds within a sample. The attractive feature of this approach is that detection limits are extremely low (nanogram to microgram range), thereby providing access to truly trace levels. The technique is in its infancy, however, and should not be considered a viable speciation method at present.

As pointed out in earlier analytical sections, several elemental analysis methods furnish fragmentary information on chemical species. For example, polarography and stripping voltammetry indicate metal oxidation states, wavelength dispersive X-ray spectrometry can distinguish between light element species, transient atomic absorption spectrometry can indicate the presence of different compounds of an element, and ESCA can establish oxidation states and electronic binding environments of elements. It has been claimed[281] that electron microprobe X-ray spectrometry can establish chemical compounds of lead based on the spatial proximity and ratios of elements present. This may be true in some cases, but such an approach should be treated with caution. Similarly, compound identification by ESCA should be recognized as being a highly sophisticated procedure and one that provides information about only the outermost 5 to 20 Å of particles.

Considerably greater success has been achieved in the development of techniques for determining the species in which an individual element exists. In the case of selenium a sequence of solvation, reaction, and analytical schemes has been designed to establish the forms of selenium present in fly ash particles.[39,282] The following chemical properties of selenium were invoked:

1. Only Se(IV) will react with 5-nitro-O-phenylenediamine to produce a thermally stable volatile piaselenol complex, which can be determined gas chromatographically.[39,283]
2. Se^o is not soluble in water or hydrochloric acid. However SeO_4^{2-} can be leached from particles by water, and $SeO_3^{2-} + SeO_4^{2-}$ can be leached by hydrochloric acid.
3. Se^o must be oxidized to SeO_3^{2-} before it will dissolve. This can be done by an oxidizing acid or Br/Br^- buffer. Use of this scheme has indicated that Se^o is the only form of selenium present in fly ash.[39] Such a sequence is also useful for illustrative purposes because it demonstrates that similar schemes may, in principle, be constructed for determination of species of other elements. However some compounds may lose their integrity on dissolution, and caution should be exercised in basing identification on species actually found in solution.

Determination of the several species of mercury[143] and arsenic[150] present in the atmosphere can also be accomplished using sequences conceptually similar to that just outlined. Since both elements occur primarily in the vapor phase, however, a series of sorption steps can be employed for species discrimination during the sample collection phase. In the case of mercury the following sequence is used:[143]

1. Incoming air first comes in contact with a plug of glass wool, which removes some particulate matter.
2. A second adsorption tube, filled with Chromosorb-W coated with 3% SE-30, adsorbs $HgCl_2$ and removes most of the remaining particles.
3. A third tube, filled with Chromosorb-W that has been washed with 0.05 *M* sodium hydroxide, adsorbs CH_3HgCl.
4. A fourth tube, filled with glass beads coated with metallic silver, adsorbs elemental mercury by amalgamation.
5. A fifth and final tube, filled with glass beads coated with metallic gold, adsorbs $(CH_3)_2Hg$.

Adsorbed species are removed by heating individual tubes and readsorbing the evolved mercury species onto gold-coated glass beads that quantitatively adsorb all forms of mercury. Thermal evolution from the gold is in the form of elemental mercury vapor, which is swept into a dc plasma for determination by measurement of the atomic mercury emission. (Alternatively, cold vapor atomic absorption spectrometry can be employed.) Detection limits are around 0.02 ng of mercury, allowing quite small air samples to be employed. This procedure has been shown to give excellent results for the quantitative determination of species in ambient air.[71] The method should be used by analytically skilled personnel, however, and it should not be selected for sampling emission sources without extensive investigation of possible interferences.

The sequence employed for determination of arsenic species in ambient air[150] involves collection of particulate matter on a filter followed by quantitative collection of volatile arsines on silver-coated glass beads. Inorganic As(III) and As(V) can be distinguished colorimetrically after dissolution, and volatile arsines are removed from the silver surface without dealkylation or disproportionation by a mild, warm alkaline wash followed by a hot water wash. The arsines are then thermally evolved into a dc plasma, as in the case of mercury described previously, and detected by observation of the atomic arsenic emission. Identification is achieved because the different arsines have different thermal retention times. Using this approach, the species As(III) and As(V) have been determined in airborne particulates together with volatile CH_3AsH_2, $(CH_3)_2AsH$, and $(CH_3)_3As$.[150] Approximately 0.5 ng of arsenic is detectable. The procedure appears to be somewhat less foolproof than that developed for mercury speciation—for example, reduction with sodium borohydride is required following air oxidation of methyl arsines during their removal from the adsorption tube; however it represents a useful approach to the speciation of airborne arsenic.

5.6.2.9. Other Methods

In addition to the methods already outlined, the field of analytical chemistry includes many approaches that can be, or have been, used to determine atmospheric trace elements. Some of these have either desirable or undesirable characteristics worthy of mention. Classically, of course, metal ions have been determined by spectrophotometric measurement of the absorbance of a specific metal complex formed by addition of a solution reagent such as dithiocarbazone.[285] Many such methods, covering essentially all elements of interest, are described in the literature.[286] Numerous colorimetric methods may provide valuable results for individual elements, and their use is very attractive when limited instrumental and financial resources make it impossible to employ recommended techniques. In general, however, colorimetric methods have rather poor detection limits (microgram to milligram range), require considerable manipulation of solutions, and are highly susceptible to interferences from metal ions other than the analyte. In fact, the complexity of solutions obtained from the digestion of airborne particles is such that interferences are more often the rule than the exception. Consequently the use of colorimetric methods is strongly discouraged except when interferences are demonstrably absent.

One variation of colorimetric analysis that partially overcomes the sensitivity limitation involves use of the ring oven.[287,288] The analyte solution (and, in some cases, the undigested particulate sample) is spotted onto the center of a filter paper placed on the flat heated surface of a ring oven. Appropriate solvents are dripped onto the sample, and analyte species are swept outward on the filter to a heated metal ring. Continuation of this process results in evaporation of the solvent and accumulation of the analyte in a narrow region (< 1 mm thick) at the circumference of the circle.

Colorimetric reagents are then added to the ring, and the intensity of color developed is compared with standards. With this technique the analyte is effectively concentrated in a very small volume (the ring) for enhanced ease of observation, the elution process provides some capability for separating interfering entities (e.g., insoluble precipitates), several elements can be determined in different portions of the ring, and rapid, inexpensive screening of the elements present can be achieved. Despite claims to the contrary,[289] however, many workers have found that use of the ring oven technique calls for considerable skill. Furthermore, the method still relies on achieving specificity of color development, which is extremely difficult for complex samples. It is recommended, therefore, that this method be used with caution and that it be employed primarily for screening purposes.

Recently it has been shown that several trace metals of interest can be determined with excellent specificity and sensitivity by derivatization gas chromatography employing selective detection.[290–292] For this method individual metals are complexed in solution with volatile organic ligands, to render them amenable to gas chromatography. Both chromatographic separation and detection characteristics can be enhanced by suitably tailoring the ligands employed. This procedure has little potential for widespread use in the analysis of atmospheric trace elements, but it is mentioned because of its capability for the determination of beryllium. This element, which is accessible by only a few techniques, can be detected at levels of 10^{-14} g.[290]

5.7. MONITORING PROGRAM DESIGN AND DATA INTERPRETATION

Throughout the last decade, information about trace elements in the atmosphere has been obtained from a variety of monitoring programs that have varied in scope from the determination of a few elements at a single site over periods of several days to nationwide measurements of many elements over a year or more.[2] Such monitoring is usually conducted in local urban atmospheres to find average ambient levels of trace elements, to establish their diurnal and seasonal variations, to indicate trace element patterns, inhalation profiles, and atmospheric enrichment, and to identify major emission sources that contribute trace elements to a local aerosol. (Determination of source emission rates constitutes a distinct monitoring operation, discussed elsewhere in this volume.)

To obtain the foregoing information, individual programs should be specifically designed to monitor all variables that can influence the measured atmospheric concentrations of each element. Unfortunately, however, the identity of such variables is not usually known prior to sampling, and the ability to establish even those which are known is limited by time, economic considerations and availability of sampling and analytical equipment. Consequently most atmospheric monitoring programs are established on a "do the best we can" basis, and the resulting data are often incoherent, fragmentary, and difficult to interpret in terms of the information required.

This situation can be improved in either or both of the following ways:

1. By careful design of monitoring strategy to include measurement of as many controlling variables as possible.
2. By deployment of multivariate statistical (pattern recognition) techniques to extract the maximum amount of information possible from the available data.

Improvement of monitoring design strategy requires recognition that the measured atmospheric concentration of a trace element depends on the following factors:

1. The location of the sampling site with respect to all sources (both point and general) from which the element is derived.
2. The physical and chemical properties of the element in question.
3. The meteorological conditions at each site, including wind direction and speed, turbulence, humidity, precipitation, solar radiation, and temperature.
4. The time interval employed for sampling and the sequence of such time intervals.
5. The method of collection (e.g., Hi Vol filtration, particle size discrimination, dustfall, wind deposition).

In particular it is important to recognize that certain elements can act as source indicators when source identification is required. For example, automobile exhaust emissions contain high levels of lead, oil-fired power plants emit substantial amounts of vanadium, and metallurgical smelter emissions contain the smelted metal. Marine backgrounds can be identified by high chlorine and sodium levels. Therefore inclusion of these elements in the analytical scheme is highly desirable.

An ideally designed general monitoring program would clearly involve many sampling sites, with numerous measurements being recorded at each site. Such a program would be prohibitively expensive and may be neither necessary nor desirable when specific information is sought. Furthermore, much of the information commonly desired can be obtained using a limited sampling program in conjunction with multivariate statistical analysis of the resulting data. This approach, which has been widely employed to interpret sociological and epidemiological survey data, takes account of the fact that a given measurement may be influenced by a large number of both independent and interdependent variables whose identity and magnitude may or may not be known.[293,294] Such is the case with atmospheric trace element analyses.

Three approaches to pattern recognition can be usefully applied to atmospheric trace element data. The first is hierarchical cluster analysis, in which the variables are divided into groups that exhibit statistically similar characteristics. This approach is particularly useful for learning which elements or environmental parameters (e.g., sampling sites) exhibit definite relationships and can be helpful in establishing specific sources. The second procedure is common factor analysis, which determines the number of common factors, or underlying causalities, that control the variance of the observed parameters (e.g., trace element concentrations). Factor analysis cannot identify the common factors, but it does supply information that enables their identity to be inferred with some confidence. For example, the relative contributions of different emission sources or meteorological conditions to a given trace element pattern can be determined. The third method is that of principal component analysis, which also enables determination of the number of underlying components that control the variability of the data. In this respect principal component analysis is effectively similar to, though mathematically distinct from, common factor analysis.

To apply multivariate analysis techniques three requirements must be met.

1. The number of samples must be equal to or, preferably, greater than the number of measured parameters associated with each sample.
2. The samples are assumed to represent a random choice from a normal distribution of the variables. Specific sampling to maximize the influence of a given variable (e.g., sampling directly downwind from a metal smelter) does not contravene this requirement, although it may distort the relative contributions of the variables.
3. All variables must be determined for each sample. This may require limitation of data sets if some sets are incomplete.

These requirements can be met, or reasonably approximated, for trace element analyses of atmospheric samples. Furthermore, computer programs suitable for performing multivariate analysis are already available in most program libraries. Multivariate analysis has not been extensively applied mainly because the utility of the approach has not been recognized, and because the data sets available are incomplete. In the latter regard it is imperative to recognize the data rquirements of multivariate statistical analysis during both the design and operational stages of a monitoring program.

Unfortunately most of the applications of multivariate analyses to date have been to existing data sets that were not generated specifically for the purpose. Nevertheless, several limited applications have provided useful information. Thus Heindryckx and Dams[63] and Bogen[64] have resolved elemental patterns into probable sources for the areas around Ghent, Belgium, and Heidelberg, West Germany, respectively. Gatz[65] has performed a similar resolution for the Chicago area using the procedure of Miller et al.[23] in which it is necessary to assume, a priori, the number and nature of the sources of particulate matter. A more sophisticated procedure has been developed by Gatrell and Friedlander,[66] taking account of gas-to-particle transformations and particle growth processes, together with an extensive source inventory. Several groups have also utilized statistical techniques to reduce their data to meaningful terms.[67,68] For example, Blifford and Meeker[67] used a principal components analysis with the data from the U.S. National Air Sampling Network (NASN) for the period 1957–1961. They were able to show that these data could be ordered into seven components, and some of these could be identified with specific emission sources.

Perhaps the most extensive application of multivariate statistical analysis has been to elemental concentration measurements of the Boston aerosol.[4,69] In this work each of the three approaches to pattern recognition was empolyed, a procedure that is strongly recommended. The data were particularly suitable to pattern recognition in that a large number of elements were determined as a function of particle size, and meteorological data were available. Common factor and principal component analysis served to distinguish six underlying causalities. These were interpreted as being due to the input of crustal dust, sea salt aerosol, emissions from residual oil burning, automobile exhaust emissions, and municipal incinerators. The sixth factor was difficult to identify but was thought to be attributable to emissions from the combustion of coal or from metallurgical processing. The results of hierarchical cluster analysis supported the foregoing interpretations. In addition, they established relationships between individual elements and delineated localized aerosol patterns in different areas of the city. The data also established unique behavior for particulate antimony, bromine, and selenium, suggesting the possibility that their initial emission is partly as vapors.

The examples above have emphasized the utility of multivariate statistical analysis for identifying the contributions of different emission sources and for establishing interelemental relationships. As pointed out earlier, however, the technique goes well beyond this in that it readily establishes statistically meaningful relationships between aerosol composition and meteorological conditions, time, and locality, as well as determining "average" trace element concentrations. In view of the considerable expense and effort involved in atmospheric monitoring, it is hoped that future programs will be designed to take full advantage of the undoubted interpretive power of multivariate statistical analysis.

5.8. CONCLUSIONS

The foregoing sections make it clear that suitable sampling and analytical techniques are available for the determination of atmospheric trace element concentrations. The range of existing methods for elemental analyses is impressive, and though no one method can be considered ideal, capabilities for obtaining precise, accurate, and semiquantitative survey data are available. Unfortunately there is a common belief that only the analytical techniques capable of the highest precision and accuracy should be employed for airborne trace element analysis. Certainly such methods are desirable; however large variabilities are introduced by sample losses during collection and pretreatment and by the variability of the samples themselves. Indeed, it can be argued that the collection inefficiencies of filter materials, the overlap of particle sizes in size discriminating devices, and the difficulties associated with defining a "representative" sample often make requirements for 5 to 10% analytical precision and accuracy rather ludicrous. In this regard it should be pointed out that multivariate statistical treatment of data sets does not require that individual measurements have high precision, provided general trends are still discernible.

An assessment of the status of current knowledge of atmospheric trace elements reveals that very little is known about their atmospheric chemistry. Extension of knowledge in this area will require performance of specific laboratory studies in conjunction with field studies in which determination of chemical species and their regional distribution in particles will play a significant role. Advances in three analytical areas will greatly benefit our understanding of the actual and potential environmental impact of atmospheric trace elements. These involve the identification and determination of chemical species, the determination of their microscopic distribution, and the deployment of multivariate statistical analysis.

ACKNOWLEDGMENTS

The authors gratefully acknowledge valuable discussions with Prof. R. K. Skogerboe, Department of Chemistry, Colorado State University. Aspects of the authors' research described herein were supported by U.S. National Science Foundation grants GH-33634 and GI-31605 and by U.S. Environmental Protection Agency grant R-803950.

SYMBOLS

C Cunningham slip correction factor

D_{aer} aerodynamic diameter of particle

D_j width of impactor slit
D particle diffusivity
E efficiency of particle collection
F flow rate of air passing through impactor or airflow in cyclone
L pore length, length of impactor slit
N_D coefficient of diffusive collection
r particle radius
R_0 initial pore radius
u face velocity
ε partial collection efficiency
ϵ filter porosity
μ viscosity
ρ density of air
ρ_p particle density
ψ inertial impaction parameter

REFERENCES

1. Rahn, Kenneth A., Marc Demuynck, Richard Dams, and Jean DeGrave, *Proceedings of the Third International Clean Air Congress*, Dusseldorf, 1973, p. C8.
2. Lee, Robert E., Jr., Stephen S. Goranson, Richard E. Enrione, and George B. Morgan, *Environ. Sci. Technol.*, 6, 1025 (1972).
3. Harrison, P. R., K. A. Rahn, R. Dams, J. A. Robbins, J. W. Winchester, S. S. Barr, and D. M. Nelson, *J. Air Pollut. Control Assoc.*, 21, 563 (1971).
4. Gladney, Ernest S., William H. Zoller, Alun G. Jones, and Glen E. Gordon, *Environ. Sci. Technol.*, 8, 551 (1974).
5. Pierson, D. H., P. A. Cawse, L. Salmon, and R. S. Cambray, *Nature*, 241, 252 (1973).
6. Struempler, A. W., *Atmos. Environ.*, 10, 33 (1976).
7. Zoller, W. H., E. S. Gladney, and R. A. Duce, *Science*, 183, 198 (1974).
8. Hoffman, Gerald L., Robert A. Duce, and Eva J. Hoffman, *J. Geophys. Res.*, 77, 5322 (1972).
9. Duce, Robert A., William H. Zoller, and Jarvis L. Moyers, *J. Geophys. Res.*, 78, 7802 (1973).
10. Blifford, Irving H. and Dale A. Gillette, *Atmos. Environ.*, 6, 463 (1972).
11. Hirao, Y. and C. C. Patterson, *Science*, 184, 989 (1974).
12. Duce, R. A., G. L. Hoffman, J. L. Fasching, and J. L. Moyers, World Meteorological Organization Report No. 368, *Proceedings of a Technical Conference on the Observation and Measurement of Atmospheric Pollution*, Helsinki, 1973.
13. Natusch, D. F. S. and J. R. Wallace, *Science*, 186, 695 (1974).
14. Carroll, Robert E., *J. Am. Med. Assoc.*, 198 (3), 267 (1966).
15. Urone, Paul, W. H. Schroeder, and S. R. Miller, *Proceedings of the Second International Clean Air Congress*, Washington, D.C., 1970, p. 370.
16. Urone, Paul and W. H. Schroeder, *Environ. Sci. Technol.*, 3, 436 (1969).

17. Pueschel, Rudolph F., Charles C. Van Valin, and Farn P. Parungo, *Geophys. Res. Lett.*, 1 (1), 51 (1974).
18. Meyer, B., and C. Carlson, *Proceedings of the Second International Clean Air Congress*, Washington, D.C., 1970, p. 835.
19. Hidy, G. M. and J. R. Brock, *Proceedings of the Second International Clean Air Congress*, Washington, D.C., 1970, p. 1088.
20. Butcher, S. S. and Rober J. Charlson, *An Introduction to Air Chemistry*, Academic Press, New York, 1972.
21. Lee, R. E., Jr. and D. J. von Lemden, *J. Air Pollut. Control Assoc.*, 23, 853 (1973).
22. Natusch, D. F. S., *Proceedings of the Second Federal Conference on the Great Lakes*, 1975, p. 175.
23. Miller, M. S., S. K. Friedlander, and G. M. Hidy, *J. Colloid Interface Sci.*, 39, 165 (1972).
24. Gordon, Glen E., Director, "Study of the Emissions from Major Air Pollution Sources and Their Atmospheric Interactions," Progress Report, University of Maryland, 1974.
25. Gordon, Glen E., Director, "Atmospheric Impact of Major Sources and Consumers of Energy," Progress Report, University of Maryland, 1975.
26. Zoller, W. H., G. E. Gordon, E. S. Gladney, and A. G. Jones, *Advan. Chem. Ser.*, 123, 31 (1973).
27. Fordyce, J. S. and D. W. Sheibley, NASA Technical Memorandum X-3054, May 1974.
28. Natusch, D. F. S., J. R. Wallace, and C. A. Evans, Jr., *Science*, 183, 202 (1974).
29. Davison, R. L., D. F. S. Natusch, J. R. Wallace, and C. A. Evans, Jr., *Environ. Sci. Technol.*, 8, 1107 (1974).
30. Kaakinen, John W., Roger M. Jorden, Mohammed H. Lawasani, and Ronald E. West, *Environ. Sci. Technol.*, 9, 862 (1975).
31. Klein, D. H., A. W. Andren, J. A. Carter, J. F. Emery, C. Feldman, W. Fulkerson, W. S. Lyon, J. C. Ogle, Y. Talmi, R. I. Van Hook, and N. Bolton, *Environ. Sci. Technol.*, 9, 973 (1975).
32. Jacko, R. B., D. W. Neuendorf, and K. J. Yost, paper presented at the Annual Meeting of the American Society of Mechanical Engineers, July 14, 1975.
33. Loh, A. and D. F. S. Natusch, unpublished results.
34. Bertine, K. K. and Edward D. Goldberg, *Science*, 173, 233 (1971).
35. White, H. J., *Industrial Electrostatic Precipitation*, Addison-Wesley, Reading, Mass., 1963.
36. Linton, R. W., A. Loh, D. F. S. Natusch, C. A. Evans, Jr., and P. Williams, *Science*, 191, 852 (1976).
37. Loh. A., doctoral dissertation, University of Illinois, Urbana, 1975.
38. Baker, J., C. F. Bauer, C. A. Evans, Jr., P. K. Hopke, R. W. Linton, A. Loh, H. Matusiewicz, and D. F. S. Natusch, *Proceedings of the International Conference on Heavy Metals in the Environment*, Toronto, Canada, 1975, in press.
39. Andren, Anders W., David H. Klein, and Yair Talmi, *Environ. Sci. Technol.*, 9, 856 (1975).
40. Fisher, G. L., D. P. Y. Chang, and Margaret Brummer, *Science*, 192, 553 (1976).
41. Friedlander, Sheldon K., *Environ. Sci. Technol.*, 7, 235 (1973).
42. Heisler, S. L., S. K. Friedlander, and R. B. Husar, *Atmos. Environ.*, 7, 633 (1973).
43. Miller, M. S., S. K. Friedlander, and G. M. Hidy, *J. Colloid Interface Sci.*, 39, 165 (1972).

44. Wesolowski, J. J., W. John, and R. Kaifer, *Advan. Chem. Ser.*, 123, 1 (1973).
45. Nifong, G. D. and J. W. Winchester, "Particle Size Distributions of Trace Elements in Pollution Aerosols," ORA Project 08903, University of Michigan, Ann Arbor, August 1970.
46. Ondov, J. M., W. H. Zoller, and G. E. Gordon, "Trace Elements on Aerosols from Motor Vehicles," paper presented at 67th Annual Meeting of the Air Pollution Control Association, Denver, June 1974.
47. Habibi, K., *Environ. Sci. Technol.*, 7, 223 (1973).
48. Habibi, K., *Environ. Sci. Technol.*, 4, 239 (1970).
49. Boyer, Kenneth W., doctoral dissertation, University of Illinois, Urbana, 1974.
50. Mueller, P. K., *Environ. Sci. Technol.*, 4, 248 (1970).
51. Boyer, Kenneth W. and H. A. Laitinen, *Environ. Sci. Technol.*, 9, 457 (1975).
52. Pierrard, John M., *Environ. Sci. Technol.*, 3, 48 (1969).
53. Laveskog, Anders, *Proceedings of the Second International Clean Air Congress*, Washington, D.C., 1970, p. 549.
54. Skogerboe, R. K., ASTM Special Technical Publication 555, American Society for Testing and Materials, 1916 Race Street, Philadelphia, 1974, p. 194.
55. Robinson, J. W., Larry Rhodes, and D. K. Wolcott, *Anal. Chim. Acta*, 78, 474 (1975).
56. Lee, Robert E., Jr., Ronald K. Patterson, and Jack Wagman, *Environ. Sci. Technol.*, 2, 288 (1968).
57. Burnham, Carole D., Carl E. Moore, Eugene Kanabrocki, and Don M. Hattori, *Environ. Sci. Technol.*, 3, 472 (1969).
58. Bara, S. S., D. M. Nelson, J. R. Kline, P. F. Gustafson, E. L. Kanabrocki, C. E. Moore, and D. M. Hattori, *J. Geophys. Res.*, 75, 2939 (1970).
59. Harrison, Paul R. and John W. Winchester, *Atmos. Environ.*, 5, 863 (1971).
60. Nifong, Gordon D., Edward A. Boettner, and John W. Winchester, *Am. Ind. Hyg. Assoc. J.*, 33, 569 (1972).
61. Ludwick, J. D., C. W. Thomas, L. L. Wendell, and T. D. Fox, Battelle Pacific Northwest Laboratories Report for 1974 to the U.S. Atomic Energy Commission Division of Biomedical and Environmental Research, Part 3, Atmospheric Sciences, February 1975, p. 1. Published in part in *J. Air. Pollut. Control Assoc.*, 26, 565 (1976).
62. Eisenbud, Merril and T. J. Kneip, "Trace Metals in Urban Aerosols," Final report to the Electric Power Research Institute, Palo Alto, Calif., October 1975, pp. 131–149.
63. Heindryckx, R. and R. Dams, *J. Radioanal. Chem.*, 19, 339 (1974).
64. Bogen, J., *Atmos. Environ.*, 7, 117 (1973).
65. Gatz, D. F., *Atmos. Environ.*, 9, 1 (1975).
66. Gatrell, G., Jr. and S. K. Friedlander, *Atmos. Environ.*, 9, 279 (1975).
67. Blifford, I. H. and G. O. Meeker, *Atmos. Environ.*, 1, 147 (1967).
68. John, W., R. Kaifer, K. Rahn, and J. J. Wesolowski, *Atmos. Environ.*, 7, 107 (1973).
69. Hopke, P. K., E. S. Gladney, G. E. Gordon, W. H. Zoller, and A. G. Jones, *Atmos. Environ.*, 10, 1015 (1976).
70. Whitby, K. T., R. E. Charleson, W. E. Wilson, R. K. Stevens, and R. E. Lee, Jr., *Science*, 183, 1098 (1974).
71. Johnson, David L. and Robert S. Braman, *Environ. Sci. Technol.*, 8, 1003 (1974).

72. Wright, W. E., G. L. Ter Haar, and E. B. Rifkin, "The Effect of Manganese on the Oxidation of SO_2 in the Air," paper presented at 67th Annual Meeting of the Air Pollution Control Association, Denver, June 1974.
73. Orr, C., Jr., F. K. Hurd, and W. J. Corbett, *J. Colloid Interface Sci.*, 13, 472 (1958).
74. Hanel, G., *Beit. Phys. Atmos.*, 43, 119 (1970).
75. Mason, B. J., *The Physics of Clouds*, 2nd ed., Clarendon Press, Oxford, 1971, p. 26.
76. Robbins, J. A. and F. L. Snitz, *Environ. Sci. Technol.*, 6, 164 (1972).
77. Committee on Biological Effects of Atmospheric Pollutants, "Airborne Lead in Perspective," National Academy of Sciences, Washington, D.C., 1972.
78. Davies, C. N., *Air Filtration*, Academic Press, New York, 1973.
79. Spurny, Kvetoslav R., James P. Lodge, Jr., Evelyn R. Frank, and David C. Sheesley, *Environ. Sci. Technol.*, 3, 453 (1969).
80. Spurny, K. R. and J. P. Lodge, Jr., NCAR Technical Note NCAR-TN/STR-77, Vol. 1, 1972.
81. Melo, O. T. and C. R. Phillips, *Environ. Sci. Technol.*, 8, 67 (1974).
82. Kubie, G., C. Jech, and K. R. Spurny, *Collect. Czech. Chem. Commun.*, 26, 1065 (1961).
83. Spurny, K. R. and J. Pich, *Collect. Czech. Chem. Commun.*, 28, 2886 (1963).
84. Skogerboe, R. K., private communication.
85. Neustadter, H. E., S. M. Sidik, R. B. King, J. S. Fordyce, and J. C. Burr, *Atmos. Environ.*, 9, 101 (1975).
86. Hwang, J. Y., *Anal. Chem.*, 44 (14), 20A (1972).
87. Dams, R., K. A. Rahn, and J. W. Winchester, *Environ. Sci. Technol.*, 6, 441 (1972).
88. Tanner, R. L., D. Leahy, and L. Newman, Paper ANAL-64, presented at the First Chemical Congress of the North American Continent, Mexico City, Mexico, December 1975.
89. Cahn, Lee, *Mater, Res. Stand.*, 3, 377 (1963).
90. Birks, L. S. and J. V. Gilfrich, EPA Report No. EPA-R2-72-063, U.S. Environmental Protection Agency, Washington, D.C.; November 1972.
91. Seeley, J. L. and R. K. Skogerboe, *Anal. Chem.*, 46, 415 (1974).
92. Heyer, J., L. Armbruster, E. Grein, and W. Stahlhofen, *J. Aerosol Sci.*, 6 (5), 311 (1975).
93. Ranz, W. E. and J. B. Wong, *AMA Arch. Ind. Hyg. Occup. Med.*, 5, 464 (1952).
94. Cohen, Jerry J. and Donald N. Montan, *Am. Ind. Hyg. Assoc. J.*, 28, 95 (1967).
95. Mercer, T. T., *Ann. Occup. Hyg.*, 7, 115 (1964).
96. Horvath, Helmuth and Augest T. Rossano, Jr., *J. Air. Pollut. Control Assoc.*, 20, 244 (1970).
97. Jaenicke, Ruprecht, *Staub-Reinhalt. Luft*, 31 (6), 1 (1971).
98. Marple, Virgil A. and Benjamin Y. H. Liu, *Environ. Sci. Technol.*, 8, 648 (1974).
99. Jaenicke, R. and I. H. Blifford, *J. Aerosol. Sci.*, 5, 457 (1974).
100. Natusch, D. F. S. and J. R. Wallace, *Atmos. Environ.*, 10, 315 (1976).
101. Flesch, J. P., C. H. Norris, and A. E. Nugent, *Am. Ind. Hyg. Assoc. J.*, 28, 507 (1967).
102. Andersen 2000, Inc., *Instructions for the Andersen Stack Sampler*, Salt Lake City, Utah.
103. Lundgren, D. A., *J. Air Pollut. Control Assoc.*, 17, 225 (1967).
104. Mitchell, R. I. and J. M. Pilcher, Fifth AEC Air Cleaning Conference, June 24, 1957, Office of Technical Services, Washington, D.C., pp. 67–83.

105. Soole, B. W., *J. Aerosol. Sci.*, 2, 1 (1971).
106. Wesolowski, J. J., B. R. Appel, and A. Alcocer, private communication.
107. O'Donnel, H., T. L. Montgomery, and M. Corn, *Atmos. Environ.*, 4, 1 (1970).
108. Dahneke, Barton, *J. Colloid Interface Sci.*, 37, 342 (1971).
109. Wesolowski, J. J., private communication.
110. Wilder, B., Mirka Fugas, J. Hrsak, and Anica Vukovic, *Proceedings of the Third International Clean Air Congress*, Dusseldorf, 1973, p. C9.
111. Dingle, A. Nelson and Bhanuprasad M. Joshi, *Atmos. Environ.*, 8, 1119 (1974).
112. Lee, R. E. and S. Goranson, *Environ. Sci. Technol.*, 6, 1019 (1972).
113. Gussman, R. A., A. M. Sacco, and N. M. McMahon, *J. Air Pollut. Control Assoc.*, 23, 778 (1973).
114. Wesolowski, Jerome J., paper presented at the Second Joint Conference on Sensing of Environmental Pollutants, Washington, D.C., December 10, 1973.
115. Pilat, Michael J., *Proceedings of the Symposium on Control of Fine-Particulate Emissions from Industrial Sources*, San Francisco, January 15, 1974, p. 709.
116. Wagman, Jack and Carl M. Peterson, *Proceedings of the Third International Clean Air Congress*, Dusseldorf, 1973, p. C6.
117. Connor, W. D., *J. Air Pollut. Control Assoc.*, 16, 35 (1966).
118. Hounam, R. F. and R. J. Sherwood, *Am. Ind. Hyg. Assoc. J.*, 26, 122 (1965).
119. Koltrappa, P. and M. E. Light, *Rev. Sci. Instrum.*, 43, 1106 (1972).
120. Lippman, M. and A. Kydonieus, *Am. Ind. Hyg. Assoc. J.*, 31, 731 (1970).
121. Yankovski, S. S. and N. A. Fuchs, *Zavod. Lab.*, 32, 811 (1966).
122. Svarovsky, L., *Proc. Soc. Anal. Chem.*, 32 (1974).
123. *Roller Particle Size Analyzer, Instruction Manual*, American Instrument Company, Inc., Silver Spring, Md.
124. Bickelhaupt, R. E., *J. Air Pollut. Control Assoc.*, 24, 251 (1974).
125. Bickelhaupt, R. E., *J. Air Pollut. Control Assoc.*, 25, 148 (1975).
126. Schroeder, H. A., *Environment*, 13 (8), 18 (1971).
127. D'Itri, F. M., *The Environmental Mercury Problem*, Chemical Rubber Co. Press, Cleveland, 1972.
128. Hatch, W. R. and W. L. Ott, *Anal. Chem.*, 40, 2085 (1968).
129. Thomas, F. J., R. A. Hagstrom, and E. J. Kuchar, *Anal. Chem.*, 44, 512 (1972).
130. Haff, L. V., private communication to F. P. Scarangelli, 1971.
131. Kenkyu, T. O., *J. Jap. Soc. Air Pollut.*, 4, 105 (1969).
132. Purdue, L. J., R. E. Enrione, R. J. Thompson, and B. A. Bonfield, *Anal. Chem.*, 45 527 (1973).
133. Linch, A. L., R. F. Stalzer, and D. T. Lefferts, *Am. Ind. Hyg. Assoc, J.*, 29, 79 (1968).
134. Oak Ridge National Laboratories, "Trace Element Measurements at the Coal Fired Allen Steam Plant," Progress report for June 1971 to January 1973.
135. Calver, S. and W. Workman, *Talanta*, 4, 89 (1960).
136. Wartburg, A. F., J. B. Pate, and J. P. Lodge, Jr., *Environ. Sci. Technol.*, 3, 767 (1969).
137. Smith, A. F., D. G. Jenkins, and D. E. Cunningworth, *J. Appl. Chem.*, 11, 317 (1961).
138. Jacobs, M. B., M. M. Braverman, and S. Hochheiser, *Anal. Chem.*, 29. 1349 (1957).

139. Bostroem, C. E., *Air Water Pollut.*, 10 (6–7) 435 (1966).
140. Moffitt, A. E. and R. E. Kupel, *At. Absorption Newslett.*, 9, 113 (1970).
141. Snyder, L. J., *Anal. Chem.*, 39, 591 (1967).
142. Joensuu, O. I., *Appl. Spectrosc.*, 25, 526 (1971).
143. Braman, Robert S., and D. J. Johnson, *Environ. Sci. Technol.*, 8, 996 (1974).
144. Wroblewski, S. C., T. M. Spittler, and P. R. Harrison, *J. Air Pollut. Control Assoc.*, 24, 778 (1974).
145. Scarangelli, F. P., J. C. Puzak, B. I. Bennett, and R. L. Denny, *Anal. Chem.*, 46, 279 (1974).
146. Monkman, J. L., private communication to S. J. Long, 1974.
147. Long, S. J., D. R. Scott, and R. J. Thompson, *Anal. Chem.*, 45, 2227 (1973).
148. ASTM Method D2681-68T, American Society for Testing and Materials, 1916 Race St., Philadelphia.
149. Snyder, L. J. and S. R. Henderson, *Anal. Chem.*, 33, 1175 (1961).
150. Braman, R. S. and C. C. Foreback, *Science*, 182, 1247 (1973).
151. Jacko, Robert B., David W. Neuendorf, and Kenneth J. Yost, *J. Air. Pollut. Control Assoc.*, 25, 1058 (1975).
152. Lapple, C. E. and H. J. Kamack, *Chem., Eng. Progr.*, 51 (3), 110 (1955).
153. Semrau, K. T., *J. Air Pollut. Control Assoc.*, 10, 200 (1960).
154. Murozumi, M., T. J. Chow, and C. C. Patterson, *Geochim. Cosmochim. Acta*, 33, 1247 (1969).
155. Prospero, Joseph M. and Tony N. Carlson, *J. Geophys. Res.*, 77, 5255 (1972).
156. von Lehmden, D. J., R. H. Jungers, and R. E. Lee, Jr., *Anal. Chem.*, 46, 239 (1974).
157. Ondov, J. M., W. H. Zoller, Ilhan Olmez, N. K. Aras, G. E. Gordon, L. A. Rancitelli, K. H. Abel, R. H. Filby, K. R. Shah, and R. C. Ragaini, *Anal. Chem.*, 47, 1102 (1975).
158. Skogerboe, R. K., A. M. Hartley, R. S. Vogel, and S. R. Koirtyohann, "Monitoring for Lead in the Environment," in *Tri-University Report on Lead in the Environment*, U.S. National Science Foundation—Research Applied to National Needs, in press.
159. Gorsuch, T. T., *The Destruction of Organic Matter*, Pergamon Press, New York, 1970.
160. Tölg, G., *Talanta*, 19, 1489 (1972).
161. Kometani, T. Y., J. L. Bove, B. Nathanson, S. Siebenberg, and M. Magyar, *Environ. Sci. Technol.*, 5, 617 (1972).
162. Burnham, C. D., C. E. Moore, T. Kowalski, and J. Krasniewski, *Appl. Spectrosc.*, 24, 411 (1970).
163. Gleit, C. E. and W. D. Holland, *Anal. Chem.*, 34, 1454 (1962).
164. Gleit, C. E., *Am. J. Med. Electron.*, 2, 112 (1963).
165. Thompson, R. J., G. B. Morgan, and L. J. Pardue, *At. Absorption Newslett.*, 9, 53 (1970).
166. Walsh, P. R., J. L. Fasching, and R. A. Duce, *Anal. Chem.*, 48, 1012 (1976).
167. Ruch, R. R., H. J. Gluskoter, and N. F. Shimp, Illinois State Geological Survey Report No. 72, 1974.
168. Olson, Kenneth W. and R. K. Skogerboe, *Environ. Sci. Technol.*, 9, 227 (1975).
169. Altshuller, A. P., "Analytical Chemistry: Key to Progress on National Problems," NBS Special Publication No. 351, Chapter 5, 1972.
170. Lisk, D. J., *Science*, 184, 1137 (1974).
171. Fundamental Reviews, *Anal. Chem.*, 48 (5), (1976).

172. Brown, R., M. L. Jacobs, and H. E. Taylor, *Am. Lab.*, 4 (11), 29 (1972).
173. Ahearn, Arthur J., Ed., *Trace Analysis by Mass Spectrometry*, Academic Press, New York, 1972.
174. Owens, E. B. and N. A. Giadino, *Anal. Chem.*, 35, 1172 (1963).
175. Morrison, George H., Ed., *Trace Analysis*, Wiley-Interscience, New York, 1965.
176. Wallace, J. R., D. F. S. Natusch, B. N. Colby, and C. A. Evans, Jr., *Anal. Chem.*, 48, 118 (1976).
177. Zoller, W. H. and G. E. Gordon, *Anal Chem.*, 42, 257 (1970).
178. Dams, R., K. A. Rahn, J. A. Robbins, G. D. Nifong, and J. W. Winchester, *Proceedings of the Second International Clean Air Congress*, Washington, D.C., 1970, p. 509.
179. Dams, R., J. A. Robbins, K. A. Rahn, and J. W. Winchester, *Anal. Chem.*, 42, 861 (1970).
180. Shah, K. R., R. H. Filby, and W. A. Haller, *J. Radioanal. Chem.*, 6, 185 (1971).
181. Shah, K. R., R. H. Filby, and W. A. Haller, *J. Radioanal. Chem.*, 6, 413 (1971).
182. Lyon, W. S. and J. F. Emery, *Int. J. Environ. Anal. Chem.*, 4, 125 (1975).
183. Olmez, I., N. K. Aras, G. E. Gordon, and W. H. Zoller, *Anal. Chem.*, 46, 935 (1974).
184. Aras, N. K., W. H. Zoller, G. E. Gordon, and G. J. Lutz, *Anal. Chem.*, 45, 1481 (1973).
185. Iddings, F. A., *Environ. Sci. Technol.*, 3, 132 (1969).
186. Kay, M. A., D. M. McKown, D. H. Gray, M. E. Eicher, and J. R. Vogt, *Am. Lab.*, 5 (7), 39 (1973).
187. Abel, K. H. and L. A. Rancitelli, *Advan. Chem. Ser.*, 141, 118 (1975).
188. Leroux, J. and M. Mahmud, *J. Air Pollut. Control Assoc.*, 20, 402 (1970).
189. Birks, L. S., *Anal. Chem.*, 44, 557R (1972).
190. Macdonald, G. L., *CRC Crit. Rev. Anal. Chem.*, 4, 281 (1974).
191. Müller, R. O., *Spectrochemical Analysis by X-Ray Fluorescence*, Plenum Press, New York, 1972.
192. Liebhafsky, H. A., P. A. Pfeiffer, E. H. Winslow, and P. D. Zemany, *X-Rays, Electrons, and Analytical Chemistry*, Wiley-Interscience, New York, 1972.
193. Woldseth, R., *X-Ray Energy Spectrometry*, Kevex Corp., Burlingame, Calif. 1973.
194. Giauque, R. D., L. Y. Goda, and R. B. Garrett, Report No. LBL-2951, Lawrence Berkeley Laboratory, Berkeley, Calif., 1974.
195. Kneip, T. J. and G. R. Laurer, *Anal. Chem.*, 44, 57A (1972).
196. Johansson, T. B., R. Akelsson, and S. A. E. Johansson, *Nucl. Instrum. Methods*, 84, 141 (1970).
197. Giauque, R. D., L. Y. Goda, and N. E. Brown, *Environ. Sci. Technol.*, 8, 436 (1974).
198. Dzubay, T. G. and R. K. Stevens, paper presented at the Second Joint Conference on Sensing of Environmental Pollutants, Washington, D.C., December 1973.
199. Cares, J. W., *Am. Ind. Hyg. Assoc. J.*, 29, 463 (1968).
200. Bowman, H. R., J. G. Canway, and F. Asaro, *Environ. Sci. Technol.*, 6, 558 (1972).
201. Rolla, A., *Chem. Ind.* (Milan), 55, 623 (1973).
202. Rhodes, J. R., A. H. Pradzynski, C. B. Hunter, J. S. Payne, and J. L. Lundgren, *Environ. Sci. Technol.*, 6, 922 (1972).
203. Grennfelt, P., A. Akerstrom, and C. Brosset, *Atmos. Environ.*, 5, 1 (1971).
204. Camp, D. C., J. A. Cooper, and J. R. Rhodes, *X-Ray Spectrom.*, 3, 47 (1974).

205. Rhodes, J. R., and C. B. Hunter, *X-Ray Spectrom.*, 1, 113 (1972).
206. Walsh, A., *Spectrochim. Acta*, 7, 108 (1955).
207. Alkemade, C. T. J. and J. M. W. Milatz, *Appl. Sci. Res. B*, 4, 289 (1955).
208. Alkemade, C. T. J. and J. M. W. Milatz, *J. Opt. Soc. Am.*, 45, 583 (1955).
209. Winefordner, J. D., Ed., *Spectrochemical Methods of Analysis: Advances in Analytical Chemistry and Instrumentation*, Vol. 9, Wiley-Interscience, New York, 1971.
210. Hildebrand, D. C., S. R. Koirtyohann, and E. E. Pickett, *Biochem. Med.*, 3, 437 (1970).
211. Kahn, H. L., G. E. Peterson, and J. E. Schallis, *At. Absorption Newslett.*, 7, 35 (1968).
212. Delves, H. T., *Analyst*, 95, 431 (1970).
213. Matousek, J. P., *Am. Lab.*, 3 (6), 45 (1971).
214. L'vov, B. V., *Spectrochim. Acta*, 17, 761 (1961).
215. Woodriff, R. and G. Ramelow, *Spectrochim. Acta*, 23B, 665 (1968).
216. Ebdon, L., G. F. Kirkbright, and T. S. West, *Anal. Chim. Acta*, 51, 365 (1970).
217. Jackson, K. W. and T. S. West, *Anal. Chim. Acta*, 59, 187 (1972).
218. Massman, H., *Spectrochim. Acta*, 23B, 215 (1968).
219. Matousek, J. P. and B. J. Stevens, *Clin. Chem.*, 17, 363 (1971).
220. Norval, E. and L. R. P. Butler, *Anal. Chim. Acta*, 58, 47 (1972).
221. Kahn, H. L., *J. Chem. Educ.*, 43, A7 and A103 (1966).
222. Seeley, J. L., D. L. Dick, J. H. Arvik, R. L. Zimdahl, and R. K. Skogerboe, *Appl. Spectrosc.*, 26, 456 (1972).
223. Koirtyohann, S. R. and E. E. Pickett, *Anal. Chem.*, 37, 601 (1965).
224. Veillon, C., *Handbook of Commercial Scientific Instruments*. Vol. 1, *Atomic Absorption*, Dekker, New York, 1972.
225. D. L. Dick, S. J. Urtamo, F. E. Lichte, and R. K. Skogerboe, *Appl. Spectrosc.*, 27, 467 (1973).
226. Marks, G. E. and C. E. Moore, *Appl. Spectrosc.*, 26, 523 (1971).
227. Schmidt, F. J. and J. L. Royer, *Anal. Lett.*, 6, 17 (1973).
228. Larsson, Leif, *Svensk. Papperstid* (Stockholm), 74, 241 (1971).
229. Thomas, F. J. and R. A. Kuchar, *Anal. Chem.*, 44, 512 (1972).
230. Braman, R. S., L. L. Justen, and C. C. Foreback, *Anal. Chem.*, 44, 1476 (1972).
231. Miller, J. A., F. J. Schmidt, D. F. S. Natusch, and T. M. Thorpe, Paper No. 112 presented at Pittsburgh Conference, Cleveland, 1974.
232. Braverman, M. M., F. A. Masciello, and V. Marsh, *J. Air Pollut. Control Assoc.*, 11, 408 (1961).
233. Sacks, R. D. and S. W. Brewer, *Appl. Spectrosc. Rev.*, 6, 313 (1972).
234. Feldman, C., *Anal. Chem.*, 21, 1041 (1949).
235. Gordon, W. A., NASA Technical Notes D2598, 4769, and 5236, Clearinghouse for Federal Scientific and Technical Information, Springfield, Va.
236. Hambridge, K. M., *Anal. Chem.*, 43, 103 (1971).
237. Slavin, M., *Emission Spectrochemical Analysis*, Wiley-Interscience, New York, 1971.
238. *Methods for Emission Spectrochemical Analysis*, American Society for Testing and Materials, Philadelphia, 1968.
239. Skogerboe, R. K. and G. N. Coleman, *Anal. Chem.*, 48, 611A (1976).

240. Fassel, V. A. and R. N. Kniseley, *Anal. Chem.*, 46, 1110A (1974).
241. Fassel, V. A. and R. N. Kniseley, *Anal. Chem.*, 46, 1155A (1974).
242. Greenfield, S., I. L. Jones, H. M. McGeachin, and P. B. Smith, *Anal. Chim. Acta*, 74, 225 (1975).
243. Winefordner, J. D. and R. C. Elser, *Anal. Chem.*, 43 (4), 24A (1971).
244. Siegerman, H. *Chem. Technol.*, 672, November 1971.
245. Damaskin, B. B., *Principles of Current Methods for the Study of Electrochemical Reactions*, McGraw-Hill, New York, 1967.
246. Harrison, P. R. and J. W. Winchester, *Atmos. Environ.*, 5, 863 (1971).
247. Harrison, P. R., J. W. Winchester, and W. R. Matson, *Atmos. Environ.*, 5, 613 (1971).
248. Colovos, G., G. S. Wilson, and J. L. Moyers, *Anal. Chim. Acta*, 64, 457 (1973).
249. Neeb, R. and F. Wahdat, *Fresenius Z. Anal. Chem.*, 269, 275 (1974).
250. Orion Research Corp., Information Leaflet No. CAT/961, 1969.
251. Flato, J. B., *Anal. Chem.*, 44, 75A (1972).
252. Melabs Scientific Instruments, Palo Alto, Calif., Application Notes 2A and 4.
253. Christian, G., *J. Electroanal. Chem.*, 22, 333 (1969).
254. Lagrou, A. and F. Verbeek, *Anal. Lett.*, **4**, 573 (1971).
255. Shain, I., in *Treatise on Analytical Chemistry*, I. Kolthoff and P. J. Elving, Eds., Part I, Vol. 4, Wiley, New York, 1964.
256. Martin, K. J. and I. Shain, *Anal. Chem.*, 30, 1808 (1958).
257. Matson, W. R., D. K. Roe, and D. E. Carritt, *Anal. Chem.*, 37, 1598 (1965).
258. Copeland, T. R., J. H. Christie, R. K. Skogerboe, and R. A. Osteryoung, *Anal. Chem.*, 45, 995 (1973).
259. Lamb, R. E., Doctoral dissertation, University of Illinois, Urbana, 1975.
260. Siegesmund, K. A., A. Funahashi, and K. Pintar, *Arch. Environ. Health*, 28. 345 (1974).
261. Ferrin, J., J. R. Coleman, S. Davis, and B. Morehouse, *Arch. Environ. Health*, 31, 113 (1976).
262. Andersen, C. A., Ed., *Microprobe Analysis*, Wiley-Interscience, New York, 1973.
263. Everhart, T. E. and T. L. Hayes, *Sci. Am.*, January 1972, p. 55.
264. Crewe, A. V., *Sci. Am.*, April 1971, p. 26.
265. Buseck, P. R. and S. Iijima, *Am. Mineral.*, 60, 771 (1975).
266. Armstrong, J. T. and P. R. Buseck, *Anal. Chem.*, 47, 2178 (1975).
267. Anderson, C. A. and J. R. Hinthorne, *Science*, 175, 853 (1972).
268. Lane, W. C., D. E. Pease, and N. C. Yew, Paper SIA-1701 presented at the Sixth Annual Technical Meeting of the International Microstructural Analysis Society, Beverly Hills, Calif., September 1973.
269. Harris, Lawrence A., *Anal. Chem.*, 40 (14), 24A (1968).
270. Hercules, David M., *Anal. Chem.*, 42 (1), 20A (1970).
271. Hulett, L. D., T. A. Carlson, B. R. Fish, and J. L. Durham, *Proceedings of the Symposium on Air Quality*, 161st National ACS Meeting, Los Angeles, April 1971. Plenum Publishing Corp., Washington, D.C.
272. Palmberg, Paul W., *Anal. Chem.*, 45, 549A (1973).
273. Evans, C. A., Jr., *Anal. Chem.*, 47, 818A and 855A (1975).
274. Watt, J. D. and D. J. Thorne, *J. Appl. Chem.*, 15, 585 (1965).

275. Cunningham, Paul T., Stanley A. Johnson, and Ralph T. Yang, *Environ. Sci. Technol.*, 8, 131 (1974).
276. Low, M. J. D., *Anal. Chem.*, 41 (6), 97A (1969).
277. Thorpe, Thomas M., Doctoral dissertation, University of Illinois, Urbana, 1975.
278. Meyer, W. T. and D. S. Evans, *Inst. Min. Metall. Symp.*, 127, August 1973.
279. Meyer, W. T. and K. C. Y. Lam Shang Leen, *Geochem. Explor., Proc. Int. Geochem. Explor. Symp.*, 325 (1972).
280. Schuetzle, D., A. L. Crittenden, and R. J. Charlson, *J. Air Pollut. Control Assoc.*, 23, 704 (1973).
281. Bayard, M. and G. L. Ter Haar, *Nature*, 232, 553 (1971).
282. Bolton, C. E., J. A. Carter, J. F. Emery, C. Feldman, W. F. Fulkerson, L. D. Hulett, and W. S. Lyon, *Advan. Chem. Ser.*, 141, 175 (1973).
283. Talmi, Y. and A. W. Andren, *Anal. Chem.*, 46, 2122 (1974).
284. Johnson, D. L. and M. E. Q. Pilson, *Anal. Chim. Acta*, 58, 289 (1972).
285. Kudsk, F. Nielsen, *Scand. J. Clin. Lab. Invest.*, 16, Suppl. 77, 1 (1964).
286. Pinta, Maurice, *Detection and Determination of Trace Elements*, Ann Arbor Science Publishers, Ann Arbor, Mich., 1972.
287. Weisz, H., *Microanalysis by the Ring Oven Technique*, Pergamon Press, New York, 1961.
288. West, P. W., in *Air Pollution*, Vol. 2, Arthur C. Stern, Ed., Academic Press, New York, 1968, pp. 147–183.
289. West, P. W. and S. Sachdev, *J. Chem. Educ.*, 46 (2), 96 (1969).
290. Ross, W. D., R. E. Sievers, and G. Wheeler, Jr., *Anal. Chem.*, 37, 598 (1965).
291. Bayer, E., H. P. Muller, and R. Sievers, *Anal. Chem.*, 43 2012 (1971).
292. Natusch, D. F. S. and T. M. Thorpe, *Anal. Chem.*, 45, 1185A (1973).
293. Harman, H. H., *Modern Factor Analysis*, 2nd ed., University of Chicago Press, Chicago, 1967.
294. Thurstone, L. L., *Multiple Factor Analysis*, University of Chicago Press, Chicago, 1947.
295. Gilfrich, J. V., P. G. Burkhalter, and L. S. Birks, *Anal. Chem.*, 45, 2002 (1973).
296. Johansson, T. B., R. E. Van Grieken, D. M. Nelson, and J. W. Winchester, *Anal. Chem.*, 47, 855 (1976).
297. Ranweiler, Lynn E. and Jarvis L. Moyers, *Environ. Sci. Technol.*, 8, 152 (1974).
298. Woodriff, R., R. W. Stone, and A. M. Held, *Appl. Spectrosc.*, 22, 408 (1968).
299. Begnoche, B. C. and T. M. Risby, *Anal. Chem.*, 47, 1041 (1975).
300. Hwang, J. Y., P. A. Ullucci, and C. J. Mokeler, *Anal. Chem.*, 44, 2018 (1972).

6

INSTRUMENTAL METHODS FOR AUTOMATIC AIR MONITORING STATIONS

D. J. Kroon

Philips Research Laboratories
N. V. Philips Gloeilampenfabrieken
Eindhoven, The Netherlands

6.1. INTRODUCTION

An air pollution monitoring system, as an instrument for environmental policy making, must present a quantitative description of the situation. This can be obtained only through the collection of extensive data on air quality, to provide the material for reliable statistics. There are two principal methods for collecting these data: manual collection of samples with centralized analysis, and use of an automatic monitoring network.

Manual collection of air samples requires a low capital investment; it is versatile with regard to the number of components to be analyzed, and the sampling sites can be easily changed. On the other hand, manpower costs are high, the density of the network is low, and the sampling frequency is limited. The resultant information tends to lag many hours behind the actual situation.

Although the capital investment for automatic sampling and analysis networks for air pollutants is higher, manpower costs are insignificant; thus the cost per analysis is of the order of 10 to 25% compared with manual data collection. Sampling frequencies and network densities can be made as high as is required, and current information is always available. Sampling sites and the air components to be measured by an automatic network must be selected *a priori* with great care, since changing of a sampling site may be difficult.

Data collection on air quality serves three general purposes: mapping, trend investigation, and warning.

6.1.1. Mapping

"Mapping" of air contaminants or the plotting of the instantaneous level of air pollution, constitutes a survey of the distribution of air pollutants over a large area. From the distribution of these concentrations it is often possible to pinpoint individual sources of pollution. If necessary in certain cases, this can result in a specific warning that appropriate control measures should be taken. Mapping requires simultaneous collection of many data over a large area. To exclude influences from atmospheric or anthropogenic sources, replicate measurements are needed, and the pollutant data must be condensed to produce statistically reliable values. Networks operating automatically and transmitting the data directly to a computer are virtually ideal for this purpose.

6.1.2. Trend Investigation

Possible trends can be deduced from slow variations in average concentrations of pollutants. The measurement of trends can be used to investigate the effect of air pollution control and abatement procedures or to determine the contribution of certain industrial areas or residential quarters to the pollution level. Reliable trend figures must be based on a large number of measurements because air pollution levels are strongly influenced by such meteorological parameters as temperature, wind direction and speed, rainfall and solar radiation, and season. To be corrected for these external parameters, the collected data must be treated statistically, and this means that for each situation a large amount of data has to be available.[1] For example, if air pollution levels on one sampling site are to be grouped according to four wind directions, high and low wind speed, winter and summer season, high and low temperature, rainfall and sunshine, there are 128 classes. For a reasonable statistical confidence level, at least 50 measurements per class, or more than 6000 measurements per site, must be carried out. Since the meteorological parameters are never evenly distributed in time, the total number of measurements may be even tenfold this number.

6.1.3. Warning

If in a continuous measurement program a sudden increase in the concentration of one or more air contaminants is observed, a danger situation may be imminent. If such an increase is observed in an air pollution monitoring system, it may show the buildup of an inversion layer. Then based on this information preventive action can be taken—for example, advising local industries, which may be required to modify their operations and postpone air polluting activities. Such an air pollution warning system needs continuous monitoring with on-line data handling, which cannot be realized with a manual system.

6.2. REQUIREMENTS FOR INSTRUMENTS IN AUTOMATIC STATIONS

Automatic stations are unmanned; air samples are taken automatically and analyzed *in situ*, and the results are transmitted by telephone, telex, or radio to the central agency, usually a computer. Therefore instruments to be used in automatic monitoring stations have to fulfill specific requirements that differ from those for instruments used in the laboratory, where the samples are analyzed manually. Automatic instruments must use analytical techniques, which are faster, more selective, and more sensitive than laboratory methods. Furthermore, since automatic instruments operate unattended for long periods (up to a number of months), the use of chemical reagents must be avoided.

6.2.1. Speed

In the laboratory the time taken for the analysis of air samples is of little importance. The time lag between sampling and analysis always exceeds the time for analysis. However, automatic instruments must be fast for several reasons. First, external conditions, such as pollutant concentrations, change quite rapidly. For mapping and warning especially, a response time of a few minutes is required,[2] as illustrated in Figure 6.1, giving typical observations on ozone (O_3) and nitrogen oxides. Around 11 A.M. the ozone concentration is observed to drop suddenly from 40 ppb to virtually zero, while the nitric oxide (NO) concentration increases from zero to 40 ppb and the nitrogen dioxide (NO_2) concentration nearly doubles. This was correlated with a sudden change in wind direction.[3] This effect, which is

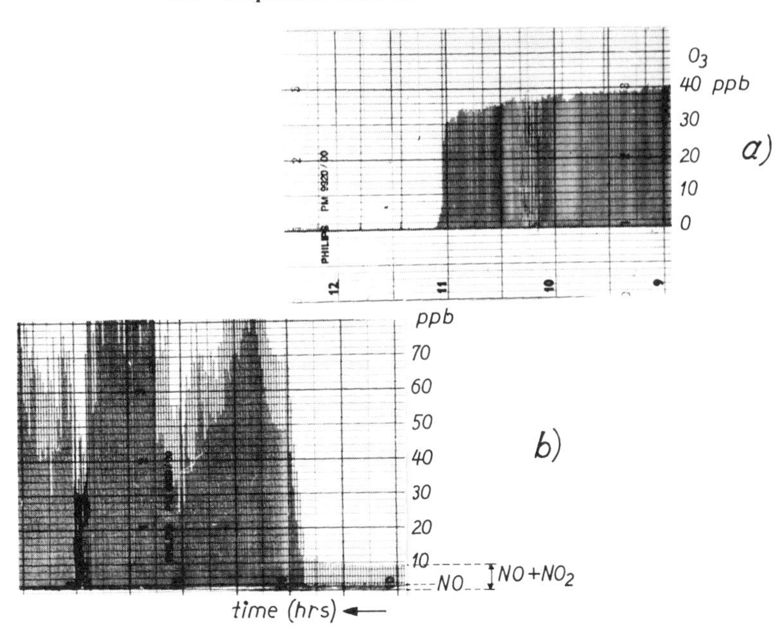

Figure 6.1. Variation of ambient O_3, NO_3 and NO_2 concentration. (*a*) At 11 A.M. the ozone concentration drops from 40 ppb to zero (rhodamine B ozone monitor, cf. Section 6.4.2.1b. The high and narrow peaks are due to the periodic calibration of the instrument.[3.26] (*b*) Rapid increase of the NO concentration (dark band) at 11 A.M. from nearly zero to about 100 ppb. The NO_2 concentration (light band) nearly doubles. Measurements were carried out with a chemiluminescent NO–NO_2 monitor. (cf. Sections 6.4.3.1 and 6.4.3.2.[3.26]

not abnormal, would have escaped observation with slow response instruments. Accuracy, too, is needed in an automatic system. In the laboratory a sample can be divided into a number of aliquot parts, and a mean value and standard deviation can be obtained by multiple analysis. In automatic monitoring a repetition of a particular measurement is not possible because the situation changes continuously. Statistical treatment of data therefore is based on a series of successive measurements. The actual frequency depends on the local situation, the season, and the air pollutant component under study. This is illustrated by the correlation function for sulfur dioxide (SO_2) obtained in an industrial and domestic area (Figure 6.2). Over 2 months the SO_2 concentration was measured every minute. Nearly 100,000 values were used to compute an autocorrelation function $F(\tau)$, using the equation

$$F(\tau) = \Sigma a(t) \cdot a(t + \tau)$$

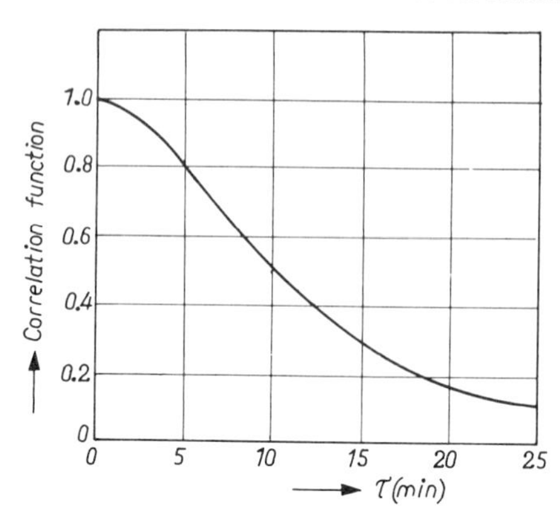

Figure 6.2. Autocorrelation function for the SO_2 concentration in an industrial and domestic area. Over a time span of 2 months, one minute values were collected. From this curve a minimum sampling time of 3 to 5 min can be deduced.

where $a(t)$ and $a(t + \tau)$ represent the modified concentrations at times t and $t + \tau$, respectively. If $c(t)$ denotes the concentration, we have

$$a(t) = \frac{c(t) - \bar{c}}{c_{\text{std}}}$$

with $\bar{c}$ the average concentration over the period of observation and c_{std} the standard deviation.

According to Figure 6.2, the correlation time of the SO_2 concentration is of the order of 15 mins. If an accurate observation of the concentration is to be made, three to five measurements should be carried out at least in one correlation time. This brings the period between sampling to 3 to 5 min. For other combinations of site, season, and component, higher or lower frequencies can be derived. Simultaneous correlation of measurements on other components may be necessary (see Figure 6.1). Thus rapid automatic analysis is essential for effective investigation of mapping, warning, and trends.

6.2.2. Selectivity

Analytical techniques to be used in automatic instruments must be more specific than those used for manual or instrumental techniques. Laboratory samples can be extensively pretreated to remove interfering components, and whether this is necessary can be decided after a preliminary semiquantitative analysis. This is not possible with automatic instruments. The air sample must be analyzed immediately, and any pretreatment is identical for all samples. The further requirement that an automatic instrument operates unattended for long periods excludes the use of large quantities of rapidly deteriorating chemical reagents. Therefore the technique chosen for automatic instrumentation should possess an inherent selectivity for the component to be analyzed or should be able to perform the analysis after a simple automated treatment, such as a filtration, scrubbing, or adsorption of an interfering compound.

6.2.3. Sensitivity

All air pollutants occur in concentrations as low as one to a few hundred parts per billion. Since it is rarely possible to obtain extensive pretreatment of the sample, which raises concentrations of contaminants, the analytical technique chosen must be capable of measuring the pollutants accurately at these very low concentrations.

6.2.4. Calibration

Whereas in laboratory instruments a number of parameters can be manually checked and the instruments adjusted, this is not possible in a field automated measurement station. Thus an automated, built-in calibration facility is essential for automatic pollutant analysis equipment, and the instrument must be safeguarded against fluctuations in temperature, voltage, gas- and liquid flows, drift of sensors, contamination of sampling lines, and so on. Since these parameters cannot be checked individually, the only possibility for adjusting the performance of the instrument is a frequent, remote control calibration of the entire analytical procedure.

Only a few analytical techniques are capable of meeting all these requirements, which is why few techniques applied in the analytical laboratory are applicable to automatic monitoring equipment. Different, very specialized techniques must be used. Examples are chemiluminescence for ozone, nitric oxide, and nitrogen dioxide, correlation spectroscopy, and coulometry.

The widely used laboratory technique using wet spectrophotometry is not really applicable to automation, because of the difficulties involved in storage and the high consumption of perishable reagents.

6.3. COMPARISON OF METHODS

In principle the particular technique employed in an automatic station is unimportant, as long as the foregoing specifications are met. If the instrument can be made sufficiently specific, sensitive, fast, and reliable (a very important requirement), a sophisticated method that is difficult to engineer may be inferior to a more conventional (i.e., "old-fashioned") method, for which engineering to an automated professional instrument is easier. In recent years a number of promising new methods have been tested for use with automatic stations and have failed. Since good engineering is both costly and time-consuming, one is faced with the paradox that instruments based on unselective methods, but well engineered, can prevail over instruments based on better principles, but for which the engineering has proved too difficult.

Because an air pollution monitoring system is set up to collect data, the techniques reviewed in this chapter are to be judged on their capability to serve this practical point of view. However, there are some difficulties in comparing different techniques. Methods for the measurement of the same component may differ in response time, range of application, lower limit of detection, and stability. Sometimes the limit of detection can be lowered by electronic integration of the signal, although this leads to a longer response time. A drift in zero level or span can be compensated for by frequent calibration and zeroing, but during these operations the instrument will not be able to supply air quality data. Instruments based on the same principle can show rather large differences in performance because of differences in the quality of their engineering.

The next section describes a number of techniques suitable for automatic monitoring of air pollution. Some of the data may be too optimistic for a certain commercial instrument, but some instruments may show more favorable characteristics. Part of the information given in the next section was collected from manufacturers' data[4] and may not be based on practical field experience. Where possible, published users' reports or other independent publications have been used. In general, such important properties of instruments as limits of detection and response times are given, indicating the range of published data. Only the principles of the methods are discussed. Commercially available instruments or published reports from laboratories may reveal many technical differences compared to the basic principle discussed here. Manufacturers' names are everywhere omitted.

6.4. EXISTING METHODS

This section describes some methods that have been found applicable to automatic monitoring of air pollutants, the most important of which are sulfur dioxide, ozone, nitric oxide, nitrogen dioxide, carbon monoxide, and dust. A criterion used was whether instruments based on the respective methods have proved able to operate unattended for a considerable period of time.

6.4.1. Monitoring Methods for Sulfur Dioxide

Five methods have been shown to be acceptable for the continuous automatic monitoring of SO_2: coulometry, flame photometry, UV spectroscopy, second derivative spectroscopy, and fluorescence.[65]

6.4.1.1. Coulometric Determination of Sulfur Dioxide

The coulometric determination of SO_2 is based on the oxidation of SO_2 by bromine, according to the reaction

$$2H_2O + SO_2 + Br_2 \longrightarrow SO_4^{2-} + 4H^+ + 2Br^- \qquad (1)$$

and the subsequent liberation of bromine by an electric current

$$2Br^- \longrightarrow Br_2 + 2e. \qquad (2)$$

The principle applied to an instrument is shown in Figure 6.3. Air containing SO_2 is passed through an electrolytic cell containing a mixture of H_2SO_4, KBr, and Br_2. In the cell there are two sets of electrodes. The indicator electrode and the reference electrode are used to determine the bromine concentration; the other set, the generating electrode and the auxiliary

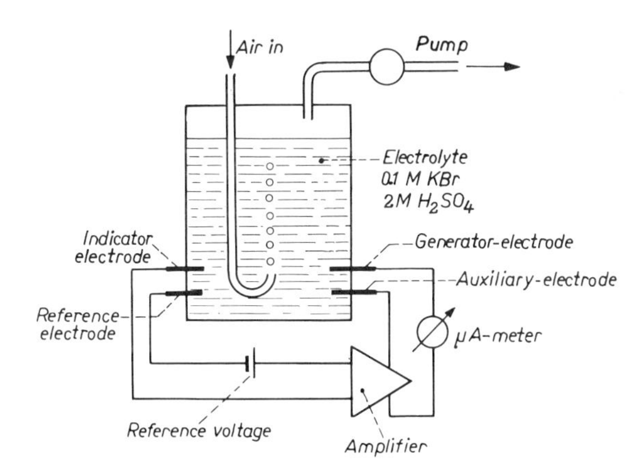

Figure 6.3. Principle of coulometric titration of SO_2. The measuring cell contains a solution of H_2SO_4, KBr_3 and Br_2 in water. The voltage between the reference electrode and the indicator electrode is compared with a reference voltage. The difference between the two voltages controls a current, which liberates Br_2 by electrolysis. This feedback system keeps the Br_2 concentration in the cell constant. If SO_2 is present in the airstream entering the instrument, it will consume bromine. Therefore the electrolysis current is proportional to the SO_2 concentration in the airstream.

electrode, serve to generate bromine by electrolysis. Indicator, generator, and auxiliary electrodes are made of platinum, the reference electrode is of the Ag-AgBr type. A voltage is developed between the indicator electrode and the reference electrode. From a certain Br_2 concentration onward, this voltage follows the Nernst law (Figure 6.4):

$$E = E_0 + \frac{RT}{4F} \ln \frac{[Br_2]}{[Br^-]}$$

This voltage is compared to a reference voltage V_0 in an electronic circuit. The difference between the Nernst voltage and the reference voltage, after amplification, causes an electric current to flow between the generating electrode and the auxiliary electrode to form bromine according to reaction 2. By this feedback system the bromine concentration in the cell is maintained at such a level that the Nernst voltage is equal to the reference voltage. If the air passing through the cell contains SO_2 (or any other pollutant that can be oxidized by bromine), bromine will be consumed according to reaction 1 and immediately replaced by way of the feedback system just described.

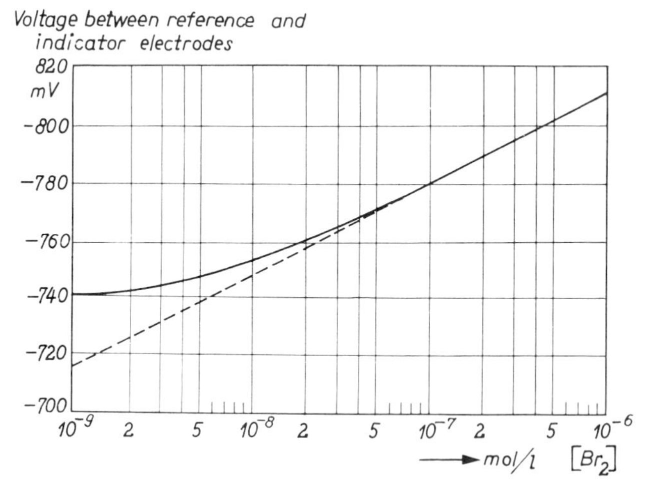

Figure 6.4. Voltage between reference electrode and indicator electrode as a function of the bromine concentration $[Br_2]$. At concentrations higher than 5×10^{-8} mole l^{-1}, the Nernst law is obeyed. At very low concentration the voltage becomes independent of $[Br_2]$ and a change in $[Br_2]$ is not measurable. Because of the logarithmic voltage–concentration relation, a change in the bromine concentration will result at high bromine concentrations in a smaller change in voltage. The optimum working point of a coulometric cell thus is where $[Br_2]$ is as low as possible (i.e., about 5×10^{-8} mole l^{-1}).

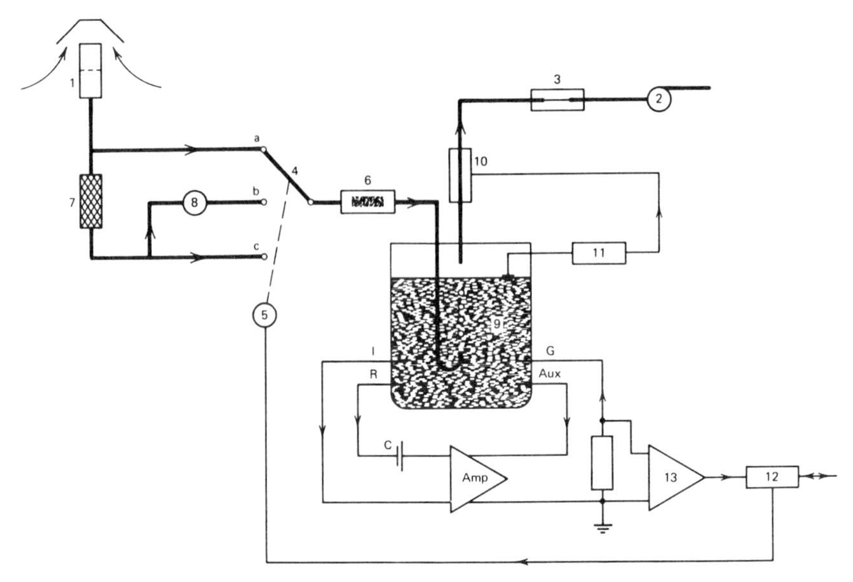

Figure 6.5. Diagram of a completely automatic SO_2 monitor based on coulometry. The pump 2 sucks the air through the instrument, and the capillary 3 keeps the airstream at a constant value. The filter 1 removes dust, insects, and so on from the airstream, and filter 6 removes many gaseous impurities in the air (except SO_2). The SO_2 in the air sample gives rise to a signal output in the cell 9, which is made compatible for transmission by telephone line by the output amplifier 13 and the telemetering unit 12. The signals from the monitor go to the central control room, where a computer is installed. Using the same telephone link, the computer sends signals to the motor 5, which controls the positions of the three-way valve 4 (shown here in measuring position *a*. In the calibration and zero mode positions of the valve, the air passes over an activated charcoal filter 7 in which all SO_2 is adsorbed. In the calibration position *b* the SO_2 source 8 adds a known quantity of SO_2 to the SO_2-free air. By comparing the sample signal with the calibration and zero signals, the computer calculates the SO_2 concentration of the input air [10, Peltier cooler]. If the level of the liquid drops under a present value, cooler is switched on, causing water vapor to condense so that the level of the liquid is maintained [11, level sensor].[5]

This means that the mass flow of SO_2 is converted into a flow of electrons through the feedback circuit, which is easily measured and recorded.

The limit of detection of this method is reported to be a low as 3 ppb and the response time of the instrument (0 to 60%) is about one minute. Although all substances that can be oxidized by bromine will give a positive signal, all oxidizing substances, such as NO_2 and O_3, will give a negative signal. Therefore a selective filter, (e.g., heated silver foil) must be inserted in front of the electrolytic cell. For SO_2 monitors the characteristics of such filters are well known, and in coulometric SO_2 monitors interference due to other gases is reported to be of the order of 1% or the less.

One commercial version of the instrument has two specific features. First, for the monitor to operate for a long time without attendance (the manufacturer claims 3 months), there is a risk that the cell will dry up, especially during winter. Therefore a level sensor has been added to the cell. When the level falls below a certain point, a Peltier cooling element is switched on. Evaporating water is thus condensed and drips back into the cell. Second, a remotely controllable calibration and zero check are supplied. Figure 6.5 is a complete diagram of the instrument.

6.4.1.2. Flame Photometric Determination of Sulfur Dioxide

In a hydrogen-rich flame (see Figure 6.6 for spectrum), sulfur compounds emit a blue light[6] at a wavelength of about 394 nm, which is due to the formation of an S_2^* molecule. In the flame the sulfur-containing molecules are first fragmented, and the S_2^* molecule is formed later. Therefore the intensity of the emission can be expected to be proportional to the square of the concentration.[7] The emission from molecules containing two sulfur atoms should The sensitivities for different compounds appear in Figure 6.7, which shows have an intensity twice that of molecules containing only one sulfur atom. that the "concentration squared" behavior is followed for all compounds; other observations are also made,[6,7] but the intensity is never proportional to the number of sulfur atoms per molecule. Since all sulfur-containing compounds in the air will contribute to the emitted light, the method is not of itself selective. However a selective filter in front of the detector or the use of a gas chromatographic separation column can overcome this problem.[8,9]

In the latter system the concentrations of all sulfur components in the air are measured separately and sequentially, and the measurement is no

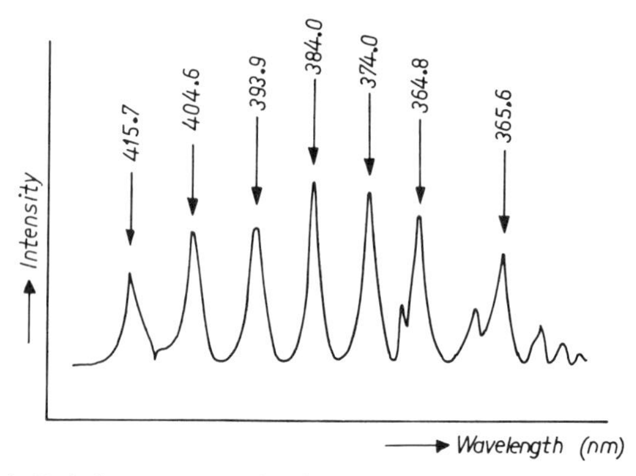

Figure 6.6. Emission spectrum of sulfur compounds in a hydrogen flame.

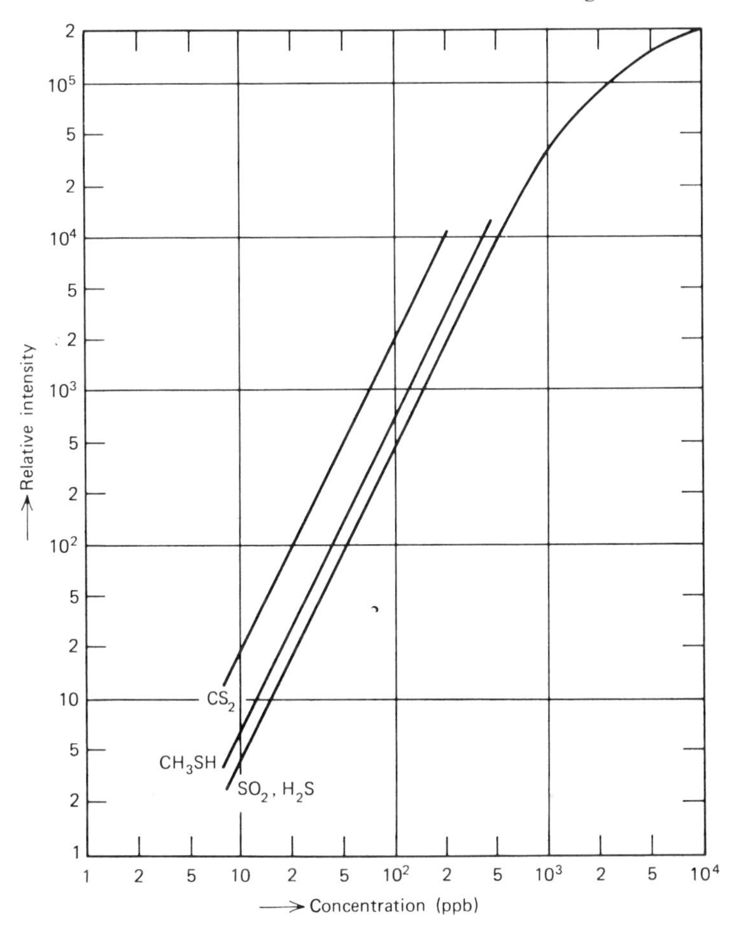

Figure 6.7. Intensity of the light emission of sulfur compounds in a hydrogen flame. As predicted by the theory, the slope of the curves is proportional to the square of the concentration.

longer continuous. Figure 6.8 is a schematic diagram of the instrument, and Figure 6.9 presents an instrument fitted with optical filters to correct for the background radiation.[7,10] Another method of increasing sensitivity employs a rotating interference filter that scans the spectrum.[6] A disadvantage of the method is the large quantity of hydrogen needed; generally about 100 ml min^{-1} or 4 m^3 per month. This introduces the hazard involved in storing such quantities of an explosive gas in a monitoring station, imposing a limitation on the period of unattended operation. In some instruments the problem is overcome by generating hydrogen by the electrolysis of water,

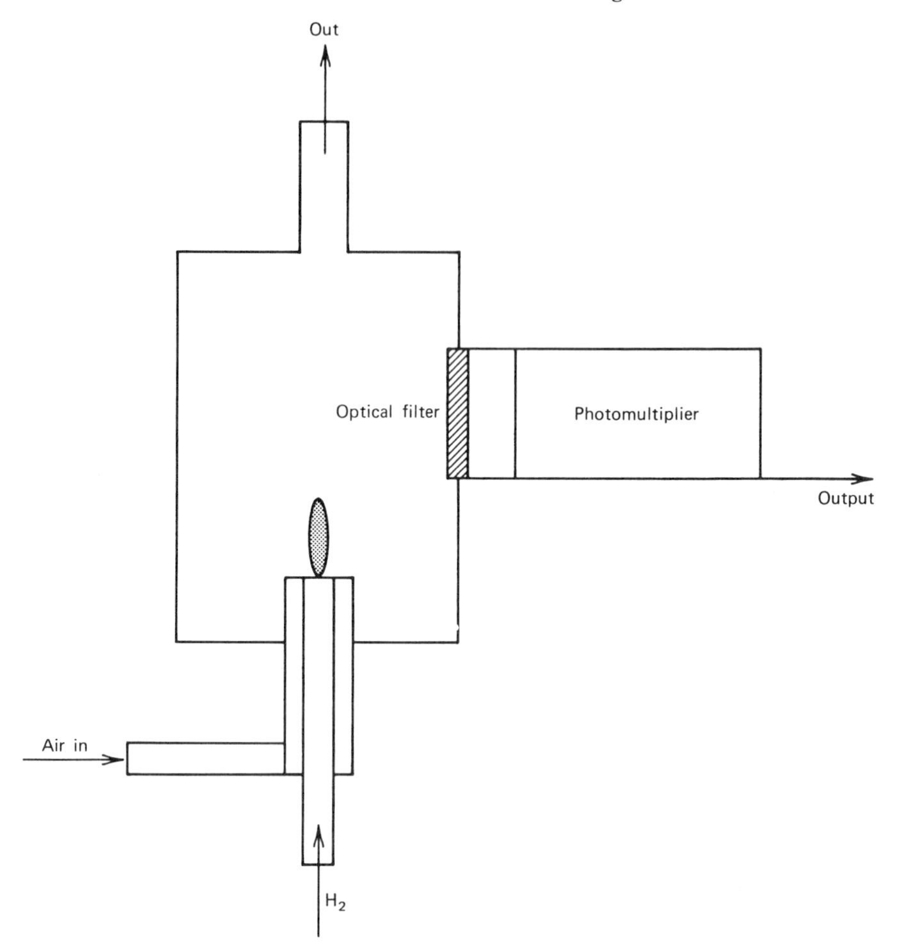

Figure 6.8. Principle of a simple flame photometric detector (FPD).

but this makes the system more complex. Detection limits for flame emission are around 5 ppb, and for total sulfur monitors responses times are about 30 sec. The combination of a gas chromatograph with flame photometric detection introduces a lag time of 5 min. Phosphorus-containing compounds may cause some interference.

6.4.1.3. Spectroscopic Techniques

Spectroscopic determination of air pollutants is based on the physical properties of the molecules rather than on their chemical activity. For most air pollutants, there are specific absorption bands in the near infrared, the

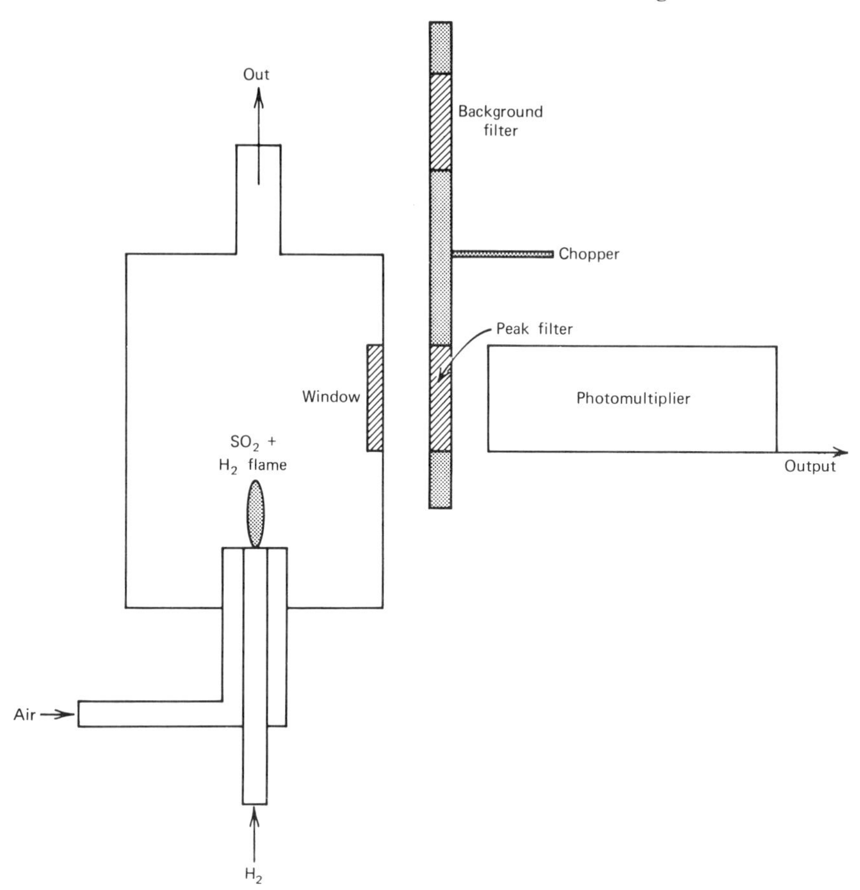

Figure 6.9. Flame photometric detector with automatic background correction. Two narrow band filters (or filter combinations) are placed in the chopper wheel. The peak filter transmits only the light of one of the peaks of the S_2 spectrum (Figure 6.7). The background filter transmits light just between two peaks. Thus the ac signal of the photomultiplier is proportional to the intensity of the S_2 radiation alone. The limit of detection of this instrument can be an order of magnitude better than can be obtained with the simple FPD of Figure 6.8.

visible, or the near ultraviolet parts of the spectrum. If for a particular pollutant these bands are sufficiently separated from the bands of other substances, a highly selective determination of the given pollutant is possible. Sulfur dioxide shows two of such absorption bands: one in the ultraviolet (200 to 235 nm)[11] and one in the infrared (7355 nm).[12] Interference due to scattering by particles can be expected in the ultraviolet band; in the infrared the light absorption of water vapor may complicate the measurement. A number of optical methods can remove interference attributable to these overlapping absorption bands or to particle scattering. Spectroscopic techniques appropriate for automatic monitoring applications include direct absorption spectroscopy, correlation spectroscopy, and second derivative spectroscopy. These techniques are also applicable to the measurement of other components.

a. Direct Absorption Spectroscopy

Direct absorption spectroscopy is the simplest type of spectroscopy and can be used only when no overlapping bands of interfering substances are present. The basic instrument (Figure 6.10) consists of a light source, emitting radiation in the required wavelength region, an optical filter, a light chopper, an absorption cell, an attenuator, a detector, and the associated electronic equipment. The chopper allows the light to pass through the absorption cell during half the period and by way of the attenuator to the detector in the other half of the period. With no pollutant gas present in the sample cell, the attenuator is adjusted to equalize the intensity of both beams. If an absorbing gas is introduced in the cell, the intensity of the light falling on the detector is decreased. The relationship between the light intensity on the detector and the properties of the pollutant is given by the Lambert-Beer law.

$$I = I_0 \exp(-\alpha c l),$$

where I and I_0 are the intensities of the light falling on the detector and emitted by the lamp, respectively, α is the absorption coefficient ($\text{atm}^{-1}\ \text{m}^{-1}$), c the concentration of the absorbing gas (atm), and l the length of the absorption cell (m). The ac signal from the detector is proportional to S, where

$$S = I_0[\exp(-\alpha c l) - 1].$$

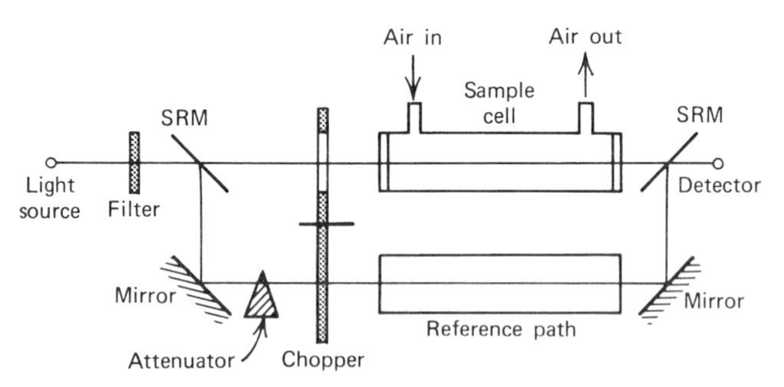

Figure 6.10. Direct absorption spectroscopy (SRM, semireflecting mirror). From the light emitted by the source, an appropriate wavelength region is selected by a filter. A rotating chopper wheel allows the light to pass alternatively through the sample cell and through a reference path to the detector. The attenuator is adjusted to equalize the two beams with the sample cell empty. The difference in intensity received by the detector in each period is proportional to the absorption of the gas in the sample cell, thus to the concentration of the pollutants.

For the weak absorptions in the case of very dilute gases, this signal can be divided by the average dc signal on the detector, thus ruling out drift in lamp intensity and detector response. In fact we should divide by the intensity I_0, not by the average dc signal on the detector; but the latter is much simpler to realize electronically. The error, however, can be estimated as follows: putting $x = \alpha c l$, the reading we obtain by dividing the ac signal by the dc signal is given by

$$R = 2\frac{I_0[\exp(-x) - 1]}{I_0[\exp(-x) + 1]} = 2\frac{\exp(-\frac{1}{2}x) - \exp(\frac{1}{2}x)}{\exp(-\frac{1}{2}x) + \exp(\frac{1}{2}x)}$$

$$= -2\tanh(\tfrac{1}{2}x) \approx -x + \tfrac{1}{12}x^3 + \cdots$$

This is equal to x within 3% for $x < 0.6$.

Since x is of the order of 0.01 to 0.10, this procedure is always safe. For sulfur dioxide long absorption cells are necessary, for instance, multipass cells (Figure 6.11).

Detection limits of 10 ppb can be achieved in the ultraviolet region with a cell length of 25 m, at 1% absorption. The response time is several seconds.

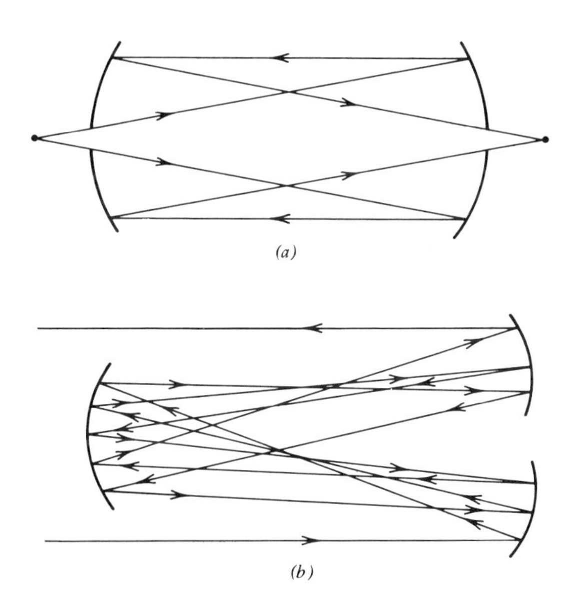

Figure 6.11. Two types of multipass optical cell. (*a*) Pfund cell. Two spherical mirrors, each having a central hole, are spaced at a distance equal to the radius of curvature of the mirrors. (*b*) White cell. The incident beam is collected by the lower spherical mirror on the right, placed at a distance of two focal lengths from the entrance aperture. The beam is focused again on the upper spherical mirror on the right. After this the procedure is repeated, the image being displaced laterally after every pass.

Aerosols and dust particles must be filtered out, especially if the UV absorption is used. Because of the short wavelength ($\lambda \approx 200$ nm), particles down to 0.2 μm contribute significantly to light scattering (cf. Appendix A6), and this effect cannot be discriminated from light absorption by SO_2 (Figure 6.12). Such scattering can be overcome by the use of two wavelengths, close together,[13] one in the SO_2 absorption band, the other just outside. The measurement on the first wavelength will give the light absorption of SO_2

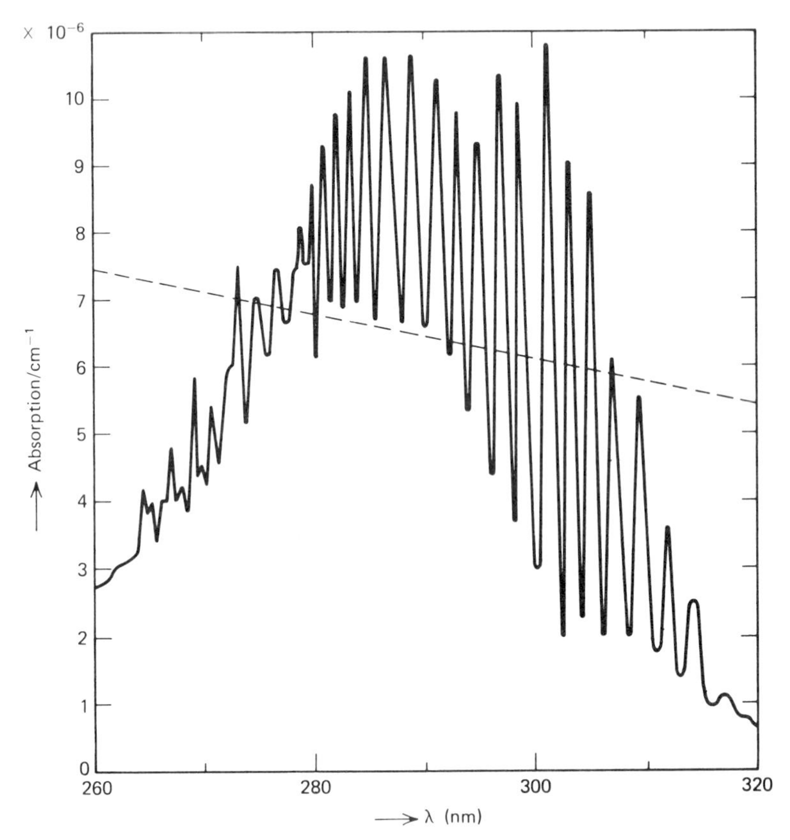

Figure 6.12. Comparison of the light losses due to particulate scattering and due to light absorption by SO_2 as a function of wavelength (260 to 320 nm). The particulate scattering (dashed line) has been derived from Figure A6.3 using the following data: average diameter of particles, 0.2 μm; concentration of particles; 100 μg m^{-3}; density of particles, 2.5 g cm^{-3}; refractive index of particles $n = 1.50$. The SO_2 spectrum is taken from the literature,[11] and an SO_2 concentration of 1 ppm is assumed. The ordinate axis represents the light absorption per centimeter path length. From this numerical example, based on frequently occurring conditions, it can be observed that particulate scattering can obscure the SO_2 absorption completely in a simple broad band light absorption instrument.

and the scattering, and the second wavelength will give the scattering only. The difference between the two absorptions gives the corrected value for the SO_2 absorption. An instrumental setup, using two light sources is depicted in Figure 6.13.

b. *Correlation Spectroscopy*

The problem of avoiding interferences due to overlap of absorption bands of other substances and by particle scattering can be tackled in a different and very elegant way by the application of so-called correlation spectroscopy. Two types of instruments are in use, one employs gas filter correlation[14,15] and the other is the correlation mask spectrometer.

GAS FILTER CORRELATION. The gas filter correlation spectrometer is a variation on the double-wavelength spectrometer, with the advantage that only one light source is employed. The instrument is represented schematically in Figure 6.14. Essentially two gas filter cells are added somewhere in the optical path to the direct single wavelength spectrometer (Figure 6.10). In Figure 6.14 these cells are placed between the light source and the chopper, but they also can be placed between the absorption cell and the detector. The upper cell is filled with pure SO_2, the lower cell with a nonabsorbing gas. The instrument is zeroed by adjusting the attenuator, with the absorption cell empty, until the intensities in both halves of the chopper period are equal.

The spectrum of the light passing through the upper cell contains all wavelengths in the selected region except the wavelengths that SO_2 absorbs, these wavelengths having been removed by absorption in the pure SO_2 in the upper cell. The light intensity received by the detector is therefore affected only by interfering absorptions, scattering by particles, and so on; but if SO_2 is present in the absorption cell, it will not be measured because the wavelengths it absorbs have been removed. The light passing through the lower cell contains all wavelengths in the selected region, and the light is attenuated by absorption due to SO_2 in the absorption cell but also by all interferences, as mentioned previously. The difference between the two signals (i.e., the ac signal from the detector) thus gives the light absorption from SO_2 only. The more complicated the absorption spectrum of the component of interest (SO_2), the better this gas filter correlation operates.

Strong interferences present over the whole width of the spectrum (e.g., water vapor) can be reduced further by adding some water to both filter cells. This attenuates the water bands in the light in both periods, and the water vapor absorption in the sample cell will be reduced.

CORRELATION MASK SPECTROSCOPY. All the foregoing spectroscopic methods were of the nondispersive type; that is, the light is not separated into the individual wavelengths by a dispersive element, such as an optical grating

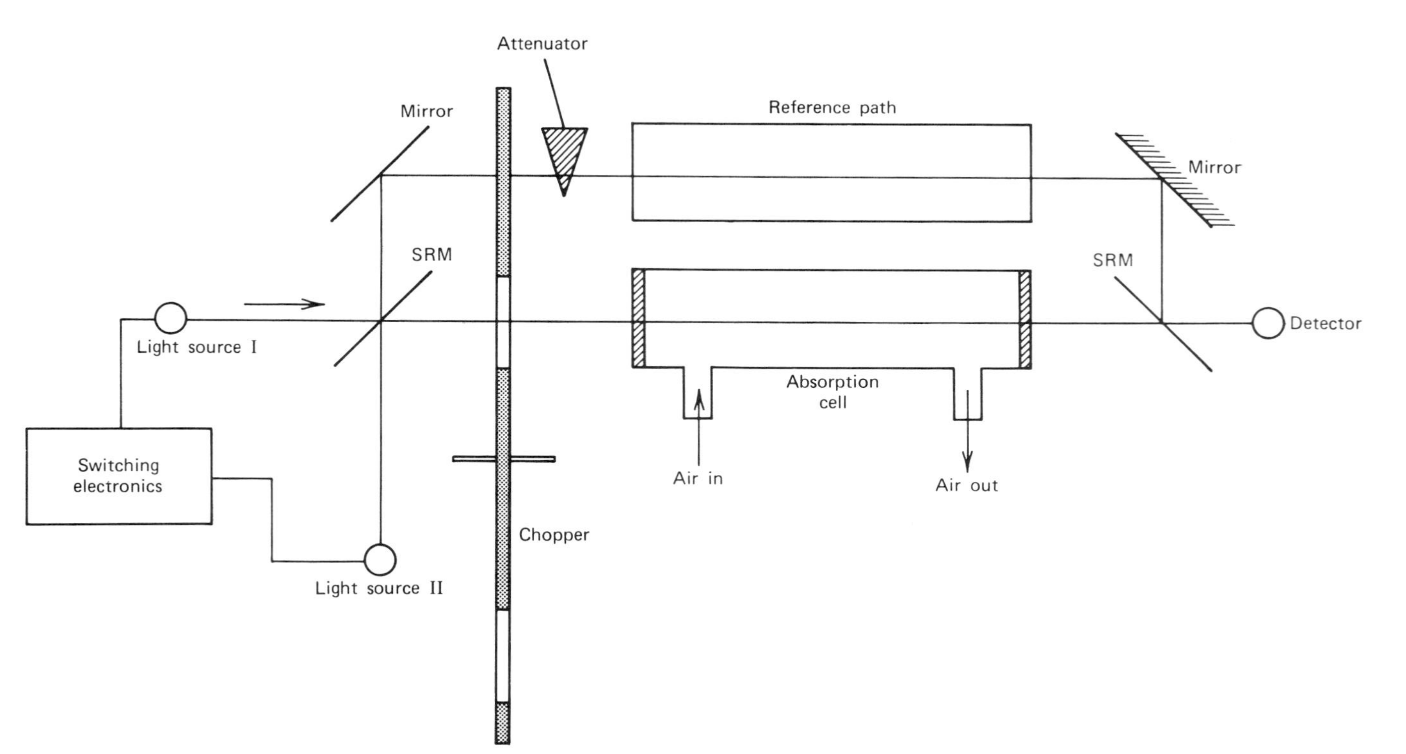

Figure 6.13. Two-wavelength double-beam optical absorption spectrometer. The principle of this instrument is the same as that of the direct absorption spectrometer of Figure 6.10. The two light sources are switched on alternatively to measure the absorption of the gas in the absorption cell at the top of an absorption line and between two absorption peaks of the gas. Interfering absorptions (e.g., due to particle scattering, (Figure 6.12) are almost equal for both wavelengths, thus can be eliminated. The reference path is used to correct for changes in lamp intensities or detector sensitivity. The attenuator is adjusted at empty absorption cell to equalize the beams.

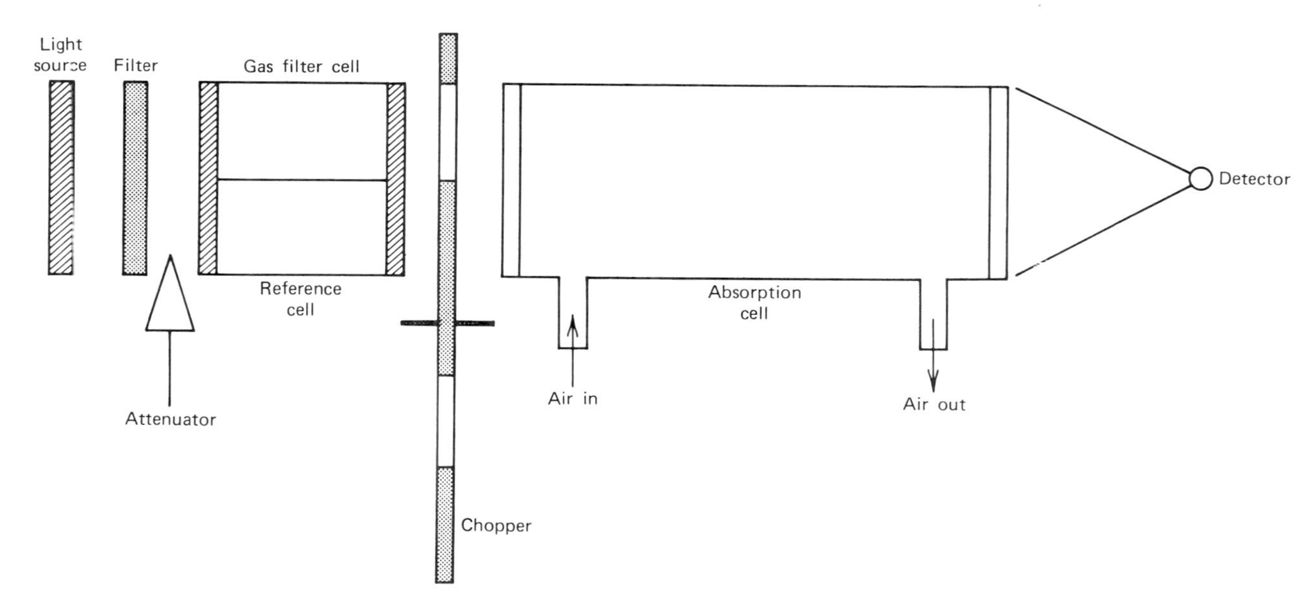

Figure 6.14. Gas filter correlation. Light from a source passes first through a filter and then, by way of a chopper, alternatively through the gas filter cell and the reference cell. In both periods the light passes through the absorption cell containing ambient air, then is focused on the detector. The difference between the two intensities is proportional to the selective absorption by SO_2 in the absorption cell.

or a prism. The following two methods, correlation mask spectroscopy and second derivative spectroscopy, involve modifications of normal spectroscopic instruments, but the equipment includes a device to distinguish a specific spectrum in a high and fluctuating background. In both instruments light passes through an absorption cell and is resolved into its different wavelengths (i.e., by an optical grating), whereupon the resulting spectrum is displayed on an exit plane. The exit plane contains one or more exit slits, and the light passing through the slits is collected on a light detector. The displayed spectrum and the exit slits are caused to vibrate with respect to each other. Thus periodically the exit slits "scan" mechanically a part of the spectrum. This mechanical device, which can be located anywhere in the optical path, may be a vibrating mirror, alternatively, the grating can be made to oscillate or the entrance slit of the spectrometer or the exit slits can be made to move.

In the correlation mask spectrometer (Figure 6.15) the exit slits are in the form of a mask in which the spectrum of the particular component (e.g., SO_2) has been etched.[16,17] If a gas mixture containing several light-absorbing components is present in the measuring cell, the spectra of all these components will be displayed on the exit mask, but only the SO_2 spectrum will match completely the pattern on the mask. The total light intensity passing through the mask is collected on a photodetector.

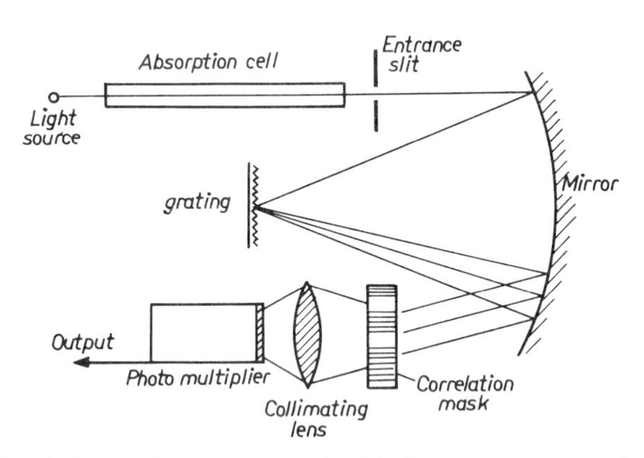

Figure 6.15. Correlation mask spectrometer. The light from a source passes first through an absorption cell filled with ambient air, then is directed, by way of an entrance slit, mirror, and grating, onto a mask consisting of a replica of the expected absorption spectrum. The "product" of the transmission of the replica mask and of the absorption cell is converted into an electrical signal by the photomultiplier. If the entrance slit is made to oscillate, the absorption spectrum oscillates over the correlation mask, and the second harmonic of the electronic signal is proportional to the absorption in the absorption-cell.

The matching of the spectrum with the spectrum etched on the mask is checked by measuring the modulation of the transmitted light due to the vibrating motion of the spectrum at the mask. If a spectral line of the light spectrum is directed exactly on a slit in the exit mask, the light will be transmitted twice during one period of oscillation of the mask with respect to the spectrum. Thus the transmitted light of the matching lines will be modulated at double the oscillation frequency. Spectral lines not exactly aligned with the slits give rise to a light modulation having the frequency of oscillation. The light detector converts the transmitted light into an electrical signal. By using an electronic circuit, such as a lock-in amplifier, tuned to the double frequency, a signal is produced that is highly specific to the spectrum of the component under observation. In principle, interfering spectra will not be observed.

The instrument operates better when the spectrum under observation is more complicated. If only a single absorption line with little structure is present, the advantage of the method is small, but this is true for all correlation instruments.

c. *Second Derivative Spectroscopy*

Second derivative spectroscopy records not the true spectrum but its second derivative,[18,19] and it has the advantage that small and narrow peaks on a high and broad background are much better resolved. If the slope of an interfering background is linear with the wavelength, the background even vanishes completely, which enables the limit of detection to be reduced by 2 to 3 orders of magnitude.

The instrument is very similar in principle to the correlation mask spectrometer except that only one exit slit is present. Furthermore the absorption spectrum of air is focused on the exit plane and made to vibrate with respect to the exit slit. A light detector coupled to an electronic circuit records the second harmonic modulation on the transmitted light. By this process the second derivative of the spectrum is recorded, as shown in the following simplified calculation.

Suppose the spectral distribution of the light at the exit plane the spectrometer is given as a function of wavelength by $I(\lambda)$. If a very narrow slit is located at λ_0 and oscillates with a frequency ω and an amplitude a, the position of the slit as a function of time is given by $\lambda_0 + a \sin \omega t$ and the transmitted light S is $I(\lambda_0 + a \sin \omega t)$. Expanding this in a Taylor series we find

$$S - \sum_{n=0}^{\infty} \frac{1}{n!} I^{(n)}(\lambda_0) a^n \sin^n \omega t,$$

where $I^{(n)}(\lambda_0)$ denotes the nth derivative of $I(\lambda)$ at λ_0. The time-dependent part of S can then be derived as

$$S(t) = \sum_{n=1}^{\infty} \frac{a^{2n-1} I^{(2n-1)}(\lambda_0)}{2^{2n-2} n!(n-1)!} \sin \omega t$$
$$- \sum_{n=1}^{\infty} \frac{a^{2n} I^{(2n)}(\lambda_0)}{2^{2n-1}(n+1)!(n-1)!} \cos(2\omega t)$$
$$+ \text{ higher frequencies.}$$

If the instrument is sensitive only to the second harmonic, the recorded signal R is proportional to

$$R \propto \frac{a^2}{4} I^{(2)}(\lambda_0) + \frac{a^4}{48} I^{(4)}(\lambda_0) + \cdots,$$

which for small amplitudes a is proportional to the second derivative.

For a Gaussian spectral line $I = \exp -(\lambda/d)^2$, and the signal becomes

$$R \propto \frac{1}{2}\left(\frac{a}{d}\right)^2 \left[2\left(\frac{\lambda}{d}\right)^2 - 1\right] \exp -\left(\frac{\lambda}{d}\right)^2$$
$$+ \frac{1}{12}\left(\frac{a}{d}\right)^4 \left[4\left(\frac{\lambda}{d}\right)^4 - 12\left(\frac{\lambda}{d}\right) + 3\right] \exp -\left(\frac{\lambda}{d}\right)^2 + \cdots$$
$$= \left(\frac{a}{d}\right)^2 F_2 + \left(\frac{a}{d}\right)^4 F_4 + \cdots$$
$$= \left(\frac{a}{d}\right)^2 \left[F_2 + \left(\frac{a}{d}\right)^2 F_4 + \cdots\right].$$

The coefficients F_2 and F_4 are plotted as a function of λ/d in Figure 6.16. For very small values of the modulation width, the second term in the expression $(a/d)^2 F_4$ vanishes rapidly because F_2 and F_4 are of the same order of magnitude. However the signal will decrease proportional to the square of the modulation width, and for small signals (low concentrations) it will be necessary to operate at a relatively broad modulation. If the peak height is used for calculating the concentration, the modulation width should not exceed 0.3 times the line width for an error of less than 5%. A more detailed theory, including the effect of final slit width is given in Reference 20. In second derivative spectrometers no built-in-spectrum recognition is present, as compared with correlation mask spectrometer. The observed spectrum is scanned, and all interfering substances (provided they have narrow peaks in the region of operation) are also recorded. For many air pollutants it is easy to choose a narrow peak, which can be used as specific "signature" (cf. Figure 6.26). Since the whole spectrum is scanned, no masks have to be

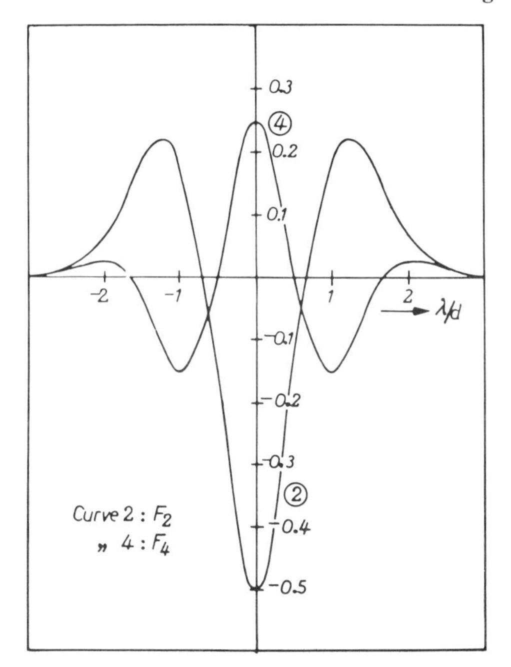

Figure 6.16. Coefficients F_2 and F_4 as a function of λ/d, the wavelength to linewidth ratio, for a Gaussian line shape. The output of a second derivative spectrometer is proportional to $(a/d)^2[F_2 + (a/d)^2 F_4]$, where a is the modulation width.

changed when switching from one component to another, and the second derivative spectrometer is more versatile.

In principle the correlation mask spectrometer is faster because the whole spectrum is displayed and checked simultaneously. The second derivative spectrometer is slower because it scans the spectrum (e.g., by slowly turning the grating), after which the spectrum recognition can be carried out.

d. Application of Spectroscopic Techniques to Sulfur Dioxide monitoring

As mentioned earlier, SO_2 shows two strong absorption bands, one in the ultraviolet (200 to 235 nm) and one in the infrared (7.4 μm). The observed detection limits are functions of the length of the measuring cell and the integration time of the electronic circuits.

An increase in cell length does not always give a proportional improvement. In multireflection cells, additional reflection losses are introduced, and an increase in cell length reduces the opening angle, thus the amount of light passing through the cell. Since there is an optimum for the number of passes through a multipass cell,[12] the minimum detectable concentration of a pollutant can be calculated for a given absorption coefficient and detector noise.

At a wavelength not coinciding with the absorption, if the transmission of the cell in one pass is P, the transmission of the cell in n passes will be P^n.

The absorption follows the Lambert–Beer law; thus the intensity of the light after one pass is given by

$$I_1 = I_0 \exp(-\alpha c l) \cdot P,$$

where I_0 is the intensity of the incident radiation, α the absorption coefficient, and l the cell length. In n passes the intensity received is

$$I_n = I_0 \exp(-\alpha c l n) P^n$$

Since the absorption is weak, we have

$$\exp(-\alpha c l n) \approx 1 - \alpha c l n$$

The difference between the intensities received at the absorption wavelength and outside the absorption wavelength, which is given by the signal S_n, is

$$S_n = I_0 P^n - I_n = n \alpha c l P^n$$

The value of S_n will be a maximum when $\ln P = -1/n$, and then

$$S_n = I_0 \, \alpha c l \frac{0.37}{-\ln P}.$$

Because P is close to unity

$$-\ln P \approx 1 - P \qquad \text{and} \qquad S_n = I_0 \alpha c l \frac{0.37}{1 - P}.$$

For a single pass cell of the same length, $S_1 = I_0 \alpha c l P$. Hence the improvement of a multipass cell over a single pass cell is

$$\frac{S_n}{S_1} = \frac{0.37}{P(1 - P)}.$$

Since P is of the order of 0.95 for soiled mirrors and 0.98 for new and clear mirrors, the optimum numbers of passes (n) will be 20 and 50, respectively, and the improvements over a single pass cell will be the factors 7.8 and 18.9, respectively. A multipass cell with 50 passes thus gives improvement by a factor of scarcely 20.

Another possibility would seem to be to increase the actual length of the absorption cell. This also requires an increase of the diameter of the cell. The relative amount of radiation entering the cell is proportional to $I_0(d^2/l)$, where d the diameter of the mirror and l the cell length; the signal for a simple pass then becomes

$$S_1 = I_0 \alpha c l P \frac{d^2}{l} = I_0 \alpha c P d^2$$

In comparing cells of different lengths, diameters, and numbers of passes, therefore, it is very difficult to decide offhand which cell is to be preferred. The following limits are reported for the different spectroscopic methods:

Direct UV Spectroscopy. At a cell length of 25 m the detection limit is 10 ppb (1% absorption). Particulates interfere. Response time is about 10 sec.

Two-Wavelength UV Spectroscopy. Detection limit, cell length, and response time are the same as for direct UV spectroscopy. The wavelengths used are 285 nm (Mg lamp) and 287 nm (Ga lamp). Particulate intereference is absent.

Correlation Spectroscopy. In the infrared (gas filter correlation), detection limits of 4 ppb are found at a cell length of 25 m at a response time of 10 sec.[21] Interferences of water and carbon dioxide can be reduced to a few percent. In the ultraviolet, correlation mask spectroscopy is used. Most of the experience here is in remote sensing, using the sun or the sky as a light source. The published limit of detection is 2 ppm m at 8 sec response time.[4] This means that for a detection limit of 10 ppb, a cell length of 200 m should be required. Although one would expect the detection limit to be lower than for single slit instruments such as the second derivative spectrometer, the results published so far do not confirm this. Probably the low intensity of the light source (blue sky) is responsible.

Second Derivative Spectroscopy. Detection limit is 4 ppb, at a cell length of 24 m and 1.5 min scanning time.[4]

6.4.1.4. Determination of Sulfur Dioxide by Fluorescence

If SO_2 is irradiated with ultraviolet light, fluorescent radiation is emitted in a continuous spectrum between 240 and 420 nm, with a maximum at about 320 nm. This effect can be used for the continuous and automatic monitoring of sulfur dioxide.[22–24]

In a typical instrument (Figure 6.17), light from a UV source is focused by a lens and passed through a filter to suppress the wavelengths that are in the range of the fluorescence spectrum of SO_2 (longer than 300 nm). This is

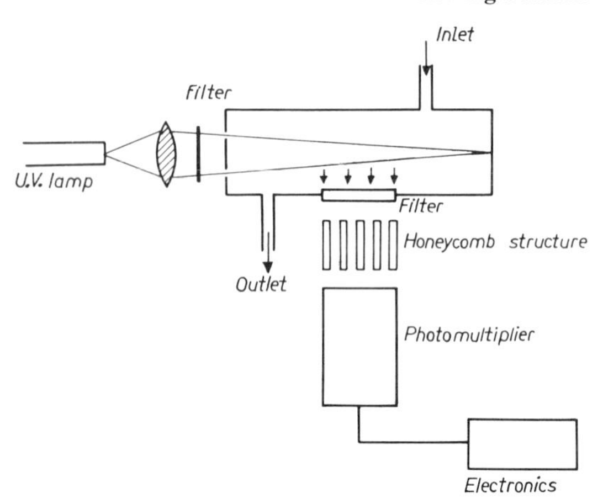

Figure 6.17. Fluorescence SO_2 monitor. Ultraviolet light from a source passes through a filter to remove light of longer wavelengths into the measuring cell. The fluorescence radiation in the direction perpendicular to the incoming beam is filtered and passes by way of the honeycomb structure (to interrupt stray light) to the photomultipler.

necessary because otherwise stray light reaching the detector would be taken to be fluorescence radiation.

Ambient air is passed through the measuring cell at a constant rate. Sulfur dioxide, if present in the air, will be excited and will emit fluorescent radiation, which is picked up by a photomultiplier provided with a filter and a honeycomb structure to remove, as far as possible, stray light that might contribute to the signal.

Fluorescence of SO_2 can be produced by irradiation at different wavelengths. Since the fluorescence intensity can be affected significantly by "quenching" (i.e., the transfer of the energy of the excited molecules not by radiation, but by collision with other molecules),[67] the excitation level should be chosen to have a very short lifetime. This minimizes the risk of collisions with other molecules.

Three absorption bands are known for SO_2:390 to 340, 320 to 250, and 230 to 190 nm, for which the lifetimes of the excited states are 8×10^{-4}, 4×10^{-5}, and 10^{-8} sec, respectively. Since the latter is best in avoiding quenching effects, the optimum excitation wavelength is 218.9 nm. Two spectral lamps are available in this region, a zinc lamp with a strong line at 213.8 nm and a cadmium lamp with a line at 228.8 nm. The values of the absorption coefficient α are about 25 atm^{-1} cm^{-1} at 213.8 nm and about 5 atm^{-1} cm^{-1} at 228.8 nm; thus the zinc lamp is preferred. Furthermore, quenching by water vapor is significantly lower when a zinc lamp is used. In the reaction vessel the following reactions are assumed to take place:

$SO_2 \longrightarrow SO_2^*$		production of SO_2 in excited state; this reaction rate is proportional to the absorption of the incoming UV radiation
$SO_2^* \longrightarrow SO_2 + h\nu$		fluorescence, reaction rate k_f
$SO_2^* \longrightarrow D$		dissociation, reaction rate k_d
$M + SO_2^* \longrightarrow SO_2 + M$		quenching, reaction rate k_q

The absorption of the incoming light I_a at low SO_2 concentrations c is given by

$$I_a = I_0 \alpha c l,$$

where I_0 is the intensity of the light source, α the absorption coefficient, and l the length of the cell. The intensity of the fluorescent radiation I_f then can be calculated:

$$I_f = \frac{k_f I_0 \alpha c l}{k_f + k_d + k_q[M]},$$

which shows that the output of fluorescent radiation therefore is proportional to the concentration of SO_2.

To avoid quenching by water vapor, the air is fed through a dryer before entering the instrument. This is one of the weaker points of this type of instrument, since memory effects are observed arising from adsorption–desorption of SO_2 on the dryer. The development of the technique is rather recent, however, and this difficulty probably will be overcome.

One commercial device uses a pulsed lamp rather than a continuous lamp. This development, when combined with a photon-counting technique, has reduced the lower limit of detection to 5 ppb, compared with 20 ppb when a continuous lamp was used. The response time is about 1 min.

The problem of stray light by particulates has only partly been solved, however, and some hydrocarbons, which also can give fluorescent radiation, also interfere.

6.4.1.5. *Summary of Methods for Monitoring Sulfur Dioxide*

Although many laboratory techniques for measuring SO_2 in air have been mechanized and promoted as automatic monitors, we have not described them in the previous sections. Among these techniques are the wet spectrophotometric method, using the color development of pararosaniline when reacting with SO_2, and the conductivity method, where SO_2 is oxidized first and the change in conductivity of a solution is measured. Although frequently employed, none of these methods meets the requirements outlined in Section 6.2.

Since both methods are rather insensitive, a relatively long sampling time is required to obtain a low limit of detection. A half-hour averaged value is the best result reported. In particular circumstances, however, a repetition time of a few minutes is necessary; thus the instruments are too slow for application in an automatic monitoring system. Furthermore, the selectivity of the conductivity method is below the standards requested.

The strongest argument against application of these pure chemical methods, is the requirement to store perishable reagents such as pararosaniline or hydrogen peroxide. The need to replenish these reagents and remove used chemicals limits the time of unattended operation to a few days.

One of the first methods used for the automatic monitoring of SO_2, coulometry, is still one of the best methods in its technical performance. The lower limit of detection is 3 ppb, and the response time is 1 min. Good filters have been devised, making the selectivity outstanding. Instruments have proved to be capable of 3 months' unattended operation. Spectroscopical methods in general suffer from lack of sensitivity, unless very long (multipass) absorption cells are used.

Flame photometry has been used in number of commercially available instruments. A detection limit of 5 ppb is reported. A disadvantage of flame photometry is the large amount of hydrogen consumed, which limits the period of unattended operation; but in principle this restriction can be eliminated by hydrogen generation *in situ* by electrolysis. Because the method responds to all sulfur components, it is suitable only in those areas where SO_2 is expected to be the only sulfur-containing pollutant. The addition of a gas chromatographic separation complicates the instrument and makes it less suited for use in unmanned stations.

It is premature to judge fluorescent methods for use in automatic stations because water vapor interference and the fluorescence of other air pollutants are problems that have not yet been completely solved. Thus although the method looks attractive, further technical development is needed.

6.4.2. Monitoring Methods for Ozone

The two principles currently being used for the automatic monitoring of ozone are chemiluminescence and UV absorption spectroscopy.

6.4.2.1. Chemiluminescence

Chemiluminescence occurs when a chemical reaction produces an excited molecule or atom that can give up its energy in the form of a quantum of ultraviolet, visible, or infrared radiation:

$$A + B \longrightarrow C^* + D \longrightarrow C + D + h\nu$$

The intensity of the emitted light is proportional to the rate of the reaction, and its spectrum is characteristic of the excited molecule or atom formed, thus generally of the reaction itself. It seldom happens that two different reactions produce the same luminescent substance.

If such a reaction is to be used for determining the concentration of one of the molecules on the left of the equation (e.g., A), B must be present in considerable excess. The reaction rate is then entirely determined by the concentration of A.

One very attractive feature of methods of measurement based on chemiluminescence is the ease with which the different air pollutants can be distinguished. First, the reaction is specific for a certain component, and second, the emitted spectrum can be used for a further characterization.

The two chemiluminescent reactions known and employed for the monitoring of ozone in ambient air use ethylene and rhodamine B, respectively.

a. The Ozone–Ethylene Reaction[25]

Figure 6.18 shows a detector used for measuring ozone by the ozone–ethylene reaction. A pump sucks a constant stream of outside air through the inlet, and ethylene ($CH_2{=}CH_2$, the auxiliary gas that enters into the chemiluminescent reaction with ozone) is supplied from a gas cylinder. The two flows meet in the reaction vessel, where the following reaction takes place:

$$O_3 + CH_2{=}CH_2 \longrightarrow H_2C\begin{matrix} O{-}O \\ \diagup \quad \diagdown \\ \diagdown \quad \diagup \\ O \end{matrix}CH_2$$

$$\downarrow$$

various decomposition products

The spectrum of the radiation emitted during this reaction is between 350 and 580 nm, with a maximum at 420 nm. The radiation falls on the cathode of the photomultiplier tube, causing an anode current to flow. No optical filter is used in this detector, because the other common air pollutants (e.g., NO, NO_2, and SO_2) do not react with ethylene, and therefore produce no radiation. In fact, no interference from other atmospheric components has

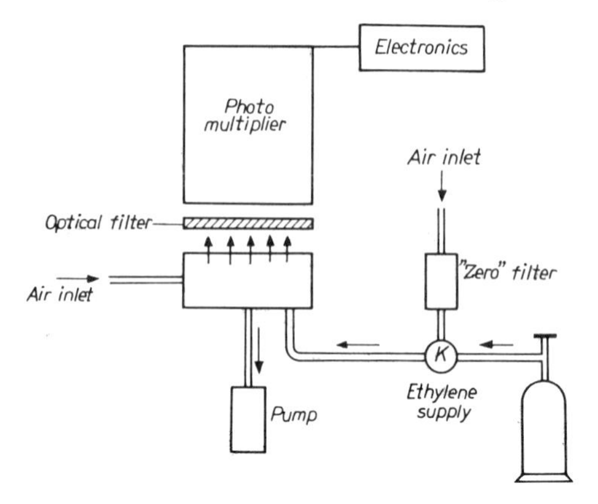

Figure 6.18. Ozone–ethylene chemiluminescent monitor for use in automatic stations. With the valve K switched to the ethylene supply, the ozone in the air will react with ethylene. The emitted light is detected by the photomultiplier. With valve K switched to the air inlet through the zero filter the "zero" level of the instrument is recorded.

ever been found. In most commercial instruments constant flows of air and ethylene are mixed, and since the ethylene consumption is about 50 to 100 ml/min, about 2.5 to 4.5 m^3 of ethylene is used up per month. To reduce the amount of ethylene to be stored in the monitoring station, intermittent operation can be envisaged. This has the additional advantage that the instrument is "zeroed" between two measurements. In the instrument appearing in Figure 6.18 there is a three-way valve K in the ethylene feed line. For a few seconds of each minute, this value is automatically turned on, to admit ethylene.[26] For the remaining period the valve shuts off the ethylene and is open to the outside air, which in this case is completely purified of air pollutants by means of a "zero filter." Under these conditions the ethylene is used up so slowly that a gas cylinder of the usual size (50 l) will last for more than a year. During the period in which purified outside air is admitted, the residual ethylene and the reaction products are dispelled, and the dark current of the photomultiplier tube is measured. Since the dark current is fairly high compared with the signal to be measured, and varies considerably with temperature, the photomultiplier is housed in a thermostat. The detection limit of this detector lies at 1 ppb. The emission intensity is proportional to the ozone concentration up to 1000 ppm. Regular calibration is necessary to allow for changes in such system parameters as the flow rates of the monitored outside air and of ethylene, and the supply voltage and sensitivity of the photomultiplier on which the output signal depends. One calibration every 24 hr limits the error in the measured values to 3%. The detector is exceptionally fast, and the measuring time can be reduced to 0.1 sec. Because of the intermittent operation, the zero drift of the photomultiplier is automatically corrected. Furthermore, the reduction of ethylene consumption makes an instrument of this type suitable for automatic stations.

b. The Ozone–Rhodamine B Reaction[27]

The reaction between ozone and rhodamine B has a lower detection limit but is somewhat more complicated and requires more frequent calibration for adequate accuracy. The detector is based on the reaction between ozone, gallic acid, and rhodamine B; the chemiluminescence extends from 550 to 630 nm and has a maximum at 560 nm. On excitation in the reaction with ozone, the gallic acid transfers its excitation energy to rhodamine B, which is thus excited in turn and emits light. Being colorless, the oxidation products of gallic acid absorb no light. The detector is shown in Figure 6.19. Gallic acid and rhodamine B, deposited on a silica-gel substrate, are contained in the reaction vessel. The silica-gel substrate has a water-repellent layer to minimize the fluctuations in the moisture content caused by varying levels of humidity in the outside air passing over the detector, because the quantity

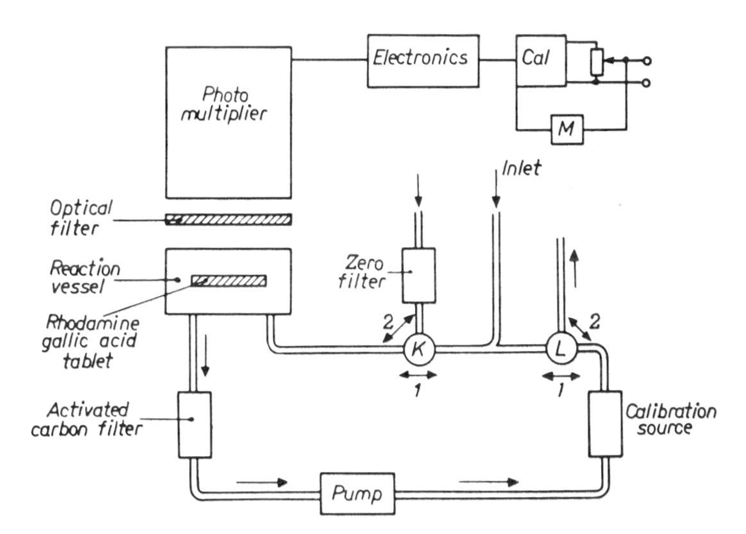

Figure 6.19. Ozone monitor using the chemiluminescent reaction between rhodamine B and ozone. With valve K in position 1 and valve L in position 2, ambient air enters the instrument, the ozone reacts with the rhodamine B–gallic acid tablet, and the light is measured by the photomultiplier (measuring mode). When valve K is switched to position 2, the incoming air first is purified, and a zero signal is observed (zeroing mode). With both valves in position 1, the air passes over a calibration source (UV lamp), where a constant concentration of ozone is formed. The signal of the photomultiplier in this calibration position is used to adjust the amplification of the electronics in the "cal" circuit. The small motor (M) drives the "cal" potentiometer.

of emitted light varies with the moisture content. Outside air is sucked through the inlet to the reaction chamber by a pump. The intake line contains a three-way valve K which admits the outside air to be monitored (position 1) for a few seconds of each minute.[26] During the remaining period, the outside air is passed through the "zero" purifying filter (position 2). In this part of the cycle only the dark current of the photomultiplier tube is measured. Regular calibration is carried out by means of the three-way valve L, using the calibration source. In this calibration source ozone is produced by irradiating a constant stream of clean outside air with a mercury lamp.

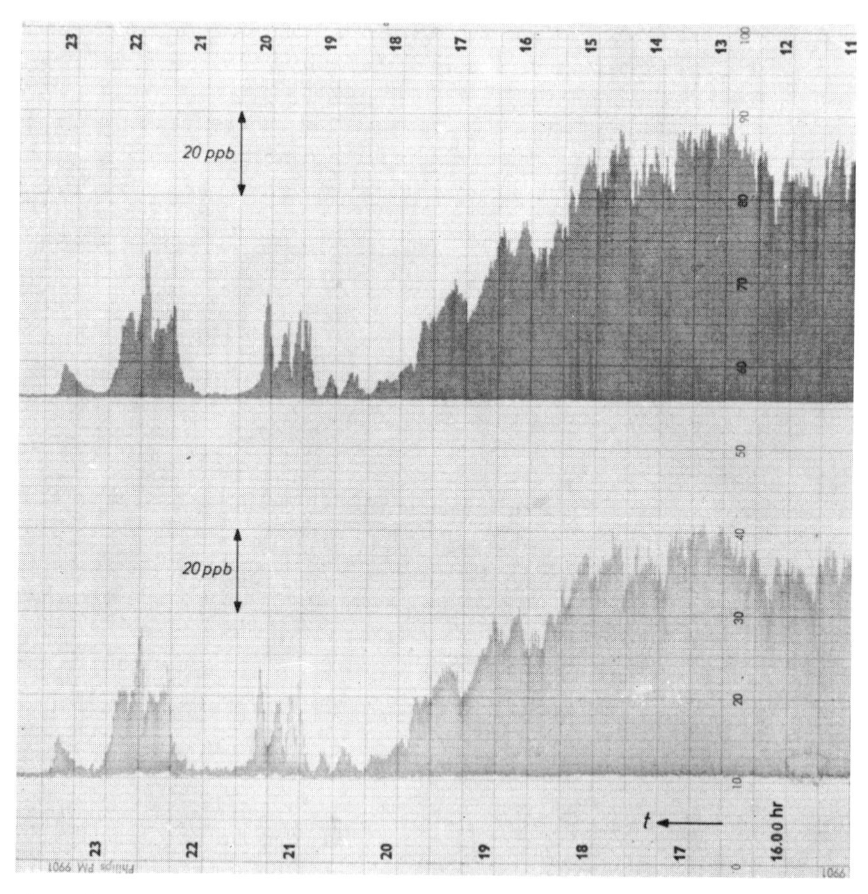

Figure 6.20. Comparison between recordings of zone measured by the chemiluminescent ozone–ethylene reaction (lower trace) and the ozone–rhodamine B–gallic acid reaction (upper trace). Apart from the baseline (more stable in the ozone–rhodamine B instrument) the two recordings are virtually identical.[26]

To ensure that the ozone formation remains constant and reproducible, the whole calibration unit is kept at constant temperature. The quantity of ozone delivered varies by no more than 1% per month. For the calibration, the two three-way valves are in position 1; the pump action is arranged to prevent the admission of further outside air during the calibration, and only the air already present in the lines of the system is circulated. All ozone is removed by a carbon filter *C*, ensuring that ozone-free air enters the calibration source. The calibration takes only a few seconds and can be carried out at regular intervals even before every measurement period. Once the two three-way valves have been set to the correct positions, a few second elapses before the stream of ozone from the calibration source reaches the reaction vessel, and the actual calibration takes place during the next 2 sec.

In this monitor an automatic correction derived from the calibration signal is applied to the measured signal. The output signal is adjusted to the value corresponding to the calibrated concentration of the ozone by a control circuit (Cal), which drives a small motor (*M*) controlling a potentiometer setting. The automatic correction for the change in the reactivity of rhodamine B, which closely depends on the previous history, is particularly important; the reactivity increases with the amount of ozone that has just been measured.

The emission intensity in this monitor is proportional to the ozone concentration up to a value of at least 400 ppb, which is sufficient, considering that even an ozone concentration of 125 ppb is very rare. The detection limit is 0.1 ppb. The life per "fill" is 500 000 ppb hr^{-1}. This means that even with an unusually high concentration of 100 ppb of ozone, day and night (a situation that never occurs), the life of one fill would be 5000 hr, or more than 6 months. This particular detector is specific to ozone and is not affected by the normally occurring concentrations of sulfur dioxide, nitric oxide, nitrogen dioxide, methane, carbon monoxide, and carbon dioxide. The results of simultaneous recordings of the measurements made by the two ozone detectors appear in Figure 6.20; the agreement is excellent.[26]

6.4.2.2. *Ultraviolet Absorption Spectroscopy*

Ozone shows a very strong absorption band between 200 and 300 nm, with a maximum around 250 nm.[28] The absorption maximum is about 300 atm^{-1} cm^{-1}; thus a cell 1 m long will have 3% absorption per ppm ozone.

Fortunately the main emission of a mercury lamp takes place at 253.7 nm, very close to the maximum of the ozone absorption band, while the absorption coefficients of other substances present in air are much lower at this wavelength (except, of course, mercury vapor). This combination of factors

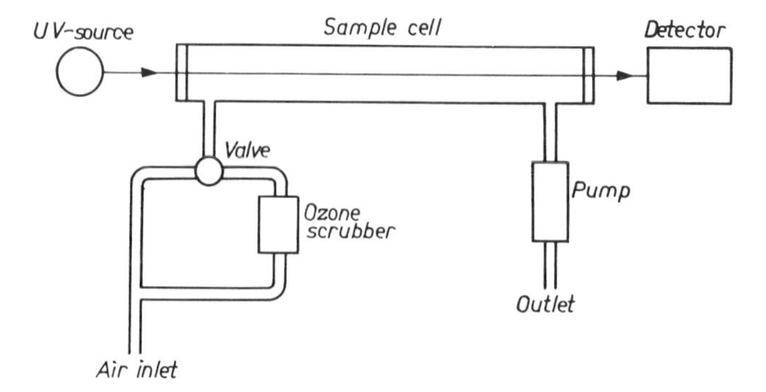

Figure 6.21. Ozone monitor based on direct UV light absorption. In the measuring position of the valve, air enters the sample cell and the light transmission of the mercury lamp is measured by the detector. In the zeroing position, the air enters the sample cell through an ozone scrubber and the transmission of the cell is measured again. The difference between the two transmissions is due to the absorption of UV light by ozone.

enables a very simple ozone monitor to be constructed.[29] In one commercial version the instrument consists of a mercury lamp, an absorption cell 0.35 m long, a switching valve, and a deozonizer or scrubber, consisting of a manganese dioxide filter (Figure 6.21). Ambient air enters either through the deozonizer or directly to the measuring cell. The difference in light absorption is due to the absorption by ozone.

A disadvantage is that two air samples with different particulate concentrations are compared, making the method very sensitive to interference by particulate light scattering (cf. Section 6.4.1.3 and Figure 6.12). For instruments like those already described, a detection limit of 3 ppb at a response time of 10 sec is reported. Mercury vapor has an absorption coefficient of about 10^6 atm^{-1} cm^{-1}. This means that very low mercury concentrations (e.g. 0.01 ppb) give a significant light absorption (equivalent to 30 ppb ozone).

6.4.2.3. *Other Methods for Ozone*

In some cases the spectroscopic methods described for SO_2 can also be applied to the monitoring of ozone. For second derivative spectroscopy, a detection limit for ozone of 25 ppb is reported, using a cell length of 24 m. Infrared spectroscopy, too, can be applied by using an absorption band located at 9.4 μm. The detection limit is 10 ppb for a response time of 10 sec and a cell 25 m long.[21]

6.4.2.4. *Summary of Ozone Monitoring Techniques*

Chemiluminescence seems to be the appropriate technique for monitoring ozone. The disadvantage of sensitivity to moisture (using the rhodamine B method) can be overcome by frequent calibration. The limit of detection is very low, which is an advantage, especially in areas where very low ozone concentrations can be expected.

The chemiluminescent method using ethylene as a reactant has an important drawback, namely, bottles containing an explosive gas must be stored in the monitoring station.* The method is less sensitive than the rhodamine B type, but the instrument is somewhat simpler. Ultraviolet absorption spectroscopy, though both simple and direct, suffers from aerosol interference, which is a problem in cities with high, varying dust levels. The sensitivity for mercury has been mentioned, but it does not appear that measurements have been carried out.

6.4.3. Monitoring Methods for Nitrogen Oxides

Methods for nitric oxide and nitrogen dioxide that are suitable for automatic monitoring include chemiluminescence (NO and NO_2 after conversion to NO), correlation spectroscopy (NO_2), second derivative spectroscopy (NO and NO_2), infrared spectroscopy (NO and NO_2), and photodissociation (NO_2).

6.4.3.1. *Chemiluminescent Monitoring of Nitric Oxide*

The operation of chemiluminescent monitors for NO determination in air is based on the reaction of NO–ozone reaction.[30,31]

$$NO + O_3 \longrightarrow NO_2^* + O_2 \longrightarrow NO_2 + O_2 + h\nu$$

Radiant energy is emitted in a very broad band starting at 600 nm, which exhibits a maximum in the infrared at 1200 nm. Since the normal photocathodes used in photomultipliers are not sensitive at wavelengths longer than 800 nm, only a very small amount of the spectrum can be measured,

* A serious gas explosion occurred in the laboratories of the New South Wales State Pollution Control Commission, Australia, in 1976. After an explosion in Yokohama, Japan, the use of the ethylene method was abandoned for mobile stations.

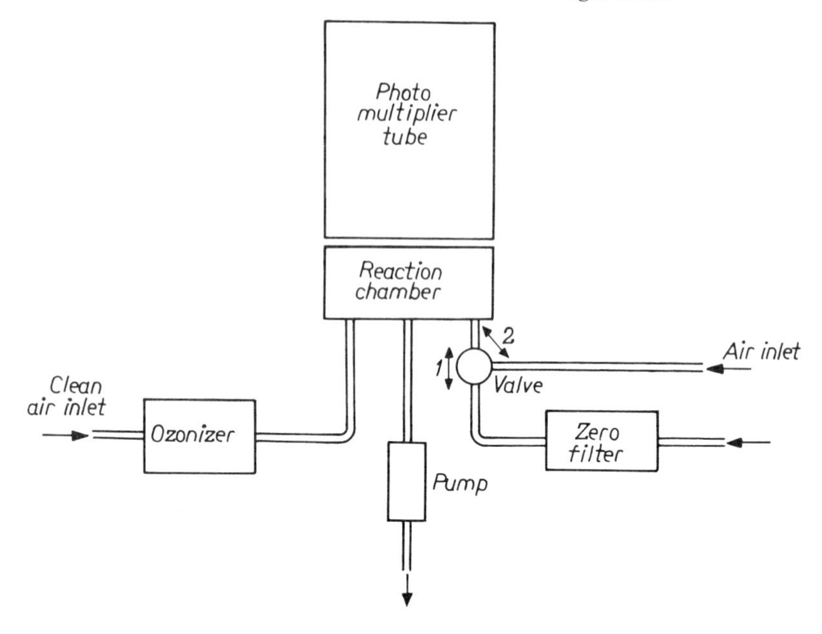

Figure 6.22. Monitor for NO based on the chemiluminescent reaction between NO and O_3. With the valve in position 2 ambient air enters the reaction chamber, where it is mixed with a high concentration of ozone, and the light from the reaction between NO and O_3 is measured by the photomultiplier. With the valve in position 2, the air is filtered before entering the instrument, NO is removed, and a "zero" signal, is recorded. The pump maintains a low pressure in the reaction chamber.

equivalent to about 3 to 4% of the energy. Nevertheless this is sufficient, and a low limit of detection can be achieved.

Another complication is that the energy of the excited NO_2^* molecule can be transferred without radiation to colliding molecules (quenching). This effect can be reduced by lowering the number of collisions (i.e., by reducing the operation pressure). The principle of the instrument is illustrated in Figure 6.22.

The air whose NO content is to be measured is mixed in a reaction chamber with clean air that has passed through an ozonizer (e.g., a silent discharge). The concentration of ozone in this latter air is very high (several percent). The radiation produced by the reaction between NO and O_3 is picked up by the photomultiplier after passing through an optical filter.

This filter passes only radiation of wavelengths longer than 600 nm; in this way possible interference from radiation generated at shorter wavelengths is avoided. Such radiation can be produced by reactions between ozone and unsaturated hydrocarbons,[33,34] hydrogen sulfide, and mercaptans.[34] A pump ensures a low pressure in the reaction chamber, where the following reactions can take place:

$$NO + O_3 \longrightarrow NO_2^* + O_2 \quad (1)$$

$$NO + O_3 \longrightarrow NO_2 + O_2 \quad (2)$$

$$NO_2^* \longrightarrow NO_2 + h\nu \quad (3)$$

$$NO_2^* + M \longrightarrow NO_2 + M \quad (4)$$

At very low pressure the quenching reaction 4 will be very slow, and the light intensity is determined by the reactions 1 and 3. Thus the light intensity will be dependent only on the NO concentration and the residence time of the molecules in the reaction chamber. At very low pressures (a few torr), the signal therefore should be proportional to the pressure. At high pressures reaction 4 becomes important, and as the number of quenching molecules increases with pressure, a light intensity proportional to $1/P$ can be expected. A more detailed analysis shows the light intensity S to be

$$S = \frac{aP}{P^2 + P_0{}^2}$$

where a and P_0 are instrumental factors determined by the volume of the reaction chamber and the flow rate.[35] The experimental verification of this pressure dependence is represented in Figure 6.23. At low pressures the

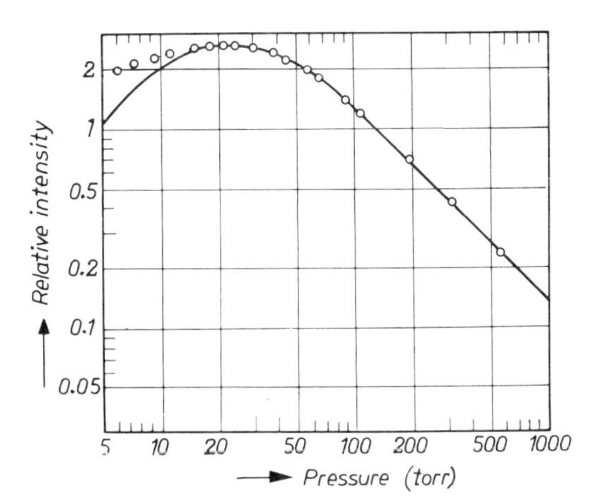

Figure 6.23. Intensity of the NO–O$_3$ chemiluminescent reaction as a function of pressure in the reaction chamber. The curve represents the theoretical behavior; the circles are experimental data points.

detection limit is 2 ppb, determined by the photomultiplier noise. When this noise is reduced by using a refrigerated photomultiplier housing and photo-counting electronics, a detection limit of a few parts per billion can be obtained at almost ambient pressure. The response time of the instrument is a few seconds.

6.4.3.2. *Monitoring of Nitrogen Dioxide by Chemiluminescence*

The same method described previously for monitoring of NO can be used for NO_2, by converting NO_2 to NO by passing the incoming air over a reducing filter. Heated metal gauzes (gold, stainless steel, molybdenum[36]) serve for this purpose, although their lifetimes and conversion rates are hardly adequate for use in automatic monitoring instruments.[37] Figure 6.24 is a block diagram for an NO–NO_2 monitor. It consists of the usual NO monitor, as described already, equipped with two valves *K* and *L*. With

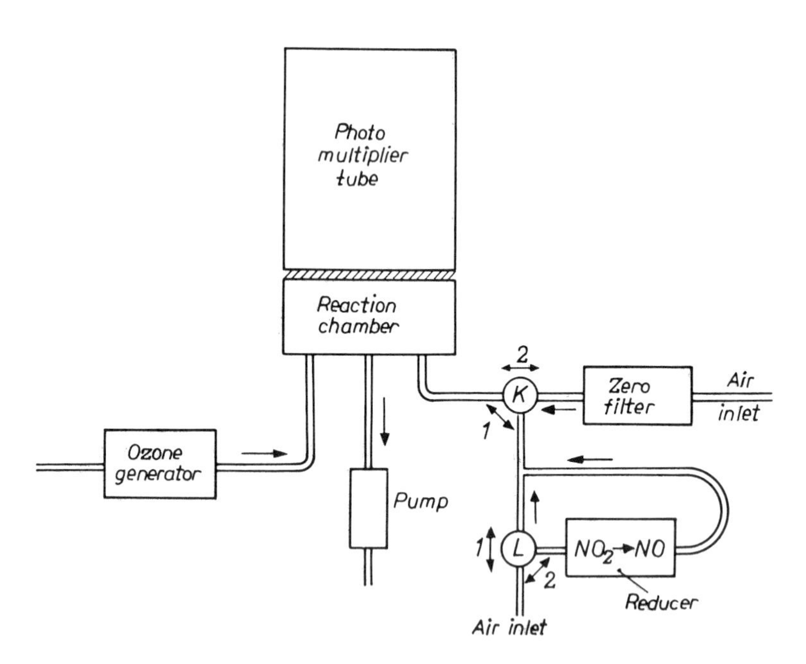

Figure 6.24. Chemiluminescent monitor for NO and NO_2 based on the preconversion of NO_2 to NO. The instrument is the same as that of Figure 6.23 except that it is equipped with an extra stage for converting NO_2 to NO. The following modes are possible: valve *K* to position 1, valve *L* to position 1, measurement of NO; valve *K* to position 1, valve *L* to position 2, measurement of NO + NO_2; valve *K* to position 2, zero measurement.

both valves in the 1 position, the NO content of the air is measured. With L switched to the 2 position, the air passes over the reducer, NO_2 is converted to NO, and the instrument measures the sum of the NO and the NO_2. If the two measurements are carried out within a very short time interval, the difference of the two readings gives a reliable value of the NO_2 concentration. A variant of this instrument uses two reaction chambers (Figure 6.25). In one chamber the sum of NO and NO_2 is measured, whereas in the other chamber only NO is recorded. By the use of a rotating chopper, the cathode of the photomultiplier is exposed alternately to the two reaction chambers. The ac signal gives directly the NO_2 content of the air; the NO content can be derived from the appropriate half-period of the signal. If the chopper blade is supplied also with a dark region for both reaction chambers, optical zeroing can be obtained. Other instruments simply employ two photomultipliers without chopper. Chemiluminescence by using atomic oxygen as reactant[38] can be used for the direct measurement of NO_2 concentrations.

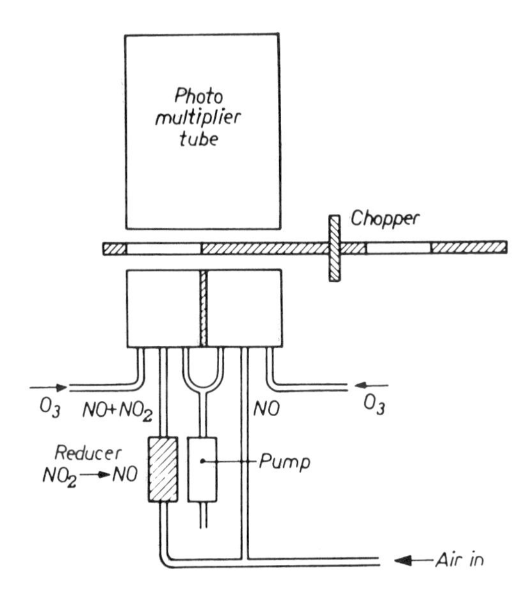

Figure 6.25. Double reaction chamber chemiluminescent NO–NO_2 monitor, employing the same principle as described in Figure 6.24, but NO and NO_2 are measured at the same time using a double reaction chamber. The air stream entering the instrument is divided into two parts. One half is led directly to the reaction chamber for NO, the other half first passes through the converter to reduce NO_2 to NO. The rotating chopper exposes the photomultipler alternately to the two reaction chambers.

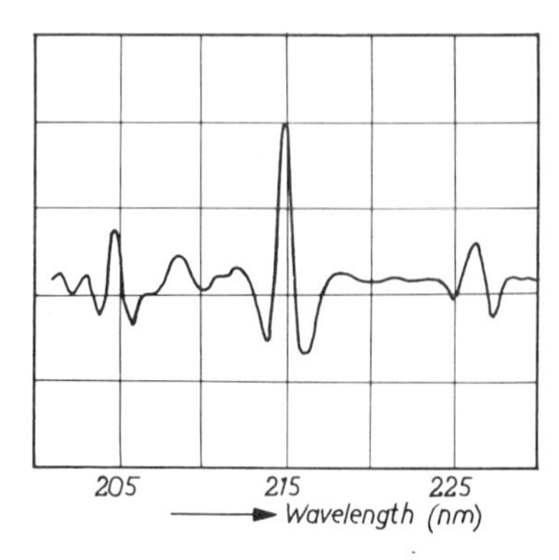

Figure 6.26. Spectrum of NO as obtained in a second derivative spectrometer. The three peaks in this wavelength region are characteristic for NO.

6.4.3.3. Spectroscopic Determination of Nitric Oxide and Nitrogen Dioxide

With a second derivative spectrometer (cf. Section 6.4.1.2*c*) NO can easily be detected in an absorption band between 200 and 230 nm. The identification of NO in the spectrum is easy because three narrow peaks appear at 204.5, 215, and 226.2 nm. Figure 6.26 gives the output signal of the second derivative spectrometer.

With an absorption cell 20 m long, the limit of detection is about 3 ppb for a response time of 1 min. For NO_2 the second derivative spectrometer offers fewer advantages than for NO. In the 440 to 460 nm band (the visible), the detection limit is about 40 ppb. The same converter used in the chemiluminescence detector for NO_2 can be used to reduce NO_2 to NO. With this converter added, the detection limit for NO_2 + NO is reduced to 3 ppb. The time for scanning and reset of the instruments is of the order of several minutes, however; thus the two measurements are carried out on two air samples taken within a few minutes of each other, and there is doubt about the accuracy of such a measurement.

Spectroscopic determination of NO_2 is also possible using the correlation mask spectrometer. The limit of detection is reported[4] to be 1 ppm m. In the infrared, when gas filter correlation is employed, NO and NO_2 can be detected using the absorption bands located at 5.3 and 6.2 μm, respectively. For a 5 sec response time and using a sample cell length of 25 m, the limits of detection for NO and NO_2 are 40 and 4 ppb, respectively.[21]

6.4.3.4. Photodissociation of Nitrogen Dioxide

Thus far photodissociation of NO_2 has been demonstrated only in the laboratory, but future instruments may be produced based on this principle.

Nitrogen dioxide dissociates under the influence of light of 300 to 400 nm wavelength.[39]

$$O_2 + NO_2 + h\nu \longrightarrow NO + O_3 \tag{1}$$

The ozone produced by this decomposition can be measured (e.g., by chemiluminescence or UV absorption), yielding a measure of the NO_2 concentration. However, other reactions consume the ozone formed:

$$NO + O_3 \longrightarrow NO_2 + O_2 \tag{2}$$

$$NO_2 + O_3 \longrightarrow NO_3 + O_2 \tag{3}$$

The respective rate constants for reactions 1, 2, and 3 at room temperature[40] are as follows:

$$k_1 = 5 \times 10^{-3}\ \text{sec}^{-1}$$
$$k_2 = 1.9 \times 10^{-14}\ \text{cm}^3\ \text{molecule}^{-1}\ \text{sec}^{-1}$$
$$k_3 = 4.5 \times 10^{-17}\ \text{cm}^3\ \text{molecule}^{-1}\ \text{sec}^{-1}$$

The rate equation for the ozone formation is

$$\frac{d[O_3]}{dt} = k_1[NO_2] - k_2[NO_2]\cdot[O_3] - k_3[NO_2]\cdot[O_3]$$

and in the stationary state, taking into account that $[NO_2] = [NO_2]_0 - [O_3]$, where $[NO_2]_0$ is the initial NO_2 concentration, this gives

$$[O_3]^2(k_3 - k_2) - [O_3](k_1 + k_3[NO_2]_0) + k_1[NO_2]_0 = 0.$$

Since k_2 is nearly $500 \times k_3$, this can be simplified to

$$[NO_2]_0 = \frac{k_2}{k_1}[O_3]^2 + [O_3]$$

which, expressed in ppb, becomes

$$[NO_2]_0 = 0.098[O_3]^2 + [O_3]$$

For NO_2 concentrations of 1 ppb, an ozone concentration of 0.9 ppb is measured, whereas for 100 ppb NO_2, only 27 ppb ozone will be present. After irradiation, the mixture of ozone, NO, and NO_2 must flow to an ozone detector. During the transfer time an additional loss of ozone will occur because of the reaction of ozone with NO (cf. Section 6.5.1). Although the reading of the instrument is highly nonlinear, this can be corrected electronically. It is therefore possible to measure ozone, NO, and NO_2 consecutively on the one instrument, as indicated in the schematic diagram of Figure 6.27. A fluorescent lamp producing ultraviolet radiation (365 nm) is used for the photolysis, and a helical glass tube around the lamp contains the air mixture,

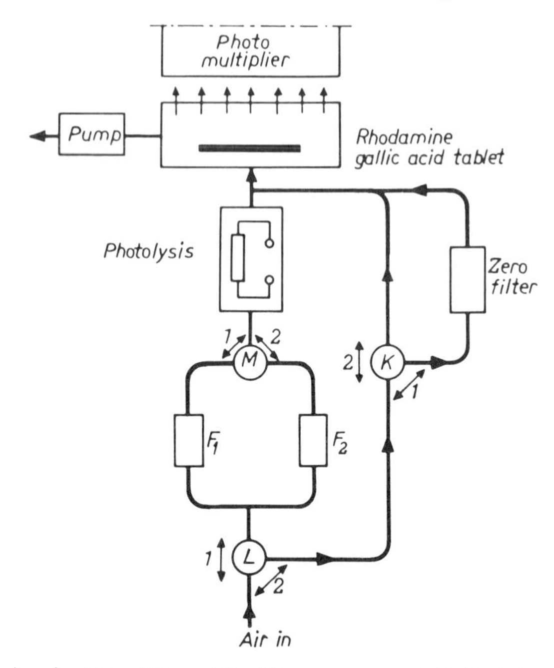

Figure 6.27. Monitor for NO_2, NO, and O_3. The measurement of NO_2 is based on photolysis of NO_2 into NO and O_3. The ozone formed is measured by a rhodamine B chemiluminescent monitor (Figure 6.19). The filter F_1 removes ozone from the air passing through it, absorbs NO_2, and oxidizes NO to NO_2. Filter F_2 removes ozone and oxidizes NO to NO_2. In the photolysis section the air mixture is irradiated and NO_2 is converted into NO and O_3. In the zero filter the air is completely cleaned.[26]

which is irradiated for about 100 sec while it remains in the tube. The ozone formed is measured with a rhodamine **B** ozone detector as described previously, although a spectroscopic detector could be used as well. The air should contain NO_2 only, and this is accomplished by passing the ozone through filter F_1 or F_2. Filter F_1 breaks down the ozone, absorbs NO_2, then oxidizes the NO to form NO_2. Filter F_2 breaks down ozone, passes the NO_2 unchanged, and oxidizes NO to form NO_2. In the first case the quantity of ozone produced is a measure of the NO concentration; in the second it is a measure of the concentration of NO and NO_2 together. The following measuring modes may be distinguished:

valve setting K_2L_2	O_3 measurement	$K{:}2$	$L{:}2$
valve setting K_1L_2	dark current measurement	$K{:}1$	$L{:}2$
valve setting L_1M_1	NO measurement	$L{:}1$	$M{:}1$
valve setting L_1M_2	NO_x measurement	$L{:}1$	$M{:}2$

Regular calibration is required among other reasons to correct for variations in the intensity of the lamp for the photolysis, in particular because hydrocarbons and atomic oxygen may react in the equipment, possibly giving rise to substances capable of oxidizing NO into NO_2. The complete measurement cycle lasts 4 min. The detection limits for ozone and for the nitrogen oxides are 0.1 and 0.5 ppb, respectively.

6.4.4. Monitoring Methods for Carbon Monoxide

Natural background levels of carbon monoxide are very high (around 100 ppb) compared to background levels of other air pollutants.

Two types of instrument are used for the monitoring of carbon monoxide. The first may be rather slow, but it must be able to detect concentrations down to 100 ppb. The second, used in the vicinity of urban traffic and industrial areas, should have a rather rapid response; but the limit of detection needs not to be much lower than 0.5 ppm. For the latter application, infrared absorption spectroscopy is currently in use.

6.4.4.1. Infrared Absorption Spectroscopy

Carbon monoxide has a characteristic absorption spectrum[41] in the infrared near 4.6 μm (Figure 6.28). Water vapor, carbon dioxide, and methane also have very weak absorptions in this wavelength region, but since their concentrations can be considerable, these compounds may interfere. However the complexity of the carbon monoxide spectrum makes it suited to be measured by gas filter correlation spectroscopy, where interference due to water, carbon dioxide, and methane can be ruled out. With an absorption cell of length 20 m, a limit of detection of 0.2 ppm can be achieved.

A special type of correlation spectrometer is used for the monitoring of carbon monoxide. The correlation cells serve also as the detector,[44] as illustrated in Figure 6.29. Infrared radiation from the source is directed

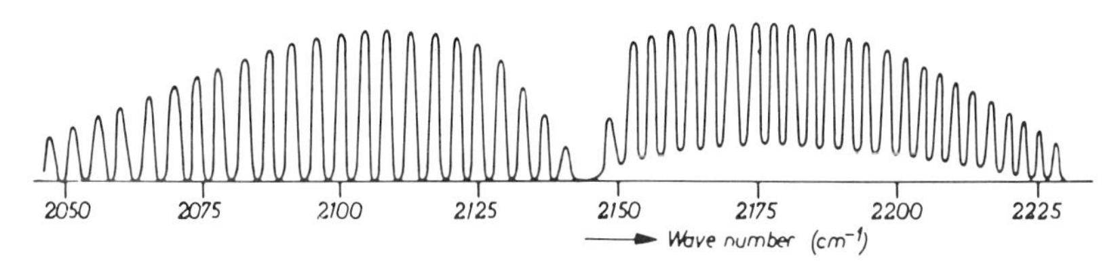

Figure 6.28. Spectrum of carbon monoxide in the infrared (near 4.6 μm wavelength). The complexity of the spectrum makes it extremely well suited for gas filter correlation spectroscopy.

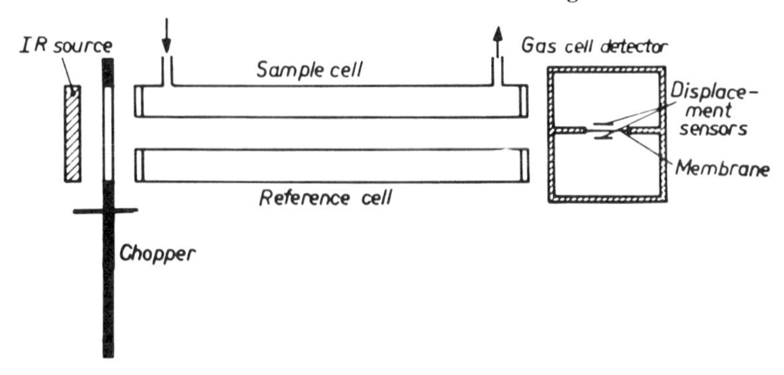

Figure 6.29. Double-beam nondispersive infrared carbon monoxide analyzer. The light of the infrared source passes through the sample cell, through which ambient air is drawn, and through a reference cell filled with clean air. The detector consists of two chambers filled with CO, separated by a thin membrane. Since the intensity of the sample beam is attenuated because of absorption in the sample cell, the intensities of the beams are unequal. Selective absorption of radiant energy in both detector cells causes the temperature of the gas in the detector cells to rise, and the pressure increases. Because the reference beam is more intense, the pressure in the lower cell is higher and the membrane is displaced. A capacitative detector converts the membrane displacement into an electric signal proportional to the pressure difference in the cells, hence to the difference in intensity of both beams. The chopper chops both beams at the same time, thus providing an instrumental "zero" signal.

onto two cells: a reference cell, filled with a gas that is nonabsorbing in the CO-absorbing region, and a sample cell. The detector, which is also the correlation cell, consists of two chambers separated by a flexible diaphragm. Both chambers are filled with carbon monoxide. The infrared radiation passing through the reference cell and the radiation passing through the sample cell are directed onto the two chambers of the detector. The gas in the two chambers is heated by the absorption of the incoming radiation because the gas in the chambers (CO) has an absorption band in the wavelength region of the infrared radiation. The heating of the gas will cause the pressure in both chambers to rise. Since the intensity of the radiation passing through the sample cell has been decreased because of light absorption by CO in the sample cell, however, the increase in pressure in the two cells will be different, resulting in a deflection of the membrane. This displacement is detected electronically. By chopping the incoming infrared radiation, the displacement of the diaphragm is made periodic, and the electronic signal can be handled with simple ac amplifiers and demodulators. As mentioned before, water vapor is the main cause of interference. With properly adjusted instruments a reduction of the sensitivity for water vapor of about 10^4 can be achieved. However the concentration of water vapor in ambient air at 100%

humidity at 20°C is about 2.3%, or about 4 to 5 orders of magnitude greater than the CO concentration, and errors of about 3 to 10 ppm may occur. Thus in infrared instruments for the monitoring of carbon monoxide, a drying stage[45] is necessary before measurements can be made. For high ranges (0 to 100 ppm), the error due to water vapor interference is negligible. A number of modifications of this type of detector are known.[46,47]

6.4.4.2. *Other Methods for Carbon Monoxide*

A slow instrument with a response time of the order of a few minutes, and very sensitive to CO, is required for the measurement of background concentrations. Instruments for this application are based on amperometry and flame ionization

a. Amperometric Monitoring of Carbon Monoxide

The amperometric monitoring of carbon monoxide depends on the reaction of CO with iodine pentoxide in which iodine is released:

$$I_2O_5 + 5CO \longrightarrow 5CO_2 + I_2$$

A stream of ambient air containing CO is passed over a column that is packed, for example, with pumice soaked in I_2O_5. Before the air is passed over this column, water vapor must be removed by drying the air in a drying stage. Some other interfering components, such as ethylene, are removed by a filter containing bromine on a granular substrate. After leaving the I_2O_5 reaction column, the air containing iodine is passed through an aqueous solution containing iodide (I^-). In the solution there are two electrodes connected by a small constant voltage source (Figure 6.30) and an ammeter. At a polarization voltage of 0.25 V, a layer of hydrogen builds up on the cathode, preventing a current from flowing through the solution:

$$2H^+ + 2e \longrightarrow H_2$$

Iodine fed in with the air flow depolarizes the electrode:

$$H_2 + I_2 \longrightarrow 2H^+ + 2I^-$$

and the sum of these reactions is

$$I_2 + 2e \longrightarrow 2I^-$$

The current measured thus is proportional to the iodine liberated by the carbon monoxide.

A detection limit of less than 0.1 ppm can be achieved with commercial instruments in a typical response time of 3 min.[48] At still lower concentrations, however, the response time may be much longer.

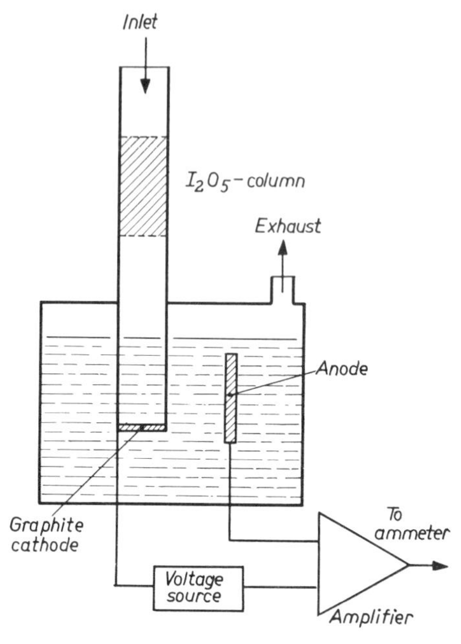

Figure 6.30. Amperometric measurement of CO. In a column packed with I_2O_5 on a granular substrate, CO is converted into CO_2 and iodine is released. The I_2 concentration is measured in an amperometric cell by the electrode reaction $I_2 + 2e \rightarrow 2I^-$. The current between anode and cathode is proportional to the iodine entering the cell, thus to the CO concentration in the air.

b. Flame Ionization Detection of Carbon Monoxide

The flame ionization detector (FID) much used in gas chromatography is based on the principle that when introduced into a hydrogen flame, hydrocarbons produce electrically charged fragments. When a voltage is applied between an electrode and the burner (Figure 6.31), a current can be measured that is proportional to the concentration of hydrocarbons in the flame.[49] The method is very sensitive, having a detection limit for methane as low as 0.01 ppm. Since the detector will not respond to CO, it first must be converted into methane by the following high temperature reaction:

$$CO + 3H_2 \longrightarrow CH_4 + H_2O$$

This reaction is fast and quantitative at 400°C, in the presence of a nickel or doped nickel catalyst,[50] but it cannot be applied directly to CO in ambient air because the 20% oxygen of the air would react more rapidly with the hydrogen than would the carbon monoxide. A preliminary separation of CO from the air is therefore necessary, as well as separation from other components, such as CO_2, CH_4, and other hydrocarbons, which would give a signal indistinguishable from that for CO. Different methods are used for this, based on gas chromatographic separation, with or without column switching. Because of the initial separation, the lag time between sampling

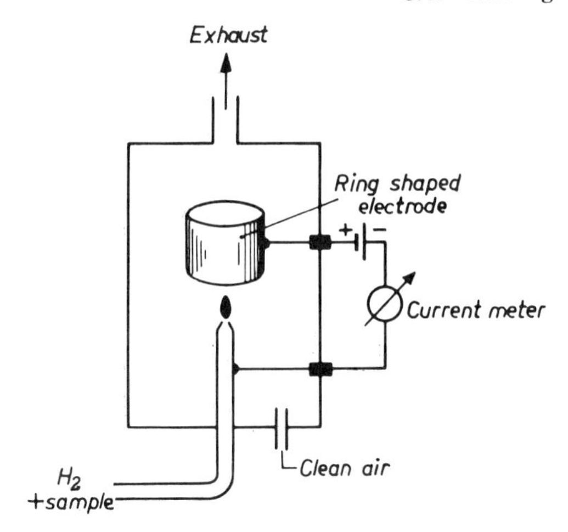

Figure 6.31. Simplified diagram of a flame ionization detector (FID). Ions are formed in the hydrogen flame, causing a current to flow between the burner and the ring-shaped electrode.

and measurement is about 5 min, and this is also the minimum time between sampling. The limit of detection can be down to 0.01 ppm. A disadvantage is the need for a supply of hydrogen for the conversion of CO to CH_4 and for the flame in the FID.

c. *Monitoring of Carbon Monoxide by Catalytic Oxidation*

Carbon monoxide can be oxidized specifically by a catalyst made of a mixture of metal oxides, (CuO and MnO_2) known as Hopcalite.* The heat of combustion causes the temperature of the air to rise, and this temperature increase can be used to measure the amount of CO present[51] (Figure 6.32). Air containing carbon monoxide passes through a column packed with the Hopcalite, and the temperature of the air is measured by two thermistor temperature sensors or by platinum thermometers before and after the column. The sensors are arranged in a bridge circuit to ensure that only the temperature difference is measured, and the temperature of the incoming air has little effect. The temperature increase of the air due to the selective combustion of CO is nearly 0.01°C per ppm CO; and since temperature differences of 0.001°C can be measured easily, this would give a theoretical limit of detection of about 0.1 ppm. In practice, because of thermal losses and thermal instabilities, the limit of detection is about 1 ppm at a

* Hopcalite is named after Johns Hopkins University, where it was developed as a filter material in gasmasks during World War I. It is also available from Hopkin and Williams Ltd, Chadwell Heath, Essex, England.

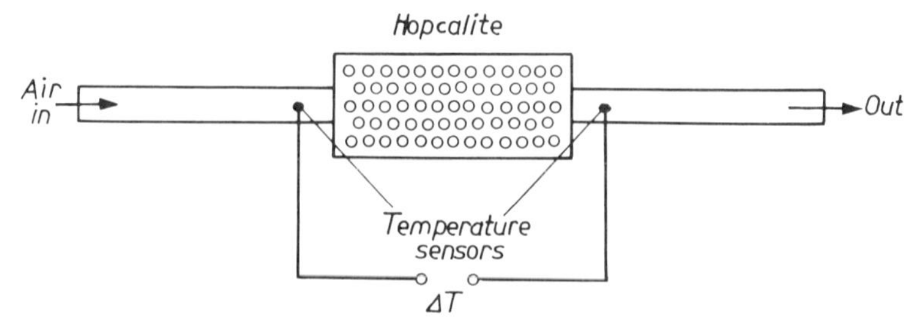

Figure 6.32. Detector for CO based on selective combustion on a "Hopcalite" catalyst. Air passes over a heated "Hopcalite" bed, where CO is oxidized to CO_2. This combustion warms the air slightly, and the temperature difference before and after the bed is proportional to the original CO content of the air.

response time of 1 min. Water rapidly deactivates the catalyst,[30] and either the whole column must be kept above 100°C, or water vapor must be removed[52] from the air entering the column. Interferences due to other combustible gases are small because of the selectivity of the catalyst and can be neglected for ambient air measurements.

6.4.5. Monitoring Methods for Particulate Matter

The foregoing sections have discussed methods for the determination of the concentrations of gaseous components. For gases, analytical techniques can be compared on the basis of simple criteria such as the lower limit of detection and the response time. For particulates, such a comparison is more difficult. For gaseous components the concentration of the pollution is a single well-defined parameter, which can be related directly to the effect on the environment, albeit in some cases in combination with other substances or meteorological circumstances, such as sunlight.

For particulate matter, the concentration, expressed as micrograms per cubic meter, is not a single valued parameter because other properties of the particulates may be important. The size of the particulates, in combination with the concentration, can be responsible for haze, effects on human health, or dustfall, and the composition of the particulates can make the particulate matter more or less dangerous. Particulates having effects injurious to health are asbestos, beryllium oxide, hazardous types of silica, and sulfuric acid. Various principles of measurement have been developed to determine size, mass, and composition.

Although in well-defined cases the results of measurements by different methods can be compared, this is not in general the case for ambient particulate measurements.

Two principles for measurement are discussed here: a method to determine the mass loading of the atmosphere based on collection of the matter on a filter and subsequent measurement by β-ray attenuation, and a method based on light scattering, which can be used to determine both the number of particles and the size distribution. The chemical composition can be ascertained only after collection on a filter, which is then analyzed in the laboratory by standard analytical techniques, such as X-ray fluorescence, atomic absorption spectrophotometry, emission spectroscopy, or X-ray diffraction.

6.4.5.1. *Mass Monitoring Instruments*

In an instrument designed to determine the mass loading of the atmosphere by solid (or liquid) particulates, the air is forced to pass through a filter (felt, glass fiber, or paper), where the solids are trapped. The increase in mass of the filter is then determined after a certain volume of air has been passed. For use in automatic stations, the mass determination of the filters by β-ray absorption has been shown to be a useful reliable method.

a. β-Attenuation Particulate Monitors

To a good approximation,[53] the attenuation of β-radiation by matter is independent of the atomic species present and depends only on the mass (in micrograms per square centimeter). The intensity I of a beam of β-particles that has passed through a layer of surface mass μ can be approximated by

$$I = I_0 \exp(-k\mu)$$

where I_0 is the intensity of the source and the constant k depends on the energy of the β-particles, thus on the β-source used. An empirical relation between k and the energy E has been found to be

$$k \approx 220E^{-1.33}$$

where k is in square centimeters per milligram and E is in kiloelectron-volts. β-Emitting isotopes used in dust monitors include

^{85}Kr	$E = 670$ keV, $k = 3.8 \times 10^{-2}$ cm^2 mg^{-1}
^{14}C	$E = 156$ keV, $k = 0.27$ cm^2 mg^{-1}
^{147}Pm	$E = 230$ keV, $k = 0.16$ cm^2 mg^{-1}

The intensity of the β-radiation is most easily measured with a Geiger counter, as in the typical instrument setup appearing in Figure 6.33.

From ambient air blown through a filter tape, the particulates are collected and form a spot on the tape. After a specified time the tape is moved forward,

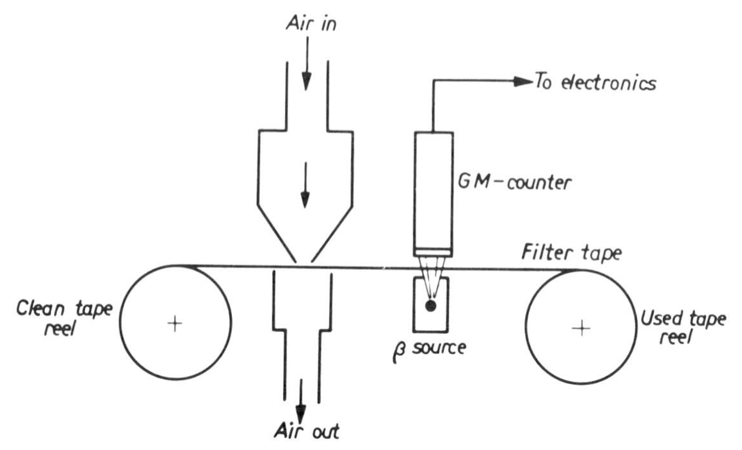

Figure 6.33. Principle of dust monitor based on attenuation of β-radiation. Particulates in the airstream are collected on a filter tape. After a certain sampling time the tape is moved forward so that the collected matter lies between a β-source and a Geiger–Müller counter. From the attenuation of the β-radiation, the total mass collected can be derived. The used tape is wound up on the reel at the right and can be used for subsequent analysis of the composition of the collected particulates.

putting the spot in the measuring position. After a sufficient number of counts, the mass collected on the tape can be calculated by subtracting the mass of the clean tape from the mass after collection. The reel with used tape can be stored and analyzed subsequently in the laboratory to find the chemical composition of the particulates as a function of time.

b. Detection Limits and Precision

If the number of β-counts detected by the Geiger counter in the case of a clean tape is N_c and that of a tape loaded with particulates N_D, then

$$N_c = A_0 \exp(-k\mu_T) \qquad \text{and} \qquad N_D = A_0 \exp[-k(\mu_T + \mu_D)]$$

where μ_T is the mass per unit area (surface mass) of the tape, μ_D the surface mass of the collected dust, and A_0 an arbitrary constant.

The value of μ_D can be derived from the two measurements:

$$\mu_D = \frac{1}{k} \ln\left(\frac{N_c}{N_D}\right)$$

The relation between μ_D and C, the mass concentration in the air, is given by

$$C = S\frac{\mu_D}{ft},$$

where S is the area of the deposition region on the tape, f the air flow during deposition, and t the sampling time. The precision and limit of detection is determined as a first approximation by the accuracy in the number of counts measured.[54,55] Since radioactive decay is a stochastic process, the standard deviation of an accumulated count number N is

$$\sigma \propto N^{1/2}$$

If $\mu_D k$ were very large, the number of counts accumulated during measurement of the loaded tape N_D, would be small because of the high attenuation; hence the relative accuracy of N_D would be low, leading to a large error in the determination of μ_D. On the other hand, if $\mu_D k$ were very small, N_c would be approximately equal to N_D, and the ratio close to unity; the accuracy also would be low. Hence it is clear that there is an optimum value for $\mu_D k$, and therefore an optimum sampling time at a given ambient particle concentration.

For the standard deviation of the *ratio* of the two counts N_c and N_D, a simple statistical analysis shows that

$$\frac{\sigma}{\mu_D} = \frac{1}{\mu_D k}[1 + \exp(\mu_D k)]^{1/2}(N_c)^{1/2}$$

The relative error will assume a minimum for $\mu_D k = 2.218$. However for the β-sources normally used (^{147}Pm and ^{14}C), k is of the order of 0.2 cm^2 mg^{-1}, which means that more than 10 mg cm^{-2} of particulates must be deposited on the tape. Assuming an air flow of 1 m^3 hr^{-1}, at a particulate concentration of 50 μg m^{-3}, 200 m^3 of air would have to be filtered, which brings the sampling time to about 100 to 200 hrs. At ambient conditions, therefore, the ultimate precision is rarely obtained. If we assume the limit of detection $\mu_{D\min}$ to be twice the standard deviation, we find for the number of counts necessary to reach this detection limit

$$N_c = \frac{4[1 + \exp(\mu_{D_{\min}} k)]}{(\mu_{D_{\min}} k)^2}$$

This relation is plotted in Figure 6.34. The left ordinate gives the product $\mu_{D_{\min}} k$, where $\mu_{D_{\min}}$ is the minimum detectable mass per square centimeter; the right ordinate gives $\mu_{D_{\min}}$, assuming that $k = 0.15$, which is correct for a ^{127}Pm source. In general, the count rate (the number of counts per second) is determined by the dead time of the Geiger counter and does not exceed 5000 counts per second.

Hence if the counting time is one hour, the count number could amount about 1.8×10^7, and a detection limit of 3 μg cm^{-2} is possible. Assuming as previously an air flow of 1 m^3 hr^{-1}, a tape surface of 2 cm^2, and a sampling time of one hour, the theoretical limit of detection would be 6 μg m^{-3}.

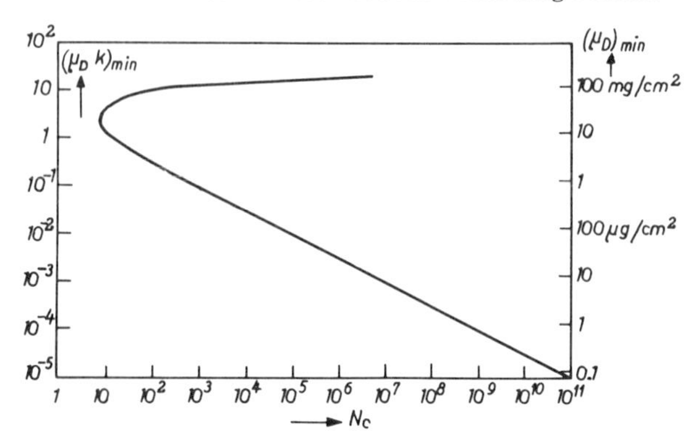

Figure 6.34. Minimum detectable concentration of particulate matter versus required number of counts in a β-ray dust monitor. The horizontal axis gives the number of counts necessary to detect amount of particulate matter of the tape. The left vertical axis gives the minimum detectable change of the exponent in the attenuation formula $(\mu_D k)_{min}$. The right vertical axis represents the minimum detectable mass when a ^{147}Pm source is used ($k = 0.16$ cm^2 mg^{-1}).

Apart from this statistical error, other parameters will influence the precision. First the density of the air between β-source and Geiger counter will depend on barometric pressure and temperature (a 1 cm air gap between source and detector is equivalent to 1.3 mg cm^{-2}). A pressure change of 3 mm Hg or a change in temperature of 1°C is equivalent to a mass change of 5 μg cm^{-2}. This is comparable to the detection limit just derived. A small source-detector gap, temperature stabilization, and frequent measurement of the clean tape will make this error negligible. Further sources of error include natural radioactivity of the collected matter and background radiation. In practice these effects on the accuracy can be neglected, but of greater importance is condensation of water on the tape or on the collected particles. To avoid errors from this source, the incoming air is usually heated to about 10°C above ambient.

Comparisons between the β-attenuation mass monitor and a gravimetric method (high volume sampler) show a good correlation between these methods.[56,57]

6.4.5.2. *Particle Counters*

Particle counters are based on the scattering of light by particulates. An air flow is passed through a light beam, and the light scattered by particles in the flow is detected by a photomultiplier. Direct illumination of the photomultiplier by the light beam is prevented by a light-absorbing structure, or stop (Figure 6.35). The number of counts registered is related to the number

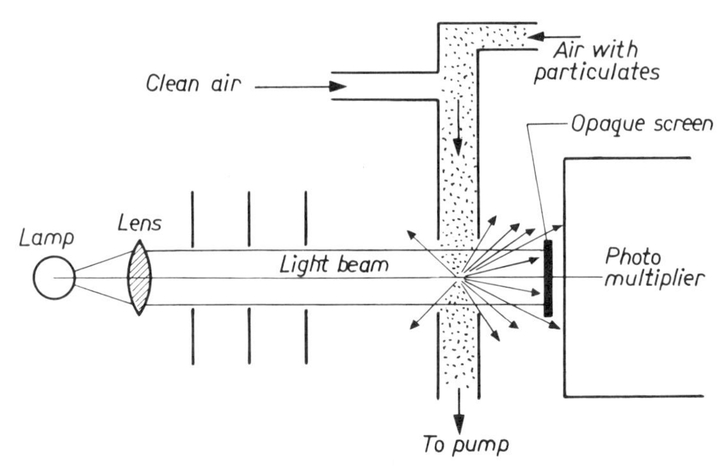

Figure 6.35. Principle of particle counter based on forward light scattering. A parallel light beam is intercepted by a stop to prevent direct light from reaching the photomultiplier. Particulates passing through the beam will scatter the light, and some of the scattered light will fall on the photomultiplier. If the concentration of particulates is high, the incoming air may have to be diluted with clean air to avoid missing counts.

of particles that have passed in a given time, the particle size can be derived from the amount of scattered light received by the photomultiplier in a pulse.[58,59] However the relation between pulse intensity and particle size depends on several factors the most important being the refractive index of the particle material. Of course, calibration of these instruments in weight per volume (micrograms per cubic meter) is impossible, although when the composition of the particles is known, a relation between weight and size can be derived.

If the particulate concentration is high, there is a risk that several particles will pass the measurement area at the same time, and some of their light scattering flashes may be missed. In this case the incoming airflow must be diluted with clean air (Figure 6.35). The main difficulty in particle counting instruments is the determination of particle size. If forward scattering is observed, the intensity of the pulse depends very little on the index of refraction of the particle, and for very small scattering angles a quite reliable pulse height versus diameter curve can be obtained. The measurement of very small angle scattering, moreover, is facilitated by the use of a laser beam as light source. The ratio between the intensities of the scattered light at two angles close together[60] is independent of refractive index and gives a good representation of particle diameter. Typical maximum values for particle concentrations are 10^6 particules per cubic meter to 10^8 particulates per cubic meter.

6.5. INSTRUMENTS IN AUTOMATIC MONITORING STATIONS

The foregoing section discussed a number of analytical techniques for measuring gaseous and particulate pollutants. Most of these techniques have been applied in commercially available instruments reliable enough to be built into automatic monitoring stations.

Additional development is needed to increase the reliability of several techniques not covered here. Although the instruments may operate well, no measurement is better than the sampling and the calibration, thus comments on these aspects in relation to the instruments should be made, if real time, reliable data are to be obtained.

6.5.1. Sampling

Air should be sampled directly from the atmosphere around the station, and it should pass without filtering to the measuring instruments. This is not always possible because sometimes the air sample must be pretreated (to remove water or other interfering substances).

The residence time in the intake tube should be minimized to a few seconds or less. For example, if an instrument needs 120 ml of air per minute for proper functioning, and the intake tube has an inner diameter of 6 mm, then for a residence time of one second, the distance between sampling point and instrument should not exceed 7 cm. For a 3 mm ID tube, the allowable length becomes 28 cm. The need for short residence time is illustrated in the following example on the monitoring of NO, NO_2, and O_3.

During daytime with sunlight, an equilibrium exists between NO, NO_2, and O_3. Nitrogen dioxide is subject to photolysis (cf. Section 6.4.3.4) and is a source for NO and O_3.

$$h\nu + NO_2 + O_2 \longrightarrow NO + O_3$$

However nitric oxide and ozone recombine to form NO_2

$$NO + O_3 \longrightarrow NO_2 + O_2$$

In the dark inlet tube only the second reaction takes place, and the NO concentration will decrease[40] according to the rate equation

$$\frac{d[NO]}{dt} = -k_1[NO] \cdot [O_3]$$

where $k_1 = 3.6 \times 10^{-4}\ ppb^{-1}\ sec^{-1}$.

The difference δ between the NO and O_3 concentrations will remain constant and equal to the difference between these concentrations outside the monitoring station:

$$\delta = [NO] - [O_3] = [NO]_0 - [O_3]_0$$

which gives the following relation for the NO concentration:

$$\frac{d[NO]}{dt} = -k_1[NO]([NO] - \delta)$$

From this differential equation one obtains

$$[NO] = \frac{[NO]_0}{[NO]_0 - [O_3]_0 \exp(-k_1\, \delta t)}$$

and

$$[O_3] = \frac{[O_3]_0 \exp(-k_1 \delta t)}{[NO]_0 - [O_3]_0 \exp(-k_1 \delta t)}$$

For residence times not exceeding 10 sec, and if $= [NO]_0 - [O_3]_0 < 100$ ppb, $k_1 \delta t < 0.3$, and we may approximate the exponential term by $\exp(-k_1 \delta t) \approx 1 - k_1 \delta t$. The losses of NO and O_3 in the tube can then be written as

$$\Delta[O_3] = \Delta[NO] = \frac{k_1 t[O_3]_0[NO]_0}{1 + k_1 t[O_3]_0}$$

For an initial NO concentration of 100 ppb, this loss is plotted in Figure 6.36. If 100 ppb ozone is present initially, about 26 ppb ozone and the same amount of NO will be lost in only 10 sec residence time in the tubing, making the NO_2 value measured too high by 26 ppb. Since the losses are proportional to the initial NO concentration, they can be calculated easily from Figure 6.36. This example illustrates only one cause of erroneous measurement, but other losses occur in the intake tube, in the filters preceding the instruments, and in the instruments themselves. This example, however, demonstrates the need for a short residence time in the intake tubes.

The sampling system in a typical automatic station is illustrated in Figure 6.37. A powerful pump draws outside air through a sampling tube from which all instruments in the station can sample. The sampling line of the upper instrument, a particulate monitor, is pointed up and the velocity in the sampling line is equal to that in the main stream (i.e., isokinetic sampling).[61] The other monitors G are monitors for gaseous pollutants, and their sampling inlets are pointed down to prevent the entry of particulates into the gas analysis instruments. The capacity of the main pump should be sufficiently large to ensure a short residence time of the air in the main sampling tube. If the

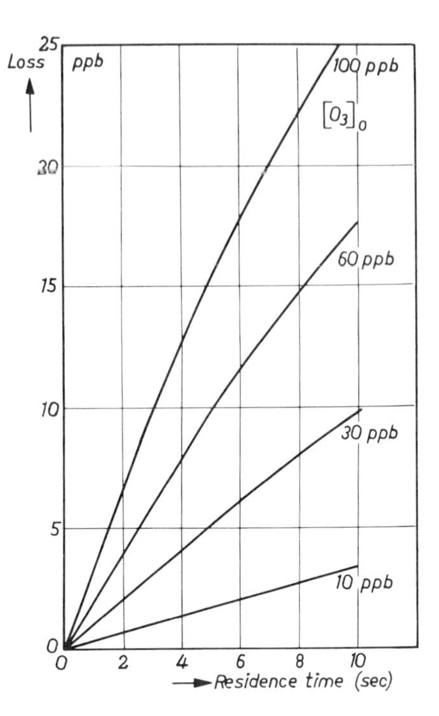

Figure 6.36. Losses of O_3 and NO in the intake tube of a monitor as a function of the residence time in the intake tube. The curves, refer to different initial ozone concentrations in the air $[O_3]_0$ indicate that even a few seconds residence time may given rise to large errors.

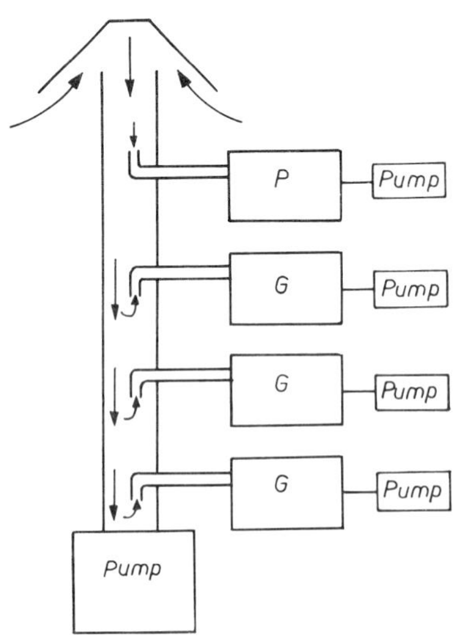

Figure 6.37. Layout of a sampling system in an automatic monitoring station. The air is sucked in at high speed by a powerful pump. The intake of the particulates monitor points up, and the velocity of the air flow in the intake line of the particulates monitor is equal to the air velocity in the main sampling tube (isokinetic sampling). The intake tubes of the monitors for gaseous components point down to prevent particulates from entering the monitors.

diameter of this tube is 3 cm, the main pump should have a capacity of 15 l sec^{-1} to ensure a maximum residence time of 0.2 sec in the sampling tube; to this must be added the residence time in the sampling tube connected to the instrument. In view of the example above, the maximum residence time in the tubing leading to the instruments should be a few seconds.

6.5.2. Calibration

As mentioned in Section 6.2, integrated calibration is necessary for instruments in automatic stations, the frequency of calibration depending on the stability of the instrument. The calibration signal can be transmitted to the central computer, stored in the memory, and used later to convert the measured signals into concentrations, or used in an internal loop to correct the signal of the instrument (see Section 6.4.2.1b).

To calibrate an instrument, clean air with a precisely known amount of the component of interest is presented to the instrument. In an automatic monitoring station a number of bottles containing the different gases are present. For some gases "permeation" tubes are available. One type consists of a short length of Teflon tube containing the gas, under pressure or liquefied, sealed with a steel ball at each end, and kept at constant temperature. Another consists of a few milliliters of the liquefied gas in a glass bottle sealed with a rubber or Teflon cap. The tube is filled with pure gas (SO_2, NO_2, etc.), and if the tube or cap is flushed continuously with clean air at a constant flow, the gas will diffuse at a constant rate through the tube or cap. After passing over the tube or cap, a precise amount of the gas will be present in the air, which then can be used as calibration gas. Since the diffusion constant is strongly dependent on temperature, permeation tubes must be thermostated. The permeation is very constant over long periods of time, and the diffusion rate can be determined by the loss of weight as a function of time.[62]

For some gases, which can be brought in liquid form under pressure at ambient temperature, a variation of the permeation tube is in use. A typical example appears in Figure 6.38. Sulfur dioxide is maintained at about 10 atm at 45°C in a stainless steel cylinder. The permeation takes place through a Teflon membrane at the top of the cylinder. The advantage of the liquid source is that the pressure in the cylinder is constant until all the SO_2 has been used up. The permeation membrane is flushed continuously by air cleaned in an activated carbon filter.

A similar device can be used for NO_2. At 45°C the vapor pressure of NO_2 is about 4 atm. An NO_2 permeation tube must be flushed with dry air. At the permeation membrane the concentration of NO_2 is high, and it may

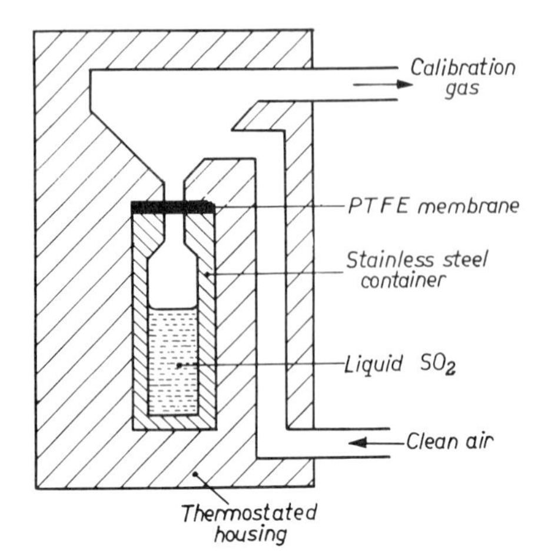

Figure 6.38. Schematic diagram of a high pressure liquid SO_2 permeation tube. At 45°C, SO_2 is in liquid form under a pressure of 10 atm. A constant amount of SO_2 diffuses per second through the PTFE (Teflon) membrane, thus providing a constant SO_2 concentration in the calibration gas. A similar device can be used for NO_2.

react easily with water vapor. To avoid the use of a large drier, the flow of flushing air is reduced. After passing over the permeation tube, the dry air containing NO_2 can be mixed with normal (wet) air. The concentration of NO_2 in the air then is so low that the reaction with water is slow enough to maintain the accuracy of the calibration.

Permeation tubes for NO must be flushed with nitrogen to avoid oxidation to NO_2. Since oxygen will penetrate by diffusion into the permeation tube, the oxidation may take place even in the tube itself. NO calibration sources for application in automatic stations can be made, using an NO_2 source followed by a reducer, such as is used in a chemiluminescent NO–NO_2 monitor (cf. Section 6.4.3.2). Since NO cannot be liquefied, the large "balloon" type of tube must be used.

A slightly different situation exists for carbon monoxide. For most calibration gases a concentration of 50 to 100 ppb is sufficient to calibrate the instrument in the region where it will operate normally. The concentrations of carbon monoxide in the atmosphere are much higher, however, which means that the concentration in the calibration gas must be of the order of 5 to 10 ppm. At a typical air flow rate of 200 to 1000 ml min^{-1}, 250 to 1500 ml of CO is used every 3 months. Since carbon monoxide cannot easily be liquefied, it is stored under pressure in a laboratory-type cylinder. The cylinder has a reducing valve and delivers CO to a silicone rubber membrane through which it can diffuse into the calibration air flow (Figure 6.39).

For ozone the standard calibration source is a pen-type UV lamp surrounded by a quartz helical tube. A constant stream of clean air passes through the tube (about 25 ml min^{-1}), and ozone is produced by the UV radiation. If maintained at constant temperature, the output of this type of

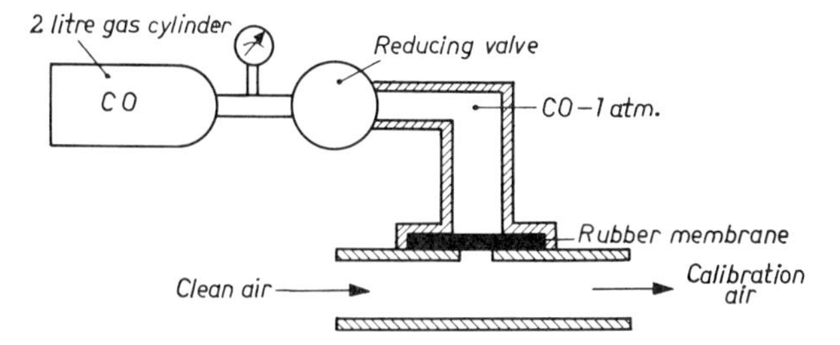

Figure 6.39. A calibration source for CO. Carbon monoxide is stored under high pressure in a steel laboratory gas cylinder. A reducing valve brings the pressure down to 1 atm in a second vessel. From here the CO diffuses at a constant rate through a rubber membrane into the air for calibration.

source has been found to be so constant (drift less than 1% per month) that it can be used as a reference source for ozone monitors.

6.5.3. Stabilization of Flow

For instruments in which the pollutant reacts chemically, the flow through the instrument must be kept constant. The response of the instrument will be proportional to the amount of pollutant received per unit of time. This is the case for coulometry, flame photometry, chemiluminescence, photodissociation, amperometry, flame ionization, catalytic oxidation, and also for the two methods for determination of dust particles. The spectroscopic techniques, including fluorescence, measure the concentration directly (i.e., the number of molecules present in the sample cell). Here a change in flow will have only a minor effect on the response time of the instrument but not on the measurement of the concentration.

For the mass-indicating instruments, however, an accurate stabilization of the flow is necessary to convert the mass measured into a concentration. A well-thermostated metering orifice operating as "critical" orifice will be sufficient, provided it is protected against corrosive vapors, dust, or moisture.

6.6 FUTURE DEVELOPMENTS

The development of air pollution instrumentation for use in automatic stations has been very rapid. Only a decade ago the most advanced methods consisted of mechanized forms of manual laboratory techniques. The minimum response was half-hour averaged values, and the stations had to be visited every few days to replenish the reagents. Special techniques have now been developed, enabling values of the pollution level to be delivered instantaneously on telephones lines connected to a computing center for immediate action. One to three months of unattended operation is not unusual.

In such a rapidly developing field it is difficult to predict how a future monitoring station will operate and what analytical techniques will be applied, but some trends can be observed. Apart from new analytical techniques, mentioned below, there is much progress in the development of low cost data processors and control computers. A few years ago a simple computer cost about the same as five analytical instruments, but nowadays such equipment can be had for only one-tenth that price. It may be expected that the price may fall even further in the near future, because of the growth of large-scale integration techniques. This creates the possibility of giving every monitoring station its own computer and converting it into an "intelligent" station, which no longer has to be connected "on-line" to the central computing center, instead, a normal connection to the telephone exchange network becomes possible. This would reduce greatly the cost of a monitoring station because dedicated telephone lines are an order of magnitude more expensive than normal connections. The rate of information flow through the telephone line from a monitoring station is surprisingly low. If the readings of five instruments are to be transmitted each minute with three figure accuracy, 50 bits per minute is to be transmitted, or 1 bit per second! The built-in computer can serve as a buffer. Once per hour or even less frequently the central computer can dial the stations and have all data transmitted in a few seconds. In case of an unexpected combination of measuring data, the microprocessor can be programmed to call the central computer, transmit the situation, and ask for instructions. The need for this capability would arise, for example, if threshold values were passed for one or more pollutants (e.g., the combination of values from which a local photochemical smog can be expected) or if an instrument began to function improperly.

It is not only in the area of data handling and interpretation that the local computer can be useful. It can also supply "dynamic programming" to the monitoring station. The frequency of the measurements would not have to be the same for all components and at all times. The frequencies could be adapted to the situation of the moment and increased or decreased to give optimum selection of components and averaging times. The application of a local computer in every monitoring station makes it possible to add new and more complicated analytical techniques, which require more frequent adjustment. Techniques, that are now usable only in the laboratory will become available for automatic monitoring. Among these are laser spectroscopy, Fourier spectroscopy, Hadamard spectroscopy, and on-the-spot determination of the composition of dust particles by X-ray fluorescence, emission spectroscopy, or (for the organic aerosols) by gas chromatography. It even may be possible to apply less selective techniques if the local computer, by simple arithmetic, can subtract the interference from the measurements.

Which components will become of interest? There is a trend to study the behavior of combinations of individual components,[66] notably the formation of photochemical smog, which depends on nitric oxide, nitrogen dioxide, ozone, and reactive hydrocarbons. The model for smog formation can be stored in the local computer and locally preventive control measures can be taken if the critical situation is approached. The same procedure can be followed for other synergistic combinations.

The measurement of nitric and sulfuric acids will certainly be added soon to the range of capabilities of automatic instruments. The existence of "local intelligence" will facilitate the setting up of very large monitoring networks, since local networks can be linked together easily and cheaply. Such developments will contribute to our knowledge of changes in the composition of air during its movement over large distances. Of course, all this will be possible only if air pollution measurements are carried out automatically. Some techniques suitable for this purpose have been described in this chapter. Most of them are still young and carry within them the potential for improvements in the near future.

SYMBOLS

a	amplitude
c	concentration
C	mass concentration in air
C_n	spherical particles per unit volume
D	diameter of particles
E	energy (keV), voltage
f	airflow
F	Faraday number
F_i	($i = 2, 4, \ldots$) coefficient
I	light intensity
I_f	intensity of fluorescence
k	constant
k,	reaction rate constant
K	scattering coefficient
l	length of absorption cell (cm)
N_c	number of counts
R	gas constant, recorded signal

S	area of deposition, transmitted light
t	time, sampling time
T	absolute temperature
V_0	reference voltage
α	absorption coefficient, line width
δ	concentration difference
λ	wavelength
μ	surface mass
μ_D	surface mass of dust
μ_T	mass per unit surface area
σ	standard deviation
τ	time
ω	frequency

APPENDIX A6. LIGHT SCATTERING BY PARTICLES

The transmission of light through a medium containing C_n spherical particles per unit volume can be described as follows:

$$I = I_0 \exp\left[-K \frac{\pi}{4} D^2 C_n l\right],$$

where I and I_0 are the transmitted light intensity and the incident light intensity, respectively, D the diameter of the particles, l the length of the light path, and K the scattering coefficient.[63]

In this relation K depends on the particle diameter, the wavelength of the light, and the refractive index of the particles. The scattering coefficient can be derived theoretically and is in good agreement with experiments. If the particles are not uniform in diameter, the light transmission can be written

$$I = I_0 \exp\left[-\frac{\pi}{4} C_n l \int_0^{D_\infty} K N(D)^2 \, D\right],$$

where $N(D)$ is the size distribution function of the particles and D_∞ the maximum diameter present.

If we define the average scattering coefficient as

$$\bar{K} = \frac{\int^{D_\infty} K N(D) D^2 \, D}{\int_0^{D_\infty} N(D) D^2 \, D}$$

and express the volume fraction of the particles C_v in terms of the number of particles by

$$C_v = C_n \frac{\pi}{6} \int_0^{D_\infty} N(D) D^3 \, dD \qquad \text{for spherical particles only}$$

the transmission becomes

$$I = I_0 \exp\left[-\frac{3}{2} \left(\frac{\bar{K}}{D_{32}} \right) C_v l \right],$$

where

$$D_{32} = \frac{\int_0^{D_\infty} N(D) D^3 \, dD}{\int_0^{D_\infty} N(D) D^2 \, dD}.$$

If we write $\beta = \frac{3}{2}\bar{K}/D_{32}$, the light transmission takes a form that is comparable to the Lambert–Beer law:

$$I = I_0 \exp(-\beta c l).$$

The distribution function of particles in the atmosphere is known to be close to log normal;[64] that is, the logarithms of the particle diameters show a normal distribution.

In practice an "upper limit" distribution can also be used, where

$$N(D) = A \frac{\exp - [\delta \ln(aD/D_\infty - D)]^2}{D^4(D_\infty - D)}$$

Where a and δ are parameters for fitting the width and skewness of the distribution, and A is a normalizing factor.

For this distribution we can easily calculate the parameter D_{32}, which has the dimension of a diameter and also the average particle diameter $\langle D \rangle$, defined as the cube root from the average particle volume

$$\langle D \rangle = \left(\frac{\int_0^{D_\infty} N(D) D^3 \, dD}{\int_0^{D_\infty} N(D) dD} \right)^{1/3}$$

We express now the "absorption coefficient" $\beta = \frac{3}{2}\bar{K}/D_{32}$ as a function of wavelength λ, average particle diameter $\langle D \rangle$, and index of refraction of the particulate matter n. According to Dobbins and Jizmagian,[63] the ratio $\bar{K}/X_{32}$, where $X_{32} = (n-1)\, D_{32}/\lambda$ is nearly independent of the shape of the distribution function. The plot of $\bar{K}/X_{32}$ versus X_{32} (Figure A6.1) is valid for a wide range of upper limit distribution functions. For a given refractive index, we can thus derive a plot of $\beta = \frac{3}{2}\bar{K}/D_{32}$ versus λ/D_{32}. We now must express the quantity D_{32} into the average particle diameter, defined earlier.

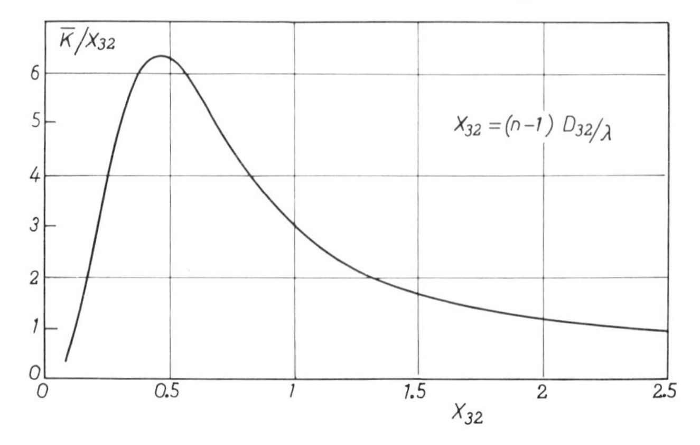

Figure A6.1. Plot of the scattering function $\overline{K}/X_{32}$, where $\overline{K}$ is the average scattering coefficient for an upper limit distribution of suspended particulates and $X_{32} = (n - 1)D_{32}/\lambda$, where n is the refractive index, λ the wavelength, and D_{32} a "particulate diameter" defined in the text; $\overline{K}/X_{32}$ depends only weakly on the parameters of the distribution function.

For a wide range of values of a and δ ($0.5 \leq a \leq 2$, $0.9 \leq \delta \leq 2.8$), the ratio $\rho = \langle D \rangle / D_{32}$ appears to be nearly constant, with the average value 0.92 ± 0.06. Hence within the inaccuracy of the assumptions of the model described, we can write

$$\rho = \tfrac{3}{2}\overline{K}/D_{32} \approx 1.38\overline{K}/\langle D \rangle \quad \text{or} \quad \beta\langle D \rangle = 1.38\overline{K} \quad \text{and} \quad \lambda/\langle D \rangle = 1.09\lambda/D_{32}.$$

These parameters, $\beta \cdot \langle D \rangle$ and $\lambda/\langle D \rangle$, are dimensionless, and the relation between the two can be plotted with the refractive index n as parameter (Figure A6.2) using the $\overline{K}$ values from Reference 63. Since $\overline{K}/D_{32}$ and $\langle D \rangle / D_{32}$ are nearly independent of the distribution function, the value of modified scattering coefficient $\beta \cdot \langle D \rangle$ will also be valid for a great variety of distribution functions.

From Figure A6.2 we can derive easily the light loss due to scattering for a fixed index of refraction as a function of wavelength, with the average particle diameter as a parameter. This has been done in Figure A6.3 for $n = 1.50$ and for average particle diameters of 0.2, 0.5, and 1 μm. The light transmission can be derived from Figure A6.2 by inserting the appropriate value of β into the transmission equation

$$I = I_0 \exp(-\beta C_v l)$$

where C_v is the volume fraction of the particulates and l the absorption path length in centimeters. Here is a numerical example. At an assumed dust concentration of 100 μg m^{-3}, with an average specific weight of the particulate matter of 2.5, the volume fraction of the matter is 40×10^{-12}. With an

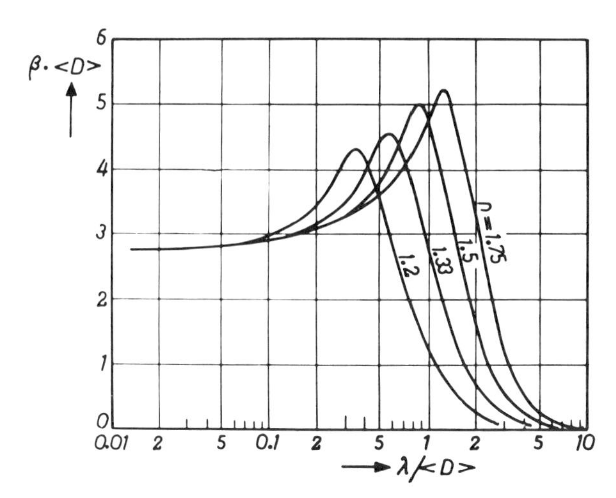

Figure A6.2. Dimensionless scattering parameter $\beta \cdot \langle D \rangle$ as a function of the dimensionless light wavelength $\lambda/\langle D \rangle$ for different values of the refractive index n of the particulate matter; $\langle D \rangle$ is the average particle diameter.

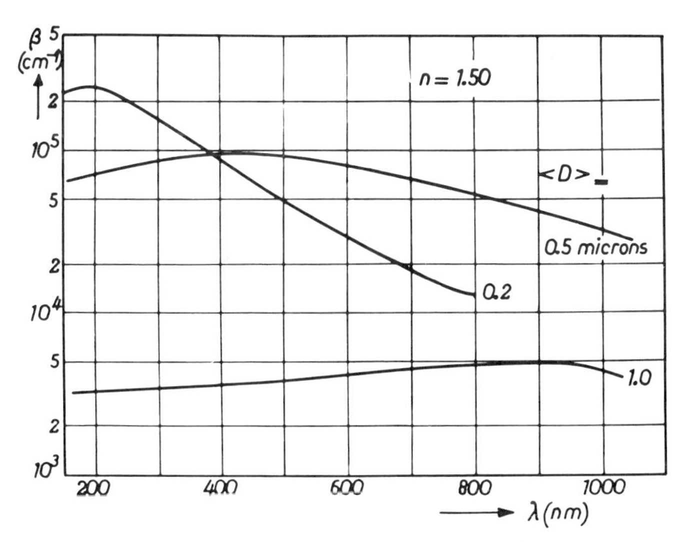

Figure A6.3. Scattering coefficient β as a function of wavelength for various average particle diameters $\langle D \rangle$. The refractive index of the particulates is taken as $n = 1.50$: For other values of n or other particle diameters, the curves can easily be derived from Figure A6.2.

average particle diameter of 0.2 μm at 300 nm wavelength, the transmission loss will be 6×10^{-6} per centimeter path length. This example was used to compute the particle scattering in Figure 6.12.

Of course the discussion above holds only for spherical particulates of uniform refractive index. For nonspherical particulates, the theoretical prediction of particulate scattering is much more complicated, and only experimental data can be used.

REFERENCES

1. Clarenburg, L. A., *Proceedings of the Second International Clean Air Congress*, Washington, D.C., 1970, p. 1193. [Academic Press, New York, 1971].
2. Bibbero, R. J. and I. G. Young, *Systems Approach to Air Pollution Control*. Wiley, New York, 1974, Section 6.1.4.
3. Observations made at the premises of Philips Research Laboratories, Eindhoven, Netherlands.
4. *Instrumentation of Environmental Monitoring: Air*. Technical Information Division, Lawrence Berkeley Laboratory, University of California, Berkeley.
5. Brouwer, H. J., S. M. de Veer, and H. Zeedijk, *Philips Tech. Rev*. 32, 33 (1971).
6. Schiller, J. W., *Bendix Tech. J*., 4, 56 (1971).
7. Hodgeson, J. A., W. A. McClenny, and R. K. Stevens, in *Analytical Methods Applied to Air Pollution Measurements*, R. K. Stevens and W. F. Herget, Eds. Ann Arbor Science Publishers, Ann Arbor, Mich., 1974, p. 43.
8. Stevens, R. K., A. E. O'Keefe, and G. C. Ortman, *Environ. Sci. Technol*., 3, 652 (1969).
9. Stevens, R. K. and A. E. O'Keefe, *Anal. Chem*., 42, 143A (1970).
10. Stevens, R. K., R. Baumgardener, H. E. Stubbs, and A. W. Berger, to be published.
11. Mettee, H. D. *J. Chem. Phys*., 49, 1784 (1968).
12. Hanst, P. L., A. S. Lefohn, and B. W. Gray, Jr., *Appl. Spectrosc*. 27, 188 (1973).
13. Boll, R. H. and J. E. Tozier, *Anal. Chem*., 47, 225 (1975).
14. Hanst, P. L., in *Advances in Environmental Science and Technology*, Vol. II, J. N. Pitts, Jr., and R. L. Metcalf, Eds., Wiley-Interscience, New York, 1971, pp. 91 ff.
15. Burch, D. E. and D. A. Gruvnak, in *Analytical Methods Applied to Air Pollution Measurements*, R. K. Stevens and W. F. Herget, Eds., Ann Arbor Science Publishers, Ann Arbor, Mich., 1974, p. 193.
16. Moffat, A. J., J. R. Robbins, and A. R. Barringer, *Atmos. Environ*., 5, 511 (1971).
17. Davies, J. H. and A. J. Moffat, *Proceedings of the Second International Clean Air Congress*, Washington, D.C., 1970, p. 552 [Academic Press, New York, 1971].
18. Hager, R. N., Jr., Paper presented at the Joint Conference on Sensing of Environmental Pollutants, Palo Alto, Calif., 1971.
19. Williams, D. T. and R. N. Hager, Jr., *Appl. Opt*., 9, 1557 (1970).
20. Hager, R. N., Jr., and R. C. Anderson, *J. Opt. Soc. Am*., 60, 1444 (1970).
21. From Reference 12, but corrected for effective cell length of multipass cell and for a signal to noise ratio of 2.

22. Schwartz, F. P., H. Okabe, and J. K. Whittaker, *Anal. Chem.*, 46, 1024 (1974).
23. Okabe, H., P. L. Splitstone, and J. J. Ball, *J. Air Pollut. Control Assoc.*, 23, 514 (1973).
24. Wolfe, C. L. and P. Giever, Paper 76-65.1 presented at the 68th Annual Meeting of the Air Pollution Control Association, Boston, 1975.
25. Nederbragt, G. W., A. van der Horst, and J. van Duijn, *Nature*, 206, 87 (1965).
26. van Heusden, S., *Philips Tech. Rev.*, 34, 73 (1974).
27. Guicherit, R., *Z. Anal. Chem.*, 256, 177 (1971).
28. Green, A. S., *The Middle Ultraviolet, Its Science and Technology*, Wiley, New York, 1966, p. 63.
29. Zafonte, L., W. Long, and J. N. Pitts, Jr., *Anal. Chem.*, 46, 1872 (1974).
30. Fontijn, A., A. J. Sabadell, and R. J. Ronco, *Anal. Chem.*, 42, 575 (1970).
31. Fontijn, A. and R. J. Ronco, Paper presented at the Joint Conference on Sensing of Environmental Pollutants, Palo Alto, Calif., 1971.
32. Stevens, R. K. and J. A. Hodgeson, *Anal. Chem.*, 45, 443A (1973).
33. van Heusden, S. and L. P. J. Hoogeveen, *Z. Anal. Chem.*, 282, 307 (1976), (in English).
34. Kummer, W. A., J. N. Pitts, Jr., and R. P. Steer, *Environ. Sci. Technol.*, 5, 1045 (1971).
35. Dzubay, T. G., H. L. Rook, and R. K. Stevens, in *Analytical Methods Applied to Air Pollution Measurements*, R. K. Stevens and W. F. Herget, Eds., Ann Arbor Science Publishers, Ann Arbor, Mich., 1974, p. 71.
36. Hodgeson, J. A., J. P. Bell, K. A. Rehme, K. J. Krost, and R. K. Stevens, Paper presented at the Joint Conference on Sensing of Environmental Pollutants, Palo Alto, Calif., 1971.
37. Winer, A. M., J. W. Peters, J. P. Smith, and J. N. Pitts, Jr., *Environ. Sci. Technol.*, 8, 118 (1974).
38. Black, F. M. and J. E. Sigsby, *Environ. Sci. Technol.*, 8, 149 (1974).
39. Leighton, P. A., *Photochemistry of Air Pollution*, Academic Press, New York, 1961.
40. Wu, C. H. and H. Niki, *Environ. Sci. Technol.*, 9, 46 (1975).
41. Whitcomb, S. E. and R. T. Lagemann, *Phys. Rev.*, 55, 81 (1939).
42. Gerritsma, C. J. and J. H. Haanstra, *Infrared Phys.*, 10, 79 (1970).
43. Pierson, R. H., A. N. Fletcher, and E. St. Clair Gantz, *Anal. Chem.*, 28, 1218 (1956).
44. Hill, D. W. and T. Powell, *Non-Disperse Gas Analysis*, Adam Hilger, London, 1968.
45. McKee, H. C., J. H. Margeson, and T. W. Stanley, *J. Air Pollut. Control Assoc.*, 23, 870 (1973).
46. Luft, K. F., *Z. Tech. Phys.* 24, 97 (1973).
47. Luft, K. F., G. Kesseler, and K. H. Zörner, *Chemie-Ingenieur-Technik*, 29, 937 (1967).
48. Falkenburg, R. and J. F. M. van Dijk, Paper presented at the ICESA Conference, Las Vegas, 1975.
49. Porter, K. and D. H. Volman, *Anal. Chem.*, 34, 748 (1962).
50. Colket, M. B., D. W. Naegeli, F. L. Dryer, and I. Glassmann, *Environ. Sci. Technol.*, 8, 43 (1974).
51. Lukaci, J. and J. Chriastel, *Anal. Abstr.*, 6, 751 (1959).
52. Bossart, C. J., Paper 75-56.8 presented at the 68th Annual Meeting of the Air Pollution Control Association, Boston, 1975.
53. Sem, G. J. and J. A. Borgos, *Staub*, 35, 5 (1975) (in English).
54. Lilienfeld, P., *Staub*, 35, 458 (1975) (in English).

55. Lilienfeld, P., Paper 75-65.2 presented at the 68th Annual Mee..ng of the Air Pollution Control Association, Boston, 1975.
56. Köhler, A. and M. Birkle, *Staub*, 35, 1 (1975).
57. Nader, J. S. and D. R. Allan, *Am. Ind. Hyg. J.*, 21, 300 (1960).
58. Seany, R. J., R. K. Holpin, and B. A. Maguire, *Staub*, 33, 213 (1973)
59. Otsuka, K., Paper 75-24.4 presented at the 68th Annual Meeting of the Air Pollution Control Association, Boston, 1975.
60. Gravatt, G. C., Jr., *J. Air Pollut. Control Assoc.*, 23, 1035 (1973).
61. Dworetsky, S. H., *Environ. Sci. Technol.*, 8, 464 (1974).
62. Salzmann, B. E., W. R. Barg, and G. Ramaswamy, *Environ. Sci. Technol.*, 5, 1121 (1971).
63. Dobbins, R. A. and G. S. Jizmagian, *J. Opt. Soc. Am.*, 56, 1345 (1966).
64. *Air Quality Criteria for Particulate Matter*, U.S. Department of Health, Education and Welfare, Washington, D.C., 1969.
65. Forrest, J. and L. Newman, *J. Air Pollut. Control Assoc.*, 23, 761 (1973).
66. Stern, A. C., *Proceedings of the Twelfth International Colloquium on Atmospheric Pollution*, M. M. Bénarie, Ed., Elsevier, Amsterdam, 1976, p. 1.
67. Jahnke, J. A., J. L. Cheney, and J. B. Homoloya, *Environ. Sci. Technol.*, 10, 1246 (1976).

7

SAMPLING NETWORK SELECTION IN URBAN AREAS

Fred M. Vukovich

Research Triangle Institute
Research Triangle Park, North Carolina

7.1. INTRODUCTION

Adverse health effects due to air pollutants motivated the development of the air quality standards. It then became obvious that urban regions, with their accompanying commercial and industrial complexes, must monitor the near-surface concentration of the air pollutants that bring about adverse health effects. The first of two main reasons for monitoring air pollutants is to provide data for the analysis of air pollution in an entire urban area to determine if, when, and where a particular pollutant concentration has exceeded the air quality standards. This information could be used to implement control measures to reduce pollution concentrations to an acceptable level and to determine the effectiveness of control measures. The second reason, as important as the first, is to furnish primary data for short-term and long-term predictions of the concentration of a particular pollutant over the entire urban region. These predictions could be used to implement control measures that would prevent pollutants from reaching or exceeding the standards, as well as to develop and test control strategies. Diagnosis and prediction of the air pollution levels have become increasingly necessary because the shortage of fossil fuels often has resulted in industries and power plants burning high sulfur coal, increasing emissions of sulfur dioxide.

Not only must the data from the network accurately depict the air pollution distribution in the urban area at any time and be dynamically consistent, to permit the accurate prediction of pollution levels for tactical purposes, the predictive capability also implies that meteorological parameters must be obtained at the stations in the network. Though this is an important factor in deciding the nature of the network, a major determinant is the economic constraints. Individual sampling stations can cost as much as $70,000 without computer systems for data retrieval and analysis. Therefore it is imperative to design a sampling network for an urban region which contains a minimum number of stations, yet yields a reasonably accurate diagnosis of the air pollution distribution and dynamically consistent data for predictions of pollutant levels.

Choosing the minimum number stations must involve the scale of the regional air pollution distribution obtained in the analysis of data. Thus the interplay between scale representation and economics is obvious. For example, phenomena such as street canyon effects in large cities[1] requires hundreds of stations because there are hundreds of street canyons. The number of stations depends on the order of variability of the constituent. This becomes small, and the number of stations not excessively large (and the cost not prohibitive) when the minimum scale represented is upper microscale or lower mesoscale (variability on scales less than 1 or 2 km is truncated). Scale representations smaller than this must be inferred through empirical results rather than through direct measurement.

An optimum sampling network implies the minimum number of stations with precise locations. Furthermore, the data from the optimum network can be used to obtain an accurate analysis of the air pollution distribution, and the accuracy of this analysis would not be significantly enhanced by adding stations. Sweeney[2] attempted to establish statistically objective guidelines for determining the number and location of sampling stations, that is, the optimum meteorological and air pollution sampling network for urban areas. He concluded that present statistical theory does not provide a basis for a general solution to the problem. In other words, an optimum network that would be universally applicable to all urban areas cannot be established. This is because the air pollution distribution and the meteorology are dynamic phenomena that not only vary in time and space in a given urban area, but also differ from one urban region to the next.

Though Sweeney's results were negative, his recommendations yielded the very important key. Generalized optimum networks that can be universally applied do not exist, but an optimum network specific for a given urban area can be determined if there is available a representative and statistically significant data set supplying the temporal and spatial variability of the air pollution concentrations and the meteorological parameters. The data set would be used to establish a statistical model to obtain the overall order of spatial variability. Through this, an optimum network could be established.

In consideration of the data set required, spatial and temporal observations of significant air pollutants and meteorological parameters must be included. For most large urban areas, there are three or more significant pollutants, and the most important meteorological parameter is the wind distribution; but including the boundary layer stability would give better estimates of diffusion coefficients. However diffusion coefficients may be inferred through indirect techniques, such as those of Turner.[3] The wind field and the diffusion coefficients are the primary meteorological parameters required for predicting air pollution levels at a later time.

As indicated, the number of stations in the optimum network depends on the overall spatial order of variability as defined by the data set for the distribution of the significant air pollutants and meteorological parameters. The overall order of variability will be markedly influenced by the parameter with the highest variability. The conservation equation shows that the air pollution distribution has the highest order of variability, rather than the distribution of the significant meteorological parameters. The conservation equation for any pollutant ψ is

$$\frac{\partial \psi}{\partial t} = -\mathbf{V} \cdot \nabla \psi + \nabla \cdot (K \nabla \psi) - \psi \sum_{n} k_n \psi_n^* + S \qquad (1)$$

The first two terms on the right of the equals sign designate the meteorological influences on the distribution of the air pollutant. The first is the advective term, where $\mathbf{V}$ is the wind vector and ∇ is the vector differential operator; and the second is the diffusion term, where K is the three-dimensional diffusion coefficient. The third term on the right is the gas phase reaction term, in which k is the rate constant specific for the second-order reaction between the given pollutant ψ and some other pollutant denoted by ψ^*. The index n of the summation indicates that more than one second-order reaction may take place. First- and third-order processes ("processes" implies techniques of removal other than gas reactions) are also possible and should be included in the equation, but it is preferable to use the simple second-order reactions to demonstrate the existence of the reaction and/or decay terms, because they are somewhat more common. The last term is the source term, S being the source strength.

The conservation equations indicate that the order of variability of any pollutant is governed to a great extent by the interaction between the source distribution and the meteorological parameters. Considering a pollution distribution that describes the variability on scales of the order of the lower mesoscale or upper microscale, the interaction between the wind field and the source distribution is seen to be the major factor influencing that order of variability. For example, in the case of one space dimension, if m is the order of variability of a pollutant in the data set, though not necessarily having the highest order of all the pollutants in that set (m is also an estimate of the number of stations required in the optimum sampling network), and if p is the number of sources in a given area, m will be less than p because the smallest scale possible has been limited to lower mesoscale or upper microscale, which will truncate higher orders of variability produced by the large number of sources. Also diffusion will smooth out some of the higher orders of variability produced by the sources. However if q is the order of variability of the wind field in one dimension, m will be greater than q because the nonlinear interaction in the advective term will produce in the pollution distribution higher orders of variability than are in either the wind field or in the initial, smoothed source distribution. If the wind field had a second-order variability along some arbitrary coordinate x and if the source distribution could also be described by a second-order function along the same coordinate, the product of the wind field representation and the gradient of the initial pollution field, as represented by the source distribution in the advective term, would yield a third-order variability for the resultant pollution field. Therefore for an optimum meteorological and air pollution sampling network, the number of stations, and to a great extent, the location of the stations is governed primarily by the statistics of the pollution distribution.

Thus general solutions to the problem of obtaining an optimum meteorological and air pollution sampling network do not exist because both the meteorological state and the air pollution distribution are dynamic. However solutions *specific* for a given area can be found, provided a representative and statistically significant data set can be obtained for the region which describes the required meteorological state and air pollution distribution in space and time over all appropriate conditions and at the required scale. Though these solutions are *specific* for a given region, they account for the spatial distribution of all pollutants and meteorological parameters, alleviating the necessity for developing sampling network criteria for each pollutant. Economic constraints suggest that the required scale is of the order of lower mesoscale and upper microscale. Higher orders of variability must be obtained through empirical relationships. The number of stations and their location in the optimum network for the given urban region is dependent on the overall order of variability of the meteorological parameters and the air pollution concentration in space as defined by the data set; and the overall order of variability is based, to a great extent, on the parameter that has the highest order of variability. In comparing the air pollution distribution with the required meteorological parameter, which is basically the horizontal wind field, the air pollution distribution would have the highest order of variability because of the larger number of sources in urban regions. Scale truncation and diffusion would reduce the gross variability produced by the sources. Nonlinear convective effects would give higher order of variations than in either the wind field or the initial, smoothed pollution field produced by the sources.

Further aspects of the all-important data set, which include the acquisition of the data set and the stability of the parameters in the data set in terms of producing an optimum network, are discussed later.

7.2. BASIC STATISTICAL PROCEDURES

This section describes a technique through which an optimum sampling network for the air pollution distribution and the meteorological parameters may be obtained, given the important data set. It is assumed that the data set is made up of observations from a lattice of points whose number is large enough to yield the required scale representation. The problem of establishing design points for a given model, especially regression models, has been explored extensively in the literature for the case of independent variables assuming a continuum of values. Goodness criteria, such as rotability, have very often been used in determining the appropriateness of the subsequent designs, and the techniques employed to constrain the design usually depend on the continuity of the independent variables. The problem of deriving such designs, which are restricted to a lattice of points, as in the case of field observations of pollutants and meteorological parameters, does not seem to have received much attention. However the general theories of least squares estimation and matrix algebra may be used in dealing with the special problem.

Consider a grid network for the given urban region containing N sampling stations having observed N^* variables. The purpose is to obtain an optimum network containing n stations, where n is much less than N, through which a reasonably accurate distribution of any one of the observed parameters can be found by sampling the parameter at the n stations in the optimum network. Polynomial models are the most applicable statistical model because of their great flexibility. These take the following general form:

$$Z(x, y) = \sum_{\alpha} \sum_{\beta} B(\alpha, \beta)P(x, y, \alpha, \beta) \tag{2}$$

where x, y are horizontal coordinates in a Cartesian coordinate system, Z represents the observed parameter, B represents the coefficients of the polynomial expansion, and the potential P represents terms of the form $x^{\alpha}y^{\beta}$ ($\alpha, \beta = 0, 1, 2, 3, \ldots$). To estimate all the coefficients in eq. 2, an infinite series, an infinite number of data points would be needed; but since the magnitude of many of the coefficients is negligible, they can be neglected with small resulting errors, considerably reducing the number of terms required in the model. The error resulting from not using all the terms in the polynomial model will be called the base error.

Among the various statistical regression procedures available for estimating the coefficients in the polynomial model, some of which could reasonably be assigned a zero value (i.e., determine which terms in eq. 2 should be retained and which should be neglected), are the following:

1. All possible regressions.
2. Forward selection.
3. Backward elimination.
4. Stepwise.
5. Maximum R^2 improvement.
6. Minimum R^2 improvement (R is the correlation coefficient).

Of these, the stepwise regression technique is most applicable because it is the most computationally efficient. From a statistical point of view, the stepwise technique also has more appealing criteria for inclusion and exclusion of terms than the forward selection or the backward elimination techniques. Recognizing that the purpose is to select a specific model that will perform reasonably well under all conditions prescribed by the initial data sets, the relative importance of terms in the general model can be judged by the frequency and the order with which they appear in the stepwise regression model. Often symmetry in the cases presented by the data set restricts the candidate model forms to those which are symmetric in x and y; that is, the terms $x^i y^j$ and $x^j y^i$ are either both included or both excluded.

From a purely statistical point of view, regression or polynomial models with large numbers of terms will, in general, perform significantly better in representing the distribution of a parameter than those with fewer terms. However it is important to note that if the resultant polynomial model contains m terms, it will take m data points (i.e., stations in a sampling network) to estimate the coefficients in the model. Furthermore, best estimates of the coefficients in the regression model are generally obtained when there are more data points than there are terms in the model. For example, Table 7.1 gives the correlation coefficient between predicted and observed wind speeds versus the number of stations used to estimate the coefficients in a 13-term, polynomial model for the wind speeds. Note the improvement when 14 stations were used instead of 13. A reasonable percentage estimate of the number of data points required for best estimate of the coefficients is 10 to 25% more than there are terms in the model. Therefore, if a 25 term model were employed, about 31 stations would be necessary to yield a reasonably accurate estimate of the coefficients in the model. Then from a practical point of view, and even more from an economic point of view, models with large number terms, calling for still more sampling stations, are not feasible.

Restricting the candidate model form to a reasonable number of terms—reasonable in terms of the creation of an economically feasible sampling network—may produce large base errors. It is then questionable whether the economic burden imposed by building a network based on a candidate model having a large number of terms (but yielding a small base error) is worth the degree of improvement obtained by using a regression model with a smaller number of terms. If there exists a quasi-discontinuous jump in the magnitude of the base error in the domain from models with large numbers of terms to models with small numbers of terms, the answer is

Table 7.1 Correlation Coefficient R Between Predicted and Observed Wind Speeds Versus Station Number[a]

Number of Stations in Optimum Network	R
13	0.50
14	0.63
15	0.65
16	0.65
17	0.66
18	0.67
19	0.67

[a] A 13 term model was used to describe the wind field.

positive. Since the regression model that would be chosen should be one far enough away from the jump not to be influenced by it, there is no choice but to absorb the economic burden if an urban or regional distribution of the pollution level is required. Otherwise, if a network were created using a model with a small number of terms, observations would have meaning only in the locale of the station, and judgments or predictions based on a distribution analysis using these observations would be extremely limited.

If it can be demonstrated that the decrease in base error is continuous in the domain from candidate models with large numbers of terms, and if the slope of the curve representing the base error estimate in that domain is small, this is more attractive. Experience has shown that often this has been the result, provided the difference between the number of terms in the model with a large number and that with the small numbers is not very great, and provided most of the dominant terms are retained in the model with the smaller number of terms (e.g., a 20 term model that may require as many as 25 sampling stations and a 10 term model that would require only about 12 sampling stations). Table 7.2 provides the correlation coefficient between predicted and observed wind speeds versus the number of terms in the polynomial model used to describe the wind field. The degree of improvement between the 13 term model and 23 term model is continuous, and the magnitude of improvement is 6 percentage points.

In some cases circumstances dictate the model chosen. For example, suppose a given urban region has determined that it can afford to set up

Table 7.2. Correlation Coefficient *R* Between Predicted and Observed Values of Wind Speed Versus Model Size

Number of Terms in Model	R
13	0.78
14	0.78
15	0.79
16	0.79
17	0.80
18	0.81
19	0.81
20	0.82
21	0.82
22	0.83
23	0.84

a sampling network with M stations. Furthermore, it has been established that the regression model with the smallest number of terms that will yield reasonably accurate distribution of all required parameters has s terms, and M is greater than s. Then it would be better to choose a model with s^* terms, where $s < s^* < M$, because more terms in the model may reduce the base error. The number of terms in the model should be maintained at a level below the number of sampling stations because errors different from the base error arise in estimating model coefficients. Better estimates of the coefficients can be had if there are more data points (sampling stations) than there are terms in the model. Models can be chosen where s^* equals M, if it can be shown that the net error is not significantly different from the base error, and if at some future time more sampling stations will be added, which will improve the estimate of the model coefficients.

The limitations of choosing a model under the condition $m < s$ (i.e., a model having fewer terms that m and a large base error) have been discussed. If an unsatisfactory model has been selected, improvement of the estimates of the model coefficients simply by adding more stations in the future is no solution to the problem, because cost of the additional stations will outweigh the improvement of the analysis of the parameter; that is, adding stations will not reduce the base error. Base error can be reduced only by adding terms in the model. Unfortunately, if terms are added when the number of sampling stations increases, an optimum sampling network obtained from the inadequate model may not be optimum for the improved model. However there may be significant improvement on the base error, which may override the optimality criteria.

Selection of the model form is clearly an important aspect in determining an optimum meteorological and air pollution sampling network. Equally important is choosing a design selection algorithm. This requires a set of candidate points for potential sampling networks (i.e., designs), and the statistical criterion for evaluating these designs. The optimality criteria for design selection are generally concerned with one, or both, of two sources of error: bias and variance. Since the model selection strategy was directed toward a model form with a small and/or acceptable bias, the design selection criterion should be aimed basically toward minimizing the variance source of error. Protection against large bias can be maintained by restricting the domain of interest to an area smaller than that established by the network used to obtain the initial data set. This is important because it shows the necessity of collecting data for the data set used to determine the optimum network over a large area. If the design area is not smaller than that covered by the initial data set, using the variance criterion to obtain optimality will result in unrealistic designs in which many of the design points are selected at the extremities of the allowable domain. The extent of bias can be evaluated further by comparing predicted data, where the predictions are made through the polynomial model and the coefficients are estimated using only data from points in the optimum network, with the remaining observed data coming from the initial data set.

Consider a fitted version of the polynomial model in eq. 2, which is given in the following form:

$$\hat{Z}(x, y) = \mathbf{x}'\mathbf{b}' \tag{3}$$

where $\mathbf{x}'$ is a vector of the form

$$\mathbf{x}' = (1, x, y, x^2, y^2, xy, x^3, y^3, x^2y, y^2x, \text{etc.})$$

and $\mathbf{b}'$ is a vector of the form

$$\mathbf{b}' = (b(0, 0), b(1, 0), b(0, 1), b(2, 0), b(0, 2), \text{etc.})$$

The vector b' is a least squares estimate of the corresponding parameter vector

$$\mathbf{B}' = (B(0, 0), B(1, 0), B(0, 1), B(2, 0), B(0, 2), \text{etc.})$$

assuming an underlying model for the initial data set that takes the general form

$$Z(x, y) = \mathbf{B}'\mathbf{x}' + e \tag{4}$$

where the deviation e from the surface $\mathbf{B}'\mathbf{x}$ is assumed to be uncorrelated and randomly distributed, and has a zero mean and a variance σ^2 that does not depend on any coordinate points (x, y).

Consider a design consisting of n points which is a subset of the total number of data points N in the initial data set (i.e., $N > n$). If the coordinates of the ith point in the subset containing n points are denoted x_i, y_i, there exists the following vector corresponding to each such point; that is,

$$\mathbf{x}_i' = (1, x_i, x_i^2, y_i^2, x_iy_i, x_i^3, y_i^3, x_i^2, y_i, x_iy_i^2, \text{etc.})$$

Then the following matrices are developed.

$$X_D' = \begin{bmatrix} \mathbf{x}_1 \\ \mathbf{x}_2' \\ \vdots \\ \mathbf{x}_n' \end{bmatrix} \quad \text{and} \quad \mathbf{Z}_D = \begin{bmatrix} Z_1 \\ Z_2 \\ \vdots \\ Z_n \end{bmatrix},$$

where the subscript D indicates that each matrix depends on the specific design made up of n points. The coefficients of the model [eq. 3] are determined through the following expression:

$$b_D = (X_D' X_D)^{-1} X_D' \mathbf{Z}_D \tag{5}$$

At any arbitrary point (x_0, y_0), the predicted value of any parameter (air pollution concentration, wind speed, etc.) based on the model in eq. 3, on estimates of the coefficients using eq. 5, and on the points in design D_1, is

$$\hat{Z}_D(x_0, y_0) = \mathbf{x}'_0 \mathbf{b}_D \tag{6}$$

and the variance $v_D(x_0, y_0)$ of $\hat{Z}_D(x_0, y_0)$ is given by

$$v_D(x_0, y_0) = \sigma^2 s_D(x_0, y_0) \tag{7}$$

where

$$s_D(x_0, y_0) = \mathbf{x}'_0(X'_D X_D)^{-1}\mathbf{x}_0 \tag{8}$$

Various techniques may be used for evaluating a design of n points using the function $s_D(x, y)$. Thus in two designs consisting of n points, D and D^*, the design D may be considered to be "better" than D^* by satisfying one of the following conditions: (*a*) $s_D(x_0 y_0) \leq s_D^*(x_0, y_0)$ for a particular point, where the point (x_0, y_0) is contained in the domain ρ, defined by the data set; (*b*) the maximum value $s_D(x_a, x_a) \leq$ the maximum value $s_D^*(x_b, y_b)$ and points (x_a, y_a), which is an element of the ρ-domain, need not be identical to the point (x_b, y_b) which is also an element of the ρ-domain; and (*c*) $\overline{s_D(x, y)} \leq \overline{s_D^*(x, y)}$, where the overbar indicates the average value of the function over the ρ-domain.

Of these, criterion c should yield best results because it is more dependent on a class of n-point designs (the class must contain at least two designs), on the domain, which may be either continuous or discrete, and on the model form.

If in certain areas of the urban region for which the data set has been provided, the constituent or parameter or both consistently tend to experience higher or lower values independent of specific cases in the data set, this information can also be used to obtain better fits for the surface in question. Unless one or more design points fall within such areas, it is unlikely that the fitted surface would behave in a manner serving to reproduce that condition. Therefore it is important to incorporate this additional information in the algorithm for selection of designs. Accordingly, the criterion (criterion c) for evaluating designs of size n may be adjusted in the following manner. Design D is better than design D^* if

$$\sum w(x, y)s_D(x, y) \leq \sum w(x, y)s_D^*(x, y) \tag{9}$$

where the summation is over the ρ-domain. The weights $w(x, y)$ attached to each point (x, y) are chosen to reflect patterns that are consistent over varying conditions, that is, in the region where the constituent or parameter consistently attains high or low values (compared to an average value). The weight would be large in that region relative to the weights in other regions, where the value is near the average value. Intuitively, the form of eq. 9 suggests that the best design, relative to other designs formed from the same set of N data points, will be found when $s_D(x, y)$ is small, where $w(x, y)$ is large in order that the summation be a minimum.

Therefore the criterion c_D, for selecting the optimum network consisting of n points selected from a large number of data points N, which will yield generally good results over most cases is

$$c_D = \sum_N w(x, y)s_D(x, y) \tag{10}$$

where a standard form for the weight $w(x, y)$ is

$$w(x, y) = \frac{1}{M} \sum_M \frac{(Z(x, y) - \bar{Z})^2}{ss}$$

and M is the number of cases in the data set,

$$ss = \sum_N (Z(x, y) - Z)^2 \qquad \text{and} \qquad \bar{Z} = \frac{1}{N} \sum_N Z(x, y)$$

The optimum network is based on finding the minimum value of c_D for an n-point design.

Most design optimality criteria depend on the inverse matrix $X'_D X_D$ as in eq. 8. If *all possible designs* were to be considered, therefore, a very large number of matrix inversions would be required, demanding an intolerable amount of computer time. For example, if the total number of points N in the initial data set were 65, and if the number of points required in the optimum design were between 10 and 20, approximately 65 years of computer time would be needed to determine the optimum 15 station network if it takes one microsecond to perform one matrix inversion. Hence it is not realistic to consider a design selection algorithm based on *all possible designs* when the initial data set consists of a large number of sampling sites (data points) that would be required in the case of an air pollution data set to determine all possible orders of variability on the upper microscale or lower mesoscale.

Two attractive design selection algorithms requiring realistic computer times are the exchange method and the backward elimination method. In the exchange method, a design of n points is chosen at random from the total available points N, where n is also the number of points required in the optimum design. The initial set of n points is then improved by first adding an $(n + 1)$ point, which must be a point in the ρ-domain and is chosen to ensure that the maximum possible decrease in the criterion c_D is achieved. The second step is to remove a point in the $(n + 1)$ design that would result in the minimum possible increase of c_D. Similarly, attempts to improve the original design may also be achieved by first removing a point and then adding a point. As a result of this procedure, the criterion for the resulting design c_D would be less than or equal to that for the initial design c_D^* (i.e., $c_D \leq c_D^*$). The algorithm would continue this procedure until further iteration showed no improvement (i.e., $c_D = c_D^*$ consistently). It should be pointed out that the basic algorithm can be and has been modified to permit the replacement of more than one point in the original design at each iteration, for a gain in flexibility.[4]

In the first step of the exchange method algorithm, exactly $(N - n)$ matrix inversions must be accomplished to determine which $(n + 1)$ design produces the maximum possible decrease in the criterion c_D, since n points are always held in common. For the second step, $(n + 1)$ matrix inversions are required to find the n-design that yields the minimum possible increase of c_D. In one iteration $(N + 1)$ matrix inversions are accomplished. Though numerous iterations are required before a solution is reached, the solution should be available in minutes on modern high speed computers.

One problem in using the exchange method for determining an optimum network is that there is no a priori way to predict the number of stations n required in the optimum network. We know only that the model adopted has s terms, that $n \geq s$, and that an optimum network is determined when we have satisfied the condition that increasing the number of stations by one or more does not significantly reduce the overall error in the analysis. Therefore it will be necessary to determine a number of best designs (in which the number of points in each design is different and $\geq s$), that is,

$$n_1 = s, n_2 = s + 1, n_3 = s + 2, \ldots, n_k = s + l$$

and to compute the average error variance over all cases in the data set for each design. These data can then be used to establish optimum network by comparing the results with the base error of the s-term model to see which design satisfied the condition that the addition of one or more stations does not significantly reduce the overall error. For example, if the base error of the s-term model is e_s over all N data points and if the overall error for a design containing n_l^* points is e_l^*, where

$$e_s \leq e_l^* \qquad \text{and} \qquad n_l^* \ll N$$

the difference

$$|e_l^* - e_s| \approx 0$$

Here it must be noted that the smallest possible value for e_l^* is e_s.

A more important problem that may arise using the exchange method alone to determining the optimum network is encountered if all the best n_k designs are different in terms of location of points. For example, a certain urban region may be able to afford initially to set up only 10 stations, yet the optimum network for a 10 term model requires 12 stations, the extra two to be added at a later date. The exchange method alone does not offer a way to determine which 10 stations in the 12 station optimum network should be constructed initially for best results. The best 10 station network selected by the exchange method might include stations whose locations depart significantly from those in the optimum 12 station network. One solution is to combine this technique with the *backward elimination* method.

The backward elimination method is similar to the second step in the exchange method; but instead of starting with n randomly selected points, it begins with all the data points. Consider N data points that are candidate sites. Let $D_N^{(N-1)}$ be a class of designs containing $(N - 1)$ points [i.e., $D_1^{(N-1)}, D_2^{(N-1)}, D_3^{(N-1)}, \ldots, D_N^{(N-1)}$ is a design of $N - 1$ points formed from the set of N points by omitting the kth point]. Let $c_N^{(N-1)}$ denote a set of values for the criterion in eq. 10 corresponding to the class of designs of $N - 1$ points, $D_N^{(N-1)}$; that is, $c_k^{(N-1)}$ is the value of the criterion for the design $D_k^{(N-1)}$. Then the minimum value of $c_k^{(N-1)}$ in the set $c_N^{(N-1)}$ is determined, and the kth point, which was omitted to produce the design $D_k^{(N-1)}$ that has the minimum value of the criterion, is then removed from further consideration as a design point. Therefore the set of candidate sites now consists of $(N - 1)$ points. Further iteration will produce the best design with $(N - 2)$ points, $(N - 3)$ points, ..., $(N - l)$ points. This process can continue until $(N - l)$ is equal in number to the number (s) of terms in the model selected to represent the distribution of the constituents and/or parameter in question (i.e., when $N - l = s$, the technique must terminate). The optimum network may be chosen here, as in the case of the exchange method, by comparing the base error of the model with s terms to the overall error produced by using each of the designs $n_1 = s$, $n_2 = s + 1$, $n_3 = s + 2, \ldots, n = s + l$, to establish a distribution of the constituents and/or parameters in question. The optimum network is generally designated as that design for which the difference between the base error of the model and the overall error due to the design is small. (As previously indicated, the base error is always less than or equal to the overall error due to that design).

The advantage of this technique is that each design containing $(n - k)$ points is a subset of the design with n points. Therefore if an urban region can set up only $(n - l)$ sampling stations, where n is the number of stations in the optimum network and $(n - l)$ is greater than s, the methods provide for the establishment of the best possible network with $(n - l)$ points. If the urban region intends to construct additional stations in the future, the method provides for the establishment of the best possible design with $(n - l + 1)$ points [which contain the same initial $(n - l)$ points plus one]), with $(n - l + 2)$ [which contain the same points in the design with $(n - l + 1)$

points plus one], and so on, until the optimum network with sampling stations is finally produced.

As far as the computational aspect in terms of computer time is concerned, the method is feasible. Using the example given when computing all possible designs, if N is 65 and n is 15, approximately 2000 matrix inversions will be required. This would take less than a minute for most modern, high speed computers.

This section has demonstrated that given the initial data set, two basic elements are required in the selection of an optimum meteorological and air pollution sampling network. First is the selection of a model form that will yield a reasonable representation of the given parameter. In terms of network selection and general modeling of a given field, polynomial models are most advantageous because of their great flexibility. The model form is constrained by the number of terms in the model that can be sustained economically by the urban region in the construction of a network, that is, the number of stations an urban region can afford now and in the future. This situation arises because a model with s terms will require a minimum of $s + 1$ stations to estimate the coefficients. Usually best "fits" from a polynomial model are obtained when there are more stations than terms. Tradeoffs and value judgments are feasible with the available and deducible statistical data and with the known economic constraints. However one factor that should overrride the economic constraints is associated with the intentions of the network. Since one of the intentions is to establish an early warning system to prevent air pollution concentrations from reaching or exceeding the set standards, perhaps producing adverse health effects to the population in the urban region, it behooves local and central governments to provide the best possible observation network in every urban region that has (or will have because of urban expansion) major air pollution problems.

The second element for an optimum network is the site selection algorithm, and two independent steps must be accomplished to establish this. First is the selection of an optimality criteria. Most commonly used is the error variance (eq. 7) or the error variance function (eq. 8). The site selection algorithm then chooses a set of n design points from a set of N candidate points ($N > n$) which will yield the distribution of the parameters with minimum error variance or error variance function based on the model selected to produce that distribution. Usually best results are obtained by minimizing the average value of the criteria over the domain of interest. However major improvements could be obtained if the information available in the data set were used in a complementary manner with the chosen criteria. This is an advantage if it can be shown that there exists in the average (which is taken over all parameters and all cases in the data set), areas in the urban region in which the parameters consistently attain high or low values. Such conditions may be produced through the interaction of the meteorological state and major sources or through accumulations of emissions from minor sources resulting from the meteorological conditions. Best fits of the selected model would be obtained only if one or more of the stations in the network were located in these regions. This condition can be satisfied if the minimization is accomplished for a weighted criterion where the weights are based on the data set.

The second step in the establishment of the site selection algorithm is to determine the method of choosing points in the optimum design which are constrained by the criteria. This should provide the optimum network of n points and also should designate the best design of $(n - 1)$ points, $(n - 2)$ points, $(n - 3)$ points, and so on, where each is a subset of the n-point optimum network; that is, all best designs of $(n - l)$ points are established using only the n points in the optimum network as candidate sites. This is done to provide a basis on which urban regions may construct the optimum network over a period of time, reducing the economic burden yet maintaining the standards required in the analysis of the parameters. Of course the optimum network and its subclasses should all be based on minimization of the chosen criteria.

7.3. SOME PROBLEMS AND SOLUTIONS

We have discussed basic statistical procedure that can be used to derive an optimum meteorological and air pollution sampling network, given a representative and statistically significant set of data for the air pollutants and relevant meteorological parameters. Essentially, a variety of statistical procedures exist for application to the data set made up of discrete points in space, to obtain an optimum network. Undoubtedly the character of the optimum network will vary depending on the criterion used and the method employed to select points; but the network will be optimum for the chosen procedure. Generally, good results should be obtained from an optimum network determined using most of the feasible statistical procedures, if the suggested changes to the procedures made in the previous sections are embodied into the technique (e.g., using weighted criteria).

One basic problem encountered is the high cost entailed in the establishment of the data set. Sweeney[2] suggested that a high resolution sampling network be set up in the urban region desiring the optimum network and that data be collected over periods of years. This is not only time-consuming but very costly. The Regional Air Pollution Study (RAPS) funded by the Environmental Protection Agency provided for the construction of a 25 station sampling network in St. Louis, Missouri, at a cost of millions of dollars. It is unreasonable to ask each urban region to support such a program, even if federal funding is available.

To ease the potential economic burden, Sweeney proposed dividing the selected urban region into parts. A high resolution network would then be set up in one of the subdivisions in which data would be collected over a number of years. After sufficient data have been obtained in the initial subdivisions, the sampling network would be moved to another subdivision, and data collection would be carried out until sufficient data had been obtained in that subdivision. The iterative process would continue until sufficient data had been collected in all subdivisions. This technique is extremely time-consuming, and the cost would probably be the same as in the former case.

One technique to circumvent the problem of the high cost of data acquisition is to simulate numerically the behavior of the air pollutants and the meteorological parameters for the specific urban region. Models based on the conservation equation (eq. 1) are available, and these may be used to simulate the pollution field. Hydrodynamic models are also required to simulate the wind field. The wind field is not only a parameter required for the data set, but it is also necessary to simulate the pollution field. Though this would call for a significant amount of time on a high speed computer with large core capacity, the cost of the computer time would be much less than the cost of a 5 year field program involving a network of 65 stations.

The accuracy of the simulated data need not be very good in the absolute sense; it is only necessary that the statistical nature of the air pollution distribution be preserved. That is, the order of variability should be reasonably well represented (i.e., the location of high and low concentrations of the pollutants should be reasonably well depicted); but neither the magnitude of the high and low concentrations nor the difference between the two needs to be delineated. Such errors would be due in part to poor source information. In the applied sense, actual observed concentrations will be employed in the statistical model, making the accuracy in the absolute sense dependent on the observations only if the model combined with the optimum network accurately shows the order of variability.

However a major dilemma arises when one determines an optimum sampling network based wholely or in part on air pollution data. It has been indicated that the number of stations in an optimum network based on air pollution and meteorological data will be more dependent on the air pollution data because there should be a higher order variability in the distribution of air pollution than in the meteorological parameters, and the network must account for all orders of variability. Most urban regions are constantly in a state of flux because new industries are moving into the region, while others are moving out. Very large shopping centers requiring vast parking lots are being constructed, and new roads, sometimes major thoroughfares, are built. In terms of air pollution, these conditions may be interpreted as follows: new industries mean new point sources, new parking lots mean new area sources, and new major roadways mean new line sources. Therefore major changes in the statistical nature of the air pollution distribution may occur over 5 to 10 years. Thus an optimum sampling network established using a present representative air pollution data set will shortly become suboptimum. The removal of a major point source (e.g., a major industry moving out of the urban region) is less serious than the addition of a major point, area, or line source, since the model, combined with the original network, can account for more variability that actually exists after the industry moves out; but the addition of new sources provides variability in pollution greater than the system was designed to handle. Moreover, if an urban region were to adopt Sweeney's approach to the acquisition of the data set, this would prevent his suggested technique from being practical. In the period between the collection of data in the first and last segments, there may occur major changes in the source configuration, making the segmented data sets totally independent of one another.

From both economic and practical considerations, the basis on which the optimum sampling network is established should be stable over a reasonably long period of time. The air pollution distribution does not offer this stability; but the conservation equation, which yields the variables necessary to completely specify the air pollution distribution, does offer the key to one possible solution to the problem. For a given urban region, the source parameters—source strength, source location, source type (line, point, etc.)—needed to solve the conservation are known or can be determined. Any changes in the source parameters due to urban expansion of one form or another also can and must be specified. The rate constant for the major reactions of the significant air pollutants are known or can and must be determined. Therefore the unknowns are the distribution of the meteorological parameters, that is, the wind field and the distribution of the diffusion coefficient. Since a reasonable estimate of the diffusion coefficients can be found through the wind observations, it will be necessary to establish only the wind field.

The order of variability of the mesoscale wind field should be relatively unaffected by urban expansion of any form in comparison to the potential effect on the air pollution distribution. Changes in surface roughness that occur when a formerly rural region is transformed into an urban residential, commercial, or industrial area, will be important, but these changes will not completely alter the nature of the flow. For example, winds from the north over a formerly rural area may have a slightly westerly component over the rougher urbanized area, but the major wind component will remain from the north; that is, a significant (say, 90°) rotation of the wind vector due to the increased roughness should not occur. Furthermore, urbanization may not change the roughness of the area at all, depending on the nature of the land before urbanization; for example, a tall forested region may be just as rough if not rougher, than a residential area.

Urbanization will increase the areal dimensions and the intensity of the urban heat island, which is likely to be the major function affecting the mesoscale flow in the urban region. However changes in the areal dimensions of the heat island occur over a period of 20 to 30 years in large urban regions whose growth has stabilized. This would not significantly alter the mesoscale wind field in terms of its order of variability if the location of the center of the heat island remained relatively the same. The expansion of the suburban residential area would tend to flatten the temperature gradient in the suburbs, which would reduce the intensity of the heat island in large urban regions and would produce changes in the intensity of the heat island circulation; but this reflects changes in the magnitude of the wind, which would be accounted for in the observed data, not in the order of variability. The establishment of another major heat island center in the urban regions would be a factor that could markedly alter the flow, but the creation of such a center with an effective areal extent to produce mesoscale changes takes many years. This may be completely avoided if it can be shown a priori that the condition will occur and may seriously affect the urban investment in the sampling network by accounting for the effect of the new heat island center in the derived wind statistics.

In smaller cities changes in the areal dimensions and intensity of the heat island can be significant, and it would be unrealistic to determine an optimum network for such an urban area based on the present meteorological conditions. It would be better to develop an optimum network based on the projected urbanization when the growth has stabilized. However smaller urban regions are unlikely to have the resources to finance an optimum network based on urban projections and may not yet need such a network because they do not have the severe air pollution problem of the larger urban areas.

This indicates that the mesoscale wind field in a large urban region should be relatively stable over a long period of time, an ideal condition in which to establish an optimum network. The development of an optimum network using the wind field as a basis is also economically attractive. It has been shown that the order of variability of the wind field will be less than that for the pollution field; thus the number of stations required for an optimum network based on the statistics of the wind field will be less than that required for the network based on air pollution distribution statistics. Furthermore, a field program designed to collect the data that will be used to establish the optimum network will be considerably less expensive when only wind observations are required compared with a program also calling for data on all the significant pollutants present. It would be less expensive to employ data based on simulations of the wind distribution using a computerized hydrodynamic model, noting that the simulations would have to describe only the statistical nature of the flow. In such a model it would be an easy step to account for important factors influencing the wind field which would result from urban expansion.

If the optimum network is based on the statistics of the wind distribution, it remains to be decided how the air pollution distribution would be determined. It is important to note that when the wind field has been established through the optimum network, all the parameters in the conservation equation (eq. 1) are specified for a given urban region. Given this information, Sasaki[5–8] has shown that the air pollution distribution may be obtained using the objective variational analysis model, which is somewhat similar to regression analysis except that a constraint equation is employed. This makes it possible to treat the parameter for which the analysis is desired in terms of behavior regardless of whether that behavior is characteristically dynamic or thermodynamic, for example, rather than in terms of a static observation as is done in regression analysis. In this manner, observations of pollutants that are generally localized can be integrated to obtain the spatial distribution of that pollutant. The constraint equation is used to determine the characteristic behavior, and with regard to the air pollution distribution, the constraint equation would be the conservation equation.

In principle, an objectives analysis model simultaneously minimizes the error variance between an analysis value of the parameter and the observed value of the parameter and between the existing value of the constraint equation and a reference value for the constraint equation over the ρ-domain. The error variance, over the area J, for the distribution of a particular pollutant may be expressed in the following manner:

$$J = \iint [\tilde{\alpha}(\psi - \tilde{\psi})^2 + \alpha_T(A - A_0)^2]dx\,dy \qquad (11)$$

where $\tilde{\psi}$ is the value of the analysis concentration, ψ is the value of the observed concentration, A is the existing value of the conservation equation, A_0 is the reference value of the conservation equation, and α and α_T are similar to Lagrangian weights. The value of A is given by eq. 1, that is,

$$A = -\mathbf{V}_H \cdot \nabla\tilde{\psi} + \nabla \cdot (K\nabla\tilde{\psi}) - \tilde{\psi} \sum k_n \psi_n^* + S$$

where $\mathbf{V}_H$ is the horizontal component of the wind vector. Since the near surface distribution of the particular air pollutant is required, the horizontal component of the wind is used because the vertical velocity is approximately zero near the surface. An example of the reference value of the consideration equation is the steady state solution (i.e., $A_0 = 0$). In order for the integral in eq. 11 to be minimum in the ρ-domain, the analysis concentration must satisfy an Euler–Lagrange equation within that domain; that is,

$$\alpha_T(K\nabla^2 A - \nabla \cdot (\mathbf{V}_H A) - A \sum k_n \psi^*) + \tilde{\alpha}(\psi - \tilde{\psi}) = 0$$

where the diffusion coefficient is held constant. Normally the Euler–Lagrange equation is constrained by conditions on the flux of the air pollutant at the boundary of the ρ-domain, or in the value of the concentration of the pollutant at the boundary; that is, the boundary condition is

$$v_n A \cdot \delta\psi = 0$$

where v_n is the wind velocity normal to the boundary and $\delta\psi$ is a variation of the pollutant concentrations. Solutions are generally obtained by relaxation techniques for which convergence can be accomplished in a relatively small number of iterations.

Therefore an optimum sampling network for the wind field would suffice to determine the distribution of a particular pollutant if the concentration of the particular pollutant were also observed at the stations in the network and if a reasonably accurate source inventory were available. The author is aware that the analysis technique depends on an accurate source inventory, which is difficult to attain. If an early warning system with regard to an air pollution "critical episode" is desired, however, or if any control measure that may be implemented is to be effective, such a source inventory must be compiled. These and the wind field derived from the optimum network are required by the objective variational analysis algorithm to determine the air pollution distribution. It is important to note that the resultant pollution analysis would yield a higher order variability for the pollutant than would otherwise be obtained using simple regression analysis of the pollution data obtained from the optimum network. This is because the sources are treated through the objective analysis algorithm as additional observations of the particular pollutant. Furthermore, the resultant analysis of the wind field and the air pollution concentrations may also be used as initial input data for the prediction of air pollution distribution over a period of time.

In summary, there is the basic economic problem of attaining an initial data set that can be used as a basis for determining the optimum network. It is suggested that a method to reduce the high cost of obtaining these data through a field program lasting over a period of years, was by simulating the distribution of the parameter using hydrodynamic and diffusion models. These data may be used as a basis for an optimum network, provided care is taken in the selection of model parameters to ensure that the statistical properties of the desired parameters are maintained. If the optimum network were based on the statistical properties of the air pollution distribution, which are subject to change over the years because of urban expansion, a major dilemma could arise because an optimum network based on the present initial data set might become suboptimum. The wind field can be used as a basis for the optimum network because it is more stable with respect to urban changes in large, more stabilized cities, and because it is the only parameter missing for the complete specification of the conservation equation for a given urban region. The air pollution distribution can be obtained by incorporating into an objective variational analysis model the wind field determined from the optimum network, air pollution observations made at the stations in the optimum network, and an emissions inventory for the region. The statistical properties of the air pollution distribution can be represented by this model because it treats the sources as additional observations of the pollutant. Furthermore, as the source configuration changes because of urban expansion, it is easy to change this configuration in the objective variational analysis model.

7.4. PRESENT APPLICATIONS OF THEORY

The author is presently engaged in research in which basic points outlined in Sections 7.1 to 7.3 have been employed to determine an optimum sampling network for St. Louis, Missouri. St. Louis was chosen because the RAPS program will provide sufficient corollary data to evaluate the optimum network. The development of the models for the research project was sponsored by the National Science Foundation. The second phase of the research project, in which the optimum network is being tested and evaluated, is jointly funded by the National Science Foundation and the U.S. Environmental Protection Agency.

A three-dimensional primitive equation model was developed to simulate the flow in urban regions. The model included the factors (essentially the urban heat island, surface roughness variability, and topography, including building heights) that would significantly change the synoptic flow. Solutions were obtained for 48 synoptic cases, which included variations in wind direction, wind speed, and thermal stability. The solutions were specific for St. Louis in that the topography and parameters needed to generate the urban heat island were obtained from data specific for that city. To obtain solutions for other urban areas, it is necessary simply to supply this model with the parameters for the urban region of interest.

Two statistical algorithms were developed to obtain the optimum network for the wind field. The first algorithm establishes the regression model that adequately characterizes the wind field in a given urban region and is based on minimizing the bias. The second uses the regression model and the wind data to drive the optimum network for the wind field. The criterion used to select the optimum network is the minimum weighted error variance. The optimum network was developed through the backward elimination technique. Again, it should be pointed out that these models are generalized, thus the optimum network obtained depends on the wind data sets that are solutions for specific urban regions. For St. Louis, a 13 term regression model was chosen and a 19 station optimum network was established.

The air pollution distribution is obtained using objective variational analysis models. The model simultaneously minimizes the error variance, by comparing the observed pollution concentrations with the derived pollution concentrations, and the area variance, by comparing the constraint equation (the continuity equation) with its steady state solution. For the variances to be minimum, it was shown that a second-order Lagrange equation must be satisfied under specific boundary conditions. The Euler–Lagrange equation is solved by relaxation techniques. The model required

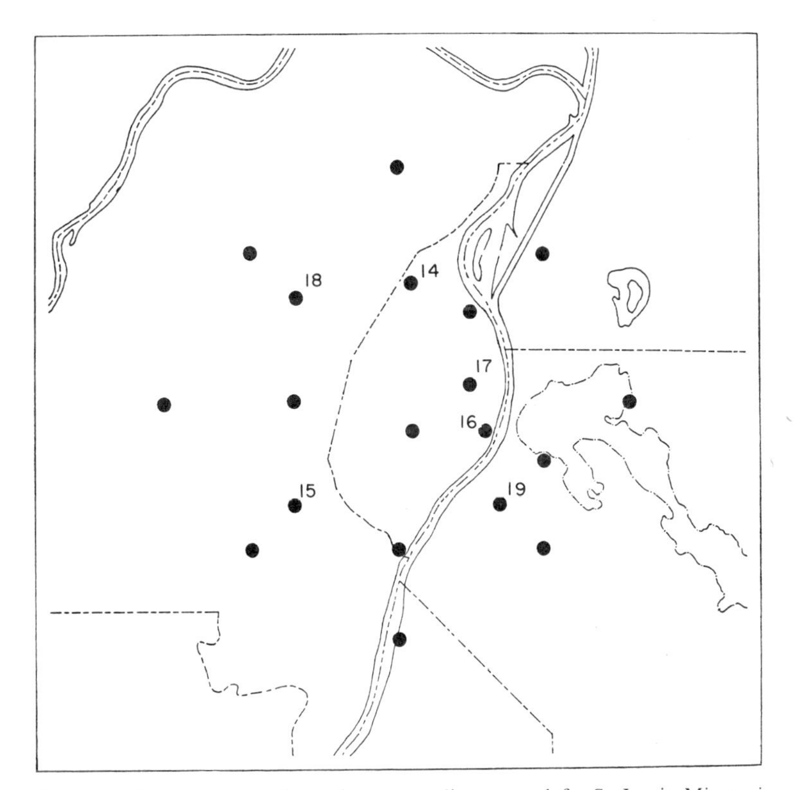

Figure 7.1. Station plot for the optimum sampling network for St. Louis, Missouri.

the following input parameters; the wind field derived from the optimum network, the observed pollution concentrations from the optimum network, and source emissions. Since this model, too, is generalized, these parameters can be specified for different urban regions.

Figure 7.1 is a plot of the stations determined for the optimum network in St. Louis. The optimum network contains 19 stations, but optimum subclasses were also determined. For instance, if the urban region could afford to set up only an 18 station network immediately, with the idea of adding the nineteenth station later, the subclass that would be created is that which includes all the stations in Figure 7.1 except the one marked "19." If 17 stations are required, delete numbers nineteen and eighteen; with 16 stations, delete nineteen, eighteen, and seventeen, and so on. Thirteen is the smallest subset that can be used because the regression model used for St. Louis has 13 terms.

We have begun the test and evaluation phase of this research project in which the optimum network is established in St. Louis and data are to be collected over a period of time. It was economically unfeasible to set up the 19 station network depicted in Figure 7.1; to reduce cost of the field program required to validate these results, many of the RAPS and St. Louis air pollution stations were used as parts of the optimum network when locations coincided or nearly coincided. A summer field program was held in July and August 1975, and a winter field program was held in February 1976. These data are now being analyzed.

7.5. CONCLUSIONS

The purpose of a sampling network of any kind is not merely to obtain data at the stations in the network, but to use the data to formulate a description of the entity over the entire region incorporated by the network. All too often a great deal of subjectivity has been employed in establishing an air pollution and meteorological sampling network in urban areas, and the consequences have been grave. Data from such networks often can be used only to give values of the concentration of a particular pollutant at a particular point in space (i.e., at the station where the observations were made). It would not be unusual for pollution levels to reach or exceed established standards in locations that lack stations. Since there is no way to detect this eventuality from a network that gives only localized observations, the problem will remain uncontrolled, resulting in adverse health effects to a segment of the population. Therefore it is important that there be procedures to combine the data from the sampling network, to yield a description of the air pollution distribution over the entire area. One basis for performing this task has been discussed. Although other techniques may exist or may be developed that are as good as or superior to the one described, the important point is the existence of an urgent need for urban regions to establish networks that can yield an analysis of the air pollution distribution over the entire urban region, not just specific points in that region.

ACKNOWLEDGMENTS

This chapter characterizes much of the research performed in the project entitled "Optimum Meteorological and Air Pollution Sampling Network Selection in Urban Areas: Phase I," which is being supported by the National Science Foundation—Research Applied to National Needs under grant GI-34345 and for which the author is the project leader. The major contributors to that project are Dr. Walter Bach, Jr., Andrew Clayton, J. W. Dunn 3rd, Dr. Kenneth Poole, and Bobby Crissman.

SYMBOLS

ψ	concentration of a pollutant (observed)
$\mathbf{V}$	three-dimensional wind vector
∇	vector differential operator
K	three-dimensional diffusion coefficient
k_n	rate constant for second-order gas phase reactions
S	source strength for a given pollutant
m	order of variability of a pollutant in one-dimensional space
p	number of sources
q	order of variability of the wind field in one dimension
N	number of sampling stations used to obtain the initial data set
N^*	number of variables (pollutants, etc.) observed for the initial data set
n	number of stations in the optimum network
Z	an observed parameter
x, y	horizontal coordinates in a Cartesian coordinate system
α, β	exponents of a polynomial series
B	coefficient of a polynomial series
P	potential of a polynomial series
R	correlation coefficient
M	number of stations in a given sampling network
s	number of terms in a polynomial model

ρ	area in an urban region covered by a sampling network most often referred to the network used to gather the data for the initial data set
$\hat{Z}$	estimate of the observed parameter Z
$\mathbf{b}'$	row vector of the least squares estimates of the coefficient B
$\mathbf{x}'$	row vector of the independent variables in the polynomial model
$\mathbf{B}'$	row vector of the coefficient B
e	the error induced in using a truncated series
σ^2	the variance of e
X_D	a matrix whose elements are the vector $\mathbf{x}'$ defined at precise points for a design D
X'_D	the transpose of X_D
$\mathbf{Z}_D$	column vector of the observed values of Z at the points in design D
v_D	variance of the model for design D
s_D	variance function of the model for design D
w	the weight
c_D	criterion for design D
l	an integer
e_s	base error for polynomial model with s terms using N data points to estimate the coefficients
e_l^*	overall error for polynomial model with s terms using n_l^* data points to estimate the coefficients
J	error variance between the analysis and observed parameter and between the actual and reference value of constraint equation
$\tilde{\psi}$	analysis value of concentration of air pollutant
A	existing value of constraint equation
A_0	reference value of the constraint equation
$\tilde{\alpha}, \alpha_T$	Lagrangian weights
$\mathbf{V}_H$	horizontal component of the wind vector
v_n	wind velocity normal to boundary of ρ-domain
$\delta\psi$	variation of the pollution concentration

REFERENCES

1. Hotchkiss, R. S. and F. H. Harlow, University of California Report 47, 1973.
2. Sweeney, H., Final report, NAPCA, Contract CPA 22-69-57, 1969.
3. Turner, B., *J. Appl. Meteorol.*, 3, 83 (1964).
4. Mitchell, T. J., *Technometrics*, 16, 203 (1974).
5. Sasaki, Y. *J. Meteorol. Soc. Japan*, 36, 77 (1958).
6. Sasaki, Y. *Mon. Weather Rev.*, 98, 875 (1970).
7. Sasaki, Y. *Mon. Weather Rev.*, 98, 884 (1970).
8. Sasaki, Y., *Mon. Weather Rev.*, 98, 899 (1970).

8

AUTOMATIC MONITORING OF AIR POLLUTION

J. J. Wilting

N.V. Philips Gloelampenfabrieken,
Scientific and Industrial Systems Department,
Eindhoven, The Netherlands

8.1. INTRODUCTION

Air pollutants are emitted by a range of sources in different locations and at different heights. The sources vary in strength and may have an unpredictable diurnal (or other time range) pattern of emissions. The relative contributions of the sources to the concentrations at a receptor location are subject to a number of factors, and some of them are not quantifiable. In densely populated areas with intense industrial and traffic activity, the air pollution abatement strategy requires a close knowledge of the pattern of emission sources coupled with the dynamics of air movement. Such a strategy of air quality management policy has to be supported by an air pollution monitoring system.

In many countries, although not in the United States, abatement strategies are based on the "best practicable means" for controlling the sources of pollution, which must allow for economic feasibility as well as available technology, and the interests of public health and amenity. This results in a balance of technological, scientific, and political factors that in practice requires governmental decisions, which, ideally, are backed by information and scientific data obtained by a well-trained staff supported by sophisticated equipment.

To determine the configuration of a monitoring system, the primary objectives of the system should be identified. These are as follows:

1. From an air quality inventory, information about air pollutant emissions in the area under consideration must be detailed.
2. The relative contributions of the individual sources must be determined. This will not only identify the major contributors, but will help in planning control measures aimed at achieving maximum effectiveness at minimum cost.
3. Probable future trends in air pollution levels must be determined. These trends can result from new industries, extensions to present industries, and residential or commercial developments, as well as the implementation of air pollution control measures.
4. The effectiveness of the control measures must be checked.
5. Buildup of air pollutants must be detected at an early stage in areas with high emission rates, occurring with stagnation of air movements. Peak levels can be avoided by a knowledge of the dispersion patterns and by staggering the processes producing the emissions.
6. The system must act as a warning system, which immediately detects air pollutant levels exceeding set values.

The detection of individual offenders requires a dense monitoring network. Ascertaining trends in air pollutant levels calls for a regular pattern of monitors. The determination of their number and position can be aided by a simulation model. If monitoring is to be used for measuring increases in levels of air pollutants or check on the effectiveness of abatement procedures, the monitoring must be sufficiently precise to provide confidence in the data, and it must be coupled with adequate data processing facilities.

Onset of air pollution episodes can be recognized from real-time monitoring with quasi-continuous measurements and immediate evaluation of results. This requires automated measuring coupled with data processing, the latter being supervised at an observation and control center. This can result in control strategy involving cooperation between the monitoring and control authority and industry in the region.

A further benefit of an effective monitoring network is to have reliable data available in an easily handled form. The data should be presented, and perhaps displayed, to disseminate the information in the present region and to neighboring regions that may also be affected.

8.1.1. Air Quality Standards and Legislation

Concentrations of air pollutants that have been accepted, after scientific investigation, as harmless to human health and to the environment, are called "air quality standards." They are incorporated in the air pollution control legislation in a number of countries, notably the United States, Canada, and the German Federal Republic (Appendix A8.2) where they constitute legal standards that are used by the governmental control authorities. In a number of other countries, including the Netherlands, air quality standards are used for air quality management by regional authorities although they do not have the force of law. Rather, they are the basis of the emission standards adapted to local conditions in the control of planning of these authorities.

In a third group of countries, with present ambient pollution levels well below those given in the air quality standards, emission standards common to the whole country can be prescribed.

An air quality management program where legal or implied air quality standards are in use requires that present and planned emissions from major sources be integrated with ambient concentrations. This means that the organizations and industrial managements controlling the sources—large industrial complexes and power generation utilities—must cooperate with the control authority in matters of present operation and future planning.

In some industries, particularly where traditional technology is used, this can cause economic hardship as the new, and more stringent, emission standards are imposed, although the harsh effects may be mitigated by a staged introduction. However, in time, older works will have to meet the same standards applied to their more modern counterparts.

The "best practicable means" become more stringent as control technology advances and as our knowledge of the effects of air pollutants at lower concentrations in the environment improves. Air pollution controls are mostly a direct charge on production, far outweighing any returns from product recovery. Thus application of controls has an effect on the cost of goods and processed materials not only nationally but internationally. Countries with less stringent controls may be able to produce goods and semiprocessed materials more cheaply than those with higher control standards. If air quality and emission control standards differ internationally, lower standards, which reduce the quality of life and of the environment, may be used competitively in lowering the cost structure.

8.1.2. Air Quality Measurements and Modeling

Correlation of ambient air pollutant concentration measurements with source emissions forms a useful foundation for developing a model on which to base a control strategy, increasing the usefulness of the measurements, as well. A number of such models have been studied experimentally and theoretically in recent years.[1,2] The efficiency of air quality management can be further enhanced by simulating the dispersion of air pollutants by a mathematical or physical model. If the observed concentrations can be attributed to particular sources, abatement action will be facilitated. However present mathematical models for distributions and dispersion of pollutants from single and multiple sources are suitable only for calculating long-term averages and frequency distributions.

Another aspect of air pollution management strategy is the prediction of pollution levels[3] as a function of meteorological conditions. When *short-term* prediction of pollutant is attempted, however, models should be based on statistical procedures that will attain greater precision.

Such a statistical method uses basically an experimental probability that certain observed trends will lead to undesirable or unacceptable concentration levels. These trends may be in one of the following factors or a combination of them: wind speed, direction, turbulence, inversion patterns, and source emissions. Indeed, the most useful indications are observations of significant and simultaneous changes in the observed values of the ambient concentrations themselves, provided the source emissions are either statistically constant or follow a known pattern of change, or their influence can be some distinguished.

In such a model, local circumstances dictate the choice of component (e.g., sulfur dioxide, oxidants, or airborne particulates), to be used as a trend-indicating tracer.

The application of any model to a region necessitates the detailed study of that region. One complicating factor is geographical location; coastal regions have frequent (diurnal) changes in wind direction from sea to land, and vice versa, while some urban areas have intricate air movements resulting from the configuration of hills and valleys. It is essential in all these cases to test the designed dispersion model with observations and measurements.

Since errors in assessing the emission sources and meteorological parameters also restrict the validity of a model, these characteristics should not be used permanently without ambient concentration measurements to confirm the model predictions. Implementing a costly abatement strategy on such a model cannot be justified unless the model is supported by ambient concentration measurements of sufficient precision.

Dispersion models tend to be least reliable under the extreme atmospheric conditions that produce high ambient pollution levels. These occur at low, random (stochastic) frequencies, and are difficult to model. Credibility can be achieved with continuous monitoring.

Long-term trends (i.e., over a period of years) in air quality are a consequence of changes in local or regional activities, such as traffic patterns and densities, and industrial developments or control measures. Following these trends may be part of the objective in monitoring, and it is necessary to eliminate meteorological factors, the most difficult being the low frequency stochastic influences. Such trends can be found either by using a regression model[4] or, for example, a statistical model based on a log-normal distribution of daily averages under comparable weather conditions. The regression model has not been satisfactory for air pollution trend computation, but the statistical model can readily cope with the numerous factors and effects that influence emissions and their dispersion. But like all statistical calculations involving a number of parameters, a large data population is needed, and this in turn means a large number of measurements.

Appropriate siting of monitors enables typical urban and rural air pollution characteristics to be distinguished. Monitoring different component gases and particulates can pinpoint the influence of specific sources. Positioning monitoring sites at different heights can help in determining variations in concentrations caused by meteorological factors.

8.1.3. Air Pollution Control Strategy Based on Monitoring

When a large amount of air pollution data is available for a region or country —as, for example, in the Netherlands—it is possible to use this information for an inventory of air quality over the critical region. Data for this purpose are given as averages over equal time intervals, namely, one hour or one day. These values can be used for developing control strategies. The daily averages, for most sites, can be approximated by a log-normal frequency distribution when taken over a year.[5,6] Since statistically there is a finite possibility that values up to the actual stack emission concentration can occur, it is not practical to set a value that must never be exceeded; rather, one sets a lower standard that must not be exceeded more than a fraction of time in a given period, say 1 year. This has lead to proposed air quality standards for sulfur dioxide in the Netherlands according to which SO_2 concentrations should not exceed 250 $\mu g\ m^{-3}$ (24 hr average) for more than

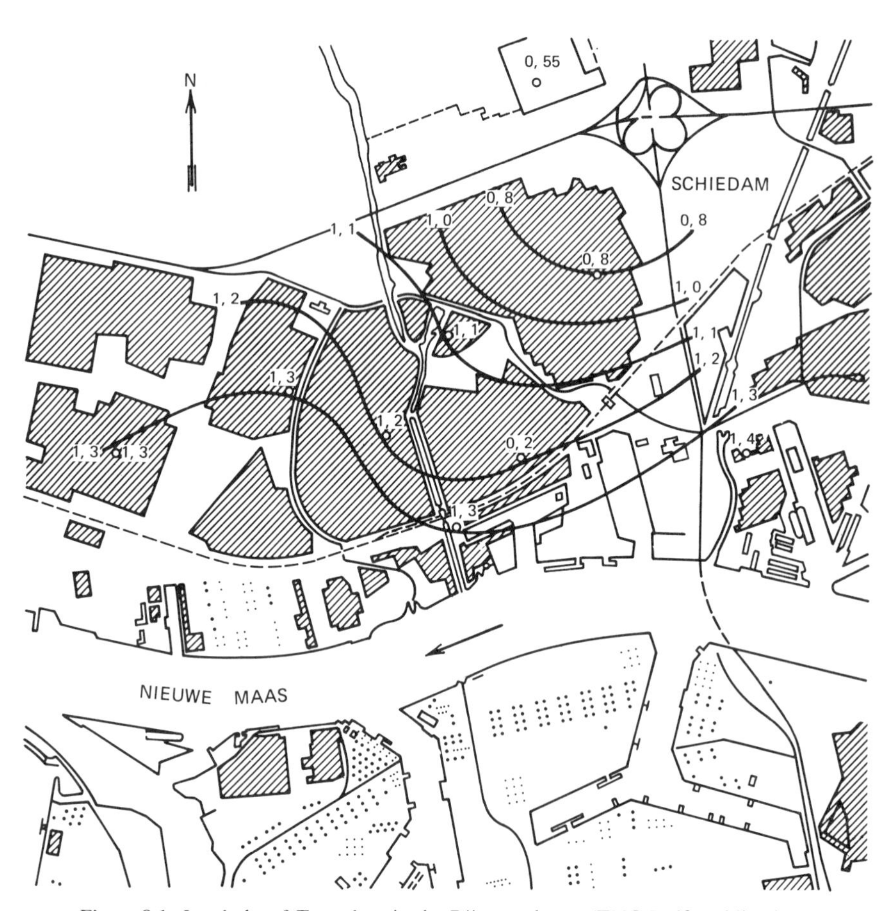

Figure 8.1. Isopleths of T_{98} values in the Rijnmond area (TNO Delft publication).

2% of each year (7 days) (Appendix A8.2). Other countries have regulations of a similar type (see appendices).

To evaluate air quality, test indices have been introduced,[6] one for the 50th percentile and one for the 98th percentile concentrations of the cumulative frequency distribution of each measuring site.

These test indices are defined by the ratio of the observed 50th and 98th percentile concentrations and the proposed corresponding standards. Hence

$$T_{50} = \frac{\text{observed } 50\% \text{ concentration}}{\text{standard } 50\% \text{ concentration}}$$

$$T_{98} = \frac{\text{observed } 98\% \text{ concentration}}{\text{standard } 98\% \text{ concentration}}$$

In practice, preference could be given to using a T_{95} index rather than T_{98}, since occasional equipment failures would give less interference. The test indices can be considered to be indicative of the amount by which emissions in a region would have to be reduced to enable standards to be met. Moreover this method provides an interesting and simple means of presenting air quality, since computed test indices for all the measuring sites can be mapped as isopleths for the region or even a whole country.[7]

For the area near Rotterdam (Figure 8.1), isopleths of the T_{98} indices for SO_2 can be drawn, based on the temporary Netherlands standards of 350 $\mu g\ m^{-3}$ SO_2 for the 98th percentile and 150 $\mu g\ m^{-3}$ for the 50th percentile (see Appendix A8.2). When cumulative frequency distributions of observed daily averages of SO_2 concentrations at a particular site are plotted on a graph with log-normal coordinates is produced (Figure 8.2). In the same diagram a standard cumulative frequency distribution is drawn (broken line) based on 250 $\mu g\ m^{-3}$ for the 98th percentile and 75 $\mu g\ m^{-3}$ for the 50th percentile (see Appendix A8.2).

The use of T indices can also be illustrated for this site by

$$T_{98} = \frac{278}{250} \div 1.1$$

$$T_{50} = \frac{44}{75} \div 0.6$$

Because T_{50} is below and T_{98} is above the air quality standard, the perturbation of high and low concentrations should be flattened at the site for which the data are shown; this is achieved by reducing peak values either in number or in intensity. High values are caused by weather conditions (stagnation periods) or by polluting industrial activities, which can occur intermittently. An investigation would reveal which strategy is best fitted to reduce the T_{98} values to the standard level.

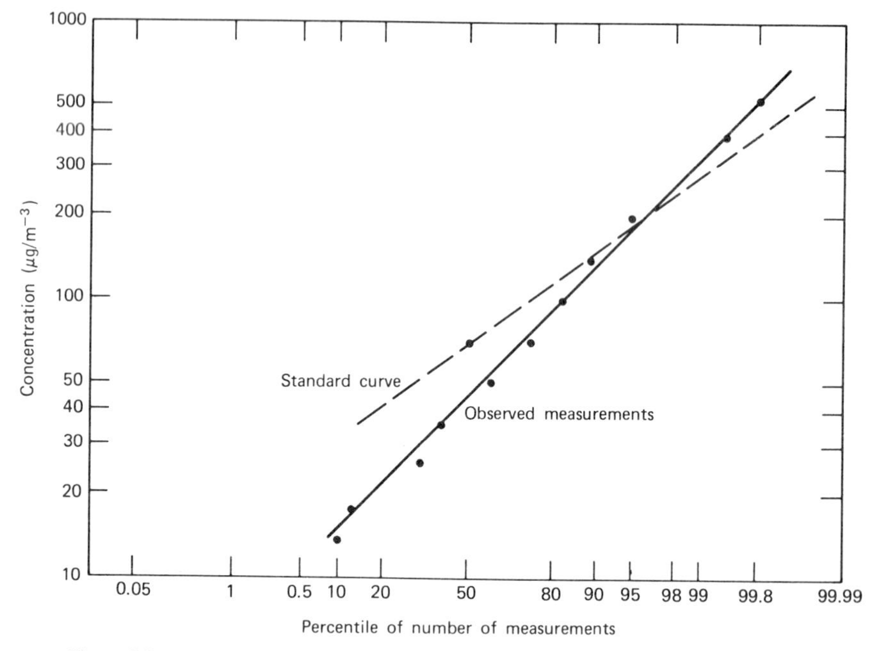

Figure 8.2. Example of cumulative frequency distribution of daily averages of SO_2.

When there is insufficient dispersion of air pollutants with adverse meteorological conditions, one control strategy, which is used by the Rijnmond Authority, is to reduce emissions from major sources. The conditions are deduced from the buildup of SO_2 concentrations monitored in the industrial and residential areas of Rotterdam, Vlaardingen, and Schiedam (Section 8.2.2.4).

8.1.4. International Monitoring

Transport of air pollutants at different altitudes depends on source height and turbulence. At levels close to the ground, air pollutants are absorbed and adsorbed by the soil, vegetation, and other surfaces. The air pollutants that are transported over long distances tend to come from the tall stacks of industrial plants and power generation facilities.[8,9] Although tall stacks solve local air pollution problems, they do not eliminate undesirable conditions regionally, nationally, or internationally.

Long-range transport simulation and monitoring on an international scale is being used to study the contribution of distant as well as local sources to observed concentrations. This is being carried out by the World Meteorological Organization (WMO), the North Atlantic Treaty Organization Committee on the Challenges of Modern Society (NATO–CCMS), and the Advisory Group for Aerospace Research and Development (AGARD).

Italy, Germany, and the Netherlands are three European countries with extensive national networks, and Belgium has one under construction. France, Switzerland, Austria, Sweden, and Denmark have smaller networks with local orientation. In the Netherlands ground level monitoring has been supplemented by higher altitude studies using a scanning correlation spectrometer mounted in a light airplane. This enables the tracking of the long-range transport of air pollutants from Germany and Belgium toward the Netherlands. The Swedish and Norwegian Institutes for Air Research are conducting similar studies[8,9] for northern Europe, as is the federal Environmental Protection Agency in the United States.[12]

Consultation and cooperative research is being undertaken under the auspices of a number of international bodies; the World Health Organization (WHO), the Food and Agriculture Organization (FAO), the United Nations Environmental Programme (UNEP), the Organization for Economic Cooperation and Development (OECD), and the European Economic Community (EEC), as well as NATO–CCMS and WMO.[13,14] This has led to discussion of standardization in air quality measurement by a committee of the International Organization for Standardization (IOS).[15]

WHO has a program for air pollution monitoring networks implemented through international reference centers in London and Washington. There are also three regional reference centers (Moscow, Nagpur, and Tokyo), seven national reference centers, and some 30 laboratories, including the Pan-American Sampling Network.[16]

A difficult problem in establishing ambient air quality standards involves the international competitiveness of goods and semifinished materials produced by inherently polluting processes. International agreement on acceptable levels is a necessary basis for successful abatement policies. This must be coupled with monitoring in the different countries to verify compliance with international agreement. Although such monitoring, nationally or internationally, does not necessarily imply central data acquisition and storage, it does require exchange of data and reporting to an international or national organization.

Global monitoring of gases and particulates that may be air pollutants if present in sufficient concentration is being started for ecological, meteorological, health, and other scientific reasons. This will establish sources and sinks, background levels as well as global trends, and "half-lives" of the materials. This effort is being carried out as part of the United Nations special agency activities such as UNEP, WHO (Earth Watch),[17] and the WMO base line stations in the Global Environmental Monitoring System (GEMS).

8.2. DATA HANDLING IN MONITORING NETWORKS

8.2.1. Selection of Components and Sites

As indicated in the preceding section, the configuration of a pollution monitoring system is a function of its objectives. The precision and time resolution of concentration measurements jointly govern the sample density and extent of a monitoring network.

Sampler locations are generally better determined by practical experience than by dispersion models, which have only limited precision until a large amount of data is available to validate the model. In practice, distances between measuring sites may be from 0.5 to 20 km, depending on source activity in the area. Monitors should not be located less than 5 times the stack height from a source, and the components to be measured at each site are a function of the pollutants expected. Thus combustion processes produce SO_2, NO, NO_2, fly ash, and soot, and specific pollutants such as H_2S, fluorides, and heavy metals may result from certain industries. Transportation produces mainly NO, CO, and unburned hydrocarbons. These can react with ozone to produce NO_2 and complex photochemical oxidants. Domestic heating with coal produces SO_2 and soot, although relatively few oxides of nitrogen (because of the low flame temperatures), and relatively little CO is given off.

In residential areas pollution from domestic heating is reduced by changing the fuel from coal and oil to gas.[18] Pollution monitoring from these sources can be implemented by measuring SO_2 only.

In built-up areas transportation is the most significant, since pollutants are emitted at ground level. Concentrations can be lowered by reducing emissions, through improved automotive engines and the regulation of traffic densities. The latter measures can be coupled to meteorological conditions (prediction of expected pollution levels) or related to the crossing of threshold levels. Controlling traffic lights for altering traffic flow can be used to modify traffic densities.

Monitoring sites where CO, NO/NO_2, ozone, and hydrocarbons are measured, must be located in streets with high traffic densities, but unfortunately the possibility of damage to the samplers by vandals prevents the placing of samplers at optimum heights.

Industry can be induced (or legally required) to monitor air quality in the areas directly affected, to enable it to control emissions sufficiently that air pollution regulations are not violated. Curtailing some emissions during adverse meteorological conditions conducive to high concentrations of air pollutants often provides an economic alternative to the installation of expensive control equipment such as scrubbers. This requires a monitoring system with real-time predictive capability.

Real-time prediction of meteorological parameters either by dispersion models or by statistical models based on meteorological classification is too inaccurate and often too slow. A better approach has been the processing of actual measured pollutant concentrations.[3] This implies that most of the (costly) meteorological towers are replaced by pollution monitoring devices, but relatively high precision in the measurements is also necessary.

An example of how this is related to the number of monitors and the averaging (time) interval appears in Figure 8.3, obtained from selected SO_2 measurements for unstable weather conditions by Stratmann.[19,20] The area of measurement is approximately homogeneous; thus statistical results are

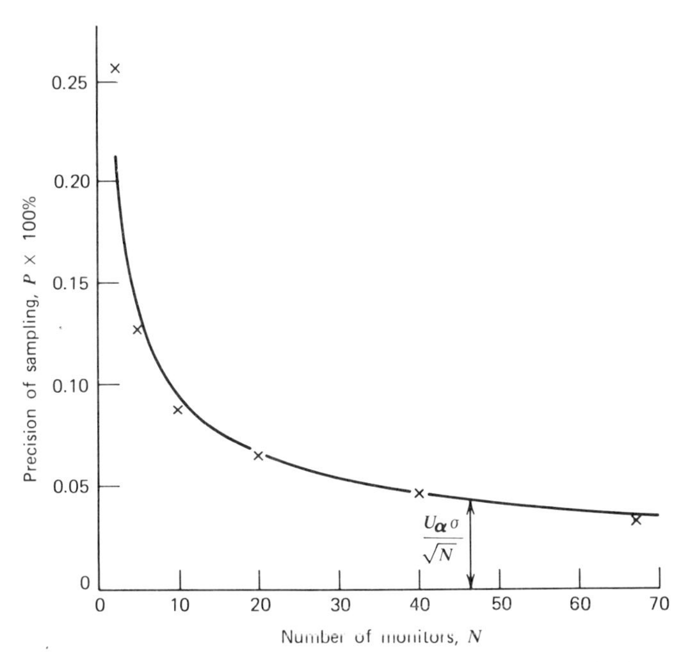

Figure 8.3. Precision of sampling P as a function of the number of monitoring stations N; two-sided interval for 90% confidence level. Crosses indicate experimental result using Student t-distribution, $ts/\sqrt{N}$. Curve fit using normal distribution, $U_\alpha \sigma/\sqrt{N}$, with $\sigma = 0.18$, $U_\alpha = 1.65$ for 90% confidence level.

not influenced systematically by wind direction, and all measurements belong to one population. The sampling interval is 10 min. From the graph it can be deduced that for this area a precision of 10% around the expected average concentration for a 90% confidence level would require a minimum of eight monitoring stations for a standard deviation $\sigma = 0.18$.

Normally an area is divided into different sectors representing different wind directions. If the area in a 45° sector is considered to be statistically homogeneous for unstable weather conditions, the same eight monitors would be needed for this sector, assuming similar conditions to those described previously.

Taking a larger sampling interval, say one hour, and assuming the autocorrelation between 10 min averages to be small at unstable weather conditions, the number of monitors required would decrease by a factor of 6; thus two monitors would be deduced as a conservative number.

A study of the measurements, preferably with the short time intervals under unstable weather conditions, will provide information about the homogeneity of the area. On the other hand, the results obtained can be made homogeneous by standardizing the distribution of sample per site, taking only the data belonging to unstable weather conditions, and approaching the information with a log-normal distribution. Hence for the concentrations at one site, for one sector of wind direction, we write

$$U = \frac{\log C_i - \log C_{mi}}{\sigma_i} \tag{1}$$

where C_i = measured concentration at site i and given wind direction, averaged over the sampling interval
C_{mi} = median of a sufficiently large number of C_i values
σ_i = standard deviation of $\log C_i/C_{mi}$ distribution

Here U is normally distributed around an average of zero and $|U| < 1.96$ with 95% probability, in good approximation independent of the site. Therefore, each hour, the average of U over N sites will be close to zero, within the precision interval $\pm 1.96/\sqrt{N}$.

A simplified normalizing procedure can be derived from eq. 1:

$$U\sigma_i = \log \frac{C_i}{C_{mi}} \tag{2}$$

or

$$\frac{C_i}{C_{mi}} = \exp(U\sigma_i) \tag{3}$$

$$= 1 + U\sigma_i + \frac{(U\sigma_i)^2}{2} + \cdots \tag{4}$$

Generally $U\sigma_i$ will be small enough compared with unity to allow for the approximation using the first two terms in the series:

$$\frac{C_i}{C_{mi}} = 1 + U\sigma_i \tag{5}$$

or

$$U\sigma_i = \frac{C_i}{C_{mi}} - 1 \tag{6}$$

Instead of using the median C_{mi}, the arithmetic mean C_{0i}, which is slightly larger, can be chosen. This compensates in a way for the omission of the higher order terms. Hence the reduced relative value becomes, as σ_i—to have only limited dependence on the site and can be replaced by σ:

$$R = \frac{C_i}{C_{0i}} - 1 \tag{7}$$

Here geographical variations due to topography are largely eliminated.

The average of all R over N monitors ought to be expected in a not quite symmetrical interval of $\pm 1.96\sigma/\sqrt{N}$ on a 95% probability. Clarenburg[3] worked with this value and found, as a first estimate, that the interval in the Rijnmond area with eight monitors in the downwind position was ± 0.32.

The limits of detection of the instruments and their inherent inaccuracies at low concentrations of pollutants in ambient air will require accurate specification of instruments related to their application. For low values of C_{0i}, say in a particular wind direction and for some sites, the inaccuracy in the ratio C_i/C_{0i} could make it useless. Adequate siting and high quality instrumentation should reduce the occurrence of inaccuracies in values of C_{0i}.

With stable weather conditions the observed values behave differently and the relations above cannot be used. Here three source categories, suited to local frequencies and inversion heights, should be distinguished. A typical division allowing for local inversion conditions and their frequency, is less than 40 m, 40 to 120 m, and more than 120 m height. Low level sources could give rise to concern during stable weather conditions, whereas the emissions from high level sources will remain aloft while diffusing over long distances. Also nonrandom switching of source activity gives rise to different distributions and may cause multipeak distributions.

Finding values of U or R that do not belong to the distribution defined previously for unstable weather conditions, within the limits of significance, indicates deviating conditions. With appropriate processing of the observed values, development of stable weather conditions may be detected at an early stage (Section 8.1.2). Real-time monitoring used for tactical operations based on information from several measuring instruments demands automated data handling equipment for ergonomic reasons.

Monitoring devices that have been developed include telemetry equipment for the immediate transmission of signals to an observation center, using either leased private lines, radio links, or connections using the public switched telephone network.

Automated monitoring is also beneficial because it excludes interference by human manipulation, enhancing accuracy and reliability. The far greater flexibility, particularly with regard to data processing, is a further incentive for automation.

Several studies[10,11] have shown that automated monitoring is generally more economic than comparable manually operated systems. This is especially true for large-scale networks using high quality monitors needing little maintenance. Monitors now available commercially have remotely controlled calibration, requiring no attendance or maintenance for periods of 3 months or more. Figure 8.4 presents such a monitor working on the principle of coulometry.

8.2.2. Transmission

Different media are available for data transmission.

Leased Telephone Lines. This mode enables continuous transmission of analogue or digital signals with high reliability. Investment costs are low, but rentals are high, although they vary from country to country.

Radio Links with Mobile Phone Systems. Radio links work in the 160 and 460 MHz bands, and the carrier replaces the leased telephone line. Generally continuous transmission of analogue and digital signals is feasible, but reliability is slightly less than with fixed lines. Investment costs are considerably higher than with leased lines, but in the long run the system tends to be more economical. Radio links are flexible with regard to site location; but their availability depends on the policy of the telecommunication authority of the country.

Public Switched Telephone Network. This system can be used with the aid of automatic dialed link systems. It permits computer-controlled calls to individual monitoring stations. This is the most economical method, especially for long distance transmission, provided batch mode transmission of digital data blocks at relatively long time intervals is acceptable. For different reasons, this medium is the least suitable for the transmission of analogue signals. Automatic dialed line equipment requires some initial investment, but subscription rates are low and tariff charges for calls can be kept at modest levels.

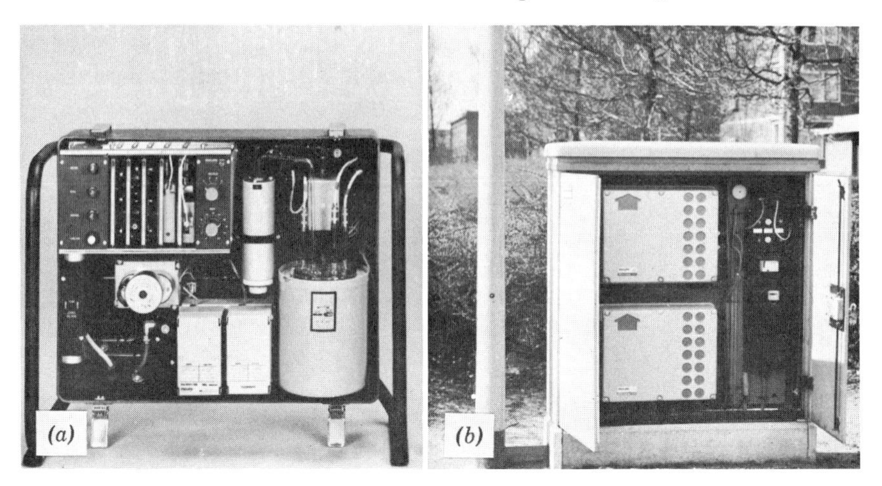

Figure 8.4. (*a*) Internal view of automated SO_2 monitor. (*b*) Open cabinet of outdoor SO_2 monitoring station.

8.2.2.1. *Analogue Transmission*

Multichannel analogue transmission of slowly varying signals can be accomplished with a so-called frequency division multiplexing (FDM) system.[21] This enables a number of analogue signals to be transmitted simultaneously through one telephone line. Each analogue signal has a frequency channel bandwidth of 120 Hz out of the 3000 Hz nominal bandwidth of a telephone channel. Hence 25 analogue channels can be provided simultaneously. The analogue signal amplitude is used to control the frequency of a square wave, constrained to remain within the limits of 5 (zero input) and 25 Hz (maximum input). This square wave is in turn the modulating signal applied to a carrier signal at the central frequency of the 120 Hz channel (Figure 8.5*a*). Figures 8.5*b* and 8.5*c* illustrate how selectivity of the channels is obtained. A block diagram of the complete transmission system is given in Figure 8.6.

To match the analogue signal transmitted to the restricted bandwidth per channel to prevent the loss of accuracy, the measuring equipment should have time constants of the order of 100 msec or more.

8.2.2.2. *Digital Transmission*

In contrast to analogue signals, digital signals, being discrete values represented by a binary code, can best be combined for transmission using time

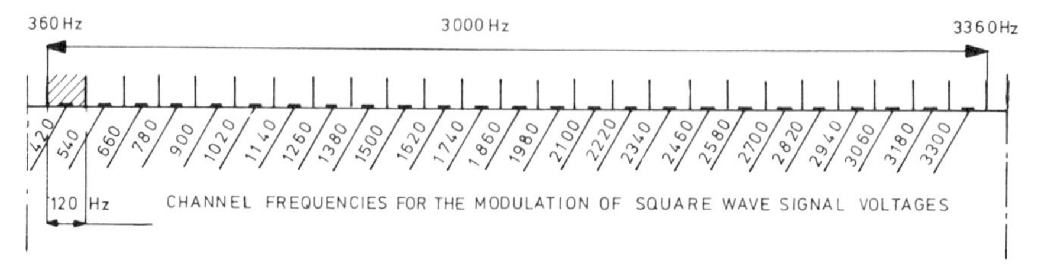
FREQUENCY AND SELECTIVITY OF CHANNELS IN ANALOG TRANSMISSION
360 Hz
3000 Hz
3360 Hz
420
540
660
780
900
1020
1140
1260
1380
1500
1620
1740
1860
1980
2100
2220
2340
2460
2580
2700
2820
2940
3060
3180
3300
120 Hz
CHANNEL FREQUENCIES FOR THE MODULATION OF SQUARE WAVE SIGNAL VOLTAGES

(a)

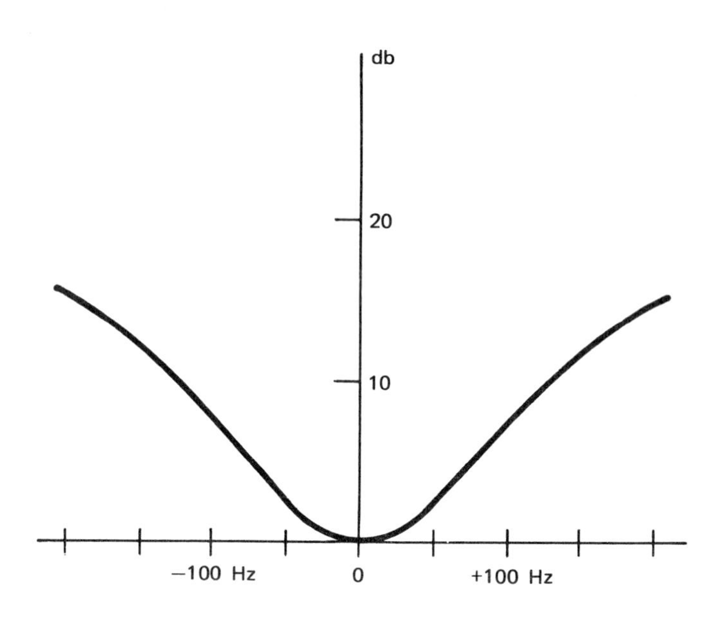
db
20
10
−100 Hz
0
+100 Hz
CARRIER FREQUENCY BAND WIDTH

(b)

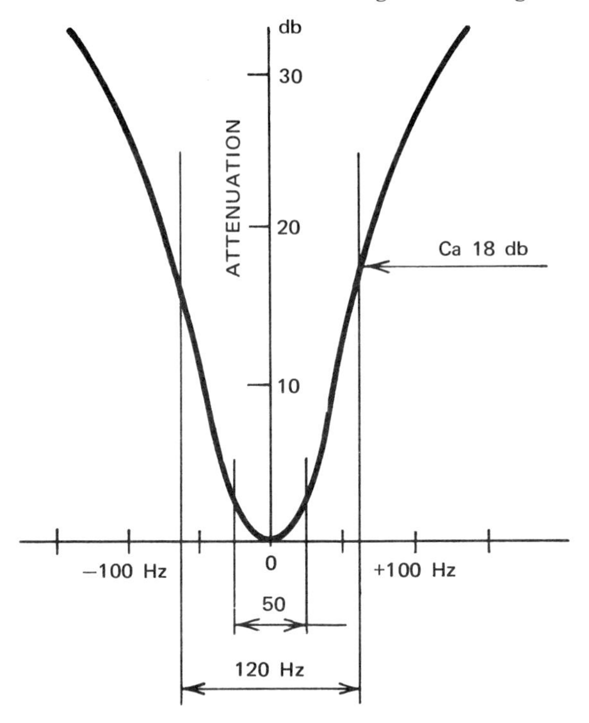

Figure 8.5. (*a*) Frequency and Selectivity of Channels in analogue transmission. (*b*) Selectivity of transmitter filter. (*c*) Selectivity of receiver filter.

division multiplexing (TDM). The principle is that a number of digital signals are transmitted sequentially in time, each occupying a short period, rather than by simultaneous transmission.

The system has considerable flexibility with respect to defining the number of signals transmitted per unit time, enabling high transmission speeds. For pollution measurements, however, transmission speed is not critical. In this system analogue measuring signals are sampled and converted to an equivalent binary number by an analogue to digital converter (ADC). The discrete values obtained represent average analogue signals over the sampling intervals. Measuring instruments should therefore have corresponding time

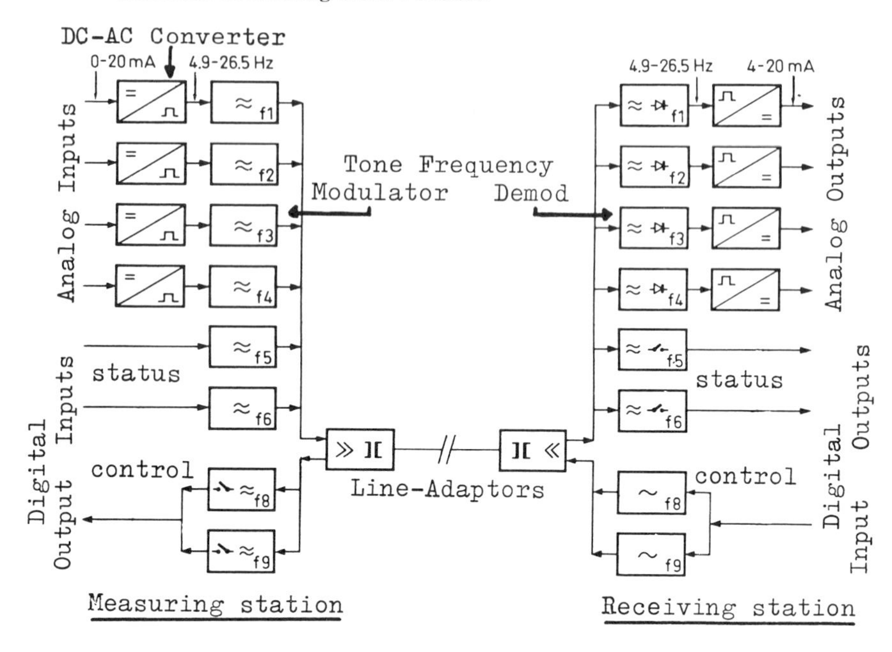

Figure 8.6. Block diagram of an FDM system. Each monitoring station on the measuring site needs a dc converter plus tone-frequency modulator and one common line adaptor per measuring signal, and at the receiving end a (central) line adaptor and a demodulator per signal. Square wave output voltages may be converted back into dc levels or evaluated digitally by the computer by way of a suitable interface.

constants. Figure 8.7 explains the principle of a TDM circuit. In this implementation digital signals are expressed in binary digits of which the "0" position is represented by a frequency f_A and the "1" position by a frequency f_Z, generated in the "modem" (modulator–demodulator).[22]

In the case of simultaneous incoming and outgoing signals (full duplex), there is a second pair of frequencies f_A and f_Z. A 200 bit sec^{-1} modem uses for the first channel (from measuring station to receiving center) $f_A = 1180$ Hz and $f_Z = 980$ Hz, and for the second channel (from receiving center to measuring station) $f_A = 1850$ Hz and $f_Z = 1650$ Hz. This configuration enables independent transmission of signals in both directions on a single telephone channel. In some countries the modems are supplied by the telecommunication authorities. A module of a Philips modem for 200 bits sec^{-1} (type RL 200) appears in Figure 8.8.

Digital data transmission normally incorporates coded checks on the transmission performance. Thus there is high confidence in the data received. Furthermore, analogue signals may be converted to binary number form with

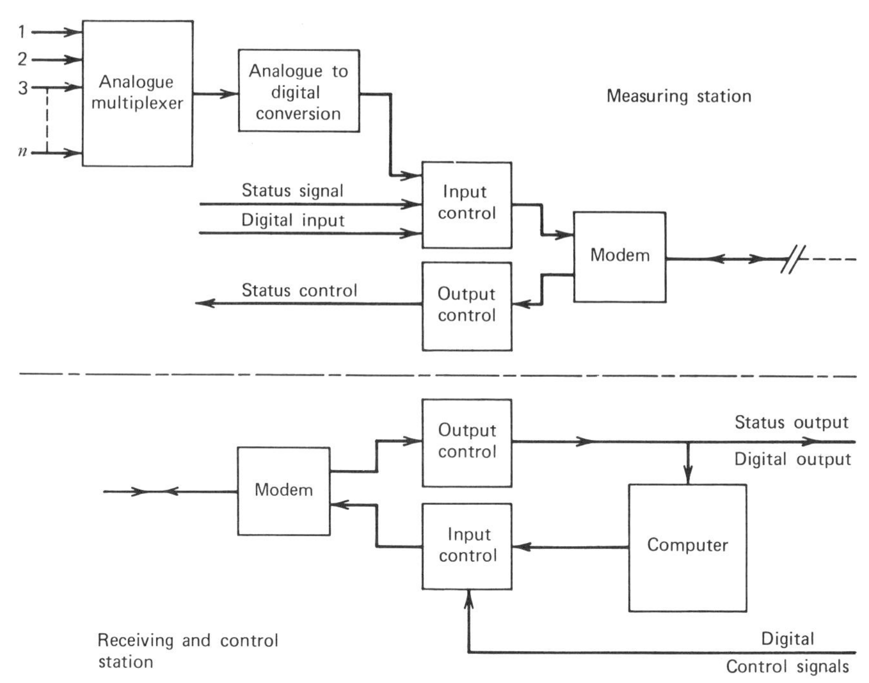

Figure 8.7. Block diagram of a TDM circuit.

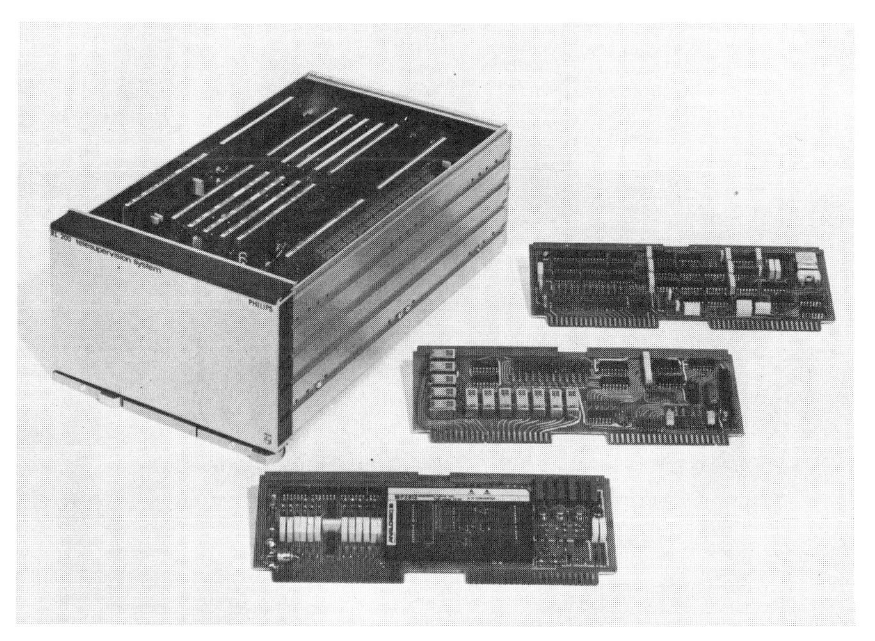

Figure 8.8. Modem module for digital transmission.

an accuracy of 0.01% or better. TDM transmission has the potential for a higher performance than FDM, but at greater cost, and this performance is not required for pollution monitoring. Total transmission cost must be considered for deciding the preferred system. A qualitative survey is given in Table 8.1.

For a small number of analogue signals, FDM is preferred because of lower cost, whereas in some instances the possibility of using remote analogue strip charts is decisive in the choice of this form of transmission.

For large numbers of signals TDM becomes more economical, although the intermediate situation may feature a hybrid system combining FDM transmission with a remotely controlled analogue scanner. Automatic dialing may be used in combination with TDM. This is available in a module (Figure 8.9) that can be mounted as an integral part of the equipment at the measuring site.

Large-scale integrated semiconductor logic in the module performs the preprocessing and storage of data. The module reacts as a normal telephone set and actively connects itself to the public switched telephone network when it is called by the computer in the central control room (receiving site).

The computer arranges the appropriate dialing and communication with the measuring site. Acknowledgment and address codes certify reliable

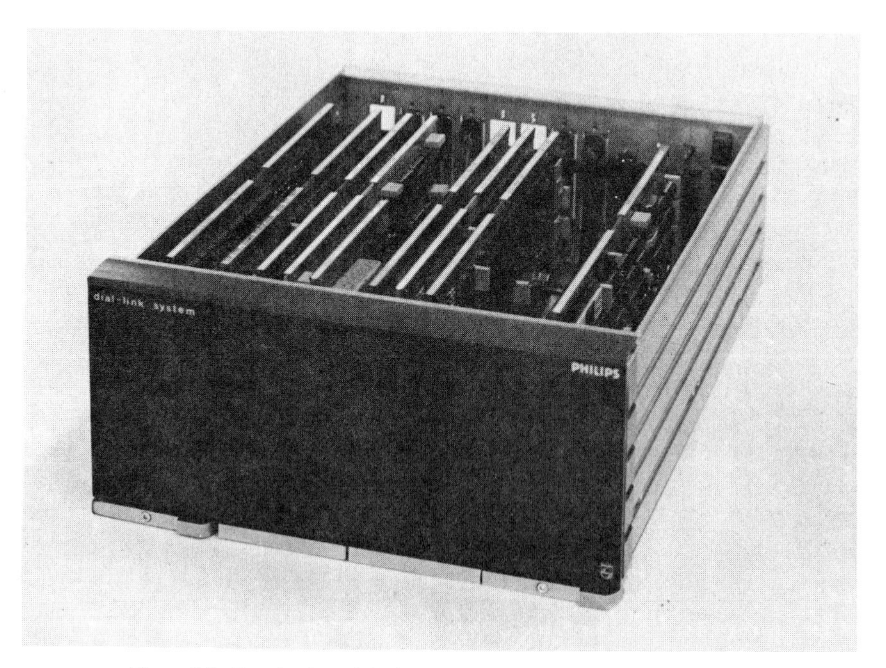

Figure 8.9. Standard module for automatic dialed-link transmission.

Table 8.1. Qualitative Comparison[a] Between Some Transmission Systems

System[b]	Transmission Speed per Channel (bits)	Data Channel Capacity per Line	Economy (number of signals per site)			Application		Order of Transmission—Error or Inaccuracy (%)
			≤2	5–8	>20	Analogue Transmission	Digital Transmission	
FDM	50	25	+++	+	–	++ Continuous	+	1
FTDM	50	150	–	+++	++	+ Discontinuous	+	1
TDM	50–1200	>100	–	++	+++	—	+++	0.01

[a] Plus and minus symbols indicate grading for applicability.

[b] Systems are abbreviated as follows: FDM, frequency division multiplexing; FTDM, hybrid system with simple time division multiplexing; TDM, time division multiplexing.

transmission of data together with data block checks. The modem at the receiving end has to be provided with a dialing circuit interfacing the computer with the public switched telephone network.

8.2.3. Data Processing

The data processing system must meet a number of requirements concerning the following areas: (*a*) the availability of data and their mutual relation, at

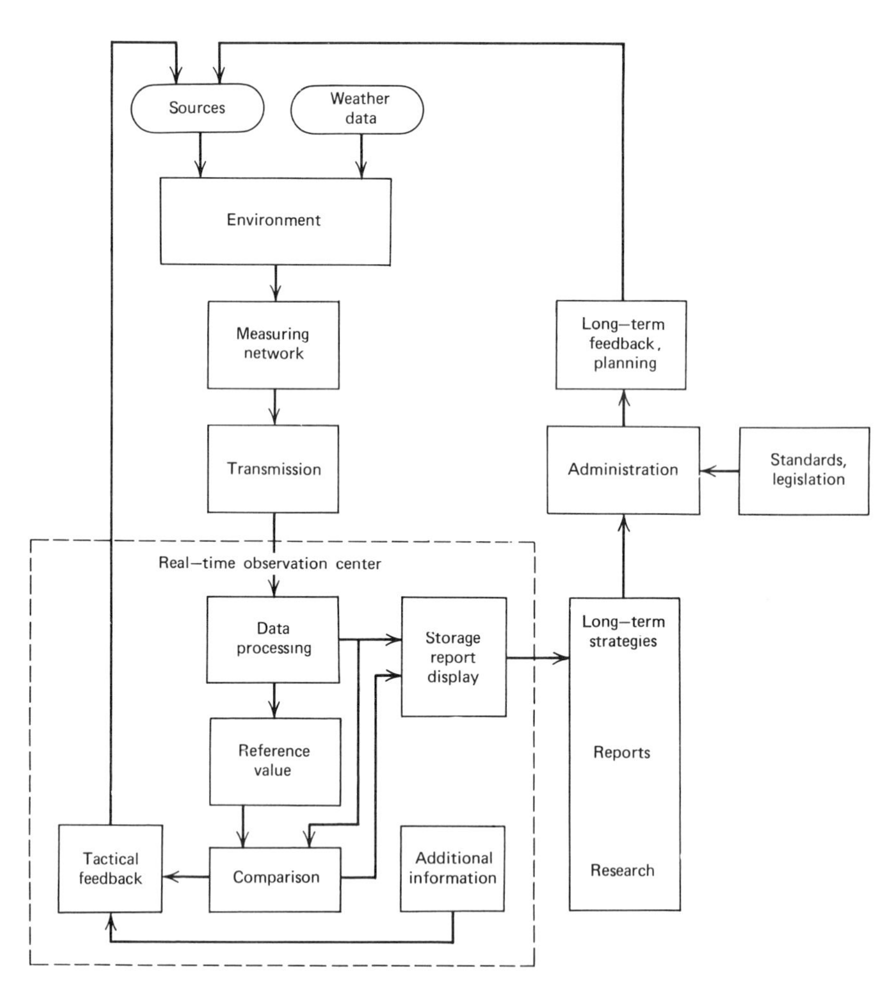

Figure 8.10. Functions in a system with pollution management, including tactical and long-term strategies.

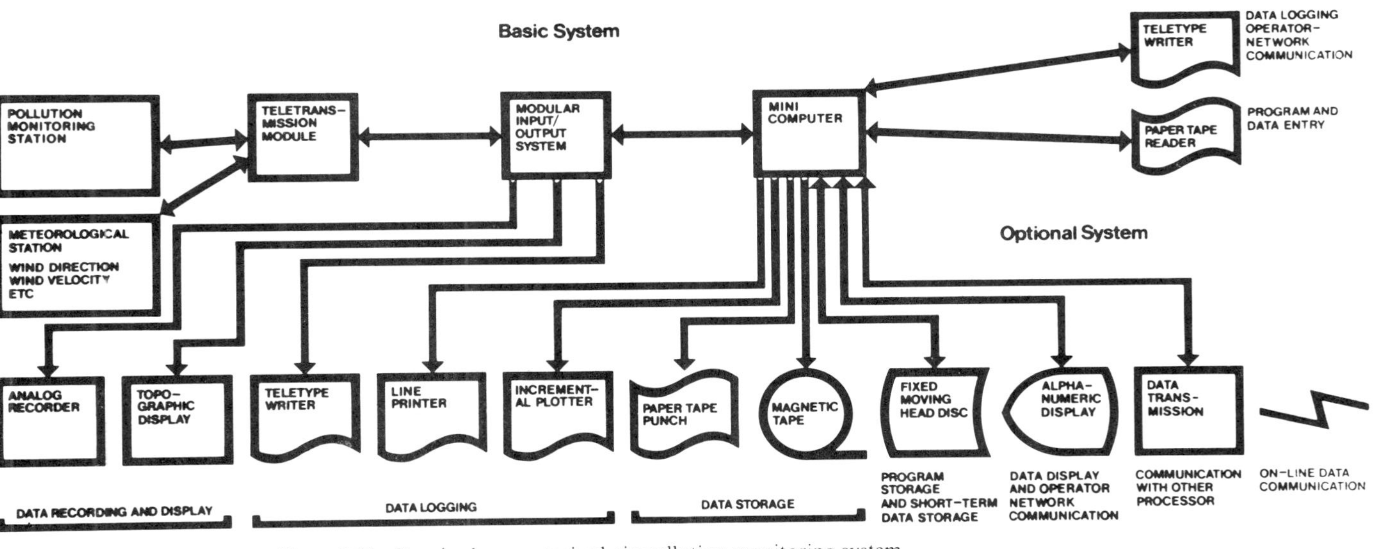

Figure 8.11. Standard computerized air pollution monitoring system.

specified times and in a specified format, and in an analogue or digital presentation; (*b*) the accuracy of the data, and their significance and reliability; (*c*) a survey of measured concentrations and their display in real time, as well as real-time statistical and other data evaluation; (*d*) statistical computation, correlation of, and research on data.

The first three can be satisfied by using an on-line, real-time data acquisition system with dedicated hardware and software. The items listed under *d*, however, should not be covered by dedicated equipment because they involve the study of models and the statistical interpretation of observed concentrations over long time intervals (months, seasons, or years). The methods depend on a knowledge of local conditions and on individual concepts. They require computer facilities with high level computer languages (PL-1, Algol, Fortran, or structured programming). Thus it is best to separate the items under *d* and optimize the others, using standardization of that part

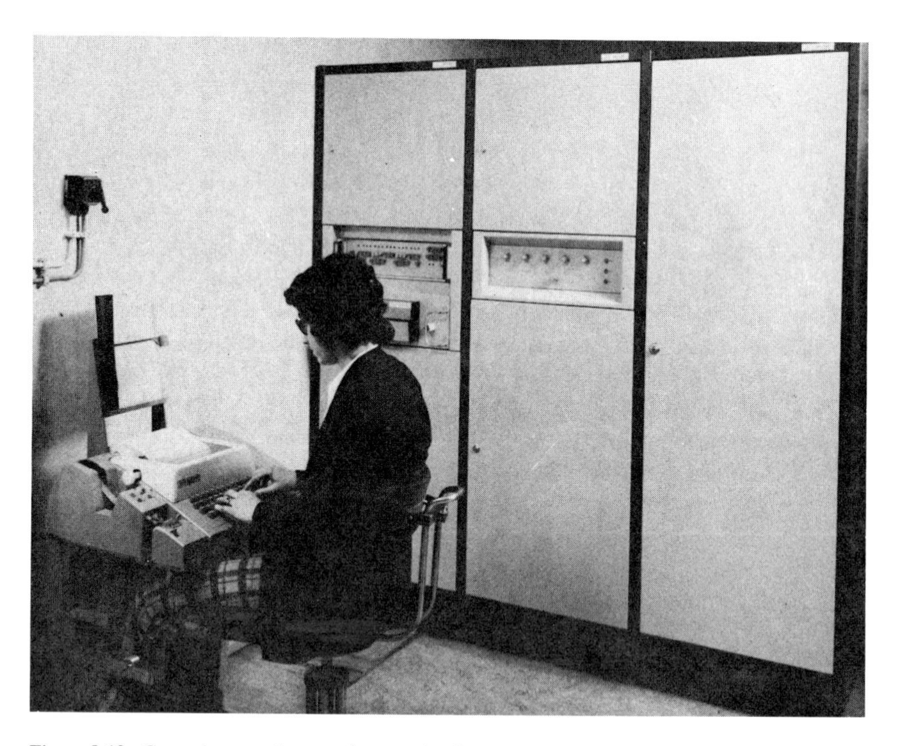

Figure 8.12. Central control room for monitoring networks. Basic configuration with mini-computer.

of the total system which is dedicated to the functions related to real-time (or quasi-real-time) data handling and processing.

Such a dedicated system can serve both as an autonomous instrument delivering the information needed to support the pollution management program and as an interface for a higher level system. The latter can be used for statistical calculations and models or long-term abatement strategies. Figure 8.10 is a synoptic diagram of the total system, showing the main functions mentioned. The dedicated part of the system can still have a variety of configurations (Figure 8.11). Such a dedicated system in a central control room appears in Figure 8.12.

Instead of a minicomputer in the central control room, even more modest is a microcomputer system. A microcomputer is built from large-scale integrated logic, constituting the processor and the semiconductor memory units. Programming of the (read-only) memory units cannot be implemented in the field, like the core memories of the minicomputers; therefore these systems are far less flexible. After acceptance of standard configurations, however, the microcomputer systems provide a more economical approach

Figure 8.13. Microcomputer system for small monitoring networks. Registration on cassette tape teletype and/or on strip printer.

to data handling. Such standard data handling units serve as a subcenter in a hierarchical configuration, with one main center containing the more flexible minicomputer. An example of an autonomous microcomputer system putting data on a cassette tape is presented in Figure 8.13. The basic standard system is made for eight monitoring devices, but up to four times as many monitors can be handled by the extended system.

8.3. THE RIJNMOND NETWORK

In the Rijnmond area an automated SO_2 monitoring network, comprising 31 SO_2 monitoring stations and two wind direction sensors, has been in operation since October 1969. The sensors are joined by a data transmission system using frequency division multiplexing to a central control room with minicomputer, typewriter, synoptic map display, and some other devices (Figure 8.14*a*). This network provides substantial support to the air pollution management program for the area (Figure 8.14*b*).

Apart from checking continuously the SO_2 levels, which is a measure of the activity of some major industries, the system is set up primarily to predict meteorological conditions unfavorable for dispersing air pollutants. In practice, therefore, SO_2 concentrations are used as a tracer to provide information about the dispersal of pollutants by the atmosphere in the area. Actual measured hourly averages are compared with expected levels occurring under neutral meteorological conditions and the ratios are evaluated statistically.[3] The information is then used for air pollution abatement tactics, especially to combat odors in residential areas, the most serious problem affecting the population in the Rijnmond area.

The odors, coming from mercaptans and other compounds, are produced mainly by oil refineries and petrochemical plants; desulfurization, combustion of refining wastes, and the innumerable small leaks and spills that occur during processing and transportation.

A mathematical model of the nuisance of odors in the environment has been developed[23] and tested statistically. The experimental nuisance is quantified by the number of complaints, by telephone calls to a special number, taking into account an estimate of the probability of response to the perception of odors. It implies that the public living in the affected area does not remain passive but can complain by telephone to the central control room at any time of the day. These calls provide a further input for evaluating the situation, by their number (frequency), the time, and the place from which the complaint is coming.

To improve the quality of life in the Rijnmond area, action has been taken, and is continuing, to reduce air pollution emissions as much as possible.

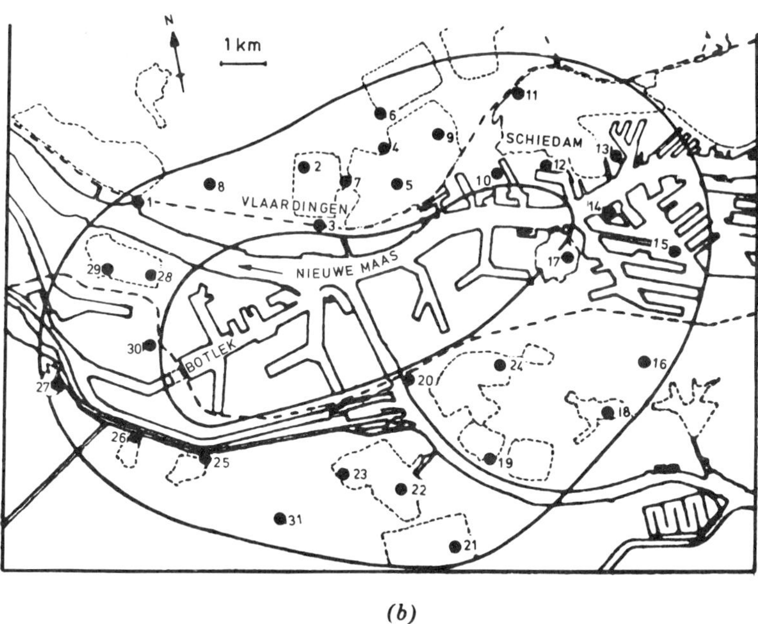

(b)

Figure 8.14. (*a*) Central control room of the Rijnmond network at Schiedam, with minicomputer. Complaints from the public are enabled by way of five telephone connections. The national meteorological service (KNMI) routinely delivers by telex information on the meteorological situation. In the background is a synoptic display of the network in the area; lamps light up at sites where set threshold levels are exceeded. (*b*) Monitoring sites in the Rijnmond area.

Licensing of new activities and extension of existing licenses is carefully considered and controlled. A preliminary study involving all industries in the area was used as a basis for the control strategy. It included questionnaires about possible countermeasures to be taken by each industry in the region when unfavorable weather conditions occur. The proposed countermeasures included the postponement of some maintenance and repair tasks; cleaning, steaming, draining, or filling of storage tanks; use of regenerating catalysts, loading and process modifications, and stopping and starting up. as indicated. Some of these activities are costly to implement; thus reliable weather information about unfavorable conditions is a prerequisite for successful industrial cooperation on a long-term basis.

The automated monitoring networks directs attention to such conditions with a special program. At Rijnmond the network configuration has been established with a model using statistical computation for the required number of monitors for SO_2 in each of the eight wind direction sectors.[3] Nonautomated monitoring for air pollutants other than SO_2 is also carried out, but this is gradually being replaced by automated stations. The actual siting of the monitoring stations has been based on practical considerations and experience (Figure 8.14*b*).

The monitoring network is of course only part of the air pollution control activity, it is, in effect, a tool used in the air quality management program being implemented by the Central Environmental Management Service, Rijnmond. This has a staff of about 120 persons: 80 are concerned with the application of the Netherlands Nuisance Act and Air Pollution Act, 20 deal with laboratory work and sample analyses, and 20 are administrative personnel. Monitoring, including the mobile station, data interpretation, alerting, supervision, and dealing with telephone complaints, requires four persons recruited from the groups just named. Network maintenance and servicing is covered by a comprehensive contract with Philips Industries of Eindhoven.

8.4. THE NETHERLANDS NATIONAL NETWORK

The Netherlands National Network was put into operation officially in June 1973 under the management of the National Institute of Public Health in Bilthoven. It has 9 regional networks, each with its own computerized control center, connected to the National Control Center in Bilthoven (Figure 8.15). The complete network has 220 monitoring stations, mainly measuring SO_2. About 100 of these are set up in a country-wide network.[6a] Each region controls an appropriate part of the network, the stations being connected by fixed telephone lines and FDM transmission with the regional

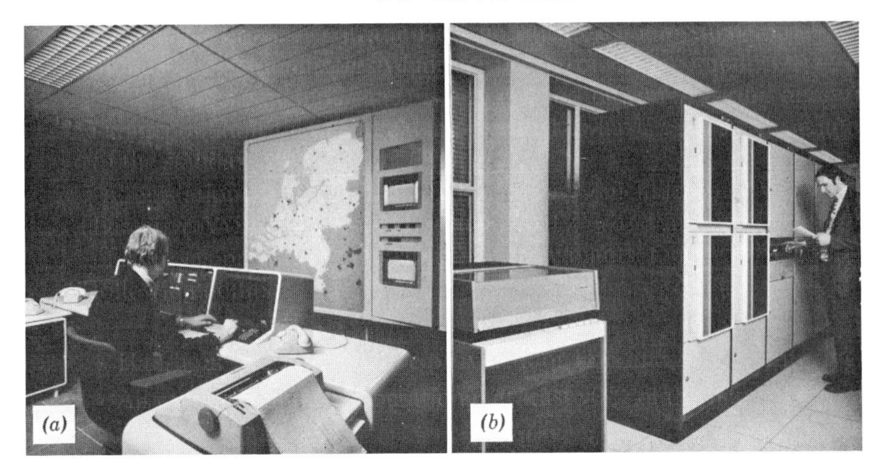

Figure 8.15. (*a*) Central control room of the Netherlands national network, controlling 9 regional centers with 220 monitoring stations. (*b*) Computer room of the Netherlands national network system with 16K computer and a.o. four magnetic tape units and one large moving head disk unit.

computer center. In turn, the regional computers are linked to the National Center by telegraph channels in a half-duplex mode for quasi-continuous information exchange. A simple channel for a 50 baud transmission rate is satisfactory. Modems are not required.

National Air Quality Management for the Netherlands comprises of the following units:

1. A measuring center, equipped for remote sensing, initial equipment service and supervision, center operation, and data processing.
2. A computer center staffed by programmers and other personnel for dealing with external inputs (mobile stations, etc.) and related activities.
3. An air pollution measuring service, for monitor supervision in the field.
4. An air chemistry unit for instrument evaluation and investigation.

Laboratory services are provided by the central facilities of the institute. Air pollution strategies that can be applied at the regional as well as national level are under investigation.

The Royal Netherlands Meteorological Institute (KNMI) works in cooperation with the National Institute of Public Health, studying air pollution dispersion. A 218 m mast used for simultaneous SO_2 and meteorological measurements at 0, 100 and 200 m, is shown in Figure 8.16.

Figure 8.16. A super 218 m high snifferpole used by the Royal Netherland Meteorological Institute (KNMI) to study the processes in the lower atmosphere. Variables measured are wind direction and velocity, temperature, humidity, and visibility; at 0, 100, and 200 m levels automated SO_2 monitors are installed.

8.5. OPERATIONAL RESULTS

Although air pollution management has been practised in the Rijnmond area for many years, the past 5 years are of special interest because abatement activities have been intensified with the installation of the automated SO_2 monitoring network.

The presence of the monitoring network also has a psychological effect, since it indicates to people in the region their personal involvement in better pollution control. The most valuable result is the ready availability of sound data that enable tactical decisions to be made regarding pollution management at any time. Thus data from the network are used for deciding when alarm code warnings are issued to industry. During 1973 an alarm code was in force 12.3% of the time, and during 1974 for 7.2%. These codes also show when unfavorable meteorological situations exist from year to year.

Industries report intended pollution-producing activities, as well as incidents and accidents, to the Regional Control Center. The number of incident reports up to 1969–1970 has been increasing, indicating that enhanced cooperation is being obtained. On the other hand, the number of

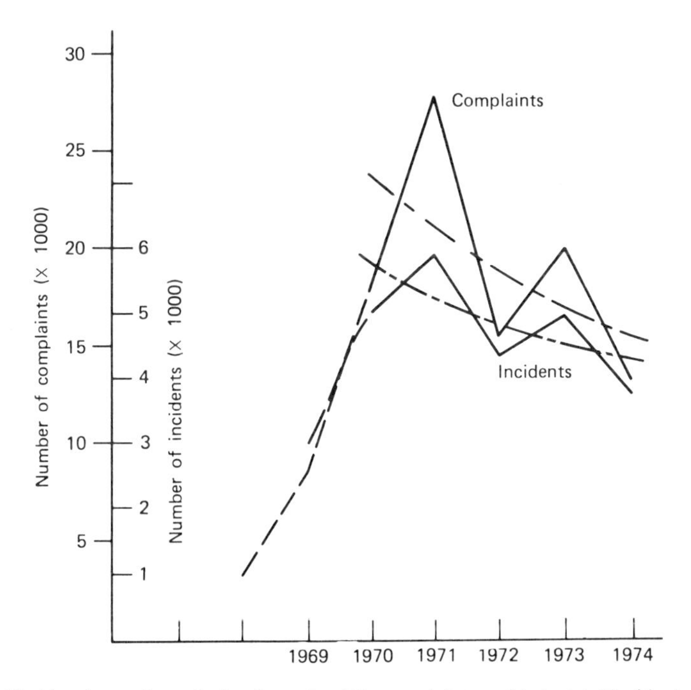

Figure 8.17. Number and trend of registered public complaints and industrial incidents over the period 1970–1975.[24]

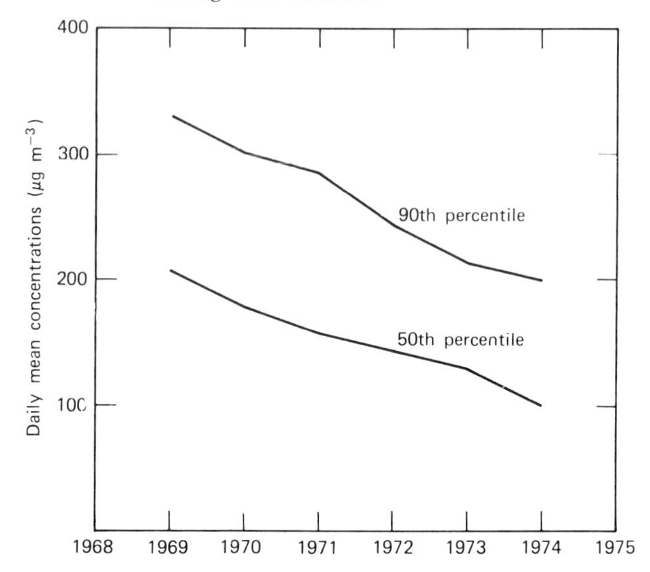

Figure 8.18. The 50th and 90th percentile values of SO_2 concentration during the winter, 1969–1975 in Vlaardingen (Rotterdam).

public complaints registered has decreased. This may be partly due to more favorable weather conditions in 1974, but it may also indicate a decrease in the real number of pollution-producing accidents. Both circumstances are plotted in Figure 8.17.

All the air pollutants measured in the Rijnmond area (SO_2, NO_x, hydrocarbons, smoke, dust, and heavy metals) have shown decreases in the past 10 years. This has been most significant for SO_2, being 12 to 15 $\mu g/m^{-3}$ $year^{-1}$ for most of the area (Figures 8.18 and 8.19). The decrease must be attributed to control measures such as the substitution of gas for oil in domestic heating

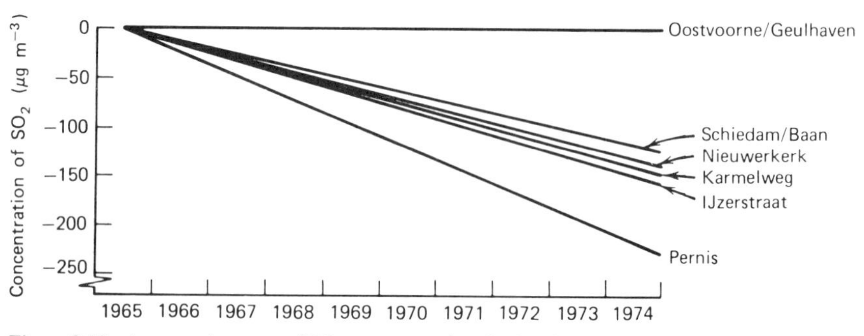

Figure 8.19. Average decrease of SO_2 concentration during the winter at six measuring stations in the Rijnmond area.

and industrial processing, and the use of process modifications when unfavorable meteorological conditions occur. These downward trends have been maintained even though industrial production has been increasing at a rate of $2\frac{1}{2}\%$ per year and population by 3% for the same period. Clarenburg[18] has shown a marked negative correlation between the number of gas connections and measured SO_2 concentrations, confirming the strong influence of domestic heating on ambient SO_2 concentrations.

A8.1. ALLOWABLE CONCENTRATION LIMITS

The allowable limits of concentration of dangerous or irritating gases depend on many factors. Various combinations of the following factors lead to different allowable limits:

1. The composition of population with respect to their susceptibility:
   a. Selected factory workers under regular medical care.
   b. Mixed population in a city.
   c. Unhealthy people (e.g., asthma patients).
2. Duration of exposure:
   a. In cases of emergency, single exposure (e.g., for 0.5, 1, or 2 hr).
   b. At the workplace, usually 8 hr day^{-1}.
   c. In residential areas, 24 hr day^{-1}.

   The allowable limits will depend on the exposure time in relation to the destruction or secretion of the pollutant in the living organism. If the pollutant is cumulative, a longer period must be taken—exposure time per year, or even exposure time per whole lifetime.
3. Combined effects of several pollutants.
4. Criteria, or dose-effect relationship:
   a. Fatal.
   b. Illness or increase of illness, irreversible or with slow recovery.
   c. Adverse effects on adequate and coordinated behavior.
   d. Illness or irritation, but reversible, with fast recovery and no lasting adverse effects on health.
   e. No adverse effects to public welfare.

Some typical combinations of the four factors are as follows:

1a, 2a, 4d. Emergency Exposure Limits: *EEL* (United States) for factory workers.

1b, 2a, 4d. Population Emergency Limits: *PEL* (United States) or EPEL (Netherlands) for mixed population.

1a, 2b, 4d. Maximum Allowable Concentration: *MAC* or MAK (Germany) or Threshold Limit Value: *TLV* for factory workers.

1b, 2c, 3, 4d. Mixed population in residential areas with adequate margin of safety to protect public health: Primary Air Quality Standard in the United States (National Ambient Air Quality Standards) or Tolerable Air Quality in Canada.

1b, 2c, 3, 4e. Mixed population in residential areas to protect the public welfare from any known or anticipated adverse effects: Secondary Air Quality Standard in the United States or Acceptable Air Quality in Canada.

1b, 2c, 3, 4e. As previously, but with a margin of safety: Desirable Air Quality in Canada; MIK values in Germany.

A8.2. METHODS OF ABATEMENT RELYING ON LEGISLATION

Ambient air quality standards are based on different situations.

1. a. Limitation of a number of pollutants separately.
   b. Limitation of combinations of pollutants, for example:
      - The product of the concentrations of SO_2 and particulates or a division in two classes:
      - More SO_2 with less smoke.
      - Less SO_2 with more smoke.
2. Division of a country into:
   - Polluted areas.
   - Urban areas.
   - Rural areas.
3. Spreading attainment of standards over a number of years, with lowering limits.
4. Accepting statistical fluctuations in concentrations due to meteorological phenomena, and the influence of exposure time, by limiting the averages (e.g., per day as well as per year).
5. Distinction of a number of levels, for example,
   a. Handled by warning systems:
      - Adverse, alert stage 1.*
      - Serious, alert stage 2.*
      - Emergency.

* Abatement to avoid the emergency level or to optimize air quality.

b. Handled by the authority:
- Desirable.
- Acceptable (secondary standard).
- Tolerable (primary standard).

A8.2.1. Legislation on Air Quality in the United States[25]

The Clean Air Act 1970 places responsibility for setting air quality standards on the Environmental Protecting Agency (EPA), formerly the National Air Pollution Control Administration (NAPCA).

Primary ambient standards are those which are requisite to protect the public health, with allowance for an adequate margin of safety.

Secondary ambient standards are those which are necessary to protect the public welfare from any known or anticipated adverse effects associated with a given pollutant in the ambient air.

Table A8.2.1 gives a survey of the National Ambient Air Quality Standards adopted from the EPA publication.[25]

Table A8.2.1. U.S. National Ambient Air Quality Standards Summary[25]

	Limits[a]			
Pollutant	Maximum ppm	Maximum Concentration per Cubic Meter	Maximum Value	Note
Sulfur oxides				
Primary standard	0.03	80 μg	Annual arithmetic mean	
	0.14	365 μg	per 24 hr	—[b]
Secondary standard	0.1	260 μg	per 24 hr	—[b]
	0.5	1300 μg	per 3 hr	—[b]
Particulate matter				
Primary standard		75 μg	Annual geometric mean	
		260 μg	per 24 hr	—[b]
Secondary standard		60 μg	Annual geometric mean	
		150 μg	per 24 hr	—[b]
Carbon monoxide	9	10 mg	per 8 hr	—[b]
	35	40 mg	per 1 hr	—[b]
Photochemical oxidants	0.08	160 μg	per 1 hr	—[b]
Hydrocarbons	0.24	160 μg	per 3 hr, 6–9 A.M.,	—[b]
			Corrected for CH_4	—[b]
Nitrogen dioxide	0.05	100 μg	Annual arithmetic mean	

[a] Reference temperature, 25°C; reference pressure, 1013 mbar.
[b] Not to be exceeded more than once per year.

A8.2.2. Legislation in Canada[26a]

"National Air Quality Objectives" were issued on October 21, 1971, based on the Federal Clean Air Act. The act distinguishes three levels of air quality objectives:

- Desirable (this defines the long-term goal).
- Acceptable (corresponds to the secondary air quality standards of the United States).
- Tolerable (indicates an imminent danger, requiring abatement action).

The values in Table A8.2.2 are to be attained and extended to include nitrogen oxides.

Standard reference methods are as follows: SO_2, the West–Gaeke method (pararosaniline method); particulate matter, the high volume method; total

Table A8.2.2. Canadian 1971 National Air Quality Objectives Summary

		Limits[a]		
Pollutant	Level	Maximum Concentration per Cubic Meter	Maximum ppm	Maximum Value
Sulfur dioxide	Acceptable	60 μg	0.02	Annual arithmetic mean
		300 μg	0.11	per 24 hr
		900 μg	0.34	per 1 hr
	Desirable	30 μg	0.01	Annual arithmetic mean
		150 μg	0.06	per 24 hr
		450 μg	0.17	per 1 hr
Particulate matter	Acceptable	70 μg		Annual geometric mean
		120 μg		per 24 hr
	Desirable	60 μg		Annual geometric mean
Carbon monoxide	Acceptable	15 mg	13	per 8 hr
		35 mg	30	per 1 hr
	Desirable	6 mg	5	per 8 hr
		15 mg	13	per 1 hr
Total oxidants, as ozone	Acceptable	30 μg	0.015	Annual arithmetic mean
		50 μg	0.025	per 24 hr
		160 μg	0.08	per 1 hr
	Desirable	30 μg	0.015	per 24 hr
		100 μg	0.05	per 1 hr
Hydrocarbons	Acceptable	—	—	

[a] Reference temperature, 25°C.

oxidants, iodometric titration method (neutral potassium iodide); carbon monoxide, the nondispersive infrared spectrometry method.

"Equivalent method" means any method of sampling and analyzing for an air pollutant that can be demonstrated to have a consistent relationship to the standard reference method.

Legislation in the Province of Alberta[26] includes the Air Monitoring Directive AMD-73-1, issued February 20, 1973, as draft. Valid for the vicinity of sour gas processing and sulfur recovery plants, the directive contains the following points:

- Half-hour averages shall be provided for H_2S and SO_2 monitoring.
- All monitors must be operational at least 90% of the maximum time each month.
- SO_2 instruments must be calibrated at three or more concentrations, and a legible photocopy of the calibration chart shall be submitted every month.
- Exceeding of the maximum levels shall be reported, together with meteorological data (not specified).
- Peak concentration of 0.05 ppm (132 $\mu g\ m^{-3}$) or greater shall be listed (SO_2), at the same time the latest average over not more than 30 min shall be listed. If the peak exceeds 0.5 ppm SO_2(1325 $\mu g\ m^{-3}$), the average time must not exceed 15 min.
- Any exceeding of concentrations of the Clean Air Regulations must be listed on a separate paper.
- Each hour, the average of the running last 24 hr is to be represented.
- An exceeding of 0.2 ppm SO_2 (0.5 hr) must be reported, however, also a shorter peak with the same dosing (e.g., 1 ppm in 6 min or 0.5 ppm in 12 min). The same holds for H_2S.
- ppm $SO_2 \times 2650 = \mu g\ m^{-3}$ SO_2 for reference pressure 1013 mbar
- ppm $H_2S \times 1400 = \mu g\ m^{-3}$ H_2S and reference temperature 25°C

Some points from the Clean Air Regulations of October 1973 are as follows:

- "Density" means the shade or opacity of an air contaminant at or near the point of emission to the atmosphere determined in accordance with the chart (prepared by the Director of Standards and Approvals).
- "Particulates" means any material, except uncombined water, having definite physical boundaries at standard conditions (25°C, 1013 mbar).
- The concentration of particulates in an effluent stream, adjusted to 50% excess air for products of combustion, is limited. The limit depends on the kind of plant, and it is to be lowered in 1975, when 0.2 to 0.6 g of particulates

per kilogram of effluent is allowed, and not more than 50% shall be retained on a 325 mesh screen.

Maximum permissible concentrations of air contaminants in the ambient air are listed in Table A8.2.3.

Table A8.2.3. Legislation In the Province of Alberta, Canada[26b]

Pollutant	Average Maximum Concentration		Average Period
	per Cubic Meter	ppm	
Sulfur dioxide	30 μg	0.01	Annual arithmetic mean
	150 μg	0.06	24 hr
	450 μg	0.17	1 hr
	525 μg	0.20	$\frac{1}{2}$ hr
Hydrogen sulfide	4 μg	0.003	24 hr
	14 μg	0.010	1 hr
	17 μg	0.012	$\frac{1}{2}$ hr
Total oxides of nitrogen, as NO_2	60 μg	0.032	Annual arithmetic mean
	200 μg	0.106	24 hr
	400 μg	0.213	1 hr
Carbon monoxide	6 mg	5	8 hr
	16 mg	13	1 hr
Total oxidants, as ozone	30 μg	0.015	24 hr
	100 μg	0.05	1 hr
Suspended particulates	60 μg		Annual geometric mean
	100 μg		24 hr
Total dustfall, due allowance for normal background levels	5.3^a g	0.18^b g/(m^2 day^{-1})	Residential and recreational areas, in 30 days
	15.8^a	0.51^b	Commercial and industrial areas, in 30 days

[a] Expressed in grams per square meter per month.
[b] Expressed in grams per square meter per day.

A8.2.3. Air Quality Standards in Germany[27,28]

German air quality standards for SO_2, H_2S, Cl_2 and HCl, and NO_2 and HNO_3 can be found in Reference 27, in the following sheets, respectively: VDI—2108, 2107, 2106, and 2105.

The definitions used in Table A8.2.4 are as follows:

MIK_D = maximum allowable concentrations in ambient air with continuous exposure
MIK_K = maximum allowable ambient concentration with intermittent exposure (specified times)
MAK = maximum allowable concentration at the workplace.

Measurements are half-hour mean values. Reference conditions for one cubic meter of air at 0°C and 1013 mbar (1 atm).

Table A8.2.4. Some Maximum Allowable Concentrations of Pollutants in the Federal Republic of Germany

	MIK_D[a]		MIK_K[a]			MAK
Pollutant	ppm	$mg\ m^{-3}$	ppm	$mg\ m^{-3}$	Period	(ppm)
NO_2	0.5	1	1	2	3 times per day	5
HNO_3	0.5	1.3	1	2.6	Once in 2 hr	10
Cl_2	0.1	0.3	0.2	0.6	3 times per day	1
HCl	0.5	0.7	1	1.4	Once in 2 hr	5
H_2S	0.1	0.15	0.2	0.3	3 times per day	20
SO_2	0.2	0.5	0.3	0.75	Once in 2 hr	5
CO						50

[a] Reduced values for Air Quality in *Technische Anleitung zur Reinhaltung der Luft* (*TA-Luft*) (Technical Guide to Clean Air).[28]

The air quality limits for nontoxic dust (*TA-Luft*) are as follows:

$0.42\ g\ m^{-2}\ day^{-1}$ average over 1 year
$0.65\ g\ m^{-2}\ day^{-1}$ average over 1 month
} outside the industrial center

$0.85\ g\ m^{-2}\ day^{-1}$ average over 1 year
$1.3\ g\ m^{-2}\ day^{-1}$ average over 1 month
} in industrial center

Table A8.2.5 summarizes recently proposed air quality limits of the German Engineering Society (VDI).

Table A8.2.5. **Air Quality Limits Proposed for Germany by the *VDI-Kommission Reinhaltung der Luft* (1973)**

Pollutant	Concentration	Limits[a,b]		Average During
		ppm	mg m^{-3}	
Sulfur dioxide	MAK	5	13	8 hr
	MIK	0.35	1.0	$\frac{1}{2}$ hr
	MIK	0.105	0.3	24 hr
		0.035	0.1	1 year
Nitrogen dioxide	MAK	5	9	8 hr
		0.10	0.2	$\frac{1}{2}$ hr
	MIK	0.05	0.1	24 hr
		0.025	0.05	1 year
Nitric oxide	MAK			
	M	0.75	1.0	$\frac{1}{2}$ hr
	MIK	0.37	0.5	24 hr
		0.075	0.1	1 year
Carbon monoxide	MAK	50	55	8 hr
		40	50	$\frac{1}{2}$ hr
	MIK	8	10	24 hr
		8	10	1 year
Ozone	MAK	0.1	0.2	8 hr
		0.07	0.15	$\frac{1}{2}$ hr
	MIK	0.023	0.05	24 hr
		?	0.01 (?)	1 year
Particulate matter		—	0.3	$\frac{1}{2}$ hr
below 10 μm	MIK	—	0.1	24 hr
		—	0.05	1 year
Zinc	MAK	ZnO	5	8 hr
	MIK	Zn	0.5	$\frac{1}{2}$ hr
	MIK	Zn	0.1	24 hr
			0.05	1 year
Cadmium	MAK	CdO	0.1	8 hr
	MIK	Cd	0.00005	24 hr
Lead	MAK	—	0.092	8 hr
	MIK		0.010	$\frac{1}{2}$ hr

Table A8.2.5. (Continued)

Pollutant	Concentration	Limits[a,b] ppm	$mg\ m^{-3}$	Average During
	MIK		0.002	24 hr
			0.001	1 year
Cryolith (Na_3AlF_6)	MIK		0.5	½ hr
			0.3	24 hr
			0.1	1 year
Aluminum fluoride (AlF_3)	MIK		0.5	½ hr
			0.3	24 hr
			0.1	1 year
Sodium fluoride (NaF)	MIK		0.3	½ hr
			0.1	24 hr
			0.1	1 year
Calcium fluoride (CaF_2)	MIK		1.0	½ hr
			0.5	24 hr
			0.2	1 year
Hydrogen fluoride (HF)	MAK		2	8 hr
	MIK		0.1	½ hr
			0.1	24 hr
			0.05	1 year
Ammonia (NH_3)	MAK	50	35	8 hr
	MIK		2	½ hr
			1	24 hr
			0.5	1 year
Sulfuric acid (H_2SO_4)	MAK		1	8 hr
	MIK		0.2	½ hr
			0.1	24 hr
			0.05	1 year
Nitric acid (HNO_3)	MAK		25	8 hr
	MIK		0.2	½ hr
			0.1	24 hr
			0.05	1 year

Table A8.2.5. (Continued)

Pollutant	Concentration	Limits[a,b] ppm	mg m^{-3}	Average During
Trichloroethylene (C_2HCl_3)	MAK	50	280	8 hr
	MIK	2.7	16	½ hr
		0.85	5	24 hr
		0.34	2	1 year
Dichloromethane, methylene chloride (CH_2Cl_2)	MAK	5000	1750	8 hr
	MIK	45	157.5	½ hr
		15	52.5	24 hr
Tetrahydrofurane (C_4H_8O)	MAK	200	590	8 hr
	MIK	60	180	½ hr
		20	60	24 hr

[a] ppm = μmole $mole^{-1}$.
[b] MAK for 20°C; air quality (MIK) probably for cubic meters reduced to 0°C.

A8.2.4. Legislation in Russia

Table A8.2.6. Legislation in Russia[29]

Pollutant	Limits mg m^{-3}	ppm	Time (hr)	Remark
SO_2	0.05	0.02	24	Maximum
	0.5	0.2	½	Maximum
NO_x (as NO_2)	0.085	0.045	24	Maximum
CO	1	0.9	24	
	3	2.7	⅓	
HCl	0.2	0.15	⅓	Maximum
Particulate matter	0.15	—	24	Maximum
	0.5	—	½	Maximum

A8.2.5. Legislation in the Netherlands

Air quality standards are not yet incorporated in the law. However there is an advice of the Board of Health (Gezondheidsraad) to the Minister of Social Affairs and Public Health (dated April 2, 1971), summarized in Table A8.2.7.

Table A8.2.7. Proposed Air Quality Standards in the Netherlands for SO_2 (DESO[a] standard smoke)

Limits ($\mu g\ m^{-3}$)[b]	Values
75 <30[a]	Median of 24 hr values
250 <90[a]	24 hr values, not to be exceeded more than 2% per year (about 7 days)

[b] Cubic meters refer to 0°C and 1013 mbar.

The proposed quality standards cannot be attained at this time. Therefore the Board of Health proposes to admit, for a limited time, higher values in industrial urban areas (B) and in obviously polluted areas (C) as given in Table A8.2.8.

Table A8.2.8. Netherlands Temporary 24 hr Values for SO_2

		Limits ($\mu g\ m^{-3}$)	
	Amount of Smoke	Median	2% Exceeding
Area B	Little	120	350
	Much	90	275
Area C	Little	150	350
	Much	125	275
Indication of amount of smoke	Little	<30	<90
	Much	>40	>120

REFERENCES

1. Stern, A. C., Ed., *Proceedings of the Symposium on Multiple Source Urban Diffusion Models*. Publication No. AD-86, U.S. Environmental Protection Agency, Research Triangle Park, N.C., 1970.
2. Deininger, R. A., Ed., *Models for Environmental Pollution Control*, Pt. III, Ann Arbor Science Publishers, Ann Arbor, Mich., 1974, p. 219.
3. Clarenburg, L. A., A System to Predict Unfavourable Weather Conditions. Rijnmond Authority Publication No. 6842, Vasteland 96/104, Rotterdam 3001, 1968.
4. Schmidt, F. H. and C. A. Velds, *Atmos. Environ.*, 3, 455 (1969).
5. Brasser, L. J., A New Method for the Presentation of a Large Number of Data Obtained from Air Pollution Survey Networks. *Proceedings of the Second International Clean Air Congress*, December 1970, H. M. Englund and W. T. Beery, Eds., Academic Press, New York, 1971, pp. 62–65.
6. Schneider, T., Ed., Automatic Air Quality Monitoring Systems. *Proceedings of the Conference Held at the National Institute of Public Health*, June 5–8, 1973, Bilthoven, Netherlands. Elsevier Amsterdam, 1974.

6a. Ibid., pp. 107–118.

7. van Egmond, N. D. and H. J. Elskamp, Annual Report 1973–1974 (in Dutch). National Measuring Network for Air Pollution in the Netherlands. National Institute for Public Health, Bilthoven, Netherlands, 1975.
8. Nördö, J., A Study of Mass Transport for Air Pollutants. Proposals for a European Project. OECD A/DAS/C51/487, Annex II, September 24, 1974.
9. Brosset, C. and Å. Åkerström, *Atmos. Environ.*, 6, 661 (1972).
10. Wilting, J. J. and H. van den Berge, *Computer*, 4 (4), 23 (1971).
11. Kellner, K. H. and J. Landbrecht, Das Lufthygienische Landesüberwachungssystem Bayern, *Gesundheits-Ingenieur*, No. 10, 297 (1974).
12. Morgan, G. B., E. W. Bretthauer, and S. H. Melfi, *Instrum. Tech.*, 21 (2), 27 (1974).
13. United Nations General Assembly, Conference on the Human Environment, Stockholm, June 1972. A/Conf. 48/11 and Add. 1, January 1, 1972; A/Conf. 48/INF. 2, February 18, 1972.
14. OECD Environmental Directorate, Record of Seminar on Economic and Legal Aspects of Transfrontier Pollution, 1972.
15. General Aspects of the Planning of Air Quality Monitoring. Working Document of 150/TCM6/SGA. 150-Central Secretariat, Geneva, Switzerland.
16. WHO, *International Air Pollution Monitoring Network, Data User's Guide*, June 1972.
17. UNEP Pollution Watch, *Financial Times*, No. 8, February 20, 1974.
18. Clarenburg, L. A., Is It Possible to Conduct a Rational Air Pollution Management? (internal report). Rijnmond Authority, Rotterdam, Netherlands.
19. Stratmann, H., M. Buck, U. Hölzel, and D. Rosin. *Schriftenreihe der Landesanstalt für Immissions- und Bodennutzungsschutz des Landes Nordrhein–Westfalen*, Vol. 1, Verlag W. Girardet, Essen, 1965, p. 25.
20. Stratmann, H., and M. Buck, *Schriftenreihe der Landesanstalt für Immissions- und Bodennutzungsschutz des Landes Nordrhein–Westfalen*, Vol. 3, Verlag W. Girardet, Essen, 1966, p. 7.

21. CCITT Recommendations R35 (50 band) for 25 channels, R37 for 12 channels (100 band) and R36A for 6 channels (200 band). Published by the International Telecommunication Union 1969, Geneva, Switzerland.
22. CCITT Recommendations V21. 200 Band Modem.
23. Milieubelasting door stank, Parts 1, 2, 3, 4. Report of the Rijnmond Authority, Rotterdam, Netherlands, 1973.
24. Annual Report 1974 (in Dutch). Central Environmental Management Service, Rijnmond, Rijnmond Authority, Schiedam, Netherlands.
25. *Federal Register*, 36 (84), part II, Title 42 Public Health Ch. IV, part 410, EPA publication, Washington, D.C., 1971.
26a. *Canada Gazette*, Part I, pp. 3736–3739, December 16, 1973.
26b. The Clean Air Act. Clean Air Regulations 10/73 part I, Government of the Province of Alberta.
27. *VDI-Handbuch, Reinhaltung der Luft*, Book 1, sheets VDI 2105–2108, VDI 2310, VDI-Kommission Reinhaltung der Luft, Düsseldorf, 1971–1974.
28. Dreissigacker, H. L., F. Surendorf, and E. Weber, *Staub-Reinhalt. Luft*, 34 (12), 431–478 (1974).
29. Syrota, J., *Pollut. Atmos.* pp. 115–117, July 1972 (in French).

9

HYDROGEN SULFIDE AS AN AIR POLLUTANT

D. F. S. Natusch* and Barbara J. Slatt

University of Illinois
Urbana, Illinois

* Professor Natusch is now Professor of Chemistry, Colorado State University, Fort Collins, Colorado.

9.1. INTRODUCTION

The importance of hydrogen sulfide (H_2S) as an air pollutant is due primarily to its toxicity, its unpleasant odor, and its reactivity with metals and metal salts. The extreme nature of these effects has, however, resulted in development of fairly effective control methodology, and only small amounts of H_2S are emitted to the atmosphere. Indeed, the atmospheric lifetime of the gas is so short that significant concentrations, although common in many industrial situations, are highly localized. For this reason air pollution standards for H_2S are based mainly on occupational rather than environmental considerations. Yet interest in H_2S as an air pollutant is widespread, and more than 1000 related articles have appeared in the literature since 1950. Currently emphasis is being placed on establishment of environmental standards and on development of more appropriate and sensitive analytical methodology for determining low H_2S concentrations in urban atmospheres.

9.2. OCCURRENCE

It has been estimated[1,2] that about 100 million tons of H_2S is produced annually from natural sources, and about 3 million tons comes from industrial sources. It is difficult to confirm such estimates, which are based on the quantities of sulfur required to balance the sulfur cycle, since few reliable measurements of global atmospheric background levels of H_2S are available.[3–5] Despite some disagreement about the absolute contributions of natural and industrial sources,[2] however, it is fairly certain that man's contribution to global H_2S is very small.

9.2.1. Natural Sources

Table 9.1 summarizes current estimates of H_2S emissions to the atmosphere. There is considerable disagreement about the amount of H_2S that enters the atmosphere from the oceans,[1–3,6] and estimates vary from 30 to 200 million tons annually.[2] There is no direct evidence that ocean waters (other than coastal waters[6]) constitute a significant source of H_2S at all; indeed the solubility and reactivity of H_2S and HS^- in saline waters suggest that very little gaseous H_2S should be released to the atmosphere.[7,8] Until reliable measurements of H_2S concentrations in maritime atmospheres are available, the role of the main bulk of the oceans as an H_2S source must remain in question.

Table 9.1. Annual Amounts of H_2S Introduced into the Atmosphere[2]

Source	Amount (tons)
Biological decay in the oceans	30×10^6
Biological decay on land	$60–80 \times 10^6$
Pollutants	3×10^6

Total annual land emissions amount to an estimated 70 million tons.[2] The main biological source is thought to be the action of the bacterium *Sporovibrio desulfuricans* in reducing sulfates to H_2S.[9] This process is particularly important in polluted tidal inlets,[10] in lakes and inland seas,[11,12] and in certain muds[13,14] and soils.[15] Certain fully or partially landlocked saline waters have a well-defined "H_2S zone" thought to arise partly from the microbiological reduction of sulfate to H_2S.[16,17] The action of bacteria such as *Escherichia coli* on sulfur-containing amino acids (notably cysteine and cystine) also produces considerable H_2S[9,18,19]—a process that can present a significant air pollution problem near refuse heaps containing large amounts of organic (particularly animal) material.

Large concentrations of H_2S often occur in regions of geothermal activity where reducing conditions exist.[20–22] The geothermal areas of New Zealand are noteworthy in this respect, and bore gases commonly contain between 150 and 11,000 ppm H_2S.[22,23] Hydrogen sulfide from geothermal sources ordinarily occurs accompanied by 15 to 40 times its concentration of carbon dioxide.[22] The resulting high density gas mixture tends to remain at a low level, giving ambient concentrations as high as 25 ppm under certain meteorological conditions. For this reason air pollution by geothermally produced H_2S can be a considerable problem in localized areas.[24]

Additional minor natural sources of H_2S include mineral waters,[25,26] inclusions in sedimentary rock,[27,28] stomach gases[29] (particularly of ruminants), and leakage of natural gas.[30] Certain natural gases can contain as much as 42% of H_2S, although levels of the order of 1 to 2% are most common.

9.2.2. Major Industrial Sources

There is no demonstrably reliable estimate of the total annual emission of H_2S from industrial sources. The figure of 3 million tons mentioned earlier is based on the results of an air pollution study in Jacksonville, Florida, where a detailed emission inventory was compiled for sulfur dioxide (SO_2).[31] The estimate for H_2S was obtained on the assumption that measured SO_2 levels were commonly about 50 times higher than those of H_2S. A number of similar studies have been conducted in localized areas.[32–41]

9.2.2.1. Kraft Pulp Mills

Because of the wide variety of operating procedures and conditions currently in use, no two kraft mills exhibit the same pattern of H_2S emissions. Sources of H_2S in the kraft industry have been widely studied,[42–59] however, and the following remarks are generally true.

In a conventional pulping and recovery operation[53] wood chips are cooked with an alkaline sulfide solution in digesters from which wood pulp and spent cooling liquor (black liquor) are continuously removed. Pulp is separated from black liquor by successive washes, and the black liquor is evaporated to a solids content of 50% in multiple-effect evaporators. Concentrated black liquor is then further evaporated to a solids content between 55 and 70% by direct contact with hot flue gases in the recovery furnace and finally is burned in this furnace. Inorganic sulfur is there reduced to sodium sulfide (Na_2S). The smelt from the recovery furnace, which contains molten sodium carbonate and Na_2S, is dissolved in water and recausticized with lime. The resulting "white liquor" is then returned to the digesters.

In most mills the main sources of H_2S are the digesters, the direct contact evaporators, and the recovery furnace, with the last two, in combination, accounting for up to 85% of the total reduced sulfur emissions (predominantly as H_2S[50,53]). Other sources of H_2S include the pulp brownstock washers, the washer filtrate seal tank vent, the knotter hood vents, and general ground level sources such as drains, sewer manholes, storage tank vents, pump leaks, and spills. Table 9.2 indicates the range of H_2S concentrations likely to be encountered from the above-mentioned sources in a representative kraft mill.[48,50,51,53,54] It should be emphasized, however, that actual levels in a given mill will depend markedly on the operating conditions and degree of odor control in force.

The concentration of H_2S in digester off gases and blow-by gases depends on a variety of factors, including the type of wood, the temperature and duration of cooking, and the sulfide concentration.[50,51,55] Emissions from direct contact evaporation are influenced most by the degree of black liquor oxidation practiced. In direct contact evaporation CO_2 and SO_2 in the hot

Table 9.2. Range of H_2S Concentrations Encountered in Kraft Mill Sampling[50]

Source	Concentration (ppm)
Digester vent	16–18,800
Blow gases	0–782
Pulp washer	0–12
Evaporator, noncondensable	907–32,600
Recovery furnace	14–1140
Smelt dissolving tank	10–44
Lime kiln	0–254
Tall oil cooking	5400–101,000

flue gases react with Na_2S in black liquor to liberate H_2S by the following reactions:[47,51]

$$Na_2S + H_2SO_3 \longrightarrow H_2S + Na_2SO_3 \quad (1)$$

$$Na_2S + H_2CO_3 \longrightarrow H_2S + Na_2CO_3 \quad (2)$$

If, however, the black liquor has been previously oxidized, the Na_2S will have been converted to sodium thiosulfate according to the following overall reaction:

$$2Na_2S + 2O_2 + H_2O \longrightarrow Na_2S_2O_3 + 2NaOH$$

and reactions 1 and 2 cannot take place. These reactions indicate that the amount of H_2S formed during direct contact evaporation should depend on the pH and the Na_2S content of the black liquor. These dependences are observed in practice.[47] Other factors include the design of the direct contact evaporator, the nature of the black liquor solids,[44] the amount of H_2S in the flue gases contacting the black liquor[47] and the design and operating conditions of the recovery furnace.[43]

Emissions of H_2S from the recovery furnace (as distinct from combined emissions from the furnace and direct contact evaporator system) are profoundly influenced by furnace operating conditions. Essentially the furnace contains three reaction zones:[43]

1. A reducing zone in the region of the smelt bed.
2. A drying and pyrolysis (oxidizing and reducing) zone above and below the nozzles introducing black liquor.
3. A gas-burning (oxidizing) zone in the superheated region above the spray nozzles.

In the reducing zone the oxygen content of the gases must be kept low to promote reduction of sodium sulfate (Na_2SO_4) to Na_2S by the reaction

$$Na_2SO_4 + 2C \longrightarrow Na_2S + 2CO_2$$

Although this reaction is essential to the kraft recovery process, the necessary reducing conditions produce H_2S by pyrolysis of any organic material containing sulfur. In the intermediate zone (2), reducing reactions produce large amounts of H_2S by pyrolysis of black liquor solids. Promotion of oxidizing conditions by increasing the total and secondary air flows to the furnace greatly reduces H_2S formation in this region. In the gas burning zone (3) H_2S is converted to SO_2, provided sufficient temperature, contact time, oxygen, and mixing of reactants are achieved. In fact, the degree of H_2S oxidation in zone (3) depends largely on the degree of oxidation occurring in the intermediate drying and pyrolysis zone (2). Overall optimization of the operating conditions of the recovery furnace is claimed to reduce H_2S emissions to less than 1 ppm.[44]

It has been established that the direct contact evaporators and the recovery furnace produce more than 23 sulfur-containing compounds[51,57,58] and that they convert a high percentage of the total sulfur to volatile products. Ordinarily H_2S makes up 30 to 75% of these products.[50,51] If the recovery furnace is overloaded beyond its design capacity, the reducing pyrolysis of black liquor solids (thus H_2S production) increases dramatically (Figure 9.1).[48,50,51,53,59] The degree of furnace overload is thus one of the major factors determining the concentration and amount of H_2S emitted from a kraft pulp mill.[52,60]

The odor from kraft mills comes from a variety of sulfur compounds in addition to H_2S.[50,51,55,56,61] Often this odor can be detected up to 30 miles from a mill, and local concentrations of H_2S of the order of 1 ppm are not uncommon.

9.2.2.2. *Petroleum Refineries*

Almost all crude oils contain sulfur, ranging in amount from 0.4% in the better Pennsylvanian crudes to 4.5% in some Mexican crudes.[62] Between 20 and 40% of the total sulfur is present as the element, and the rest is mainly dissolved H_2S.[62,63] Both can be released as H_2S in the refining process, where reducing conditions exist.[64,65] The primary sources of H_2S in a refinery are the distillation, cracking, and hydrogenation processes.[65–67]

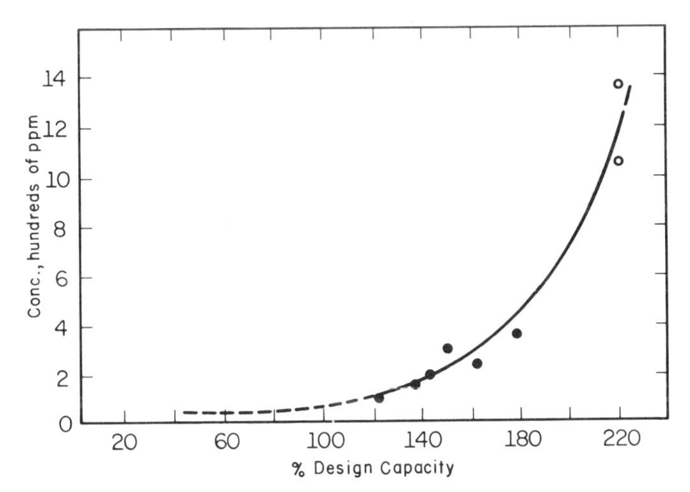

Figure 9.1. Effect of recovery furnace overload on H_2S emissions.[50]

In the distillation process H_2S may be either discharged to the atmosphere from the vacuum jet or concentrated in noncondensable gases (some of which may be discharged).[65] However ammonia is often added to the overhead reflux system of a crude oil distillation column,[66] and the majority of H_2S from this source is present as ammonium bisulfide (NH_4HS) in aqueous steam condensates.

A small amount of H_2S is discharged in the plume of catalytic cracking units; however the majority is again present as NH_4HS in aqueous steam condensates from both thermal and catalytic cracking and from hydrogenation. All these processes produce both NH_3 and H_2S. (A typical gaseous effluent from a catalytic cracking unit consists of 55% CH_4, 42.8% N_2, 0.85% NH_3, and 0.85% H_2S.[68]) Nowadays, most H_2S produced in refinery operations is recovered either from gaseous or aqueous effluents, and only relatively small amounts are vented to atmosphere. Some H_2S is often released, however, from aqueous effluents (even after recovery operations) in the sewage system.[65,66,69]

Table 9.3 gives an example of the range of H_2S concentrations determined in the vicinity of a group of oil processing and petrochemical plants consisting of three refineries, a synthetic alcohol plant, a chemical plant, and three oil-fired power stations. Even 12 miles from the main sources (the refineries), H_2S levels constitute significant pollution.[70]

9.2.2.3. *Natural Gas Processing*

The amount of H_2S in natural gas varies widely from a fraction of 1 ppm in so-called sweet gas to as much as 42% in some gas fields in Wyoming.[71,72] Natural gas containing more than about 15 ppm (1 grain 100 ft^{-3}) of H_2S must be purified before distribution and domestic consumption,[72] while gas containing high percentages of H_2S is used as a commercial source of sulfur.[71,73] Sulfur recovery operations, such as that practiced on natural gases from the Lacq field in France, where the H_2S content is initially 15.3%,

Table 9.3. H_2S Concentrations in the Vicinity of a Group of Oil Processing and Petrochemical Plants[35]

Location	H_2S (ppm)
Immediate neighborhood	0.0112–0.0988
2–5 km	0.0053–0.046
20 km	0.0006–0.032

usually exhaust quite high levels of H_2S to the atmosphere. Even after incineration, the tail gas from the Lacq works contains 0.005% H_2S.[74]

9.2.2.4. Gasification of Coal

Considerable quantities of H_2S are produced in the manufacture of coal gas and producer gas for household and industrial consumption and in the coking of coals in the steel industry.[36,75–77] Most coals contain between 0.8 and 4% of sulfur, of which about half is present as iron sulfides and half as organic sulfur compounds.[78] The manufacture of coal gas typically converts between 30 and 50% of the total sulfur to H_2S, leaving the remainder in the coke. As a result, untreated coal gas usually contains 0.3 to 2% of H_2S.[75,78–80] Producer gas, which is manufactured from coke, contains somewhat less H_2S (~0.3%).[75] In both cases the H_2S levels depend on the method of gasification employed.[80]

Most of the H_2S that is introduced into the atmosphere from gas manufacturing operations enters during recharging of the retorts.[81] Considerable amounts are also released when hot coke is cooled by spraying it with water,[76] especially in some older works where the water is recirculated to an open cooling frame.[75] Other sources include aqueous condensates in drains and sewers and gas leaks prior to the H_2S purification stage. In many works iron oxide (Fe_2O_3) is used to remove H_2S from the gas before distribution. The spent oxide is piled in the open air while reoxidation takes place, and these piles usually lose some of their sulfur as H_2S.

Hydrogen sulfide is only one of many air pollutants produced in coal gasification, as indicated by the typical inventory of air pollutants presented

Table 9.4. Inventory of Air Pollutants from Coke Plants[36]

Pollutant	Annual Emission[a] (tons)
Sulfur dioxide	13,000
Dust	9,900
Carbon monoxide	3,030
Benzene	1,660
Hydrogen sulfide	990
Phenol	990
Ammonia	950
Cyanide	99

[a] Total emissions from eight plants in northern Moravia, Czechoslovakia.

in Table 9.4.[36,82] Nevertheless, considerable quantities of H_2S are emitted, and the odor is readily apparent in the vicinity of gasification plants.

9.2.2.5. *Coal Refuse Disposal*

Approximately 20% of the bituminous coal and anthracite mined is rejected because of its small size, high ash content, or mineral impurities. This refuse is collected in large heaps near the minehead or coal processing plant. Oxidation of iron sulfides and other reducing materials by air in the presence of moisture is exothermic and often produces sufficient heat to ignite the coal refuse, which emits large quantities of H_2S and SO_2.[83,84] Some refuse heaps are also ignited either deliberately or through carelessness.

The exact mechanism of H_2S production in a burning refuse heap has not been established; however the reactions

$$H_2SO_4 + FeS \longrightarrow FeSO_4 + H_2S$$

and

$$3C + 2SO_2 \longrightarrow CS_2 + 2CO_2$$

followed by

$$CS_2 + 2H_2O \longrightarrow CO_2 + 2H_2S$$

have been suggested.[83]

Concentrations of H_2S and SO_2 in the ranges 0.05 to 0.4 ppm and 0.5 to 5.0 ppm, respectively, are commonly encountered in communities adjacent to refuse heaps. Although such levels are undesirably high, it is extremely difficult to extinguish the fires. There are more than 500 burning coal refuse heaps in the United States, and some have been burning for more than 20 years.

9.2.2.6. *Viscose Processing Plants*

During the manufacture of viscose rayon yarns, commercially pure cellulose is steeped in alkali, ground into crumbs, and aged; carbon disulfide (CS_2) is then added to form cellulose xanthates, which constitute the viscose solution. The viscose is mixed with caustic soda and, after blending, filtering, and deaerating, is extruded through nozzles in a spinning bath containing sulfuric acid, sodium sulfate, zinc sulfate, and sugars. At this stage solid fibers form and the process is termed "spinning." The fiber is then washed and twisted into yarn.[85]

During the formation of cellulose xanthate, considerable amounts of CS_2 are converted to H_2S[86–88] (about 25% of which is oxidized to sulfur in the spinning bath[88]). Approximately 90% of the H_2S remaining is evolved from the spinning bath,[87] and most of this is exhausted by high velocity fans. Concentrations of CS_2 and H_2S of the order of 300 to 800 ppm are common in this exhaust,[86,89,90] and H_2S levels around 1 ppm can occur in the vicinity of the precipitating and spinning baths. Approximately 4% of the H_2S produced is evolved during finishing of the thread, particularly from the washing cylinders.[87]

The actual amount of H_2S evolved can vary widely and depends on the operating temperatures of the various stages, the geometry of the equipment and, to a small extent, on the fiber size. There is no apparent dependence on process rate or on the amount of residual cellulose xanthate in the fibers.[87] Concentrations of H_2S measured in the vicinity of viscose plants can be of the order of 0.1 ppm at 2000 yards distance under suitable wind conditions.[37–41,91]

Atmospheres of viscose plants constitute health hazards,[72,92–95] and there is considerable emphasis on control of both H_2S and CS_2 in the industry.[96]

9.2.2.7. *Meat Processing*

The thermal breakdown of protein under the reducing conditions that apply during cooking, rendering, and smoking of meat, fish, fat, blood, and vegetables produces highly offensive odors.[97–99] Predominant odor constituents include H_2S, mercaptans, disulfides, ammonia, amines, and certain fatty acids.

It has been shown that the rate of evolution of H_2S from meat during cooking increases with time and with temperature but reaches a maximum at 120°C.[100] Effluents from dry rendering of animal wastes (bones, skin, meat, hair, and offal) at 125°C commonly contain H_2S concentrations in the range 20 to 120 ppm even after steam condensation.[101] The amount of H_2S produced, however, depends markedly on the design of the rendering equipment.

The onset of bacterial decay greatly increases the amount of H_2S liberated during rendering,[18,99,101,102] presumably because both processes achieve protein breakdown. Also, when small amounts of keratin in the form of wool or hair are present, there is a significant increase in the amount of H_2S evolved.[101] This effect is most pronounced in high temperature (~370°C) blood driers where wool mixed with the blood is incinerated.

Exhaust gases from rendering plants are classed among the very worst industrial odors, and the smell is usually apparent up to a mile from a plant even on calm days. If H_2S is selectively removed from rendering exhausts, the intensity of the odor is greatly reduced, indicating that H_2S is indeed a major odor constituent.

9.2.2.8. *Sewage Disposal*

The two main constituents of sewage gases are hydrogen sulfide and methane (CH_4). A variety of compounds such as mercaptans, amines, ammonia, putrescine, cadaverine, indole, and skatole[103] are also present in small amounts, and it is primarily skatole that in synergism with H_2S, gives sewage its characteristic odor.

Hydrogen sulfide is formed from sulfates and proteinaceous material by microorganisms under anaerobic conditions.[104–106] These organisms, which exist in the slimes formed below the water level in sewer pipes, include *Spirillum desulfiricans*, *Microspira alstanrii*, and *Vibro desulfuricans*.[107] The reactions by which they form sulfides are not known.[108]

It has been shown that H_2S formation occurs after the reduction of oxygen and nitrogen compounds in sewage.[105] Consequently, H_2S production increases with biological oxygen demand (BOD)[109,110] as well as with the concentration of sulfates and proteinaceous material in solution. Formation of H_2S can be markedly reduced by diluting sewage to reduce BOD, retain H_2S in solution, and prevent deposition of organic material. Indeed, when the sulfate concentration is below 25 mg l^{-1}, generation of H_2S is often inhibited.[105]

Sewage microorganisms thrive in an atmosphere of high humidity and temperature, and the magnitude of H_2S generation increases about 7% per degree celsius rise in sewage temperature up to 30°C (86°F).[106,110] Generation of H_2S in sewers is thus greatest in hot countries and during the summer months. The pH of sewage affects H_2S generation in two ways: the optimum range of bacterial operation is 7.5 to 8.0,[104,110] and the proportion of total sulfide as H_2S increases with decreasing pH. Maximum generation of H_2S occurs between pH 6.5 and 7.5.[105]

The velocity and turbulence of flow of the sewage also affects the evolution of H_2S.[105] Low velocities promote sludge deposition and slime growth, whereas high velocities cause turbulent conditions that release H_2S from solution. Another important factor is the retention time of the sewage in the sewerage system. In general, short, direct flow sewerage systems have much lower H_2S levels than more complex systems in which sewage is exposed to bacterial action for long periods.

Most sewerage systems handle industrial as well as domestic wastes, and the sewage content can be highly variable. It has been shown that addition of sulfates, sulfides, and sulfites greatly enhances H_2S production, and there is evidence that ground garbage may also contribute.[107,111] Highly alkaline industrial wastes tend to reduce the amount of H_2S evolved, as do sulfide precipitating agents such as metal ions.

Levels of H_2S in sewer pipes are commonly around 1 ppm, although levels of several hundred ppm can occur. Atmospheric pollution from sewage gases is, however, normally restricted to the vicinity of pumping or treatment stations, where H_2S concentrations up to 1 ppm can be encountered. The gas is also evolved through street drainage gratings, particularly in cold weather when the temperature of the sewage gases is higher than that of the air.

9.2.3. Minor Sources

Ambient concentrations of H_2S up to 1.3 ppm have been reported in the vulcanization sections of plants processing natural and artificial rubbers and ebonite based on artificial rubbers.[112–114] However the presence at more toxic levels of other gases and vapors (notably butadiene, styrene, oil fog, formaldehyde, methanol, acreolin, aromatic amines and nitriles, and carbon monoxide) makes H_2S a relatively minor pollutant. Significant health hazards can result from vulcanization in poorly ventilated areas.

In the manufacture of phosphoric acid by the thermal processes, H_2S, Na_2S, or NaHS is added to the acid to precipitate trace metal impurities. Excess sulfide is blown off as H_2S. The emissions are intermittent but may range from 10 to 2500 ppm.[115] Hydrogen sulfide is second only to phosphoric acid mist as an air pollutant from this industry.

Fumes emitted from varnish cookers (particularly the open pot variety) during the manufacture of varnish contain significant quantities of H_2S.[116] Sulfur compounds present as impurities in tall oil or rosin are converted to H_2S during esterification with glycerine and pentaerythritol and during depolymerization of some natural resins. Some H_2S is also evolved from varnish (and oil-based paint) solvents after the cooking stage. The predominant odor from varnish cooking is that of acreolin; however H_2S and some allyl sulfide, butyl mercaptan, and thiophene also contribute because of their low odor thresholds.

The manufacture of synthetic alcohol from petroleum products containing small amounts of sulfur compounds produces H_2S in an essentially similar manner to the cracking and distillation processes in a refinery. Most of the sulfur is emitted as SO_2, but some H_2S is also present. Measurements made in the vicinity of one synthetic alcohol plant indicated that ambient levels of H_2S 300 yards from the plant were the order of 0.06 ppm. The corresponding SO_2 level was 4.4 ppm.[117]

Manufacture of chemicals from coal tar produces H_2S by reduction of the sulfur compounds remaining in tar after coking of the coal.[118] Normally H_2S is accompanied by a variety of other noxious gases and vapors (notably phenols, ammonia, and carbon monoxide). These gases are also emitted with H_2S from smelting operations and blast furnaces,[119,120] where reducing conditions convert sulfur compounds present in ores (particularly pyritic ores) to H_2S. In one case measurements made 1500 ft above a steel works indicated H_2S concentrations up to 0.3 ppm.[120]

A small but significant fraction of the sulfur compounds used in the manufacture of sulfur dyes and other sulfur-containing chemicals is converted to H_2S. Nowadays gaseous effluents from such industries are fairly effectively controlled, but local concentrations of H_2S are usually readily detectable by smell.

From the earliest times H_2S (or "stink damp") has been known in mines. In most cases it is produced by the action of the bacterium *Desulfovibrio desulfuricans* on pyritic material under moist anaerobic conditions.[121] Seepage of natural gas containing H_2S also occurs in a few mines. The concentrations of H_2S found in mines vary widely and are poorly documented; however a number of cases of both fatal and nonfatal intoxication of miners have undoubtedly been due (at least in part) to H_2S.

The gas also can present a lethal hazard in tanneries, where it is formed by bacterial and occasionally acidic action on spent sulfide lime liquors used in the dehairing process.[122] High H_2S levels are often encountered in tannery drainage canals and sewers from disposal of sulfide liquors, and corrosion of concrete sewers by H_2S is common near tanneries.

Fermentation reactions involving yeast produce small quantities of H_2S whose odor is often apparent in breweries[123] and during dough mixing in bakeries.[124] The amount of H_2S emitted depends on the grain or flour used and on the temperature and pH.

Although H_2S can scarcely be classed as a major household pollutant, its presence in the home[125] constitutes a definite nuisance (e.g., as sulfide tarnishing of silver and copper). Common sources include cooking vapors (especially from chicken, pork,[100] fish,[102] cabbage, and onions), resin linoleum fumes,[126] fresh oil-based paint and varnish fumes, body odors, decaying garbage, and unsanitary kitchen and bathroom facilities.

Bacterial action on excreta, straw, and waste food in animal rearing and fattening units produces a highly odorous effluent whose major constituents are H_2S, NH_3, CO_2, and CH_4.[127] When cleaning and ventilation are inadequate, the effluent can reach lethal levels.[128]

Although only a minor air pollutant, some H_2S is produced in cement kilns where the compounds SiS, SiS_2, and SiOS are formed from sulfates or pyritic material at the high temperatures ($\sim 1100°F$) employed. Hydrolysis of these compounds produces H_2S.[129]

Incineration of household waste and home or warehouse fires can produce lethal concentrations of H_2S if natural fibers such as wool, hair, silk, feathers, or textiles containing these materials, are burned. Wool is particularly dangerous in this respect, and H_2S liberated from burning wool has contributed to facilities in a number of warehouse fires. Tobacco smoke has also been shown to contain H_2S. The amount liberated depends on the length of tobacco smoked and on the rate of smoking.[130]

9.3. AEROCHEMISTRY

One of the major scavenging reactions for H_2S in the atmosphere is thought to be that with ozone[2,131,132]

$$H_2S + O_3 \longrightarrow H_2O + SO_2$$

In the laboratory this reaction is at least partly heterogeneous, since its rate depends on the surface area of the reaction vessel. The rate equation is

$$-\frac{d[H_2S]}{dt} = k[O_3]^{1.5}[A]^{0.5} \tag{3}$$

where $[A]$ is the specific area of the reaction surface. The reaction is thus zero order in H_2S and the rate constant $k = 200$ cm mole$^{-0.5}$ sec^{-1} at 300K when ozone concentrations are expressed in moles per cubic centimeter.

If it is assumed that airborne particles provide the necessary surface for reaction between H_2S and O_3 in the atmosphere, the lifetime of H_2S can be estimated from eq. 3. For particle concentrations (in continental atmospheres) of 15,000 per cubic centimeter near the earth's surface and 200 per cubic centimeter in the upper troposphere, and calculated specific areas of 10^{-4} and 5×10^{-7} cm^2 gram^{-1}, respectively, the lifetime of 1 ppb of H_2S in the presence of 0.05 ppm of O_3 turns out to be 2 hr near the earth's surface and 28 hr in the troposphere.[1]

These calculated lifetimes take no account of the huge available reaction surface at ground level where H_2S predominates; nor do they include contributions from reactions other than that with ozone. Nevertheless such considerations make it clear that H_2S levels in areas remote from any significant source will be very low. The few available measurements of atmospheric H_2S in remote areas indicate levels of 20 to 60 parts per trillion (ppt: 1 in 10^{12}) in a remote area close to the North American Continental Divide[5] and in maritime atmospheres off the Florida coast,[133] with somewhat higher levels (100 to 350 ppm) in rural areas of Illinois and Missouri.[134] Based on these measurements, calculated lifetimes are in general accord with those just indicated.

Clearly, man's total contribution of H_2S to the atmosphere has a negligible effect on global background levels of H_2S. Since H_2S is converted ultimately (via SO_2) to sulfate aerosols, its primary contribution is to particulate matter in the atmosphere.[135] However considering the amounts of industrially produced H_2S involved ($\sim 3 \times 10^6$ tons $year^{-1}$) and the wide distribution of sources, it is doubtful whether this contribution is of any significance.

9.4. MEASUREMENT

At present there is no single analytical method for the determination of H_2S that incorporates all the desirable characteristics of speed, low cost, ruggedness, high sensitivity, wide concentration range, freedom from interferents, ease of manipulation, and automatic operation. Fortunately, however, one seldom requires all these characteristics for any given application, and the available methods go a long way toward meeting most specific analytical requirements.

Many of the published methods for the determination of H_2S have been strongly criticized in the literature for shortcomings in sensitivity, specificity or (most commonly) instability of the collected sulfide. Such criticisms are often valid, but it must be stressed that they are generally only important where low H_2S levels (<1.0 ppm) are involved. The following sections emphasize the conditions necessary for obtaining reliable analyses. Analytical procedures are divided into gas collection and analytical determination stages, and a small number of methods are recommended for the determination of H_2S with respect to specific situations.

9.4.1. Collection and Preconcentration

Until recently research has been heavily weighted toward developing methods for estimating H_2S after it has been collected from a given volume of air. This is unfortunate, since in common with most other environmental analyses, collection of H_2S is the limiting step. Ideally, one would like to determine H_2S *in situ* without recourse to a collection step, but this is not usually possible unless high concentrations (~ 100 ppm) are present. The following remarks are intended as a critique of existing methods of H_2S collection with respect to their efficiency, selectivity, and ability to stabilize the collected sulfide.

9.4.1.1. Absorption

The efficiency of mass transfer of H_2S from air bubbling through an aqueous solution to the solution phase depends on the gas–liquid contact time, the diffusion coefficients of H_2S in the gas and liquid phases, the bubble size, the

H_2S concentration, and the solubility of H_2S in solution. Manipulation of these parameters by suitable choice of bubbler design, air flow rate, and collecting solution can usually result in quantitative collection.

Calvert and Workman[136] have studied the mass transfer characteristics of a number of bubbler designs and have produced a method for predicting their efficiency of collection. This method should be recognized as predominantly qualitative, but it can serve as a useful guide. Some of the more efficient designs include that of Wartburg et al.,[137] the petticoat bubbler, and the Greenberg–Smith impinger.[4,138] Midget impingers,[138] Dreschel bottles,[4] and packed columns[139] are useful where low flow rates are involved. Care must be taken to prevent carryover of solution at high flow rates, but in practice the efficiency of collection of H_2S should not be limited by lack of a suitable bubbler design.

Collection efficiency is greatly improved by choosing an absorbing solution in which H_2S is highly soluble. Consequently, a number of analytical methods rely on the collection of H_2S as HS^- in alkaline solutions.[140–143] This procedure lacks specificity in that a wide variety of acidic (and even some nonacidic) gases will be collected. Also, sulfide ions in alkaline solution are readily oxidized by air passing through the solution,[78,144,145] and the resulting $S_2O_3{}^{2-}$ and $SO_4{}^{2-}$ ions are not detected by most subsequent analytical techniques. Collection in alkaline solution is acceptable for sampling H_2S concentrations above about 1 ppm, but care should be taken to keep the collection time short and to perform the sulfide analysis immediately.

The majority of solutions for collecting H_2S employ a dissolved or suspended metal salt that stabilizes the gas as an insoluble sulfide. The most reliable and widely used is a suspension of cadmium hydroxide.[4,138,143–146] (Mercaptans are also collected as insoluble cadmium mercaptides, and the solution retains some acidic gases, notably SO_2.)

There is very convincing evidence that cadmium sulfide suspensions are susceptible to photooxidation.[4,147–150] Irreproducible sulfide losses as high as 80 to 90% have been reported in both laboratory[147] and field studies.[150] Methods proposed to eliminate photooxidation include the addition of antioxidants such as STRactan 10 (arabino-galactan) to the collection solution,[147] the exclusion of light,[4,147] and the use of alternative light-resistant collecting reagents such as the zinc amine complex[149] or a mixture of cadmium acetate, cadmium sulfate, sodium citrate, and triethanolamine.[151] The most satisfactory procedure is to exclude all light.

Although it is possible to prevent photooxidation of cadmium sulfide, recovery of the sulfide from the collecting solution is still not quantitative and decreases with storage time.[147] The reason for this additional sulfide loss is not clear. It does not depend significantly on pH in the range 7 to 12 nor on the presence of antioxidants; however losses are much greater from darkened collecting solutions through which air has been aspirated than from identical unaspirated solutions that have stood for the same time. Oxidation by air seems unlikely, since the amount of sulfide lost is the same whether nitrogen, oxygen, or air is aspirated through the solution (the same is true when photooxidation is allowed to take place).[147] When zinc acetate collecting solutions are used, however, oxidation by air definitely takes place and produces very large measurement errors.[152]

It has been suggested[147] that the loss of sulfide by mechanisms other than photooxidation is fairly constant and that 85% recovery of the collected sulfide can be achieved independent of H_2S concentration. Contrary to these findings, Smith et al.[4] report quantitative recovery of known weights of H_2S (6 to 58 μg) generated rapidly by the action of H_2SO_4 on sodium sulfide, and it is clear that sulfide losses from cadmium sulfide suspensions warrant further study. Nevertheless, cadmium hydroxide suspensions undoubtedly provide a reliable absorbing medium for collecting concentrations of H_2S above about 0.1 ppm. Lower concentrations may be sampled, provided calibrations are based on standard H_2S concentrations rather than on nonuniform H_2S generation or on sulfide standards.

Information regarding the reliability of collecting solutions containing reagents other than cadmium hydroxide or alkali is sparse. Zinc acetate,[149,153] silver nitrate,[4,154] and ammonium arsenite[4,155] have been used to collect H_2S concentrations substantially lower than 1 ppm, but there is little to recommend them for such low concentrations. Both zinc sulfide and silver sulfide are rapidly oxidized in solution (the latter to silver oxide), with consequent loss of sulfide.[4,147] The arsenious sulfide sol is stated to be stable and insensitive to SO_2, yet Smith et al.[4] were unable to achieve better than 60% recovery of known weights of sulfide in the 190 to 360 mg range. Nevertheless, these instabilities do not produce significant H_2S losses when high H_2S levels (>1 ppm) are involved.

In addition to the above-mentioned reagents, potassium bisulfate–silver sulfate mixtures,[156] ammonium molybdate,[157] lead oxalate,[158] iodine solution,[141,159,160] sodium bromide,[161] hydrazine solutions,[162] and dimethylformamide[163] have all been used for collecting high concentrations of H_2S (>1 ppm) in solution. Some of these reagents enable direct evaluation of the collected product, thus reducing the number of steps in the analysis and enabling direct continuous monitoring of high H_2S concentrations. Their use in the field of air pollution is, however, very limited.

9.4.1.2. *Adsorption*

Many collection procedures involve reaction between H_2S and a metal salt that has been impregnated onto a solid substrate such as a filter paper,[5,24,154,164–167] silica gel,[168–170] or porous ceramic tiles.[171] Quantitative collection of H_2S can be obtained by passing the air sample through impregnated filter paper or through a tube packed with impregnated granules. In the case of impregnated tiles, however, only that fraction of H_2S reaching the surface by atmospheric transport processes is collected. This procedure is normally used only for qualitative or rough integrated determinations of H_2S and is not considered further.

The efficiency of mass transfer of H_2S from air to a solid surface depends on the gas–solid contact time, the diffusion coefficient of H_2S in air, the H_2S concentration, and the nature and amount of the impregnated reagent. Quantitative collection can generally be achieved by manipulating the gas flow rate and the depth and density of the impregnated filter.[164] In these respects it should be noted that the fraction of H_2S adsorbed increases exponentially with the depth of the filter.

Low concentrations of H_2S can be quantitatively removed from air passing through filters impregnated with silver nitrate,[4,5,24,164] potassium argentocyanide,[24,164,165] mercuric chloride,[24,164,166,167] lead acetate,[4,24,140,164,172] or zinc acetate.[173] A large number of filter types have been used, but the most common is Whatman No. 4 paper, whose low density makes it possible to employ high air flow rates. The dependences of collection efficiency on the nature of the impregnating reagent, the impregnating time, and the H_2S concentration are illustrated as a function of flow rate in Figures 9.2, 9.3, and 9.4, respectively. Recommended[164] impregnating solutions and impregnating

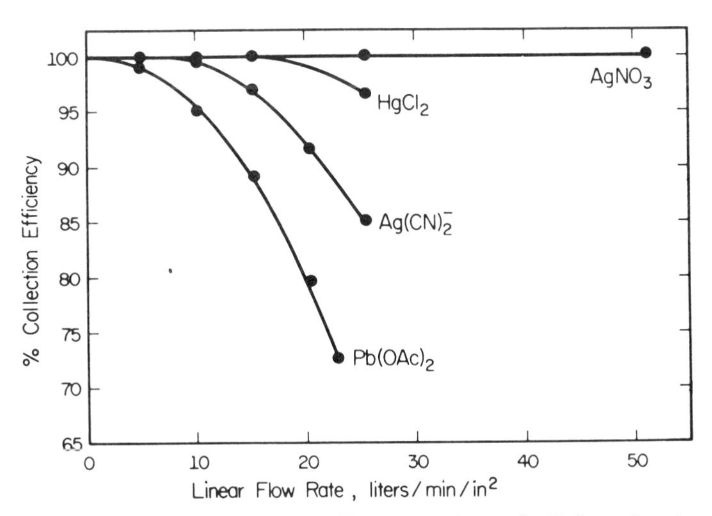

Figure 9.2. Percentage collection efficiency of impregnated tapes for H_2S as a function of linear flow rate; $[H_2S]$, 1.2 ppm, soaking time, 2 hr.

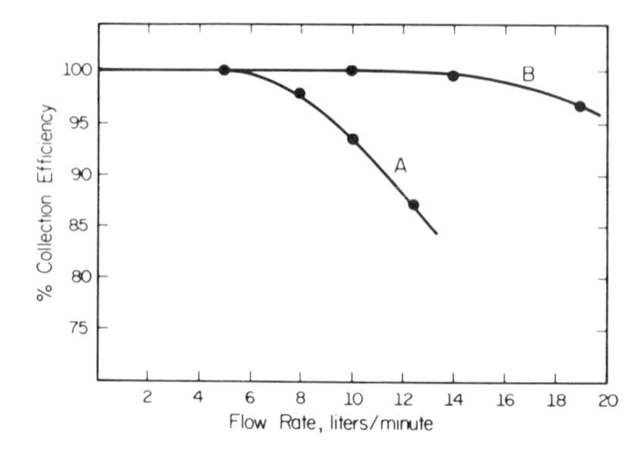

Figure 9.3. Influence of soaking time on percentage collection efficiency of $AgNO_3$-impregnated tapes for H_2S as a function of flow rate: *A*, soaking time, 1 min; *B*, soaking time, 2 hr, $[H_2S]$, 1.2 ppm.

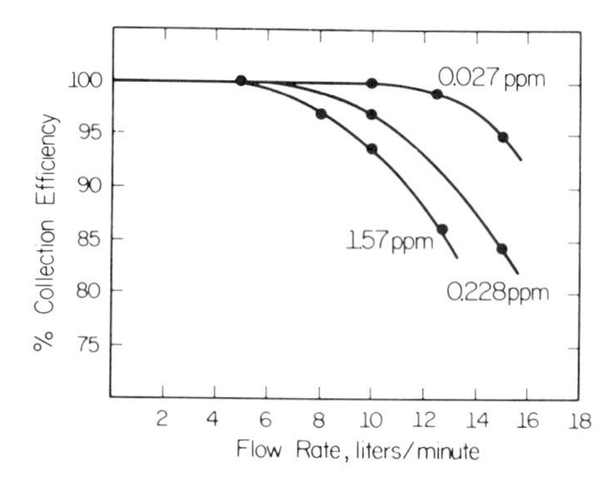

Figure 9.4. Influence of H_2S concentration on percentage collection efficiency of $AgNO_3$-impregnated tapes for H_2S as a function of flow rate.

times for Whatman No. 4 filter paper are given in Table 9.5. Filters prepared in accordance with the data in Table 9.5 will achieve 100% collection of 0.1 ppm H_2S up to the air flow rates listed.[164]

Impregnated solid materials require the presence of a small amount of moisture for efficient reaction with H_2S. In this sense an impregnated filter should be considered chemically as a solution that contains a small amount of solvent and a large amount of predominantly precipitated reactant. Therefore it is reasonable to expect the efficiency of H_2S collection to be low when air of low relative humidity (RH) dehydrates the filter. This effect is of little importance for air above about 15% RH, but progressively lower flow rates must be used[5] to achieve quantitative collection for RH values below 15%. The use of a prehumidifier is not recommended for precision low level work

Table 9.5. Operational Comparison of Impregnated Paper Tapes, Maximum Flow Rates, Sensitivity Factors, and Detection Limits

Tape Reagent	Amounts and Time	Maximum Flow Rate ($l\ min^{-1}\ in.^{-2}$)	Sensitivity Factor K ($ppm\ l^{-1}\ in.^{-2}$)	Detection Limits	
				1 hr Sampling Time (ppm)	1 m^3 Sample (ppm)
Silver nitrate		60	71	0.0002	0.0007
$AgNO_3$	2 g				
HNO_3 (1 N)	1 ml				
Ethanol	20 ml				
Water	80 ml				
Soaking time	1 hr				
Argentocyanide		10	57	0.0010	0.0006
$AgNO_3$	15 g				
KCN	15 g				
Water	100 ml				
Soaking time	2 hr				
Mercuric chloride		15	57	0.0007	0.0006
$HgCl_2$	10 g				
Urea	2 g				
Ethanol	100 ml				
Soaking time	1 hr				
Lead acetate		5	45	0.0016	0.0005
$Pb(OAc)_2$	10 g				
Acetic acid	2 ml				
Water	100 ml				
Soaking time	1 hr				

because the device absorbs some H_2S. Humidity effects can be better overcome by addition of 1 ml of a humectant such as glycerol to the impregnating solution.[5]

Metal sulfides resulting from the collection of H_2S by filters impregnated with silver nitrate, potassium argentocyanide, or mercuric chloride are extremely stable to the passage of large quantities of air. The presence of 200 ppb ozone (i.e., ~4 times the natural atmospheric concentration) causes oxidation of about 18% of the silver sulfide formed from collecting 1 ppb H_2S,[5] but with this minor exception none of these sulfides is affected by any of the normal constituents of polluted air. Silver nitrate impregnated papers are light sensitive, but the light reaction has no demonstrable effect on the stability of the sulfide.[5]

By contrast, the lead sulfide formed on lead acetate impregnated filters is unstable to light, and the color can be leached out by pure air and by air containing SO_2, ozone, or nitrogen oxides.[4,147,164,174,175] The stability of lead sulfide is greatest when the impregnating solution has a pH lower than 2. Quantitative collection and stabilization however, can be achieved at low flow rates (<5 l min^{-1} in.$^{-2}$ for 0.1 ppm H_2S) and in darkness. It should be noted that many automatic impregnated tape samplers employ flow rates much higher than this and that certain instruments continuously illuminate the spot during its formation. Results obtained using lead acetate tapes in this manner should be treated with considerable caution[164] for H_2S concentrations <1 ppm.

Mercaptans will react with the impregnated filters listed in Table 9.5 to form metal mercaptides, which may interfere with later sulfide analyses. Heterogeneous reactivity with mercaptans increases in the order silver < mercury < lead, and in all cases collection efficiencies are less than for H_2S.[164] Selectivity for H_2S is thus promoted by employing high flow rates and using silver nitrate impregnated filters.

The undoubted limitations of lead acetate impregnated filters have resulted in recommendations against their use in air pollution control work.[154,176–178] This is unfortunate because the shortcomings of lead acetate have been reflected onto silver nitrate as an impregnating reagent. This reflection is completely unjustified. Indeed, silver nitrate filters are to be strongly recommended, since they allow quantitative collection of H_2S at high flow rates, completely stabilize the collected sulfide, and have excellent selectivity for H_2S.[4,24,164] By comparison, the other impregnating reagents are much less desirable, although all can be used provided their limitations are clearly recognized and avoided.

Impregnated silica gel is normally used as the adsorbing material in indicator tubes[168,169] to collect high concentrations (>1 ppm) of H_2S. Common impregnating reagents include silver cyanide,[168,169] bismuth nitrate,[179] copper sulfate,[180] lead acetate, or a mixture of these reagents.[168] Iodine is used to collect both H_2S and SO_2 in admixture.[168] The collecting and stabilizing abilities of reagents impregnated onto silica gel are essentially similar to those for impregnated paper filters.

9.4.1.3. *Direct Sampling*

Ideally, it is desirable to eliminate the collection stage of any analytical procedure and to determine the pollutant *in situ*. A number of analytical procedures employing such direct sampling of H_2S have been reported.[181–184] With the notable exception of gas chromatography,[183] however, few detection methods are sufficiently sensitive for use without preconcentration of H_2S. This is not true, of course, where industrial levels greatly in excess of 1 ppm are involved.

A variation of direct sampling involves collection for later analysis of a small volume of air in a gas-tight syringe or other suitable container. However there are very few materials that will not adsorb or react with H_2S.[184–188] For example, Koppe and Adams[187] report extensive adsorption losses from low H_2S concentrations (<1 ppm) on glass, Teflon, and stainless steel. Thick-walled Mylar laminates have proved suitable for containing H_2S concentrations around 1 ppm for several weeks,[164] and Saran containers can also be used for periods of a few hours.

9.4.2. Evaluation

The literature describes a wide variety of methods for evaluating H_2S either directly or as a collected sulfide. The great majority of these methods are useful only for concentrations in excess of 1 ppm H_2S, and relatively few are applicable to air pollution control work. The following sections discuss the advantages and disadvantages of a number of evaluation methods with respect to their sensitivities, selectivities, and areas of application for determining H_2S. Other reviews are also available.[141,175]

9.4.2.1. *Volumetric Methods*

Volumetric methods for the determination of sulfide in solution require the manipulation of solutions and glassware and are best suited to laboratory analyses. Specificity is determined by the nature of the reaction between sulfide ions and the chosen reactant. Titration end points can be determined with high precision, thus making volumetric analyses particularly suitable for calibration purposes.

The most widely employed volumetric determination of sulfide involves oxidation of sulfide with a standard iodine solution.[141,159] The reaction

$$S^{2-} + I_2 \longrightarrow S + 2I^-$$

is quantitative in weakly acid solution; in alkaline solution, however, part of the sulfide is oxidized to sulfate by the reaction

$$S^{2-} + 4I_2 + 8OH^- \longrightarrow SO_4^{2-} + 8I^- + 4H_2O$$

Iodine titrations therefore must be carried out in acid solution, and precautions must be taken to prevent loss of any H_2S transferred to the gas phase. One microgram of sulfide can be detected.

In common with many oxidation-reduction titrations, the iodometric method also indicates the presence of other oxidizable species (e.g., SO_2 and mercaptans) which, together with a small amount of oxidation of iodide ions by air, may cause significant end point errors. For these reasons the method should be restricted to the determination of fairly high concentrations of H_2S (>1 ppm) in the absence of alternative reducing agents.

9.4.2.2. *Gravimetric Methods*

Gravimetric methods involve precipitation of the sulfide as an insoluble salt (normally a metal sulfide such as cadmium sulfide), which is collected and weighed.[141,189] The need for filtration equipment and a sensitive balance restrict these analyses to the laboratory. Sensitivities in the milligram to microgram range can be achieved, but the method is best employed to determine high concentrations of H_2S such as occur in natural gas. Since all insoluble material will be detected, selectivity can be achieved only by suitable choice of the precipitating reagent.

9.4.2.3. *Spectrophotometric Methods*

Direct observation of the characteristic electronic absorption peak of H_2S at 2537 Å is of little use in evaluating H_2S concentrations because of the poor sensitivity afforded.[190] Sulfur dioxide, however, has an intense characteristic absorption band at 2850 Å and as little as 2 ppm H_2S can be detected, after oxidation to SO_2, at this wavelength.[187,191,192] Interferences are minimal, and the method is particularly suitable for continuous monitoring of H_2S and SO_2 in flue gases, although the necessary equipment is expensive.

Much greater sensitivity can be achieved by colorimetric determination of a suitable reaction product generated by collected sulfide in solution. This procedure is widely used for the determination of H_2S.[4,141,175,178,193,194] The most sensitive method involves the formation of methylene blue by reaction between sulfide ions and *p*-amino dimethylaniline (PADMA) in the presence of acid and ferric chloride.[153,182] As little as 0.04 μg sulfide can be detected.[4] Mercaptans and SO_2 are common interferents in this reaction. Manipulation of solutions and glassware makes colorimetric estimation of sulfide in solution most suitable for laboratory use, although the methylene blue procedure has been automated for the determination of H_2S in industrial atmospheres.[195]

Measurement of the turbidity or optical density of a suspended sulfide or sulfide reaction product provides a simple and inexpensive method for evaluating the amount of sulfide collected in solution.[4,141,155,196,197] Since turbidimetry detects all suspended material, selectivity must be achieved through choice of a specific precipitating reagent. Arsenious sulfide, formed by reaction between sulfide ions and ammonium arsenite, is detected in this way, and the reaction is unaffected by SO_2.[4,155,197] One microgram of sulfide ions can be detected. The simplicity of turbidimetry makes it suitable for field use, and the arsenious sulfide method is recommended in the USSR for widespread use in air pollution control work.[155,197] However Smith et al.[4] have reported difficulties in using the method for H_2S concentrations below 1 ppm.

Optical density measurements can also be used to determine the amount of a colored metal sulfide collected on a filter paper impregnated with a metal salt.[4,24,141,164–167,174,178] For a half-inch diameter sulfide spot area, as employed in Bendix instruments, the detection limits for filter paper impregnated with silver nitrate, potassium argentocyanide, mercuric chloride, and lead acetate are, respectively, 0.23, 0.18, 0.18, and 0.14 μg of H_2S.[164] Although optical density measurements are not selective in that they register the presence of any solid reaction product, the selectivity of the first three reagents for H_2S is sufficiently great to ensure extremely high selectivity for the overall method. The apparatus and procedure are simple, and the method is very suitable for field operation. Indeed, automatic sampling and evaluation instruments can operate unattended for several days.

Observation of the quenching of fluorescence by sulfide ions is one of the most sensitive methods for detecting soluble sulfide.[5,140,198] Selectivity is achieved by the choice of a fluorescent species that has a fluorescence-sensitive reactive site highly specific for sulfide. In this respect fluorescein mercuric acetate is so selective for sulfide that mercaptans and disulfides are only minor interferents, and as little as 2 ng of sulfide can be detected.[5,140] The method is ideally suited for the determination of H_2S levels below 1 ppb, although at such levels evaluation of the collected sulfide should be carried out in the laboratory to minimize contamination.

9.4.2.4. *Electrochemical Methods*

Electrochemical methods have the advantage that they are often readily adaptable to continuous monitoring. Many such methods have been reported, but their use in air pollution control is not widespread. In general, most electrochemical procedures are extremely sensitive to small changes in the ionic character of the solution and are not suitable for absolute measurements without careful and repeated calibration.

The simplest potentiometric procedure involves the determination of pH changes resulting from sulfide formation in solution.[141] Only gross pH changes are meaningful; thus the method is limited to the determination of high H_2S concentrations (near 1%) by utilizing solution reactions of the form

$$H_2S + CdSO_4 \longrightarrow CdS + H_2SO_4$$

Clearly, any basic or acidic species will interfere with the measurement.

Vibrating silver–silver sulfide electrodes have been used to detect as little as 4 ppb H_2S present as sulfide in an alkaline collecting solution.[156,199,200] Greater selectivity is achieved than by pH indication, but the electrode also responds to HCN and a number of other ions in solution. Thus the method is analytically useful only where the concentrations of interfering species remain fairly constant. Recently developed sulfide ion selective electrodes have somewhat better selectivity and may find future use as sulfide sensors for continuous monitoring in industry.

The concentration of sulfide ions in solution can be determined with sensitivity by measuring either the absolute or relative electrical conductance of the solution.[141,163] Also H_2S concentrations in the range 0.01 to 1.0 ppm have been determined in air, using a thin film electroconductivity detector.[201] Conductivity measurements are poorly selective, since all ionic species present contribute to the observed conductance. For reliable results, care must be taken to ensure that all ionic species other than those derived from H_2S remain constant. In particular, hydrogen ions, with their high equivalent conductance, can produce large errors.

Coulometric measurements detect the flow of electrical charge resulting from a given amount of sulfide reacting at a silver, bromine, or iodine electrode.[141,160,161,173] The silver electrode reaction will indicate mercaptan and disulfide ions as well as sulfide ions, but the bromine and iodine electrodes (which involve oxidation reactions) will register any species that is electroactive at the applied potential. Microcoulometers are now receiving attention as H_2S detectors in gas chromatography, where their lack of specificity is of little account. The bromine microcoulometer can detect 1 ng of sulfide, making it one of the most sensitive detectors for H_2S determination.

Polarographic determination of H_2S in air that is bubbled vigorously over a carbon indicator electrode at an operating potential near -0.6 V has been reported by a number of workers.[141,162] Specificity is high, and sensitivity is of the order of 5 μg of sulfide; although lower limits are attainable with pulsed polarography. Considerable care must be taken to avoid contamination of the solution by other electroactive species that are discharged close to -0.6 V. The method has been employed for monitoring H_2S in industrial effluents, but susceptibility to interferences seems likely to limit its use.

9.4.2.5. Radiochemical Methods

The detection of beta or gamma radiation can be employed to detect very small amounts of H_2S that have been collected as the insoluble metal sulfide on a filter impregnated with $^{203}HgCl_2$ or $^{110}AgNO_3$.[5,202] Alternatively, an equivalent amount of soluble radioactive material can be displaced by the reaction

$$2\,Ag^{131}I + S^{2-} \longrightarrow Ag_2S + 2^{131}I^-$$

on an impregnated paper filter.[203] In both cases the radioactive material of interest is separated on the basis of solubility, and its activity is determined. High selectivities for H_2S can be achieved, although detection limits are governed by the relative solubilities of the reacted and unreacted radioactive species. Nevertheless detection limits of 5 ng (equivalent to 0.05 ppb) have been reported.[5,203] Radiochemical detection requires extremely careful analytical procedures and specialized equipment, and at present this technique shows no particular advantages over alternative procedures.

9.4.2.6. Gas Chromatographic Methods

Gas chromatography is potentially one of the most promising techniques for determining H_2S.[141,183,184,187,204] Attractive features include complete specificity for H_2S, direct determination without prior collection, and ability to determine other gases and vapors in the same air sample. Detection limits are governed primarily by adsorption losses on column materials,[183,187,205] although certain gas chromatographic detectors respond poorly to H_2S (Table 9.6). In this respect flame photometric, microcoulometric, and electron capture detectors are the most sensitive. Recent refinements aimed at reducing adsorption losses[183,206,207] indicate that H_2S concentrations of

Table 9.6. Maximum Sensitivity to H_2S of Gas Chromatographic Detectors

Type of Detector	Approximate H_2S^a (ppm)	Reference
Thermal conductivity	100	175, 205
Thermionic emission	25	205
Flame ionization	10	175
Electron capture	1	101, 205
Microcoulometric	0.1	187, 205
Flame photometric	0.002	175,183

[a] These approximations, based on injection of a 5 ml sample, greatly depend on column conditions, thus they should not be considered to be absolute sensitivities.

2 ppb can be detected in a 10 ml air sample using flame photometric detection.[183] This demonstration clearly foreshadows the increased use of gas chromatography in determining ambient H_2S concentrations in lightly polluted atmospheres. To date, however, the main applications have been for determining relatively high concentrations (>1 ppm) of H_2S in the kraft pulp,[184,187,208–211] natural gas,[212] and petroleum[213,214] industries.

9.4.3. Major Analytical Procedures

It is not implied that the following methods for determining H_2S are the only acceptable ones. Collectively, however, they cover most common situations where H_2S measurements are required, and individually they represent either the most fully documented or the most potentially promising procedures.

9.4.3.1. Indicator Tubes

Indicator tubes contain a molecular sieve (usually silica gel) impregnated with a chemical that reacts with H_2S to produce a visible color change when air is drawn through the tube.[168–180,215] The length of the color change can be related to the concentration of H_2S in the air sample. The collection characteristics of a number of common impregnating reagents are discussed in Section 9.4.1.2.

Argentocyanide is the only impregnating reagent whose reaction product (Ag_2S) is not oxidized by air. Silver sulfide stains remain unchanged for at least a year if the tube is resealed after use and stored in the dark. To prevent bleaching of the colors of copper, lead, and bismuth sulfides, prolonged passage of air should be avoided.[168] These sulfides also fade if stored for long periods. In all cases air flow rates less than 350 ml min^{-1} are required to achieve quantitative collection of H_2S with presently available commercial tubes.

Prolonged passage of dry air ($<20\%$ RH) will impair the color reaction, although the presence of a humectant like glycerol enables operation at low RH. The color reaction is unaffected by high levels of water vapor, carbon dioxide, ammonia, benzene, methane, or carbon monoxide. However SO_2 concentrations above 25 ppm cause significant irregular interferences. High concentrations of hydrochloric acid vapor can convert some metal sulfides back to H_2S, but the net effect is only to displace the zero point of the color, and this does not necessarily affect the tube indication.[168]

Most indicator tubes have an unused storage life of about 2 years at temperatures below 30°C. Storage at 50°C leads to appreciable formation of the metal silicate, however, and performance may thereby be impaired.

Commercially available indicator tubes can be obtained to cover the range 1 ppm to 7% H_2S by volume;[168,215] however improvements by Stepankova and Juranek[169] have resulted in tubes with sensitivities as low as 1 ppb. One of the most sensitive, specific, and stable of the commercial tubes is the Dräger type 1/a. This tube has a standard deviation of $\pm 6\%$ at 20 ppm. General purpose tubes, however, normally deviate $\pm 12\%$ in the 7 to 80 ppm range and $\pm 17\%$ between 70 and 900 ppm. Concentrations of H_2S above 100 ppm tend to produce systematic positive errors as high as 20%.

Indicator tubes have the advantages that they give a very rapid reading, cover a wide range of concentration, and are ideal for rugged field work. The results obtained are only semiquantitative, however. Nevertheless, indicator tubes are strongly recommended for spot measurements aimed at determining whether potentially dangerous concentrations of H_2S are present.

9.4.3.2. The Methylene Blue (PADMA) Method

The methylene blue method is widely considered to be one of the most sensitive and reliable for determining H_2S.[4,138,146,147,153] For best results, recommended procedure[4,138,147] is to draw air containing H_2S through an aqueous suspension of cadmium hydroxide in a darkened collection vessel, or series of vessels, as discussed in Section 9.4.1.1. For high H_2S levels, midget impingers can be used. The PADMA reagent (0.6 ml) and 1 drop of 1% ferric chloride solution are added to 25 ml of the collecting solution with shaking after each addition, and the solution is allowed to stand for

30 min. The amount of methylene blue produced is then estimated spectrophotometrically at 667 nm. Recommended[4,147,153] reagent preparations are as follows.

a. Absorption Mixture

Dissolve 4.3 g of $CdSO_4 \cdot 8H_2O$ in deionized water and add a solution containing 0.3 g of NaOH. The whole is diluted to 1 l with deionized water, stored in the dark, and shaken well before use.

b. PADMA Solution

Add 50 ml of concentrated H_2SO_4 to 30 ml of deionized water, cool, and add 12 g of *N,N*-dimethyl-*p*-phenylene diamine (*p*-amino dimethylaniline). Dilute 25 ml of this solution to 1 l with 1:1 H_2SO_4.

c. Ferric Choride Solution

Dissolve 100 g of $FeCl_3 \cdot 6H_2O$ in deionized water and dilute the solution to 100 ml.

d. Comparison of Reagents

The chief disadvantage of the methylene blue procedure is the unexplained loss of sulfide from cadmium hydroxide collecting solutions (see Section 9.4.1.1). By comparison with the collection stage, the evaluation stage of the overall method is relatively foolproof, and the following considerations are important only when high precision and sensitivity are sought.

Vigorous shaking of the collecting solution after addition of the acidic PADMA reagent transfers a small (but rarely constant) amount of H_2S to the gas phase. The flask should therefore be stoppered immediately, and gentle mixing employed. On addition of the ferric chloride solution, however, vigorous shaking should be commenced immediately to facilitate reaction of the gaseous H_2S. Low yields result if shaking is not begun directly.[153]

The yield of methylene blue increases with increasing acidity, but its extinction coefficient at 667 nm decreases. The optimum acidity is 0.3 *M* H_2SO_4 (or 0.5 *M* HCl[216]), which gives a yield of approximately 66%.[153] This yield is fairly insensitive to small variations in the concentrations of other reactants, but large excesses of $FeCl_3$ cause the 667 nm absorption maximum to decrease with time. Since increasing the temperature produces an increase in this absorption maximum but decreases the yield, the overall analysis does not depend critically on temperature.

For trace determination of H_2S it is essential to use deionized rather than distilled water and to eliminate traces of heavy metals from reagents and glassware. Both Hg^{2+} and Ag^+ (but not Pb^{2+}) will remove sulfide from solution at all stages of the reaction sequence and Cu^{2+} present *before* the addition of PADMA (but not afterward) will do likewise.[153] Large and erratic errors are produced by SO_2 at concentrations greater than about 5 times that of the H_2S present. The methylene blue procedure is not suitable, therefore, for determining H_2S in the presence of SO_2.[4]

The methylene blue reaction also produces small amounts of the dyes Wurster's red, sulfide green, leucomethylene blue, and methylene red.[153] The first three readily convert to methylene blue, but methylene red normally introduces a systematic negative error of 2%. Also, dimerization of methylene blue causes deviation of the 667 nm absorption maximum from Beer's law and introduces a negative error of 0.7% for 20 μg of sulfide.

Clearly, a well-defined analytical procedure must be adhered to if high precision is required. Provided the stated precautions are taken, however, the reproducibility of color development is $\pm 2.5\%$ and the overall method has a reproducibility of $\pm 6\%$. The detection limit is 1 ppb, and the method can be used to determine H_2S concentrations up to several percent.

The advantages of the methylene blue method are its reliability, its high precision, and the wide range of H_2S concentrations accessible. On the other hand, the method is relatively slow, and selectivity for H_2S in the presence of SO_2, nitrogen oxides, and mercaptans is poor. Also, the loss of sulfide during collection[147] renders determinations of low H_2S concentrations (<0.1 ppm) somewhat suspect unless careful calibrations are achieved.

The methylene blue procedure is most suited to laboratory determinations of H_2S and individual determinations in the field. The procedure can be successfully employed in a commercial autoanalyzer,[195] although little if any time advantage is gained over manual operation.

9.4.3.3. Impregnated Paper Densitometry

In impregnated paper densitometry a known volume of air is drawn at constant rate through a filter paper impregnated with silver nitrate, potassium argentocyanide, mercuric chloride, or lead acetate.[4,5,24,154,164–167,174,177,202] Hydrogen sulfide reacts to form a colored metal sulfide, and the optical density of the reaction area is then compared with that of a similar area of unreacted filter. The concentration of H_2S in the volume of air V, which passes through the filter, is determined from the relation

$$\text{ppm } H_2S = \frac{KAD}{V} \tag{4}$$

where D is the optical density of the colored sulfide spot of area A, and K is a constant that depends on the impregnating reagent and the design of the optical densitometer. Deviations from this relation generally become significant for $D > 0.6$.[164]

The collection characteristics of a number of impregnated filter papers have been discussed in Section 9.4.1.2, which stressed the favorable properties of filters impregnated with silver nitrate—high efficiency of collection, stability of the collected sulfide, and selectivity for H_2S. This impregnating reagent is undoubtedly the best for general purpose use, although impregnated materials must be stored in the dark and shielded from direct light during use.

Paper densitometry is most commonly employed with an automatic sampler-evaluator system that automatically advances a tape of impregnated filter paper after each timed sample and reads the optical density of the sulfide spot formed. Instruments that continuously illuminate the sulfide spot as it is being built up should be avoided because both silver nitrate and lead sulfide are light sensitive (the former darkens and the latter fades), and erroneous optical densities result.[164]

Mercuric chloride tapes have the disadvantage that sulfide spots must be developed by exposure to ammonia fumes in a desiccator.[164] These spots fade rapidly but can be repeatedly developed to their original intensity for at least 2 years after their original exposure to H_2S.[164]

The optical densities of sulfide spots on filter paper are *diffuse* optical densities as opposed to the *specular* optical densities obtained when both incident and emergent beams are colinear. Diffuse optical densities depend markedly on the optical arrangement of the densitometer and may vary considerably even between apparently identical instruments. It is essential, therefore, that the value of K in eq. 4 be obtained for each impregnated tape-densitometer combination. Thus the values of K (obtained on a Gelman model 14103 densitometer[164]) in Table 9.5 indicate only the relative sensitivities of the individual tapes.

The detection limits attainable with a one hour sampling period at maximum allowable flow rates are listed for four tapes in Table 9.5. For such low levels, however, it is desirable to remove any particles from the air with a Gelman glass fiber type A or Nucleopore prefilter.[5] Most commercial paper tape samplers employ flow rates such that H_2S concentrations in the range 1 ppb to 10 ppm can be most readily determined. In principle, however, the method can be extended to higher concentrations. With careful operation, reproducibilities of $\pm 7\%$ can be achieved using silver nitrate tapes.

Paper tape densitometry employing silver nitrate impregnated tapes is well suited for rapid continuous automated determination of H_2S levels in both heavily and lightly polluted atmospheres, even in the presence of SO_2 and mercaptans. The equipment is inexpensive, rugged, and portable, and it can be battery operated in remote areas without loss of necessary performance. Furthermore, the sensitivity, selectivity, and precision of H_2S determinations are comparable to those obtained with much more sophisticated, but less versatile, methods.

9.4.3.4. Trace Determination of Hydrogen Sulfide

Only one demonstrably reliable analytical method for determining H_2S concentrations well below 1 ppb has been developed.[5] By the trace method H_2S is collected as silver sulfide on a disk of Whatman No. 4 filter paper impregnated for 2 min in 0.01 *M* HNO_3 containing 2% $AgNO_3$ and 20% ethanol. The filters must be dried in a desiccator, and great care taken to exclude body odors or tobacco smoke, which contain H_2S. Gelman glass fiber type A or Nucleopore prefilters should always be used to collect dust.

Exposed filters are soaked in, and washed repeatedly in a Buchner funnel with aliquots of 0.1 *M* NaOH/0.1 *M* NaCN solution. This dissolves all silver in the form of argentocyanide [mainly $Ag(CN)_2^-$] and releases free sulfide ions in solution. Next 1 ml of 10^{-6} *M* fluorescein mercuric acetate (FMA)[5,140] in 0.1 *M* NaOH is added to 9 ml of the solution, and the resulting fluorescence quenching is determined by comparison with sulfide standards and unexposed impregnated filter blanks.

As discussed earlier (Section 9.4.1.2), collection of H_2S by this method is highly selective, although high ozone levels may cause a small amount of sulfide oxidation. On the other hand, complete blackening of exposed filters by strong sunlight introduces only a minimal error ($\sim 10\%$) in the determination of collected sulfide.

Precipitation of free sulfide by traces of heavy metal cations can cause significant errors in the analyses, and the use of deionized water is imperative. Also, for sulfide concentrations greater than 8×10^{-8} *M*, the variation of fluorescence intensity with sulfide concentration becomes nonlinear (Figure 9.5), apparently because of the formation of a labile complex involving sulfide and cyanide ions. This can be overcome by dilution. Mercaptans and disulfides are interferents in solution; however the collection stage discriminates against these species, particularly at high flow rates.[5]

The quenching of fluorescence by 2 ng of sulfide can be detected, and this amount corresponds to a detection limit of 5 ppt in a 1 m^3 air sample. Ultimate detectability is limited by the amount of sulfide in the filter blanks. With care, reproducibilities of $\pm 5\%$ can be obtained for H_2S concentrations of 20 ppt.

At present the major application of the fluorescence procedure appears to be in the determination of atmospheric background levels of H_2S (~ 20 to 50 ppt). A significant advantage of the method is that filters exposed in remote

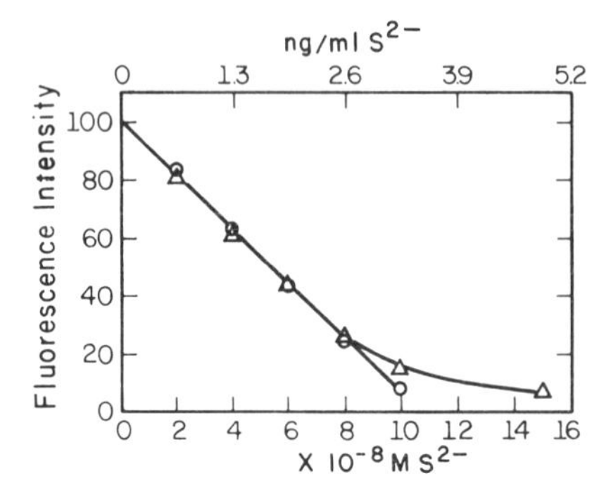

Figure 9.5. Effect of cyanide on fluorescence intensity of FMA-S^{2-} solutions: circles, $C_{CN^-} = 0$, $C_{OH^-} = 0.1\ M$; triangles, $C_{CN^-} = 0.1\ M$, $C_{OH^-} = 0.1\ M$.

areas can be transported to the laboratory (wrapped in Mylar film) for processing many days later.

9.4.3.5 *The Hydrogen Sulfide–Sulfur Dioxide Ultraviolet Analyzer*

The H_2S–SO_2 ultraviolet analyzer was developed recently for the *in situ* determination of H_2S and SO_2 in stack gases.[181,192] It is particularly noteworthy in that direct sampling is employed.

Part of the stack effluent is processed to remove entrained solids and diverted through a heated bed of iron filings, where the H_2S is oxidized to SO_2. The sample then passes through a quartz cell in which the total concentration of SO_2 is determined from its ultraviolet absorption at 285 nm. A similar sample passes directly to a second cell without oxidation (Figure 9.6). The ultraviolet absorption from the second sample gives the concentration of SO_2, and the concentration of H_2S is determined from the difference between the two absorption readings.

A detection limit of 2 ppm is claimed for H_2S,[192] although since the final reading is the difference between two experimental measurements, the precision is low. Given this difference determination, the absorption and scattering of radiation by foreign species has a minimal effect on the accuracy of the H_2S measurement. Difficulties do arise, however, because of the long-term deposition of solids on the cell windows.

At present the H_2S–SO_2 ultraviolet analyzer seems to be one of the most promising instruments for continuous in-plant monitoring of H_2S concentrations in the range of 2 to 2000 ppm. For this purpose simplicity and speed of continuous operation are of greater consequence than high precision or absolute accuracy. Furthermore, the simultaneous determination of H_2S and SO_2 can aid in optimization of oxidation-reduction conditions in the plant.

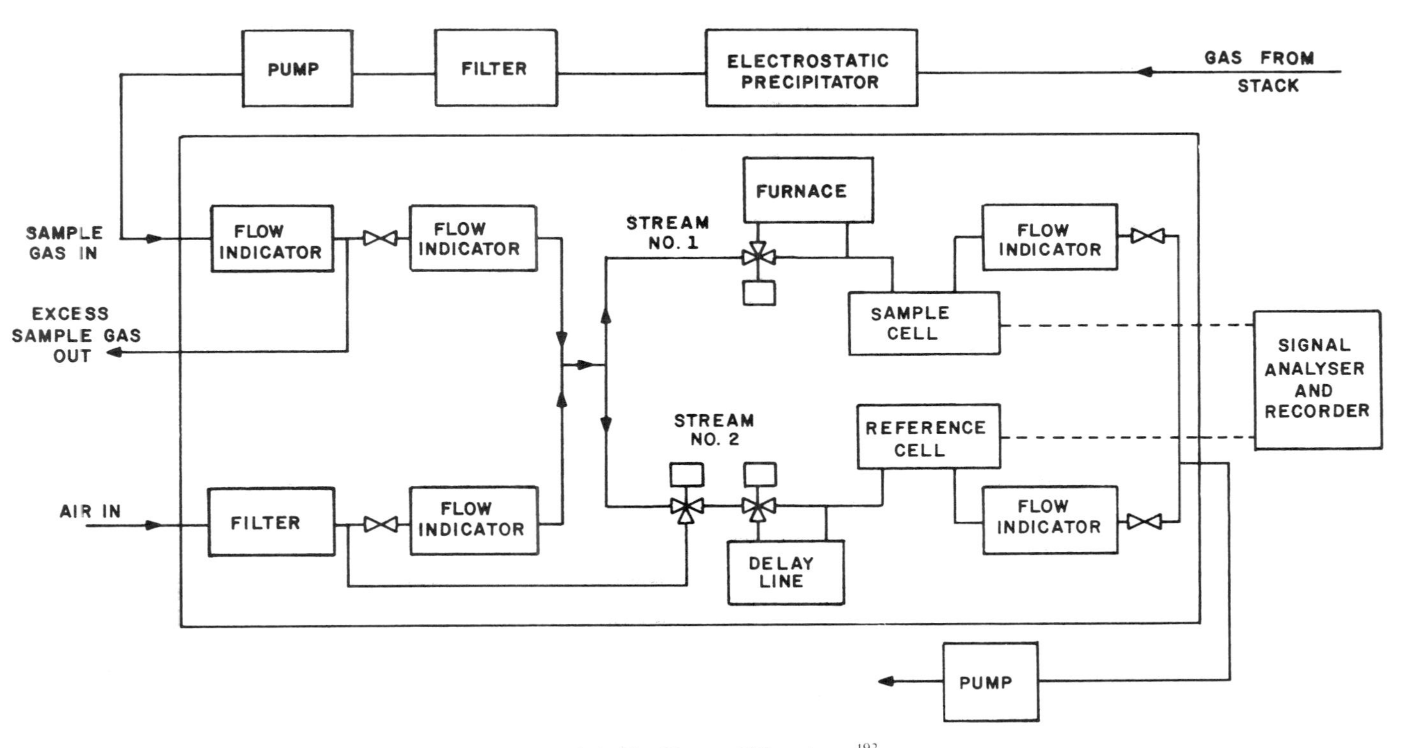

Figure 9.6. The Murray UV analyzer.[192]

9.4.3.6. Gas Chromatographic Analysis

A major attraction of gas chromatographic analysis is that a number of gaseous pollutants can be determined concurrently in the same sample while still retaining specificity for H_2S. This may represent a disadvantage when a large number of components are present, however, unless a sulfur-selective detector[217] is employed to discriminate against unwanted compounds.

To date the greatest single factor limiting the application of gas chromatography to H_2S determination has been poor sensitivity. Recently, however, it has been recognized that sensitivity is limited mainly by irreversible adsorption onto column materials (indeed, as indicated in Table 9.6, gas chromatographic detectors have extremely high sensitivities for H_2S). An elegant study by Stevens et al.[183] has demonstrated that adsorption losses can be dramatically reduced by suitable choice of column packing and constructional materials. As a result, gas chromatography now can be applied not only to the determination of H_2S in industrial effluents but also to ambient air analyses.

The literature describes a number of major variations of the gas chromatographic method for determining H_2S in widely differing situations.[141,183,184,187,205,218,219] Most of these variations have been designed specifically to achieve separation of other air pollutants, to exclude unwanted species (e.g., water vapor), or to improve detection limits. Relatively few procedures have been optimized for H_2S alone.

a. Ambient Air Analyses

The best available gas chromatographic procedure for determination of H_2S in ambient air has also been designed for the simultaneous determination of SO_2, methyl mercaptan (CH_3SH), and dimethyl mercaptan (CH_3SCH_3).[183] For rapid repetitive analyses, elution of sulfur compounds containing three or more carbon or sulfur atoms is prevented by passing the air sample through an initial 24 in. stripping column whose packing material is identical to that of the analytical column (Figure 9.7): H_2S, SO_2, CH_3SH, and CH_3SCH_3 pass rapidly through the stripping column, which is then backflushed to remove any remaining unwanted species. The whole operation is readily automated.

For optimum sensitivity and separation, the following apparatus and experimental conditions are recommended:[183]

Sample size	10 ml
Carrier gas flow rate	100 ml min^{-1}
Carrier gas pressure	<60 psi
Analytical column	36 ft, $\frac{1}{8}$ in. × 0.085 ID fluorinated ethylene propylene

Analytical column packing	40–60 mesh Haloport F coated with polyphenyl ether (a 5-ring polymer) and orthophosphoric acid
Column and detector temperature	105 $\pm 3^\circ$C
Detector	Flame photometric
Detector gas flow rates	H_2, 80 ml min^{-1}; O_2, 16 ml min^{-1}
Stripping column	24 in. × 0.085 ID (construction and packing materials same as for the analytical column)
Stripping time	60 sec
Backflushing time	8 min

It is important that the column and delivery tubing be connected to minimize contact between the gases and any metallic parts that will adsorb H_2S. Also, carrier gas pressures should not exceed 60 psi, since higher pressures rapidly produce leaks in Teflon fittings. Temperatures above 130°C cause gas losses by thermal decomposition. A small amount of orthophosphoric acid mixed with the polyphenyl ether stationary phase considerably reduces peak tailing, which is otherwise severe.

A flame photometric detector utilizing a hydrogen–oxygen flame is the most sensitive for sulfur compounds. A narrow band filter giving 56% optical transmission at 394 nm (bandwidth 5 nm) has been employed successfully in this detector.[183] High sensitivity is also achieved with microcoulometric and electron capture detectors (Table 9.6). The flame photometric detector will respond to 2×10^{-4} ng of H_2S entering the flame per second,[220] which for a 10 ml sample and a carrier flow rate of 100 ml min^{-1}, corresponds to a theoretical sensitivity of about 0.1 ppb. The experimental detection limit[183] is 2 ppb, indicating that some H_2S is still being irreversibly adsorbed. The amount is small, but considerable care should be exercised in establishing calibrations below about 20 ppb.

Although gas chromatographic determination of H_2S in ambient air is still in its infancy, the technique is of importance because of its high sensitivity, excellent selectivity, and direct sample presentation. The instrument can hardly be considered portable for field use, but its future for automatic continuous monitoring of sulfur compounds at selected sites in the vicinity of kraft mills, refineries, viscose plants, power plants, and other sources of sulfur gases seems assured.

b. Industrial Effluent Analyses

Industrial effluent gases normally contain many more components of interest than does ambient air. Consequently gas chromatographic separation requirements are more stringent. Detection of H_2S concentrations below about

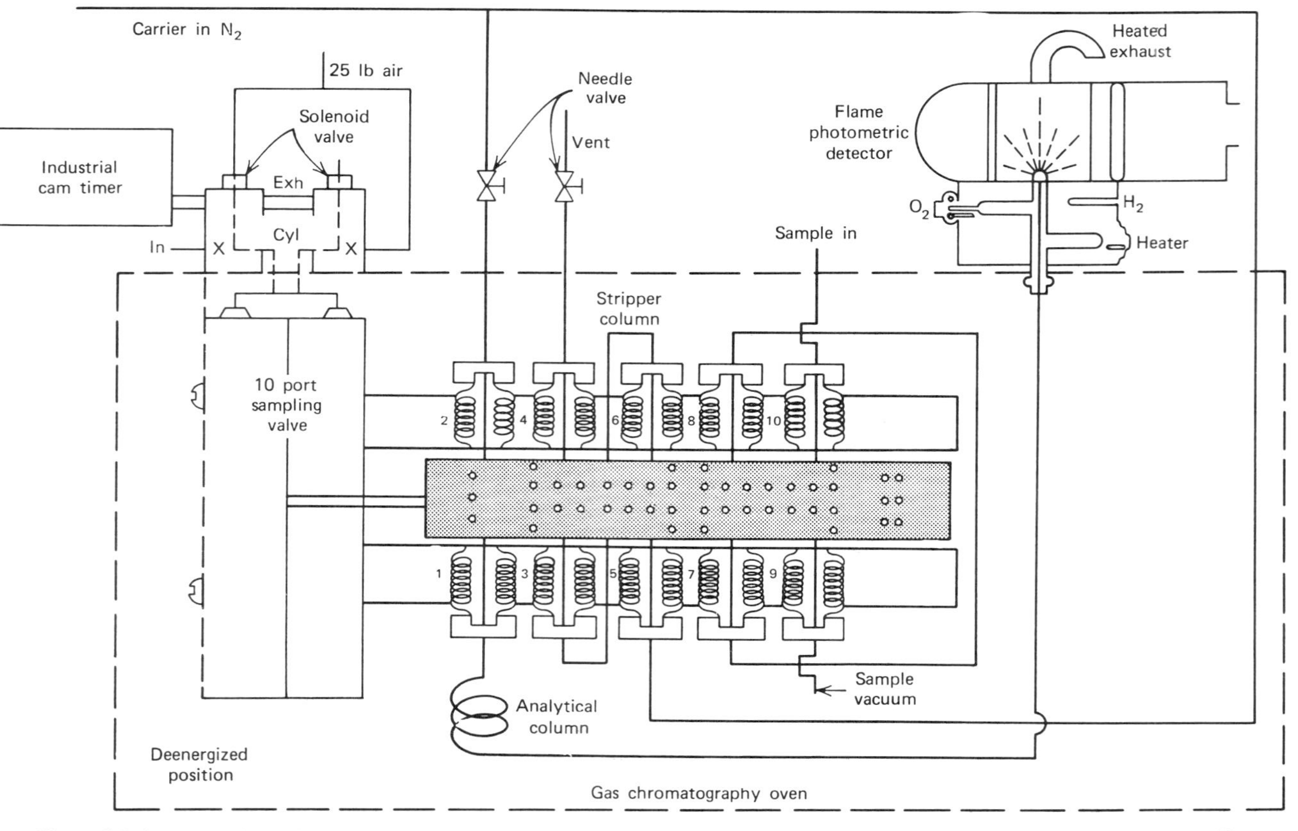

Figure 9.7 Automated gas chromatographic–FPD sulfur gas analyzer equipped with precolumn and backflushing modifications.[183]

1 ppm is not generally required; however most published methods suffer such large adsorption losses that detection even at this level is marginal.

A number of procedures overcome the sensitivity problem by employing a preconcentration step (e.g., cold trapping or adsorption onto a precolumn).[221–223] Such procedures can hardly be recommended; they are cumbersome, they involve significant losses, and they forego the direct sampling advantage of gas chromatography. Furthermore, they are not necessary if column adsorption losses are minimized. It is, however, often necessary to remove particulate material and water vapor before analysis.[210,224] Water vapor is usually removed by cold trapping or by adsorption onto calcium sulfate. The latter is preferable, but both methods involve losses of H_2S. Where the gaseous effluent temperature is greater than 100°C, it is advisable to use steam jacketed delivery lines to prevent condensation of water vapor.[53,210]

Separation of a large number of sulfur-containing gases is most commonly achieved using Triton X305 as the stationary liquid phase on silanized Chromasorb P or G.[187,209,218,224] A number of other partitioning agents and supports also achieve separation,[187,222,225–227] although significant amounts of H_2S are lost in all cases.

Therefore quantitative interpretation of results within a factor of 10 of the sensitivity should be treated with caution.

Either constant temperature conditions or linear temperature programming can be used to separate compounds of predominantly similar polarity. In the presence of compounds of widely differing polarity (e.g., H_2S, SO_2, mercaptans, hydrocarbons, CO, CO_2, H_2O), separation is best achieved by employing nonlinear temperature programming[218,228] or by using two different analytical columns in series.[223]

Flame ionization and thermal conductivity detectors are most useful for determining a wide variety of compounds, but microcoulometric and flame photometric detectors offer the best sensitivity and selectivity for sulfur compounds.[217,220] For a sample size of 10 ml, H_2S detection limits of about 10 ppm can be achieved even without minimization of adsorption losses. With reproducible sample injection, precisions of $\pm 5\%$ can readily be achieved.

Potentially the gas chromatographic determination of H_2S (and other air pollutants) in industrial effluents is extremely attractive. The selectivity of the method is without equal, and the precision, direct sampling, and ease of adaptation to automatic monitoring offer significant advantages for industrial air pollution control applications. Equipment cost is moderate, and it is worth noting that the very short retention times of H_2S and SO_2 may enable development of simplified instruments designed specifically for the determination of these gases.

9.4.4. Calibration

Absolute calibration of analytical procedures for the determination of H_2S presents considerable difficulties because of the high adsorption of H_2S onto most constructional materials (see Section 9.4.1.3) and to the high reactivity of S^{2-}. The following calibration procedures have been used.

9.4.4.1. Standard Sulfide Solutions

Standard sulfide solutions can be employed only for evaluation of the S^{2-} ion. The H_2S collection stage is not included in this type of calibration, and errors result from inefficient collection or from losses that occur before the evaluation of collected S^{2-}. Normally $Na_2S \cdot 9H_2O$ serves as the primary sulfide standard. This compound leaves much to be desired because it decomposes spontaneously, is deliquescent, is oxidized by air to thiosulfate and sulfate, and polymerizes to polysulfide. However $Na_2S \cdot 9H_2O$ performs adequately with careful preparation,[4] and no better solution standard has yet been proposed.

9.4.4.2. Hydrogen Sulfide Generation

The best procedure for quantitative generation of gaseous H_2S involves addition of a concentrated standard solution of sodium sulfide to 25% H_2SO_4 by means of a micrometer syringe. The solution is then stripped with clean air or nitrogen.[4] The action of acid on a sulfide, however, does not generally produce quantitative generation of H_2S. Also, this method does not produce a uniform concentration of H_2S, and some collection methods may not be quantitative for the relatively high concentrations presented.

9.4.4.3. Static Dilution of Standard Hydrogen Sulfide Concentrations

The determination of high concentrations of H_2S (>50 ppm) can be achieved with reasonable accuracy using a volumetric technique such as the iodine titration.[141,159] The standardized concentration can then be diluted by transferring the contents of a gas-tight syringe to a large volume, laminated Mylar bag.[4,164] With care, H_2S concentrations reproducible within $\pm 5\%$ can be achieved, but adsorption losses in delivery tubing and in the Mylar bag, though probably small, are difficult to establish.

9.4.4.4. Dynamic Dilution of Standard Hydrogen Sulfide Concentrations

A third method involves dilution of a known high H_2S concentration (from a gas cylinder) with a continuously flowing stream of clean air. The reproducibility of such gas dilution techniques[229] is the order of $\pm 5\%$, but errors occur from adsorption of H_2S onto glassware.

9.4.4.5. Permeation Devices

When enclosed in an inert plastic tube, a liquefiable gas such as H_2S escapes at a constant temperature-dependent rate that can be determined by measuring the weight loss of the tube over a long period of time. The concentration C of H_2S (ppm) in air flowing over the permeation tube at a rate of L (l min^{-1}) is given by the relation[220]

$$C = \frac{RG}{ML}$$

where R = rate of weight loss at constant temperature (μg min^{-1})
M = molecular weight of H_2S
G = gas constant

A variation of this procedure is to allow pure H_2S from a gas cylinder to diffuse at a constant rate through a ceramic frit into the airstream.[230]

Permeation tubes provide a simple and accurate method for preparing known H_2S concentrations in air. However adsorption losses again restrict the lower limits at which H_2S concentrations can be considered reliable. To achieve precise calibrations, the temperature of the permeation tube must be carefully controlled.

Although the foregoing calibration methods can provide meaningful calibrations for H_2S concentrations above about 10 ppb, lower concentrations should always be treated with caution because of the significance of adsorption losses. At present, permeation tubes clearly provide the most simple and reliable method of calibration.

9.5. EFFECTS

Hydrogen sulfide almost always occurs in admixture with other undesirable gases, and its individual effects are often difficult to isolate from those of other gases and from synergisms. The undesirable effects of H_2S can be divided into three categories in terms of its influence on living things, its unpleasant odor, and its ability to corrode materials such as metals, paint, and concrete.

9.5.1. Biological Effects

There are many cases on record[92,122,231,232] of fatal intoxication of humans by H_2S. There is no exact threshold above which fatal intoxication occurs, but exposures to concentrations in the range of 660 to 1400 ppm have proved fatal.[92,233] People who have respiratory complaints appear to be most susceptible, and there is evidence that the toxicity of H_2S is enhanced by the presence of dust and other gases such as CS_2, CO, CO_2, NH_3, and SO_2.[233–235] There is also some evidence that H_2S may be adsorbed through the skin.[236] Lethal concentrations of H_2S for cats, dogs, and cattle[127] are similar to those for humans; but smaller animals (rats, mice, and birds) can succumb to lower concentrations ($\sim$60 ppm).[233] Dissolved H_2S in water has also been known to kill fish.[237,238]

High concentrations of H_2S have been shown[232,239] to have a direct effect on the respiratory center of humans and animals. Chronic poisoning can give rise either to stimulation of the nervous system with consequent hyperventilation, hypocapnia, or apnea, or to depression and paralysis. Respiratory failure and asphyxiation result in both cases. There is no specific treatment known for H_2S poisoning, and medical practice is aimed at relieving the observed symptoms by applying artificial respiration or administering oxygen. Recent experiments involving intermittent exposure of rabbits to 300 ppm H_2S have indicated, however, that the resulting disruption of mineral metabolism, acid–base balance, and kidney function can be considerably reduced by administration of vitamins and minerals.[240]

A number of subacute physiological effects have been noted[241–243] and studied in humans and animals exposed to low levels of H_2S. Rats exposed to 6.6 ppm H_2S for 12 hr daily over a 3-month period exhibit marked changes in the functional state of the central nervous system, irritation of mucous membranes in the trachea and bronchii, and morphological changes in the brain cotex.[244,245] These effects were negligible at concentrations of 0.013 ppm under the same experimental conditions. Low H_2S levels have also been shown to paralyze the olfactory response,[244] to cause irritation of conjunctival membranes of the eyes,[232] and to depress reflex responses[93] and heart function.[246] Many of these effects are accentuated in physically active individuals.

The demonstrated loss of sensitivity of the eyes to light is probably of most significance.[244,247] This effect is apparent at H_2S concentrations above 0.01 ppm (i.e., below the odor threshold). (Somewhat ironically, exposure to H_2S will reverse eye damage produced by mercury compounds.[248,249])

Although the general mode of action of H_2S poisoning in animals is fairly clear, the actual biochemical processes that occur are not well understood. The action of a number of enzymes (notably phosphatases and DNAase II) has been shown to be impaired,[250–252] and changes in mineral balance and the histochemistry of the central nervous system have been noted.[252]

At present the maximum allowable concentration (MAC) of H_2S to which a person can be exposed for 8 hr is set at 20 ppm in the United States.

The standard in the USSR is 6.6 ppm (10 mg m^{-3}). Such levels undoubtedly give rise to headaches, vertigo, nausea, anemia, general debility, and other subjective sensations. Furthermore, adverse health effects have been demonstrated[94,112,117] in people (particularly children) living in the vicinity of industries that produce H_2S. Thus there may be a case for lowering MAC levels[232,235,243,253] in situations of repeated or continuous exposure.

By comparison with its effect on animals, the toxicity of H_2S to plants is minor[254] and is normally manifest only at concentrations of several hundred ppm. For humans and animals the order of decreasing toxicity of the most dangerous common air pollutants is HCN, H_2S, Cl_2, SO_2, NH_3. However this order is almost reversed for plants[255] (Cl_2, SO_2, NH_3, HCN, H_2S). Prolonged exposure of plant life to several hundred ppm of H_2S will result in leaf damage, particularly at the growing tips of young leaves, although significant root damage from dissolved H_2S is observed at much lower levels.[256,257] Common garden plants are affected to different degrees, but leaf damage is most commonly encountered in rice grown on soils high in sulfide[258] and in apple and citrus trees that have been sprayed with lime sulfur or sulfur dust.

9.5.2. Odor

A surprisingly large number of unpleasant odors contain H_2S as a constituent (e.g., those from sewers, meat rendering plants, tanneries, refineries, kraft mills, viscose plants, and decaying animal or vegetable matter). The odor threshold for pure H_2S varies markedly among individuals;[4,259,260] for most people the minimum detection limit lies within the range 0.025 to 0.100 ppm.[259] This range may vary when other odors are also present. For example, 0.006 to 0.02 ppm H_2S can be detected in the presence of oil vapor.

It is well established[243] that H_2S concentrations above about 20 ppm can paralyze the human olfactory response—an effect that is potentially dangerous when such concentrations are encountered. Normally the odor problem is of major concern for H_2S concentrations above 0.1 ppm, although prolonged exposure produces olfactory fatigue, which somewhat reduces the problem. This fatigue can last for several weeks after individuals cease their exposure to H_2S.

9.5.3. Corrosion

The corrosiveness of H_2S is due to its extremely high reactivity with a variety of materials. The most common corrosion products are metal sulfides, whose high color and undesirable electrical and mechanical properties make sulfide corrosion undesirable on both aesthetic and economic grounds. Discoloration of domestic utensils and fittings and of minor artworks usually can be remedied by cleaning and is more an annoyance than a problem. Corrosion of metals, paints, plastics, and concrete however, can have considerable economic importance. In almost all cases, significant corrosion occurs only in close proximity to sources of H_2S.

9.5.3.1. Metals

Most metals react with H_2S to form metal sulfides. However the greatest reactivity is exhibited by heavy metals whose stable ionization states have a d^9 or d^9s^2 electronic configuration. Thus although sulfide corrosion of iron and steel, aluminum, nickel, zinc, chromium, cobalt, palladium, and tantalum has been widely reported, its degree is usually much less than that of silver, copper, cadmium, mercury, thallium, lead, and bismuth.[261–266] In fact, major problems resulting from metallic sulfide formation by atmospheric H_2S are usually encountered only with silver and copper or their alloys such as brass and bronze.[24,247,266,267]

Exposure of silver or copper or their alloys to H_2S results in the formation of a dark brown or black powdery layer of sulfide on the metal surface. Copper sulfide forms a somewhat better protective coating than silver sulfide, but neither adheres sufficiently well to prevent further attack. The presence of moisture (particularly acidic moisture) on the metal surface greatly increases the rate of sulfide formation by promoting formation of the metal and S^{2-} ions.

The rate of sulfide formation also depends markedly[266–268] on crystal orientation in the metal, on temperature, and on the relative humidity of the surrounding atmosphere (Figure 9.8). Indeed, silver and copper can be effectively protected from sulfide corrosion by lowering the relative humidity, even when high H_2S concentrations are present.[266] Sulfidation proceeds very slowly, however, even in atmospheres that are nominally at 0% relative humidity.[267]

It is evident from Figure 9.8 that there are two ranges within which the rate of sulfide formation depends markedly on relative humidity.[267] Above the critical RH (about 75%), water vapor condenses rapidly on the metal surface and causes fast nucleation of the metal sulfide. This in turn promotes further condensation, with a consequent rate increase as RH increases. Below the critical relative humidity, however, the initial rate of sulfide formation is almost independent of RH until the crevices between sulfide nuclei become small enough to promote capillary condensation of water according to the Kelvin equation.[267] This produces a rate acceleration that is dependent on

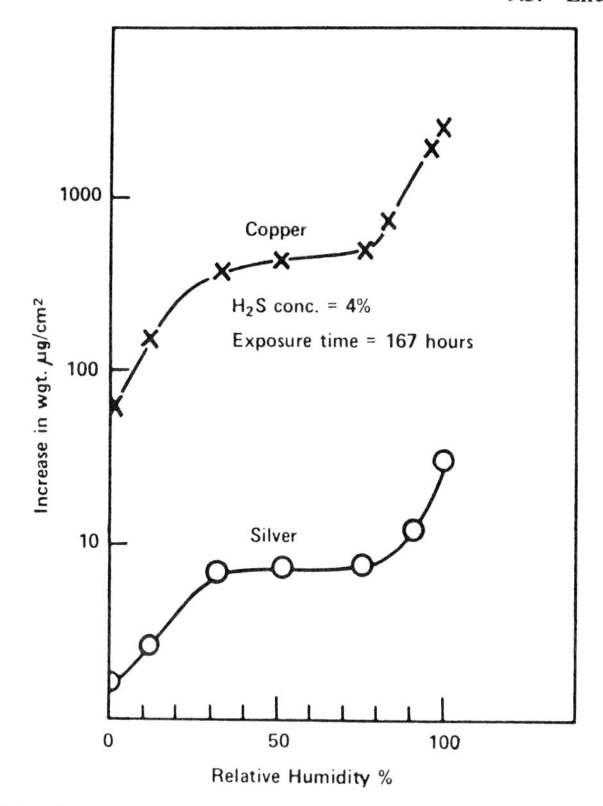

Figure 9.8. Effect of humidity on the rate of sulfide corrosion of copper and silver.

relative humidity. At about 30% RH, all condensation sites are occupied, and additional increase up to the critical relative humidity does not further enhance the rate of sulfide formation. It should be noted that accelerated studies of sulfide corrosion at relative humidities below the critical value will give erroneous results because of the rapidly increasing initial sulfidation rate with time.[266]

There is evidence that sulfidation of copper is influenced by the partial pressure of oxygen in the atmosphere.[269] However the significance of this finding with respect to practical corrosion problems appears to be minimal. Of greater importance is the enhancement of sulfide formation by surface contaminants such as dusts or metallic particles,[269] which either act as nucleating sites for moisture or provide a galvanic couple.

It is commonly observed that H_2S corrosion predominates in areas of strain[270] in the metal, and in the region of an electrical contact or discharge (Figure 9.9). This is because the controlling step in sulfide corrosion appears

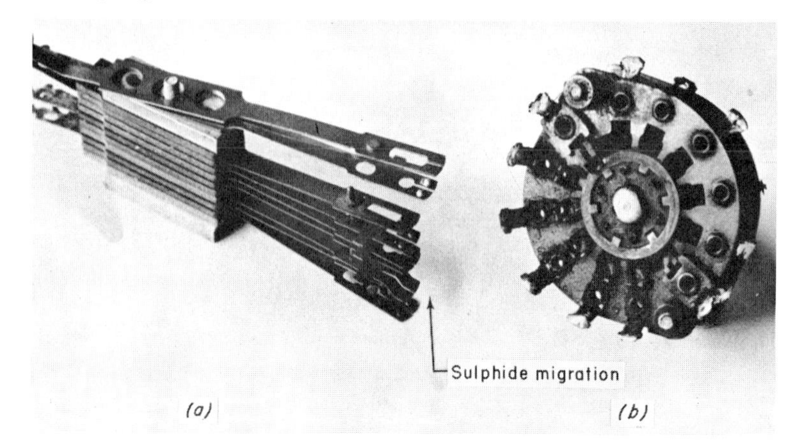

Figure 9.9. Examples of sulfide corrosion. (*a*) Telephone relay contacts showing localized corrosion of contacts and migration of the sulfide. (*b*) Rotary switch contacts.

to be the removal of metal ions from their surface environment[270]—a process that is promoted by the presence of moisture, by surface strain, and by galvanic or electrical action.

The mechanical properties of metals are impaired only by extensive sulfide formation. This situation is encountered, for example, with copper spouting and down-piping in the vicinity of H_2S sources. Invariably the parts that are continually damp disintegrate first. Mechanical weakening or failure is of greatest significance when the susceptible metal is extremely thin. This occurs in the cases of fine silver wires[271] connecting ferrite cores in memory circuits of computers, fine springs and hair springs (particularly those made of copper or phosphor-bronze) in electrical switch gear and meters, thin copper foils, and copper sensing tubes in pressure gauges.[24]

The effects of sulfide formation on the electrical properties of metals are generally of greater economic importance than the effects on mechanical properties. Metal sulfides have high electrical resistance,[272] and formation of sulfide layers or sulfide "whiskers"[267,273,274] on copper, silver, or brass contact points of electrical relays[261] or switches[24] greatly impairs their performance. Sulfide formation takes place very rapidly at such contact points, and cleaning of the contact material is not usually effective in restoring the conductance to its original value.

Breakdown of electrical contacts is commonly encountered in telephone relays, in industrial switch gear, in rotary switches on electronic equipment, and in printed circuitry or soldering that is exposed to H_2S attack (Figure 9.9). Generally contacts that have a wiping action are less susceptible to malfunction than those relying on pressure contact. However sulfide formation on copper commutators in electric motors and generators exposed to H_2S can result in both high resistance at the brush–commutator interface and rapid wearing of the commutator surface, due to the wiping action of the brushes. This has been a considerable problem at the geothermal power station at Wairakei, New Zealand.

In addition to the factors already mentioned, the extent of metal sulfidation depends on the rate of delivery of H_2S molecules to the surface. Thus metals exposed to a constant stream of air containing H_2S undoubtedly corrode more quickly than those exposed to the same air which is not moving. In practice, this effect makes it extremely difficult to establish meaningful data on the exposure time required to produce malfunction of a metal component under given conditions of H_2S concentration, temperature, and relative humidity. Furthermore, the nonlinear rate of corrosion discussed earlier renders the results of accelerated corrosion tests highly suspect. As a rough guideline, however, Table 9.7 lists H_2S concentrations at which electrical contact malfunction might be experienced within 5 to 10 years of operation. These results[24] are based on observation of existing equipment in geothermal areas where the humidity range is 35 to 75% RH, with mean indoor temperatures between 20 and 25°C.

Susceptible metal components can be protected from H_2S attack by substituting alternative corrosion-resistant materials,[275–277] by coating the metal with a film that is impervious to H_2S,[278–280] by enclosing the components in a box and removing H_2S and water with a tray of activated carbon and desiccant,[24,266] or by supplying the unit with air from which H_2S has been removed by passage through a filter such as activated carbon.[24] Any of these procedures can be extremely effective, but all have certain disadvantages. Thus components made of resistant materials such as gold, platinum, palladium, rhodium, or tin[275,277] are expensive because they

Table 9.7. Minimum Average H_2S Concentrations Likely to Give Rise to Malfunction of Components Listed Within 5 to 10 Years

Component	RH %	"Average" conc. limit
Silver relay pressure contacts (low voltage)	55	0.002 ppm
Silver relay pressure contacts (low voltage)	Uncontrolled	0.001 ppm
Copper relay pressure contacts (high voltage)	Uncontrolled	0.007 ppm
Silver switch wiping contacts (low voltage)	55	0.005 ppm
Silver switch wiping contacts (high voltage)	Uncontrolled	0.005 ppm

must be custom-made; materials impervious to H_2S such as Mylar, cyclo- or chlorinated rubbers, epoxy resins, and cellulose[280,281] do not normally conduct electricity; low relative humidity often causes dehydration problems in other materials such as nylon; and enclosure may not be possible because of physical construction or the requirements of heat dissipation. Finally, continuously flowing air must contain extremely low levels (<0.001 ppm) of H_2S, thus requiring highly efficient filtration.

9.5.3.2. Paints and Plastics

Discoloration of paints by H_2S results from reaction of the gas with salts of metals (e.g., lead, mercury, cobalt, iron, tin, and cadmium) used as coloring pigments or internal fungicides.[282] In the case of plastics, the reactive constituent is usually a heavy metal soap (e.g., lead stearate) used as an internal stabilizer or lubricant. Such discoloration is a considerable problem, particularly in the case of lead-based paints near kraft mills, petroleum refineries, and viscose plants, or near polluted tidal inlets.[10]

Lead-based exterior house paints most commonly exhibit sulfide discoloration and take on a "battleship gray" color. In extreme cases, large sections of paint can be peeled off the surface.[282] Since moisture is a major factor promoting sulfidation,[274,283,284] discoloration is most marked under eaves and in regions that retain moisture. Equally important is the condition and age of the paint,[274,283,285] since surfaces that are cracked and powdery allow better penetration of H_2S and hold more moisture than fresh paint films. Freshly painted surfaces are afforded considerable protection from the paint binder and from the coating of clear varnish often applied by artists to oil paintings. Studies of the susceptibility of different paint bases to sulfide discoloration suggest that susceptible pigments are better protected by the new acrylics than by more traditional oil bases.[283,285,286]

Oxidizing agents (e.g., SO_2, Cl_2, O_3, and O_2) and light promote rapid conversion of highly colored sulfides to pale (usually white) sulfates[4,147,164,174,177,274] This does not occur for mercury sulfides. Thus paintwork that has been discolored by an incidental exposure to H_2S will fade to its original color within 3 to 6 months under normal weathering conditions.[274] Where fairly constant H_2S levels are encountered, a true discoloration threshold exists above which the rate of discoloration exceeds that of bleaching. This threshold has been reported to lie at various points in the range of 0.001 to 0.1 ppm.[274] However this level must depend markedly on the weathering conditions and on the various parameters listed earlier. In any event, darkening of exterior paints containing metal pigments susceptible to sulfide formation may occur at H_2S levels close to or below the odor threshold.

Sulfide discoloration of paint is often confused with discoloration due to growth of the fungus *Pullularia pollulans*. The two can be distinguished in the field with the aid of a low power hand lens or by addition of a drop of 5% sodium hypochlorite solution, which will remove the mildew but not the sulfide. Sulfide discoloration can be positively identified by wiping with a 3% solution of hydrogen peroxide. This removes the color, which is then rapidly restored. X-Ray or electron diffraction techniques, of course, enable positive laboratory identification of metal sulfides in paint samples.[282]

9.5.3.3. *Concrete*

Corrosion of concrete by H_2S is not widespread, but it creates a considerable problem in sewers and drains where the gas is present.[105,106,287,288] Corrosion is normally localized just above the water line, where H_2S is capable of leaching the constituents of calcium carbonate from the moist concrete by the reaction

$$2CaCO_3 + 2H_2S \longrightarrow Ca(HCO_3)_2 + Ca(HS)_2$$

It is generally accepted, however, that corrosion of concrete in sewers is due primarily to the sulfate ion,[105] which is produced by bacterial action on H_2S under moist oxidizing conditions on the sewer wall. This process is also localized just above the water line, where the bacteria thrive.

In extreme cases where high H_2S levels occur, corrosion can be so extensive that the pipe disintegrates along the water line.[288] Corrosion can be prevented, however, by employing ceramic pipes or by introducing a tightly adhering protective film (usually plastic).

9.6. CONCLUSION

For many years the vast majority of research projects involving H_2S as an air pollutant have been directed toward occupational rather than environmental situations. Consequently there exists an extensive body of information relating to the occurrence, measurement, and effects of H_2S in specific instances where its biological or corrosive effects are known to have a demonstrable influence on human health or to be undesirable on economic grounds. For the most part, studies relating to H_2S as an environmental pollutant have been conducted only to the stage where it is clear that the gas need not be considered as a major air pollutant in terms of the amounts released to the atmosphere and of its reactions therein.

As a result of these findings, the importance of H_2S as an air pollutant was somewhat underemphasized during the 1960s, when awareness of environmental problems increased dramatically. In the last few years, however, attention has been deservedly refocused on the very large number of specific industrial situations in which H_2S is produced. It seems probable, therefore, that the importance of specific sources of H_2S will be further recognized and emphasized in the near future.

In assessing the completeness of the data presented herein, it is clear that considerably more information is required on the chemical processes that produce H_2S in each of the many industrial operations that contribute to its atmospheric emission. At present, probably most is known about H_2S production in the paper industry, and it is clear that in this instance at least, fairly minor process modifications can markedly reduce H_2S emissions. Considerable work is also required in developing methodology for automated continuous measurement of H_2S in specific industrial effluents and in the vicinity of these effluents. Improvements in methods for measuring urban concentrations of atmospheric H_2S may also be necessary if environmental H_2S standards are set. In general, the adverse effects of H_2S are fairly well characterized, although there is a need for more basic research into the fundamental processes that produce significant health effects and corrosion of materials.

REFERENCES

1. Kellogg, W. W., R. D. Cadle, E. R. Allen, A. L. Lazarus, and E. A. Martell, *Science*, 175 (4022), 587 (1972).
2. Robinson, E. and R. C. Robbins, in *Air Pollution Control*, Part II, Werner Strauss, Ed., Wiley-Interscience, New York, 1972.
3. Junge, C. E., *Air Chemistry and Radioactivity*, Academic Press, New York, 1963, Chapter 2.
4. Smith, A. F., D. G. Jenkins, and D. E. Cunningworth, *J. Appl. Chem.*, (London), 11, 317 (1961).
5. Natusch, David F. S., Homer B. Klonis, Herman D. Axelrod, Ronald J. Teck, and James P. Lodge, Jr., *Anal. Chem.*, 44 (12), 2067 (1972).
6. Jensen, M. L. and N. Nakai, *Science*, 134 (3496), 2102 (1961).
7. Avrahami, M. and R. M. Golding, *J. Chem. Soc. A*, 647 (1968).
8. Cline, Joel D. and Francis Asbury Richards, *Environ. Sci. Technol.*, 3 (9), 838 (1969).
9. Altshuller, Aubrey P., *Tellus*, 10, 479 (1958).
10. Denmead, C. F., *Proc. Clean Air Conf.*, University of New South Wales, 1962, 1, 4.1–17.
11. Schulze, Karl L., *Proc. Ann. Nat. Diary Eng. Conf.*, 16, 51–9 (1968).
12. Chebotarev, E. N., V. M. Gorlenko, and V. I. Kachalkin, *Mikrobiologiya*, 43 (2), 321 (1974).
13. Butlin, K. R., Sylvia C. Selwyn, and D. S. Wakerley, *J. Appl. Bacteriol.*, 23, 158 (1960).
14. Espino de la O., Ernesto, *Diss. Abstr. B*, 29 (5), 1710 (1968).
15. Park, N. J., C. S. Park, Y. S. Kim, B. H. Cho, and C. Y. Lee, *Nongso Sihom Yon'gu Pogo*, 10 (3), 9 (1967); *Chem. Abstr.*, 69, 18291e (1968).

16. Shimoda, Nobuo, Kozo Ishimaru, and Hirotoshi Tanaka, *Muroran Kogyo Daigaku Kenkyu Hokoku*, 6 (1), 13 (1967); *Chem. Abstr.*, 70, 90647p (1969).
17. Chebotarev, E. N., V. M. Gorlenko, and V. I. Kachalkin, *Mikrobiologiya*, 42 (3), 537 (1973).
18. Nicol, D. J., M. K. Shaw, and D. A. Ledward, *Appl. Microbiol.*, 19 (6), 937 (1970).
19. Parsky, S. and C. Billy, *Ann. Inst. Pasteur*, 103, 461 (1962); *Chem. Abstr.*, 58, 12871e (1963).
20. Takejiro, Ozawa, *Nippon Kagaku Zasshi*, 87 (9), 959 (1966); *Chem. Abstr.*, 66, 8109t (1967).
21. Mizutani, Yoshihiko, *J. Earth Sci. Nagoya Univ.*, 10, 125 (1962); *Chem. Abstr.*, 59, 6143e (1963).
22. Mahon, W. A. J., *New Zealand J. Sci.*, 5 (1), 85 (1962).
23. Ross, J. B., *New Zealand J. Sci.*, 11 (2), 249 (1968).
24. Natusch, D. F. S., *Clean Air*, 4 (4), 69 (1970).
25. Koridze, I. T., *Tr. Nauch.-Issled. Inst. Kurortol. Fizioter., Gruz, SSR*, 29, 199 (1968); *Ref. Zh. Khim.*, Abstr. No. 71232 (1969); *Chem. Abstr.*, 72, 35611p (1970).
26. Konopac, J., *Vod. Hospod. B*, 22 (2), 31 (1972); *Chem. Abstr.*, 79, 45570h (1973).
27. Anisimov, L. A., *Sov. Geol.*, 13 (3), 75 (1970); *Chem. Abstr.*, 72, 135171p (1970).
28. Le Tran Khanh, *Bull. Cent. Rech. Pau*, 5 (2), 321 (1972); *Chem., Abstr.*, 80, 5837q (1974).
29. Matsumoto, Tatsuro and Yoshitaka Shimoda, *Tohoku J. Agr. Res.*, 13, 135 (1962); *Chem. Abstr.*, 58, 2712c (1963).
30. "Harvesting Sulphur from Sour Gases and Oil," *Eng. Mining J.*, 169 (10), 85 (1968).
31. Sheehy, James P. and John J. Henderson, U.S. Public Health Service Publication No. 999-AP-3, 1963.
32. Anderson, D. O., I. H. Williams, and B. G. Ferris, *Can. Med. Assoc. J.*, 92, 954 (1965).
33. Ferris, B. G. and D. O. Anderson, *Proc. Roy. Soc. Med.*, Pt. 2, 57 (10), 979 (1964).
34. Hochheiser, Seymour, Sanford W. Horstman, and Guy M. Tate, Jr., Robert A. Taft Sanitary Engineering Center, Technical Report A62-22, 1962; *Chem. Abstr.*, 58, 11889c (1963).
35. Krasovitskaya, M. L., L. K. Malyarova, and T. S. Zaporozhets, *Gig. Sanit.*, 30 (4), 103 (1965); Chem. Abstr., 63, 2303g (1965).
36. Masek, Vaclav and Josef Sedlak, *Hutn. Listy*, 25 (3), 149 (1970); *Chem. Abstr.*, 72, 136096e (1970).
37. Jedrzejoroski, J., *Gaz, Woda Tech Sanit.*, 33, 240 (1959).
38. Rosival, L. and Strecha, M., *Smokeless Air*, 30, 194 (1960).
39. Klimek, W., *Gig. Sanit.*, 33 (8), 88 (1968).
40. Lahmann, Erdwin and Hanna J. Koerner, *Gesundheit-Ingenieur*, 90 (10), 293 (1969); *Chem. Abstr.*, 72, 15512d (1970).
41. Dutkiewicz, T., I. Kesy, and J. Piotrowski, *Panstw. Zakl. Hig. Rocz.*, 9 (6), 543 (1958); *Air Pollut. Control Assoc. Abstr.*, 7 (7), 4035 (1961).
42. Hendrickson, E. R. and C. I. Harding, *Int. Clean Air Congr. Proc.*, I, 95 (1966).
43. Murray, F. E. and H. B. Rayner, *Pulp Paper Mag. Can.*, 69 (5), 71 (1968).
44. Thoen, G. N., G. G. DeHaas, R. G. Tallent, and A. S. Davis, *Tappi*, 51 (8), 329 (1968).
45. Thomas, J. F., K. H. Jones, and D. L. Brink, *Tappi*, 52 (10), 1873 (1969).
46. Douglass, I. B. and L. Price, *Tappi*, 51 (10), 465 (1968).

47. Cave, G. C. B., *Tappi*, 46 (1), 1 (1963).
48. Landry, J. E. and D. H. Longwell, *Tappi*, 48 (6), 66A (1965).
49. Murray, F. E. and H. B. Rayner, *Tappi*, 48 (10), 588 (1965).
50. Harding, C. I. and J. E. Landry, *Tappi*, 49 (8), 61A (1966).
51. Douglass, I. B., *J. Air Pollut. Control Assoc.*, 18 (8), 541 (1968).
52. Clement, J. L. and J. S. Elliot, *Paper Trade J.*, 153 (16), 63 (1969).
53. Walther, J. E. and H. R. Amberg, *Chem. Eng. Progr.*, 66 (3), 73 (1970).
54. Tretter, Vincent J., Jr., *Tappi*, 52 (12), 2324 (1969).
55. McKean, W. T., B. J. Hruitfiord, K. V. Sarkanen, L. Price, and I. B. Douglass, *Tappi*, 50, 400 (1967).
56. Thoen, G. N., G. G. DeHaas, and R. R. Austin, *Tappi*, 52 (8), 1485 (1969).
57. Feldstein, D. L., J. Thomas, and D. L. Brink, *Tappi*, 50 (6), 258 (1967).
58. *Ibid.*, p. 276.
59. Hendrickson, E. R. and C. I. Harding, *J. Air Pollut. Control Assoc.*, 14 (12), 487 (1964).
60. Clement, J. L. and J. S. Elliot, *Paper Trade J.*, 152 (40), 59 (1968).
61. Miyamoto, Isao, Yukio Inada, Minoru Ishizaka, and Shuichi Okumura, *Kogai To Taisaku*, 9 (7), 733 (1973); *Chem. Abstr.*, 80, 51949a (1974).
62. Goldstein, R. F. and A. L. Waddams, *The Petroleum Chemicals Industry*, 3rd ed., E. and F. N. Spon Ltd., London, 1967.
63. Ahmed, L., I. K. Abdou, and B. H. Mahmoud, *J. Parkt. Chem.*, 38 (1–2), 1 (1968); *Chem. Abstr.*, 69, 11898z (1968).
64. de Sanctis, G., *Fumi Polveri*, 4, 141 (1964); *Air Pollut. Control Assoc. Abstr.*, 10 (8), 6348 (1965).
65. Ellis, F. J., *Am. Ind. Hyg. Ass. J.*, 19, 313 (1958).
66. Beychok, M. R., *Aqueous Wastes from Petroleum and Petrochemical Plants*, Wiley, London, 1967.
67. Galanina, S. V. and E. A. Kadarova, *Neftepererab. Neftekhim.* (Moscow), (3), 35 (1969); *Chem. Abstr.*, 71, 23440y (1969).
68. Pollio, Frank X., Kenneth A. Kun, and Robert Kunin, South African Patent 68 02, 028 (1969); *Chem. Abstr.*, 72, 23165h (1970).
69. Sokolov, A., A. Y. Litovchenko, *Tr., Gos. Inst. Proekt. Issled. Rab. Neftedobyvayushchei Prom.*, (10), 147 (1967); *Chem., Abstr.*, 71, 24594p (1969).
70. Polyanskii, V. A., A. N. Musserskaya, and N. E. Nesterova, *Gig. Sanit.*, 33 (11), 24 (1968).
71. Stockton, J. R., R. C. Henshaw, and R. W. Graves, Bureau of Business Research, Research Monograph No. 15, University of Texas, Austin, 1952.
72. Williams, A. F. and W. L. Lom, *J. Appl. Chem.* (London), 17 (6), 179 (1967).
73. Mallete, F. S., Ed., *Problems and Control of Air Pollution*, Reinhold, New York, 1955.
74. Bapseres, P., *Pharm. Biol.*, 6 (61), 187 (1969); *Chem., Abstr.*, 71, 128335w (1969).
75. Nonhebel, G., Ed., *Gas Purification Processes*, George Hewnes Ltd., London, 1964, Chapters 4 and 8.
76. Gilpin, A., *Control of Air Pollution*, Butterworths, London, 1963, Chapter 10.
77. Korsh, M. P., I. F. Bogdanov, and N. V. Lavrov, *Tr. Inst. Goryuch. Iskop., Akad. Nauk SSSR*, No. 16, 367 (1961); *Chem. Abstr.*, 57, 16959d (1962).
78. Smith, B., *Trans. Chalmers Univ. Technol.*, Gothenberg, Sweden, 184 (1), (1957).

79. Leidnitz, Kurt, *Gas-Wasserfach*, 106, 1204 (1965); *Chem. Abstr.*, 66, 117719v (1967).
80. Mazur, Margareta and S. Forizs, *Metalurgia* (Bucharest), 20 (10), 567 (1968); *Chem. Abstr.*, 71, 52030v (1969).
81. Brandt, A. D., *Proceedings of the Third National Conference on Air Pollution*, U.S. Department of Health, Education and Welfare, No. 1649, 236, 1966.
82. Masek, V., *Koks Khim.* (9), 61 (1973); *Chem. Abstr.*, 80, 30287t (1974).
83. Sussman, V. H. and J. J. Mulhern, *J. Air Pollut. Control Assoc.*, 14, 279 (1964).
84. Sussman, V. H., in *Air Pollution*, Vol. III, A. C. Stern, Ed., Academic Press, New York, 1968, Chapter 5, p. 129.
85. Lemming, Joseph, *Rayon. The First Man-Made Fiber*, Chemical Publishing Co., New York, 1950.
86. Pratt, D. C. F. and A. Rutherford, *Chem. Ind.* (London), No. 41, 1281 (1955).
87. Khor'kova, O. G., E. M. Mogilevskii, and G. G. Finger, *Khim. Volokna*, 52 (1968); *Chem. Abstr.*, 71, 102976f (1969).
88. Selin, A. N. and P. I. Nivin, *Khim. Volokna*, (6), 65 (1968); *Chem. Abstr.*, 70, 38743c (1969).
89. Roberts, C. B. and H. T. Farrar, *Roy. Soc. Health J.*, 76, 36 (1956); *Cbem. Abstr.*, 50, 15078a (1956).
90. Chivilikhina, M. P., A. T. Serkov, B. M. Sokolovskii, S. I. Barskaya, N. P. Shishkina, and Y. N. Kurylev, *Khim. Volokna*, 15 (4), 24 (1973); *Chem. Abstr.*, 80, 30255f (1974).
91. Vdovin, B. I., I. M. Zrazhevskii, T. A. Kuz'mina, R. I. Onikul, A. A. Parlenko, G. A. Panoilova, G. P. Rastorgueva, and B. V. Rikhter, *Tr. Gl. Geofiz. Observ.*, No. 254, 57 (1971); *Chem. Abstr.*, 78, 19888c (1973).
92. Hromadka, Miroslav, *Prac. Lek.*, 17 (2), 68 (1965); *Chem. Abstr.*, 62, 16872e (1965).
93. Petri, H., *Staub*, 21, 64 (1961).
94. Holasova, P., *Cesk. Hyg.*, 14 (7–8), 260 (1969); *Chem. Abstr.*, 72, 35459v (1970).
95. Desai, N. F., *Colourage*, 16 (2), 55 (1969); *Chem. Abstr.*, 71, 116243b (1969).
96. Bubnova, G. P. and V. I. Kostrikov, *Zh. Vses. Khim. Obshchest.*, 14 (4), 399 (1969); *Chem. Abstr.*, 72, 4429f (1970).
97. Strauss, W., *J. Air Pollut. Control Assoc.*, 14, 424 (1964).
98. *Reduction of Inedible Animal Matter*, Air Pollution Engineering Manual, U.S. Department of Health, Education and Welfare, Cincinnati, Ohio, 1967, p. 776.
99. Faith, W. L., in *Air Pollution*, Vol. III, A. C. Stern, Ed., Academic Press, New York, 1968, Chapter 40.
100. Sowa, Tedeusz, *Przem. Spozyw.*, 22 (1), 27 (1968); *Chem. Abstr.*, 71, 11859d (1969).
101. Natusch, D. F. S., unpublished data.
102. Osada, Hiromitsu and Ikuko Goto, *Eiyo To Shokuryo*, 20 (5), 387 (1968); *Chem. Abstr.*, 69, 1871w (1968).
103. White, Richard K., *Proc. Int. Symp. Identification Meas. Environ. Pollut.*, 1971, 105–9; *Chem. Abstr.*, 80, 87006e (1974).
104. Mueller, Wilhelm J., *Gas-u. Wasserfach*, 102, 986 (1961); *Chem., Abstr.*, 56, 1303e (1962).
105. Santry, I. W., Jr., *J. Water Pollut. Control Fed.*, 35, 1580 (1963).
106. Rao, G. J. Mohan, C. A. Sastry, and W. F. Garber, *J. Inst. Eng. India* (Calcutta), 46 (6), 90 (1966), *Air Pollut. Control Assoc. Abstr.*, 12 (5), 7617 (1966).

107. Tanner, F. W. and F. W. Tanner, Jr., *Bacteriology*, Wiley, New York, 1938.
108. Wagner, G. C., R. J. Kassner, and M. D. Kamen, *Proc. Nat. Acad. Sci.* (U.S.), 71 (2), 253 (1974).
109. Gloyna, E. F. and E. Espino, *J. Sanit. Eng. Div. ASCE*, 95 (SA3), 607 (1969).
110. Pomeroy, R. and F. D. Bowlus, *Sewage Works J.*, 18 (4), 597 (1946).
111. Vomos, Rezso, *Munkavedelem*, 17 (4–6), 41 (1971); *Chem., Abstr.*, 80, 18940t (1974).
112. Volkava, Z. A. and Z. M. Bagdinov, *Gig. Sanit.*, 34 (9), 33 (1969).
113. Verhaar, G., *Trans. Proc., Inst. Rubber Ind.*, 1 (2), 109 (1967); *Chem. Abstr.*, 71, 22699c (1969).
114. Kochanova, O. M., G. A. Blokh, F. S. Kohman, I. M. Strelok, S. A. Levina, E. F. Ermolenko, and L. N. Malashevich, *Kauch. Rezina*, 29 (3), 15 (1970); *Chem. Abstr.*, 73, 4704n (1970).
115. Goodwin, Don R. and Fred G. Rolater, U. S. Public Health Service Publication AP-48, 1968.
116. "Scrubbers Handle Varnish Fumes," *Paint, Oil Chem. Rev.*, 119, 14 (1956).
117. Kononova, V. A. and V. B. Aksenova, *Gig. Sanit.*, 26, 3 (1961); *Air Pollut. Control Assoc. Abstr.*, 7 (11), 4346 (1962).
118. Mel'nichenko, R. K., *Gig. Tr.*, (1), 55 (1969); *Chem. Abstr.*, 72, 47087k (1970).
119. Stern, Arthur C., Ed., *Air Pollution*, Vols. I, II, and III, 2nd ed., Academic Press, New York, 1968.
120. Basmadzhieva, K. and M. Argirova, *Khig. Zdraveopazvane*, 11 (3), 237 (1968); *Chem. Abstr.*, 69, 69522p (1968).
121. Slyusarenko, V. G., A. M. Kirichenko, V. I. Taranets, L. S. Belokrys, and G. N. Kravchenko, *Izv. Vyssh. Ucheb. Zaved., Gorn. Zh.*, 11 (8), 57 (1968); *Chem. Abstr.*, 70, 30775u (1969).
122. Strack, W., *Leder*, 18 (9), 233 (1967); *Chem. Abstr.*, 68, 22723u (1968).
123. Lawrence, William C. and Edward R. Cole, *Wallerstein Lab. Commun.*, 31 (105), 95 (1968); *Chem. Abstr.*, 70, 46151y (1969).
124. Mecham, D. K. and Maura M. Bean, *Cereal Chem.*, 45 (5), 445 (1968); *Chem. Abstr.*, 70, 2538a (1969).
125. Hoffman, E., *Aust. Paint J.*, 4 (7), 15 (1959); *Chem. Abstr.*, 54, 13509e (1960).
126. Druyan, E. A., *Mater. Nauch.-Prakt. Konf. Molodykh Gig. Sanit. Vrachei*, 11th, ed., A. P. Shitskova, Ed., Moscow Nauch.-Issled. Inst. Gig., USSR, pp. 26–28; *Chem. Abstr.*, 72, 35431e (1970).
127. Taiganides, E. Paul and Richard K. White, *Trans. ASAE*, 12 (3), 359 (1969); *Chem. Abstr.*, 71, 73738x (1969).
128. Moorman, R., *J. Air Pollut. Control Assoc.*, 15, 34 (1965).
129. Vogel, Erich, *Silikat. Tech.*, 11, 476 (1960); *Chem. Abstr.*, 55, 3949c (1961).
130. Mokhnachev, I. G. and A. A. Sirotenko, *Bulg. Tyutyun*, 8–9, 31 (1967); *Chem. Abstr.*, 68, 36836v (1968).
131. Cadle, R. D. and M. Ledford, *Int. J. Air Water Pollut.*, 10, 25 (1966).
132. Hoshika, Yasuyuki, *Kogai To Taisaku*, 9 (3), 285 (1973); *Chem. Abstr.*, 79, 57279h (1973).
133. Slatt, B. J., and D. F. S. Natusch, "Fluorescence Determination of Atmospheric H_2S and SO_2 at the Parts per Trillion Level," Paper presented at the 166th Annual Meeting, American Chemical Society, Chicago, August 27, 1973.

134. Breeding, R. J., J. P. Lodge, Jr., J. B. Pate, D. C. Sheesley, H. B. Klonis, B. Fogle, J. A. Anderson, T. R. Englert, P. L. Haagenson, et al., *J. Geophys. Res.*, 78 (30), 7057 (1973).

135. "Cleaning Our Environment: The Chemical Basis for Action," American Chemical Society Report, Washington, D.C., 1969.

136. Calvert, S. and W. Workman, *Talanta*, 4, 89 (1960).

137. Wartburg, A. F., J. B. Pate, and J. P. Lodge, Jr., *Environ. Sci. Technol.*, 3, 767 (1969).

138. Jacobs, M. B., M. M. Braverman, and S. Hochheiser, *Anal. Chem.*, 29, 1349 (1957).

139. Bostroem, C. E., *Air Water Pollut.*, 10 (6–7), 435 (1966).

140. Axelrod, Herman D., Joe H. Cary, Joseph E. Bonelli, and James P. Lodge, Jr., *Anal. Chem.*, 41 (13), 1856 (1969).

141. Karchmer, J. H., Ed., *The Analytical Chemistry of Sulphur and Its Compounds*, Part 1, Wiley-Interscience, New York, 1970.

142. Weiland, Ralph H. and Olev Trass, *Anal. Chem.*, 41 (12), 1709 (1969).

143. Pribyl, M. and Z. Slovak, *Mikrochim. Ichnoanal. Acta*, 5–6, 1119 (1963).

144. Carter, Cecil Neal, *Diss. Abstr. B*, 27 (2), 453 (1966).

145. Alferova, L. A. and G. A. Titova, *Zh. Prikl. Khim.* (Leningrad), 42 (1), 192 (1969); *Chem. Abstr.*, 70, 81365t (1969).

146. Jacobs, M. B., *J. Air Pollut. Control Assoc.*, 15, 314 (1965).

147. Bamesberger, W. L. and D. F. Adams, *Environ. Sci. Technol.*, 3 (3), 258 (1969).

148. Adams, D. F., *Health Lab. Sci.*, 7, 157 (1970).

149. Fukui, Syozo, Syoji Naito, Mikihiro Kaneko, and Saburo Kanno, *Eisei Kagaku*, 13 (1), 16 (1967); *Chem. Abstr.*, 68, 52991x (1968).

150. Lahmann, E. and K. E. Prescher, *Staub*, 25, 527 (1965).

151. Haun, R. *Z. Ges. Hyg. Ihre Grenzgebiete* (Berlin), 14 (2), 93 (1968); *Air Pollut. Control Assoc. Abstr.*, 14 (5), 10277 (1968).

152. Kovalenko, N. P., and T. V. Martynenko, *Zavod. Lab.*, 35 (9), 1051 (1969); *Chem. Abstr.*, 72, 18244d (1970).

153. Gustafsson, L., *Talanta*, 4, 227 (1960).

154. Prescher, K.-E. and E. Lahmann, *Gesundheit-Ingenieur* (Munich), 87 (12), 351 (1966); *Air Pollut. Control Assoc. Abstr.*, 14 (7), 10533 (1968).

155. Levine, B. S., Transl., "U.S.S.R. Literature on Air Pollution and Related Occupational Diseases," U.S. Department of Commerce, Office of Technical Services, Washington, D.C., 1963, p. 8.

156. Peter, F. and R. Rosset, *Anal. Chim. Acta*, 70 (1), 149 (1974).

157. Sakwai, H. and T. Toyama, *Japan. J. Ind. Health*, 5 (11), 689 (1968).

158. Kasper, Eberhard, *Chem. Tech.* (Berlin), 11, 616 (1959); *Chem. Abstr.*, 54, 17151f (1960).

159. Gershkovich, E. E., *Novoe V. Oblastic Sanit.-Khim. Anal.* (Raboty po Prom.-Sanit. Khim.), 53 (1962); *Chem. Abstr.*, 59, 12076g (1963).

160. van Langermeersch, A., *Method. Phys. Anal.*, April–June, 67 (1967); *Chem. Abstr.*, 68, 9118f (1968).

161. Adams, D. F. and R. K. Koppe, *J. Air Pollut. Control Assoc.*, 17, 161 (1967).

162. Wolf, F. and H. Langen, *Chem. Ing. Tech.*, 39 (16), 945 (1967); *Air Pollut. Control Assoc. Abstr.*, 13 (11), 9397 (1968).

163. Yoshimura, C., H. Hara, and K. Tamura, *Japan Anal.*, 18 (6), 689 (1969); *Air Pollut. Control Assoc. Abstr.*, 15 (6), 11947 (1969).

164. Natusch, D. F. S., J. R. Sewell, and R. L. Tanner, *Anal. Chem.*, 46 (3), 410 (1974).
165. Schumann, Horst, Kurt Lobenstein, and Horst Blaschke, British Patent 1,047,700 (1966); *Chem. Abstr.*, 66, 34627n (1967).
166. Paré, Jean P., *J. Air Pollut. Control Assoc.*, 16 (6), 325 (1966).
167. Hochheiser, S. and L. A. Elfers, *Environ. Sci. Technol.*, 4, 674 (1970).
168. "Information on the Dräger Gas Detector," *Drägerwerk Bull.*, Nos. 41 and 48, Lubeck, August 1960, June 1961.
169. Stepankova, Kamila and Jiri Juranek, *Chem. Prum.*, 23 (1), 36 (1973); *Chem. Abstr.*, 78, 150907a (1973).
170. Boström, C. E., and C. Brosset, *Atmos. Environ.*, 3 (4), 407 (1969).
171. Gilardi, E. F. and R. M. Manganelli, *J. Air Pollut. Control Assoc.*, 13, 305 (1963).
172. Okita, Toshiichi, James P. Lodge, Jr., and Herman D. Axelrod, *Environ. Sci. Technol.*, 5 (6), 532 (1971).
173. Chamberland, André M., Pierre Bourbon, and Robert Malbosc, *Int. J. Environ. Anal. Chem.*, 2 (4), 303 (1973).
174. Sanderson, H. P., R. Thomas, and M. Katz, *J. Air Pollut. Control Assoc.*, 16, 328 (1966).
175. Bethea, Robert M., *J. Air Pollut. Control Assoc.*, 23 (8), 710 (1973).
176. Lahmann, E., *Erdoel Kohle*, 18 (10), 796 (1965).
177. Jacobs, M. B., National Academy of Sciences–National Research Council Publication No. 652, 24 (1959); *Air Pollut. Control Assoc. Abstr.*, 5 (11), 3019 (1960).
178. Bamesberger, W. L. and D. F. Adams, *Tappi*, 52 (7), 1302 (1969).
179. Sinkevish, O. V. and A. E. Noshchenko, USSR Patent 174,002 (1965); *Chem. Abstr.*, 64, 4158a (1966).
180. Dräger, Heinrich and Bernhard Dräger, German Patent 1,206,176 (1965); *Chem. Abstr.*, 64, 10788e (1966).
181. Risk, J. B., and F. E. Murray, *Can. Pulp Paper Ind.*, 17 (10), 31 (1964).
182. Adams, D. F., W. L. Bamesberger, and T. J. Robertson, *J. Air Pollut. Control. Assoc.*, 18 (3), 145 (1968).
183. Stevens, R. K., J. D. Mulik, A. E. O'Keefe, and K. J. Krost, *Anal. Chem.*, 43, 827 (1971).
184. Bethge, P. O. and L. Ehrenborg, *Svensk Papperstid.* (Stockholm), 70 (10), 347 (1967); *Air Pollut. Control Assoc. Abstr.*, 13 (6), 8766 (1967).
185. Mason, D. McA., *Hydrocarbon Process. Petrol. Refinery*, 43, 145 (October 1964).
186. Urone, P., J. B. Evans, and C. M. Noyes, *Anal. Chem.*, 37, 1104 (1965).
187. Koppe, R. and D. Adams, *Environ. Sci. Technol.*, 1 (6), 479 (1967).
188. Schuette, F. J., *Atmos. Environ.*, 1 (4), 515 (1967).
189. Bloomfield, Bernard D., in *Air Pollution*, Vol. II, A. C. Stern, Ed., Academic Press, New York, 1968, Chapter 28, p. 521.
190. Hanson, V. H., *Ind. Eng. Chem., Anal. Ed.*, 13, 119 (1941).
191. Saltzmann, Robert S. and Elton B. Hunt, Jr., *ISA Trans.*, 12 (2), 103 (1973); *Chem. Abstr.*, 79, 127773r (1973).
192. "Two Systems Monitor H_2S in Stack Gases," *Can. Chem. Process.*, 53 (8), 58 (1969).
193. Humphrey, R. E., W. Hinze, and W. M. Jenkines, III, *Anal. Chem.*, 43, 140 (1971).
194. Naunczyk, J., *Gas, Woda Tech. Sanit.*, 45, 346 (1971); *Chem. Abstr.*, 76, 37265k (1972).
195. Grasshof, K. and K. Chan, *Anal. Chim. Acta*, 53, 442 (1971).

196. Lyubimov, N. A., *Predel'na Dopustimye Kontsentratsii Atm. Zagryazn.*, No. 5, 169 (1961); *Chem. Abstr.*, 58, 1849d (1963).

197. Polezhaev, N. G., *Gig. Tr. Tekhn. Bezop.*, 6, 38 (1937); *Limits of Allowable Concentrations of Atmospheric Pollutants*, Book 1, V. A. Ryazanov, Ed., and B. S. Levine, Transl., U.S. Department of Commerce, Office of Technical Services, Washington, D.C., 1952, p. 104.

198. Andrew, T. R. and P. N. R. Nichols, *Analyst*, 90, 367 (1965).

199. Roth, H. H., U.S. Patent 2,864,747 (1958); *Chem. Abstr.*, 53, 5022i (1959).

200. Sutherland, J. C., *Proc. Inst. Environ. Sci. Tech. Meet.*, 1970, 298–301; *Chem. Abstr.*, 74,67523u (1971).

201. Stepanenko, V. E., S. I. Krichmar, and Z. F. Khar'kovskaya, *Zh. Anal. Khim.*, 28 (11), 2136 (1973); *Chem. Abstr.*, 80, 90660u (1974).

202. Popescu, G., and Mihaela Dura, *Rev. Chim.* (Bucharest), 20 (5), 290 (1969); *Chem. Abstr.*, 71,73762a (1969).

203. Kriván, V., S. Pahlke, and G. Tölg, *Talanta*, 20, 391 (1973).

204. Kaiser, R., *Chromatographia*, 5 (39), 177 (1972); *Chem. Abstr.*, 76, 80834g (1972).

205. Applebury, T. E. and M. J. Schaer, *J. Air Pollut. Control Assoc.*, 20 (2), 83 (1970).

206. Bruner, Fabrizio, Arnaldo Liberti, Massimiliano Possanzini, and Ivo Allegrini, *Anal. Chem.*, 44 (12), 2070 (1972).

207. Greer, David G. and Thomas J. Bydalek, *Environ. Sci. Technol.*, 7 (2), 153 (1973).

208. Adams, D. F., R. K. Koppe, and W. N. Tuttle, *J. Air Pollut. Control Assoc.*, 15, 31 (1965).

209. Feurerstein, D. L., J. F. Thomas, and D. L. Brink, *Tappi*, 50 (6), 258 and 276 (1967).

210. Walther, J. E., and H. R. Amberg, *Tappi*, 50 (10), 108A (1967).

211. Robertus, Robert J. and Michael J. Schaer, *Environ. Sci. Technol.*, 7 (9), 849 (1973).

212. Drews, Bruno, G. Baerwald, and H. J. Niefind, *Monatsschr. Brau.*, 22 (6), 140 (1969); *Chem. Abstr.*, 71, 108942p (1969).

213. Jones, Caroline N., *Anal. Chem.*, 39 (14), 1858 (1967).

214. Drushel, H. V., *Anal. Chem.*, 41 (4), 569 (1969).

215. Johnson, Bruce A., *Am. Ind. Hyg. Ass. J.*, 33 (12), 811 (1972).

216. Hofman, K. and R. Hamm, *Anal. Chem.*, 232 (3), 167 (1967).

217. Natusch, D. F. S. and T. Thorpe, *Anal. Chem.*, 45 (14), 1185A (1973).

218. Adams, D. F. and R. K. Koppe, *Tappi*, 42 (7), 601 (1959).

219. Obermiller, E. L. and G. O. Charlier, *J. Chromatogr. Sci.*, 7 (9), 580 (1969).

220. Stevens, R. K., A. E. O'Keefe, and G. C. Ortman, *Environ. Sci. Technol.*, 3 (7), 652 (1969).

221. Cave, G. C. B., *Tappi*, 46 (1), 1 (1963).

222. Mikhailenko, A. S. and E. I. Koroleva, *Zavod. Lab.*, 39 (8), 944 (1973); *Chem. Abstr.*, 79, 142593a (1973).

223. Robbins, L. A., R. M. Bethea, and T. D. Wheelock, *J. Chromatogr.*, 13 (2), 361 (1964).

224. Adams, D. F. and R. K. Koppe, *J. Air Pollut. Control Assoc.*, 17 (3), 161 (1967).

225. Draganescu, A., I. Bica, S. Serban, and Ana Maria Pavlovschi, *Rev. Roum. Chim.*, 17 (10), 1779 (1972); *Chem. Abstr.*, 78, 37478y (1973).

226. Pekhota, F. N., A. I. Parfenov, and K. I. Sakodynskii, *Nov. Sorbenty Khromatogr.*, No. 16, 119 (1971); *Ref. Zh. Khim.*, Abstr. No. 3G175 (1973); *Chem. Abstr.*, 79, 44088v (1973).

227. Bollman, D. H. and D. M. Mortimore, *J. Chromatogr. Sci.*, 10 (8), 523 (1972).
228. Obermiller, E. L. and G. O. Charlier, *J. Gas Chromatogr.*, 6 (8), 446 (1968).
229. Axelrod, H. D., J. B. Pate, W. R. Barchet, and J. P. Lodge, Jr., *Atmos. Environ.*, 4, 209 (1970).
230. Axelrod, H. D., R. J. Teck, J. P. Lodge, Jr., and R. H. Allen, *Anal. Chem.*, 43 (3), 496 (1971).
231. Evans, C. L., *Quart. J. Exp. Physiol.*, 52 (3), 231 (1967).
232. Milby, T. H., *J. Occup. Med.*, 4, 431 (1962); *Air Pollut. Control Assoc. Abstr.*, 9 (3), 5341 (1963).
233. Gurinov, B. P., in *Limits of Allowable Concentrations of Atmospheric Pollutants*, Book 1, V. A. Ryazanov, Ed., B. S. Levine, Transl., U.S. Department of Commerce, Office of Technical Services, Washington, D.C., 1952, p. 46.
234. Jacobs, Morris B., *The Analytical Toxicology of Industrial Inorganic Poisons*, Wiley-Interscience, New York, 1967.
235. Baikov, B. K., *Gig. Sanit.*, 28, 3 (1963); *Air Pollut. Control Assoc. Abstr.*, 9 (8), 5692 (1964).
236. Petrun, N. M., *Farmakol. Toksikol.*, 28 (4), 488 (1965); *Chem. Abstr.*, 63, 17021e (1965).
237. Veszpremi, Bela, *Orsz. Mezogazd. Minosegvizsgalo Int. Evk.*, 6, 255–262 (1961–1963) (Pub. 1964); *Chem. Abstr.*, 65, 4326h (1966).
238. Vamos, R. and R. Tasnadi, *Fortschr. Wasserchem. Ihrer Grenzgeb.*, No. 14, 235 (1972); *Chem. Abstr.*, 79, 96556r (1973).
239. Hays, Frederic L., *Diss. Abstr. Int. B*, 34 (3), 935 (1973).
240. Kosmider, Stanislaw, Edmund Rogala, Zbigniew Szulik, and Jerzy Dwornicki, *Patol. Pol.*, 24 (1), 171 (1973); *Chem. Abstr.*, 80, 531a (1974).
241. Oseid, Donavon and Lloyd L. Smith, Jr., *Trans. Am. Fish. Soc.*, 101 (4), 620 (1972); *Chem. Abstr.*, 78, 53609x (1973).
242. Vasil'eva, I. A., *Gig. Sanit.*, 7, 24 (1973); *Chem. Abstr.*, 80, 18982h (1974).
243. Kaplun, S. Y. and E. G. Koptera, *Fiziol. Zh.* (Kiev), 19 (3), 328 (1973); *Chem. Abstr.*, 79, 62310r (1973).
244. Fyn-Djui, Duán, *Gig. Sanit.*, 24 (10), 12 (1959); *Air Pollut. Control Assoc. Abstr.*, 5 (11). 3042 (1960); *U.S.S.R. Literature on Air Pollution and Related Occupational Diseases*, Vol. 5, B. S. Levine, Transl., U.S. Department of Commerce, Office of Technical Services, Washington, D.C., 1961, p. 66.
245. Ryazanov, V. A., *Proc. Int. Clean Air Conf. London*, 1959, 175–6 (1960); *Chem. Abstr.*, 55, 13726d (1961).
246. Krekel, K., *Zent. Arbeitsmed. Arbeitsschutz*, 14 (7), 159 (1964); *Chem. Abstr.*, 62, 2168g (1965).
247. Beasley, R. W. R., *Brit. J. Ind. Med.*, 20, 32 (1963); *Chem. Abstr.*, 58, 13047g (1963).
248. Yamada, Masaru and Kenzo Tonomura, *Hakko Kogaku Zasshi*, 50 (12), 901 (1972); *Chem. Abstr.*, 78, 55219u (1973).
249. Fomicheva, L. V. and A. I. Gmyrya, *Vestn. Oftal'mol*, (2), 77 (1973); *Chem. Abstr.*, 79, 74469n (1973).
250. Kosmider, S., *Polsk. Arch. Med. Wewn.*, 36 (5), 615 (1966); *Chem. Abstr.*, 65, 20734g (1966).
251. Kaminski, Marcin and Piotr Mikolajczyk, *Med. Pracy*, 18 (1), 42 (1967); *Chem. Abstr.*, 66, 114224u (1967).

252. Kosmider, Stanislaw and K. Zajusz, *Zentralbl. Allg. Pathol. Anat.*, 109 (4), 411 (1966); *Chem. Abstr.*, 66, 114228y (1967).
253. Basmadzhieva, K., Z. Rashev, and M. Argirova, *Khig. Zdraveopazvane*, 12 (1), 33 (1969); *Chem. Abstr.*, 71, 53280p (1969).
254. Gassman, Merrill L., *Plant Physiol.*, 51 (1), 139 (1973); *Chem. Abstr.*, 78, 53307x (1973).
255. Murphy, Walter J., Ed., *Chem. Eng. News*, 33 (19), 1966 (1955).
256. Scholl, W. and H. Gehlker, *Landwirt. Forsch., Sonderh.*, 28 (1), 40 (1973); *Chem. Abstr.*, 80, 81366a (1974).
257. Ford, H. W., *J. Am. Soc. Hort. Sci.*, 98 (1), 66 (1973); *Chem. Abstr.*, 78, 132440p (1973).
258. Shimose, N. and Y. Miyake, *Nippon Dojo-Hiryogaku Zasshi*, 36 (5), 117 (1965); *Chem. Abstr.*, 64, 10358g (1966).
259. Loginova, R. A., Transl. *Predel'no-dopustimye kontsentrat. atmosfernykh zagriaznenii* (Moscow), 3, 63 (1957); *Air Pollut. Control Assoc. Abstr.*, 6 (9), 3504 (1961).
260. Pomeroy, Richard D. and Henry Cruse, *J. Am. Water Works Assoc.*, 61 (12), 677 (1969).
261. Chandler, K. A., *Gas World Coking*, 150, 7 (1959).
262. Sanyal, Bhavtosh, G. K. Singhania, and D. Bhadwar, *Tr. Mezhdunar. Kongr. Korroz. Metal.*, 3rd, N. D. Lesteva, Ed., Izd. "Mir", Moscow, 1966 (Pub. 1968), 4, pp. 464–473; *Chem. Abstr.*, 71, 127722h (1969).
263. Yocom, J. E., *J. Air Pollut. Control Assoc.*, 8, 203 (1958).
264. Greenburg, L. and M. B. Jacobs, *Am. Paint J.*, 39, 64 (1955); *Air Pollut. Control Assoc. Abstr.*, 2 (5), 961 (1956).
265. Chiarenzelli, Robert V., *IEEE, Trans. Parts, Mater. Packag.*, 3 (3), 89 (1967); *Chem. Abstr.*, 68, 52611y (1968).
266. Pope, D., H. R. Gibbens, and R. L. Moss, *Corros. Sci.*, 8 (12), 883 (1968).
267. Backlund, P., B. Fjellsala, S. Hammarbaeck, and B. Maijgren, *Ark. Kemi*, 26 (21–23), 267 (1966); *Chem. Abstr.*, 66, 78831d (1967).
268. Blaise, Bernard, Annick Genty, and Jean Bardolle, *C. R. Acad. Sci., Paris., Ser. C*, 276 (15), 1227 (1973); *Chem. Abstr.*, 79, 23891m (1973).
269. Backlund, Per, *Acta Chem. Scand.*, 23 (5), 1541 (1969); *Chem. Abstr.*, 71, 94172y (1969).
270. Allpress, J. G. and J. V. Sanders, *Phil. Mag.*, 10 (107), 827 (1964); *Chem. Abstr.*, 61, 15784a (1964).
271. Dvoraczek, Jean P., Jean C. Colson, and Denise Delafosse, *C. R. Acad. Sci., Paris, Ser C*, 268 (19), 1646 (1969); *Chem. Abstr.*, 71, 52799c (1969).
272. Hentsch, A., *Elektrie*, 16, 234 (1962); *Chem. Abstr.*, 57, 14847a (1962).
273. Draeger, Hans J., *Elektrotech. Z.*, Ausg. A, 94 (9), 562 (1973); *Chem. Abstr.*, 80, 29888h (1974).
274. Wohlers, H. C. and M. Feldstein, *J. Air Pollut. Control Assoc.*, 16, 19 (1966).
275. Haritani, Hiroshi, Ichikazu Kawanishi, and Michio Asahina, Japanese Patent 70 15,618, 1970; *Chem. Abstr.*, 73, 69325j (1970).
276. Krumbein, Simeon J. and Morton Antler, *IEEE Trans. Parts, Mater. Packag.*, 4 (1), 3 (1968); *Chem. Abstr.*, 69, 31194v (1968).
277. Elliot, J. F. E. and A. C. Franks, *Tin—Its Uses*, 8, 8 (1968).
278. Mennenoeh, Siegfried, *Hoesch. Ber. Forsch. Entwickl. Unserer Werke*, 8 (1), 23 (1973); *Chem. Abstr.*, 79, 6888y (1973).

279. Khamzin, A., *Gazov. Prom.*, (1), 10 (1973); *Chem. Abstr.*, 78, 127430f (1973).

280. Kofman, N. I. and M. I. Finkel'shtein, *Lakokras. Mater. Ikh. Primen.*, (1), 40 (1970); *Chem. Abstr.*, 72, 123021b (1970).

281. Braunisch, M. and H. Lenhart, *Kolloid-Z.*, 177, 24 (1961); *Chem. Abstr.*, 55, 25327i (1961).

282. Reffner, J. A., C. I. Harding, and T. R. Kelley, *J. Air Pollut. Control Assoc.*, 17 (1), 36 (1967).

283. Holbrow, G. L., *J. Oil Color Chem. Assoc.*, 45, 701 (1962); *Chem. Abstr.*, 57, 16781d (1962).

284. "Air Pollution and Paint," *Carnegie-Mellon Paint Tech.*, 20, 134 (1956).

285. Vanicek, O., V. Civin, and V. Taborsky, *Chem. Prumysl.* (Prague), 7 (32), 273 (1957); *Air Pollut. Control Assoc. Abstr.*, 3 (8), 1620 (1958).

286. Hoffman, E., *J. Oil Color Chem. Assoc.*, 42, 452 (1959).

287. Brebion, G. and R. Cabridenc, *Corrosion* (Rueil-Malmaison, Fr.), 17 (2), 71 (1969); *Chem. Abstr.*, 70, 117822f (1969).

288. Schremmer, Herbert, *Gas-Wasserfach, Wasser-Abwasser*, 113 (12), 591 (1972); *Chem. Abstr.*, 78, 139946s (1973).

AUTHOR INDEX

SUBJECT INDEX